Recent Progress in Wireless Technology

Recent Progress in Wireless Technology

Edited by
Gary Payton

Willford Press
www.willfordpress.com

Published by Willford Press,
118-35 Queens Blvd., Suite 400,
Forest Hills, NY 11375, USA

ISBN: 978-1-68285-595-9

Cataloging-in-Publication Data

Recent progress in wireless technology / edited by Gary Payton.
p. cm.
Includes bibliographical references and index.
ISBN 978-1-68285-595-9
1. Wireless communication systems. 2. Mobile communication systems.
I. Payton, Gary.
TK5103.2 .R43 2019
621.384--dc23

For information on all Willford Press publications
visit our website at www.willfordpress.com

Willford Press

Contents

Preface

In my initial years as a student, I used to run to the library at every possible instance to grab a book and learn something new. Books were my primary source of knowledge and I would not have come such a long way without all that I learnt from them. Thus, when I was approached to edit this book; I became understandably nostalgic. It was an absolute honor to be considered worthy of guiding the current generation as well as those to come. I put all my knowledge and hard work into making this book most beneficial for its readers.

Wireless communication undertakes the access and retrieval of data and information across two or more remote points of exchange. Wireless networking employs radio waves for long range communication. It has varied applications in a number of industries and fields such as telecommunication industry, medical industry, data communication, etc. Modern wireless tools such as WiFi have become essential systems in diverse spaces like homes, offices, schools, etc. This book attempts to understand the multiple branches that fall under the discipline of wireless technology and how such concepts have practical applications. It sheds light on some of the unexplored aspects of this domain and the recent researches in this field. This book will serve as a reference to a broad spectrum of readers.

I wish to thank my publisher for supporting me at every step. I would also like to thank all the authors who have contributed their researches in this book. I hope this book will be a valuable contribution to the progress of the field.

Editor

1

Compact UWB-MIMO antenna with metamaterial FSS decoupling structure

Xiaoming Zhu[1,2*], Xiaodong Yang[2,3], Qichao Song[1] and Baisen Lui[1]

Abstract

In this paper, an UWB-MIMO antenna with frequency selective surface (FSS) decoupling structure is proposed. To realize antenna system miniaturization and integration, silicon material is applied as substrate. The proposed antenna becomes a very compact construction with dimension of 38.2 × 26.6 × 0.4 mm^3. The proposed broadband FSS unit consists of four split rectangles and one I-shaped strip. The effective permittivity or effective permeability of FSS unit is negative, so the FSS is a metamaterial structure and has a broadband frequency band gap in the entire UWB range. Six FSS units are positioned in the middle of antenna backside, like band-stop filter to reduce the coupling between the antennas placed side-by-side. Compared to UWB-MIMO antenna without FSS, the proposed array not only keeps each antenna performance but also decreases 7.2 dB coupling. The UWB-MIMO with FSS structure provides an overall isolation of more than 16 dB in the UWB spectrum, so it is suitable for portable UWB-MIMO system applications.

Keywords: UWB-MIMO antenna, FSS, Metamaterial, Decoupling

1 Introduction

The ultra-wideband (UWB) system is a radio engineering technology with frequency from 3.1–10.6 GHz [1], which has several advantages including very large bandwidth, low power consumption, high date rate, high time resolution, resistance to interference, co-existence with narrowband systems, and so on. Such advantages enable UWB technology widely applied in communication, radar, imaging, and positioning [2–5]. To improve transmission rate and communication reliability of UWB system, multiple-input-multiple-output (MIMO) technology can be joined to become UWB-MIMO system. The combinative system needs multiple antennas coexisting in the finite space of transmitters and receivers. The individual antenna not only has broad impedance matching characteristic over the entire spectrum but also has better isolation from adjacent antenna. The coupling influence between antennas is more serious as the increase of the antenna number, so the effective decoupling method is a key technology for UWB-MIMO system. The straightforward decoupling way is to extend separation distance of antennas, but the size of each component is strictly controlled for miniaturization UWB mobile terminals. The correct scheme is to design special structure to decrease coupling without sacrificing transceiver space. For plane monopole MIMO antennas, the common decoupling structures are parasitic elements and defected grounds. The detailed structures are diversified such as T-shaped element, Y-shaped element, or combination with several long and short strips [6–8]. Another approach for designing UWB-MIMO antennas is to use slot antennas with orthogonal feeding to achieve polarization diversity and pattern diversity [9–12]. These layouts are relatively larger than the above decoupling antenna structures. Frequency selective surface (FSS) is a period electromagnetism material with frequency band gap performance. FSS structures can selectively determine electromagnetic waves to pass or prevent within specified frequency ranges.

In this paper, a UWB-MIMO structure of antenna is presented. The broadband FSS cell is composed of electric resonator and magnetic resonator with the band gap characteristic as same as metamaterial. This UWB-MIMO antenna with FSS structure obtains more than 16 dB isolation between antennas in the UWB frequency band. The rest of this paper is as follows: Section 2

* Correspondence: zhuxiaoming213@163.com
[1]College of Electrical and Information Engineering, Heilongjiang Institute of Technology, Harbin 150050, China
[2]College of Information and Communication Engineering, Harbin Engineering University, Harbin 150001, China
Full list of author information is available at the end of the article

presents the antenna configuration, Section 3 discusses the performances of metamaterial FSS unit, Section 4 analyzes the results of UWB-MIMO antenna, and finally, the paper is concluded in Section 5.

2 Antenna structure design

According to UWB technology in small mobile devices, the antenna is designed on planar microstrip transmission line with coplanar waveguide (CPW) feeder. The 400 μm thickness silicon wafer is chosen as antenna substrate. Because the relative dielectric constant of silicon is 11.9, the size of MIMO antenna can be shrunk. Then, silicon wafer is the material of most of integrated circuits, so antennas with silicon substrate are easily integrated with functional circuits to become a complete on-chip system.

The proposed UWB-MIMO antenna configuration is shown in Fig. 1. Table 1 presents the final dimensions of antenna. There are two same rectangular radiation patches and CPW feeders on the upper side of array. The shape of patches is rectangle initially, but this construction cannot cover the whole UWB frequency band. Therefore, transforming the bottom profile of patches is the simple method to increase bandwidth because the current is concentrated in the bottom edge of the antenna. The rectangle patch area is changed to staircase profile, in order to increase current path. If the antenna feeder is microstrip structure, the radiation patch and ground are on the different sides of substrate. But the proposed antenna uses CPW feeder because patch and ground are on the same plane. The coplanar characteristic makes antenna easy to manufacture with low radiation loss and dispersion. To solve current losses of right-angled grounds, arc processing is utilized for grounds next to the feeder, which can improve impedance matching and extend frequency bandwidth.

Table 1 Dimensions of the UWB-MIMO antenna

Parameter	mm	Parameter	mm
W1	38.2	L1	26.6
W2	14.88	L2	19.36
G	0.9	W3	1.2
W4	5	L3	6.01
L4	0.96	L5	4.91
L6	3	x1	2.1
x2	2.9	y1	3.95
y2	1.81	d	3.18
s1	0.4	s2	10.26
h	0.4	h1	0.2
c1	15.96	c2	19.54
c3	6.88	FW	2
FL	4	f1	0.04
f2	0.04	f3	0.03
f4	0.04	f5	0.06
f6	0.1		

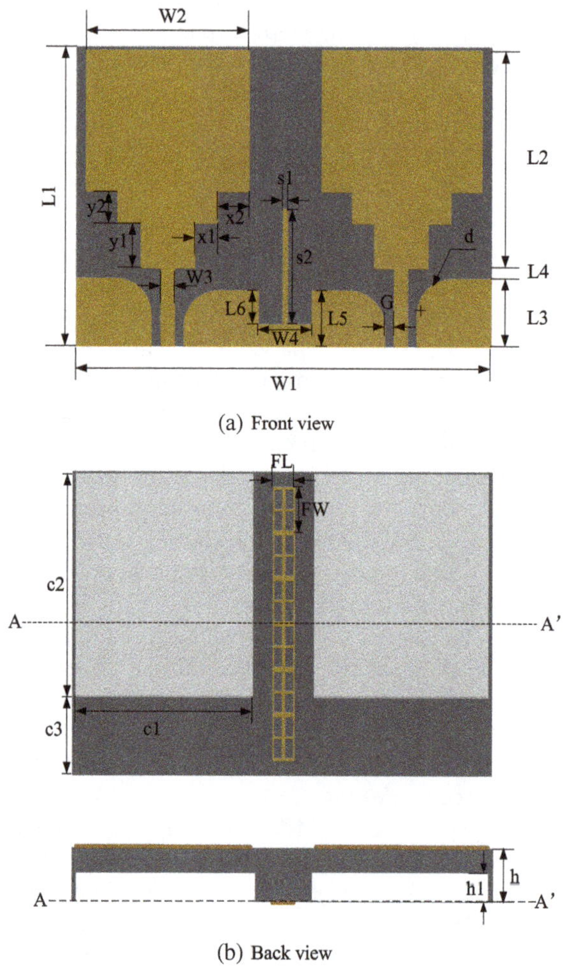

Fig. 1 Configuration of the UWB-MIMO antenna. **a** Front view. **b** Back view

There are two rectangular cavities and decoupling FSS structure on the rear side of the proposed antenna. High-permittivity substrate can reduce antenna size, but surface waves losses influence the antenna performance seriously. To restrain surface wave appearance, two vertical cavities are etched underneath the patches on the back of substrate. If the antenna substrate is formed by silicon and air cavity, the effective dielectric constant of mixture substrate can be calculated as,

$$\varepsilon_e = \frac{\varepsilon_0 \varepsilon_r}{\varepsilon_0 + (\varepsilon_r - \varepsilon_0) h_1 / h} \tag{1}$$

where ε_0 is the permittivity of vacuum, ε_r is relative permittivity of silicon, h_1 is the height of air cavity and h is

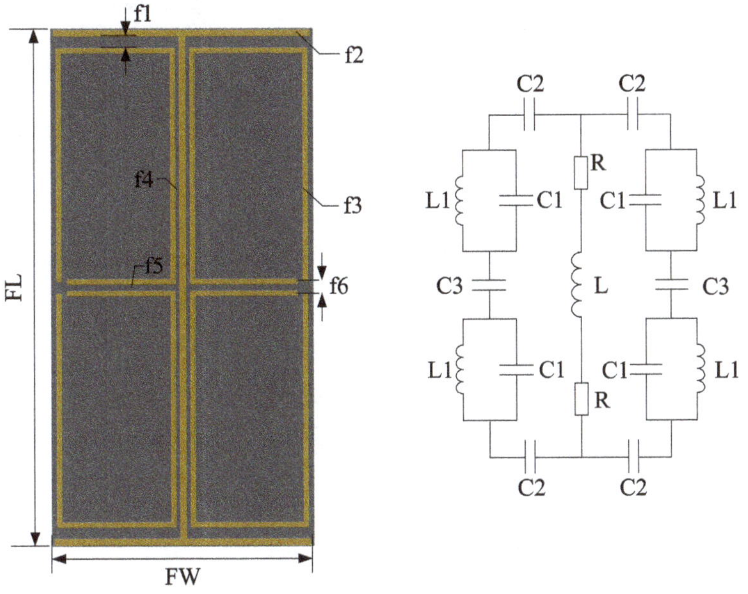

(a) FSS unit structure (b) FSS equivalent circuit

Fig. 2 FSS unit structure and equivalent circuit. **a** FSS unit structure. **b** FSS equivalent circuit

the thickness of silicon wafer. The vertical air cavities enable to decrease relative dielectric constant, solve surface wave problem, and improve MIMO antenna performances.

3 Decoupling structure design

Metamaterial is an artificial composite material with negative refractive index, negative permittivity, or negative permeability. Metamaterial belongs to subwavelength structure, which unit size is much less than working wavelength. The equivalent permittivity and equivalent permeability can be applied to analyze the physical property of metamaterial. The ε and μ indicate the interaction of electric field and magnetic field, and the propagation constant k of electromagnetic waves can be described as,

$$k = w\sqrt{\mu\varepsilon} \tag{2}$$

When k is a real number, electromagnetic waves can be propagated in the medium. While ε or μ is a negative value, the propagation constant k becomes an imaginary number. In this condition, electromagnetic waves cannot be propagated and the band gap appears. Therefore, when electromagnetic waves are incident on the surface of metamaterials, total reflection phenomenon will occur.

The UWB-MIMO antenna is proposed with novel decoupling structures, which are the front defected ground

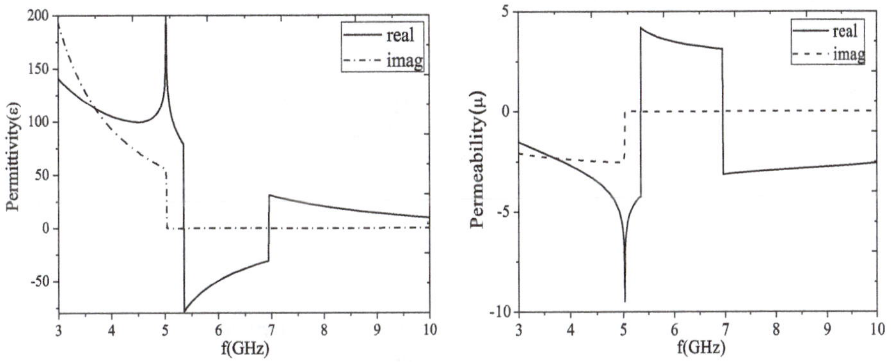

(a) equivalent permittivity (b) equivalent permeability

Fig. 3 Equivalent parameters of FSS unit. **a** Equivalent permittivity. **b** Equivalent permeability

structure and the rear FSS structure. Because of the CPW feeder of antenna, the middle grounds are connected together with mutual effect in high-frequency band of UWB. The defected ground transformation reduces coupling of connected ground, and the center short-circuited strip plays a role of reflector between two antennas. Broadband FSS structure is positioned in the middle of cavities. As compact antenna dimension, FSS should be placed on the back side to lower the direct impact on the front radiation patches.

The designed FSS unit consists of four split rectangles and one I-shaped strip, which pattern is made of metal material printed on the 400 μm silicon substrate. The FSS structure and equivalent circuit are shown in Fig. 2. L and R are the equivalent inductance and equivalent resistance of I-shaped strip. L1 and C1 are the equivalent inductance and equivalent capacitance of the split rectangle. C2 is the equivalent capacitance between the I-shaped strip and split rectangle. C3 is the equivalent capacitance between the two split rectangles. The split rectangle structure is equivalent to LC resonance circuit and belongs to a magnetic metamaterial, which can generate strong magnetic response with negative equivalent permeability. The I-shaped structure is also equivalent to LC resonance circuit and belongs to an electrical metamaterial, which can produce huge induced electric field to form electric resonance with negative permittivity. In fact, the proposed FSS cell is the combinative element with electrical and magnetic metamaterials. The equivalent permittivity and permeability curves of FSS cell are in Fig. 3. In frequency band of 5.36–6.96 GHz, FSS appears electric resonance phenomenon with negative equivalent permittivity and positive equivalent permeability. And in the other frequency bands, FSS has magnetic resonance characteristic with opposite parameter performance. The S parameters of FSS unit are presented in Fig. 4, where S11 is the reflection coefficient and S12 is the transmission coefficient. In Fig. 4, S12 values are all less than −20 dB in the whole UWB bandwidth, so the FSS unit can generate a wide band gap to restrain the electromagnetic waves propagation like a band-stop filter. Therefore, the metamaterial FSS unit is suitable for UWB-MIMO antenna decoupling.

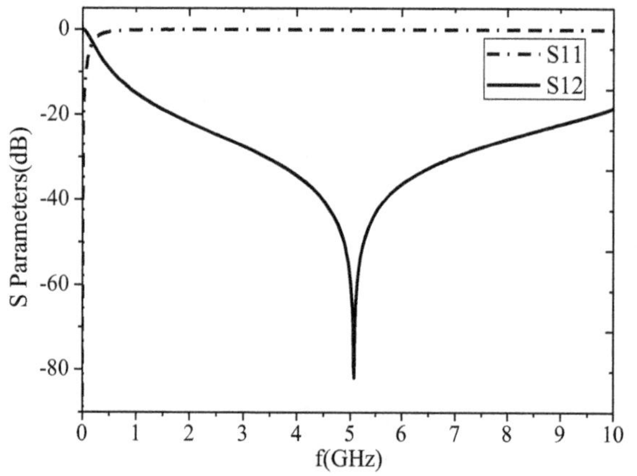

Fig. 4 Reflection coefficient and transmission coefficient of FSS unit

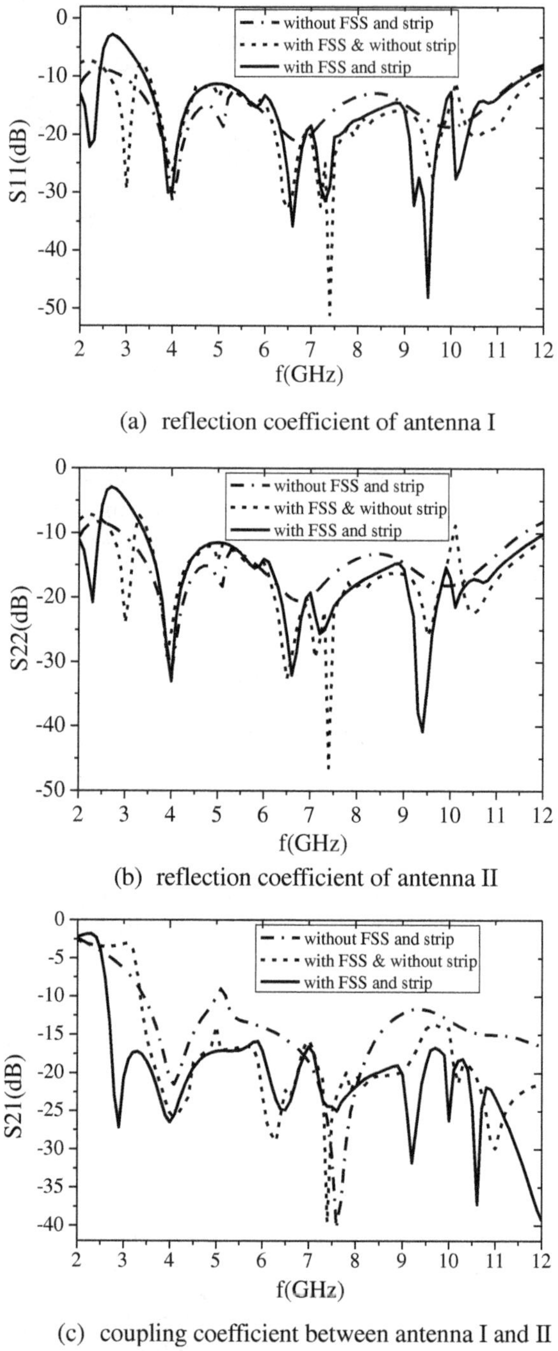

Fig. 5 Return loss and isolation of the UWB-MIMO antenna. **a** Reflection coefficient of antenna I. **b** Reflection coefficient of antenna II. **c** Coupling coefficient between antenna I and II

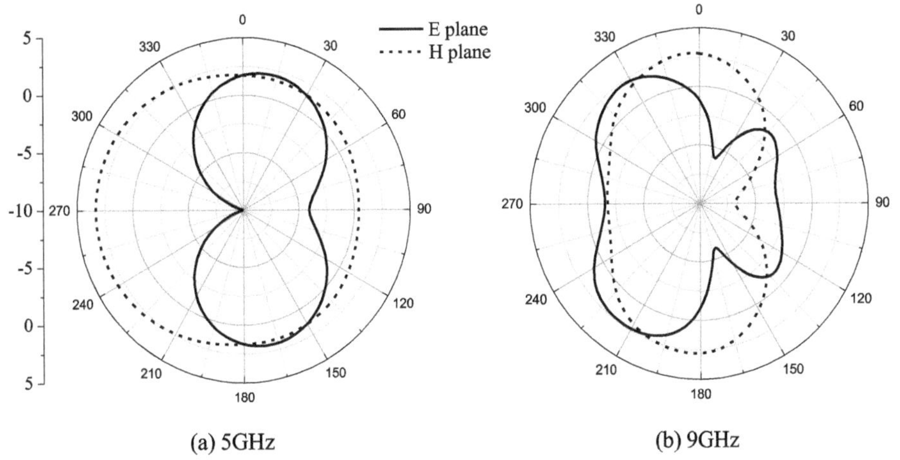

Fig. 6 Radiation patterns of the UWB-MIMO antenna. **a**5 GHz. **b** 9 GHz

4 Results and discussions

Firstly, the left antenna unit is regarded as antenna I and the other unit is antenna II in Fig. 1. S11 and S22 are respectively reflection coefficients of antennas I and II. S12 and S21 are equivalent because of symmetrical structure, and both represent coupling coefficient between two antennas. To analyze the decoupling performance, the proposed UWB-MIMO antenna is simulated with and without the decoupling structures. The reflection coefficients and coupling coefficient of the MIMO antenna are shown in Fig. 5. The S11 and S22 are less than −10 dB in UWB frequency range, so each independent antenna has better impedance matching performance. The antenna array without decoupling structure has worse isolation especially around 5 and 9 GHz. After adding six metamaterial FSS units on the rear of MIMO antenna, S21 decreases over the entire UWB range, but coupling coefficient is still above −15 dB between 9.42 GHz and 10 GHz. The center short-circuited strip of defected ground is applied to solve coupling problem around 9.7 GHz. Finally, S21 is all below −16 dB from 2.72GHz to 12 GHz, and the minimum value is up to −37.2 dB. It is demonstrated that the decoupling structure keeps good impedance matching performance and improves isolation of antennas I and II. The E-plane and H-plane patterns of the proposed antenna are obtained at frequencies 5GHz and 9 GHz in Fig. 6. The radiation pattern is nearly omni-directional at 5 GHz. But at frequency of 9 GHz, certain distortion appears in the pattern within the tolerable limit. The reason is working wavelengths become tiny close to the size of the proposed antenna at higher frequencies.

5 Conclusions

This paper proposes a novel MIMO antenna based on silicon substrate for UWB communication systems. To achieve high isolation between antennas, six metamaterial FSS units are employed in the middle of array. Finally, the isolation performance is less than −16 dB over the entire UWB frequency band. More importantly, the designed MIMO antenna becomes a very compact construction especially suitable for mobile terminals, like notebook computer or mobile, where antennas need to satisfy the demands of low profile, small dimensions, and integration with other components.

Acknowledgements

This work is supported by National Natural Science Foundation general projects, China (No.61471056), the Youth Learning Scholar Supporting Program of Colleges and Universities of Heilongjiang Province, China (No.1254G051), and the Natural Science Foundation of Heilongjiang Province, China (No. F201322).

Funding

This work is supported by National Natural Science Foundation general projects, China (No. 61471056), the Youth Learning Scholar Supporting Program of Colleges and Universities of Heilongjiang Province, China (No.1254G051), and the Natural Science Foundation of Heilongjiang Province, China (No. F201322).

Authors' contributions

XZ proposed the main idea and completed the antenna construction design. XY assisted the theory research. QS and BL assisted the simulation and analysis. All authors read and approved the final manuscript.

Competing interests

The authors declare that they have no competing interests.

Author details

[1]College of Electrical and Information Engineering, Heilongjiang Institute of Technology, Harbin 150050, China. [2]College of Information and Communication Engineering, Harbin Engineering University, Harbin 150001, China. [3]Collaborative Research Center, Meisei University, Tokyo 1918506, Japan.

References

1. Federal Communications Commission (FCC), First report and order in the matter of revision of part 15 of the com-mission's rules regarding ultra-wideband transmission systems. ET-Docket 98-153, FCC02-48(2002)
2. B Li, CL Zhao, MW Sun, Z Zhou, A Nallanathan, Spectrum sensing for cognitive radios in time-variant flat fading channels: a joint estimation approach. IEEE. Trans. Commun **62**, 2665–80 (2014)
3. B Li, MW Sun, XF Li, A Nallanathan, CL Zhao, Energy detection based spectrum sensing for cognitive radios over time-frequency doubly selective fading channels. IEEE Trans. Signal Process **2**, 402–17 (2015)
4. B Li, Z Zhou, WX Zou, XB Sun, GL Du, On the efficient beam-forming training for 60 ghz wireless personal area networks. IEEE Trans. Wirel. Commun **2**, 504–15 (2013)
5. B Li, J Hou, XF Li, YJ Nan, A Nallanathan, Deep sensing for space-time doubly selective channels: when a primary user is mobile and the channel is flat rayleigh fading. IEEE Trans. Signal Process **13**, 3362–75 (2016)
6. MS Khan, MF Shafique, AD Capobianco, E Autizi, and I Shoaib, in Proceedings of 10th international Burgan conference on applied sciences & technology. Compact UWB-MIMO antenna array with a novel decoupling structure(IBCAST 2013), pp. 347-350
7. TS See, ZN Chen, An ultrawideband diversity antenna. IEEE Trans. Antennas Propag. **57**, 1597–605 (2009)
8. B Li, SH Li, A Nallanathan, CL Zhao, Deep sensing for future spectrum and location awareness 5G communications. IEEE J. Sel. Areas Commun **7**, 1331–44 (2015)
9. H Seokjin, C Kyungho, L Jaewon et al., Design of a diversity antenna with stubs for UWB applications. Microw. Opt. Technol. Lett **50**, 1352–6 (2008)
10. M Koohestani, AA Moreira, AK Skrivervik, A novel compact CPW-fed polarization diversity ultrawideband antenna. IEEE Antennas Wirel. Propag. Lett **13**, 563–6 (2014)
11. M Gallo, E Antonino-Daviu, M Ferrando-Bataller et al., A broadband pattern diversity annular slot antenna. IEEE Trans. Antennas Propag **60**, 1596–600 (2012)
12. B Li, Z Zhou, ZHJ Zhang, A Nallanathan, Efficient beamforming training for 60-ghz millimeter-wave communications: a novel numerical optimization framework. IEEE Trans. Veh. Technol **2**, 703–17 (2014)

2

Context-aware radio resource management below 6 GHz for enabling dynamic channel assignment in the 5G era

Ioannis-Prodromos Belikaidis[1*], Stavroula Vassaki[1], Andreas Georgakopoulos[1], Aristotelis Margaris[1], Federico Miatton[2], Uwe Herzog[3], Kostas Tsagkaris[1] and Panagiotis Demestichas[4]

Abstract

Heterogeneous networks constitute a promising solution to the emerging challenges of 5G networks. According to the specific network architecture, a macro-cell base station (MBS) shares the same spectral resources with a number of small cell base stations (SBSs), resulting in increased co-channel interference (CCI). The efficient management of CCI has been studied extensively in the literature and various dynamic channel assignment (DCA) schemes have been proposed. However, the majority of these schemes consider a uniform approach for the users without taking into account the different quality requirements of each application. In this work, we propose an algorithm for enabling dynamic channel assignment in the 5G era that receives information about the interference and QoS levels and dynamically assigns the best channel. This algorithm is compared to state-of-the-art channel assignment algorithm. Results show an increase of performance, e.g., in terms of throughput and air interface latency. Finally, potential challenges and way forward are also discussed.

Keywords: Dynamic channel assignment, Channel segregation, Quality of service, Heterogeneous networks, 5G

1 Introduction

5G is characterized by the challenges of rapid growth in mobile connections and traffic volume [1, 2]. To address these challenges, the European project SPEED-5G (standing for quality of service provision and capacity expansion through extended-DSA for 5G) focuses on the efficient exploitation of wireless technologies so as to provide higher capacity along with the ultra-densification of cellular technology [3]. Under the framework of SPEED-5G, novel techniques for optimizing spectrum utilization will be developed, following three main dimensions: (i) ultra-densification through small cells, (ii) load balancing across available spectrum, and (iii) exploitation of resources across different technologies. Considering the specific three-dimensional model, which is referred to as extended-dynamic spectrum allocation (eDSA), different spectrum bands and technologies can be jointly managed so as to improve the users' quality of experience (QoE). Hence, the ultimate goal of SPEED-5G boils down to the development of a dynamic radio resource management framework, including mechanisms for interference control, coexistence of heterogeneous networks, management of spectral resources in lightly licensed bands, and other smart resource allocation schemes. It is worth mentioning that this work is an extended version of the work published by the authors in [4, 5].

One of the main scenarios addressed in SPEED-5G is the case of heterogeneous networks where a massive deployment of small cells is put into place to deliver a uniform broadband experience to the users, considering applications with different QoS requirements, such as high resolution multimedia streaming, gaming, video calling, and cloud services. A significant challenge in these networks is the efficient management of co-channel interference (CCI) that occurs due to proximity among the SBSs. Hence, given that the same channels are reused among SBSs due to the scarce spectral resources, CCI constitutes an important restrictive factor for the network performance.

* Correspondence: iobelika@wings-ict-solutions.eu
[1]WINGS ICT Solutions, Athens, Greece
Full list of author information is available at the end of the article

To confront this challenge, dynamic channel assignment (DCA) techniques have been proposed in the literature, either considering a centralized approach [6] or a distributed one [7]. In particular, in [6], a centralized DCA technique considering a heterogeneous network that consists of small cells and macro cells is investigated based on the graph approach. It should be noted that the centralized approaches have several advantages in terms of performance. Nevertheless, the high computational complexity renders them inappropriate for the case of a heterogeneous network with a massive number of small cells. Therefore, distributed DCA techniques have gained the interest of many researchers as a solution that can be applied in future wireless networks. An interesting approach of a distributed adaptive channel allocation scheme known as channel segregation has been proposed in [8], to improve the spectrum efficiency in cellular networks. According to this approach, each cell creates a priority table with the available channels and tries to use the channels with the highest priority. Using this technique, an efficient stable channel re-use pattern is formed and the system performance is ameliorated. Due to the inherent advantages of this method, various DCA mechanisms based on channel segregation have been proposed by the research community [9–12]. However, the majority of the DCA schemes in the literature consider that the SBSs do not differentiate between traffic requests from user equipment (UE) applications, even if the applications do not have the same priority from the user point of view. Considering that in 5G networks, the traffic will range from high data rates to machine type traffic, covering a variety of different applications, there is an emerging need for DCA schemes that provide differentiated QoS to each user, coping with the changing network conditions and the time-varying CCI. Based on this remark and the work presented in [9], we study a modified distributed channel segregation mechanism that takes into account the CCI and the QoS characteristics of the users. The proposed interference and QoS aware channel segregation-based DCA (IQ-CS-DCA) can be employed in order to use the spectral resources efficiently and at the same time prioritize the users with delay-constrained applications (such as video streaming).

The rest of the paper is organized as follows: In Section 2, a brief description of the scenarios considered in SPEED-5G is given, focusing on the scenario of interest. Section 3 summarizes the previous work in this research area, and Section 4 presents an algorithmic description of the proposed IQ-CS-DCA mechanism. Finally, Section 6 discusses the challenges that should be addressed in the mechanism and some future work whereas Section 7 concludes the paper.

2 Scenarios in SPEED-5G

SPEED-5G will mainly investigate indoor and indoor/outdoor scenarios (around buildings) where capacity demands are the highest and where eDSA will exploit efficiently the co-existence of different technologies. More specifically, the selected scenarios are the following:

- Massive IoT (Internet of Things): This scenario refers to the "low-end IoT" and covers devices with sporadic and delay-tolerant traffic, mainly composed of short packets. Among others, this category typically includes wearable devices, smart meters, home automation devices, healthcare, non-critical smart cities sensors, and wireless sensor networks for environmental monitoring.
- Ultra-reliable communications: This scenario refers to a network that supports services with extreme requirements on availability and reliability. Particularly, it is envisioned to have new applications based on M2M (machine-to-machine) and IoT communication with real-time constraints, enabling new functionalities for traffic safety, traffic efficiency, or mission-critical control for industrial and military applications.
- High-speed mobility: This use case considers high-mobility environments (e.g., high-speed trains and cars on highways) where broadband communications need to be achieved.
- Broadband wireless: This use case constitutes the scenario of interest, and it focuses on a mixture of domestic, enterprise and public access outdoor and indoor environments located in a densely populated urban area (see Fig. 1). In this case, a large number of small cells co-exist within a macro-cell offering an improved communication experience to the users.

In order to meet the 5G requirements, which characterize the specific use case, we propose a DCA mechanism for the efficient usage of the available spectrum, driven by the coordination of the CCI and the users' QoS requirements.

3 Related work

The concept of heterogeneous networks focuses on the improvement of spectral efficiency per unit area using a diverse set of base stations (BS), in a mix of macro cells and small cells. As highlighted in the introduction section, one of the main problems in these networks is the efficient management of CCI between the different cells to enhance the network performance. Towards this direction, one promising category of DCA mechanisms, which has been recently studied in the literature, refers to the channel segregation-based DCA (CS-DCA) mechanisms [10].

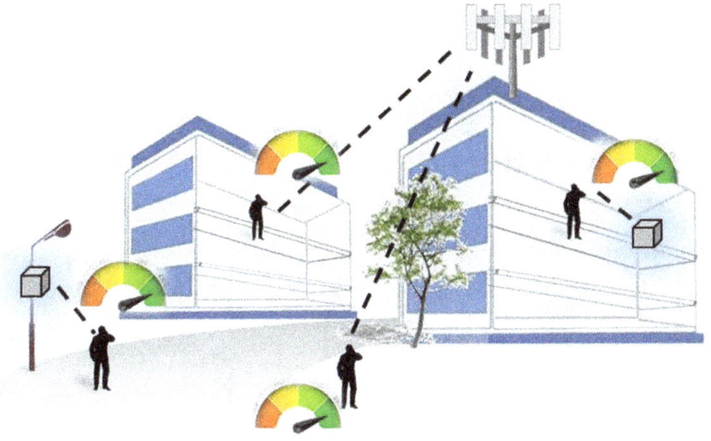

Fig. 1 Broadband wireless scenario of SPEED-5G

One of the first approaches of channel segregation appears in [8]. In this work, each BS acquires its favorite channel independently, through learning from statistical data. As a result, the process of channel re-use is self-organized, leading to an amelioration of spectrum efficiency. In the simulation results, the proposed mechanism is compared with a system without segregation and the performance improvement in terms of blocking probability and channel utilization is demonstrated.

In [9], the authors modify the previous CS-DCA mechanism for application to Distributed Antenna Networks (DANs). According to the modified scheme, known as interference-aware CS-DCA (IACS-DCA), the average CCI is computed for the available channels and a CCI table attached to each antenna is created with the first channel to have the lowest CCI value. Hence, before a transmission attempt, the user selects the closest distributed antenna and the first channel of the corresponding table is assigned to him. In the simulation analysis, the performance improvement due to the existence of DAN is confirmed. Furthermore, the superiority of the proposed algorithm compared to a fixed channel allocation (FCA) mechanism is proven. Based on this mechanism, the authors in [11] examine several metrics such as the autocorrelation function, the fairness index of channel reuse pattern and the minimum co-channel distance among the BSs. From their analysis, it can be seen that the proposed scheme forms a CCI minimized channel reuse pattern that ameliorates the signal-to-interference ratio (SIR) compared to other channel assignments schemes.

Taking into account the increasing number of wireless terminals, the authors in [12] propose a modification of the IACS-DCA, named as 'multi-group IACS-DCA.' According to this mechanism, the available channels are divided into multiple groups and the initial IACS-DCA is applied to each channel group. As they highlight in their simulation analysis, the specific mechanism results not only in a more stable reuse pattern in case of multiple users but also in amelioration of the SIR compared to the single-group IACS-DCA scheme.

An energy-efficient approach of the IACS-DCA scheme is presented in [13]. In this work, the use of a transmission power control scheme in combination with IACS-DCA is studied and the stability of the channel reuse patter is verified. From the simulations analysis, some indicative positive results for the transmission power reduction are presented. A similar energy-efficient approach is also proposed by the authors in [14]. In their work, they propose a modified approach of IACS–DCA mechanism in combination with a learning game-theoretic algorithm for the BS sleep process. The performance of the proposed mechanism is investigated through simulation and its superiority in terms of energy and spectral efficiency is proven.

4 Algorithmic approach for RRM/MAC

In this section, we describe the proposed IQ-CS-DCA mechanism that is based on the IACS–DCA mechanism presented in [9]. At first, we present an abstract formulation of the considered optimization problem whereas in the second subsection a more algorithmic approach of the proposed mechanism is given.

4.1 Abstract form of the optimization problem

In our approach, we consider the scenario of a heterogeneous network with one macro-cell and multiple small-cells, similar to Fig. 2. Considering the network elements, an abstract formulation of the studied resource allocation problem can be given as follows:

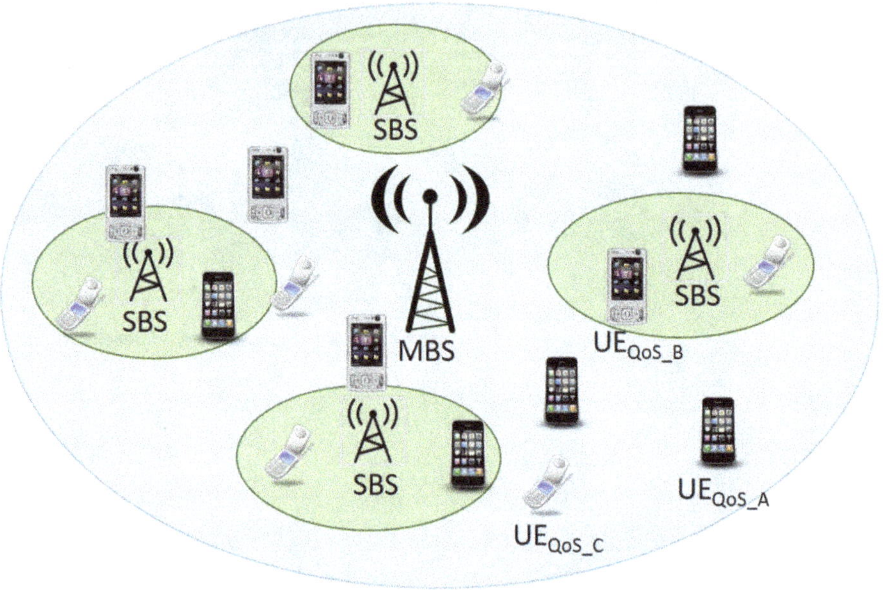

Fig. 2 Heterogeneous network scenario

4.1.1 Given:

- The large number of SBSs
- The diverse QoS requirements of the UEs
- The time-varying network conditions (due to various traffic characteristics, changing propagation environment, power control, etc.)
- The limited number of spectrum channels

4.1.2 Find:

- An efficient association of UEs to SBSs
- An efficient channel assignment to UEs

4.1.3 So as:

- To maximize the spectral-efficiency (via an adequate re-use channel pattern)
- To satisfy the communication quality of the UEs

4.2 High level description of IQ-CS-DCA mechanism

The proposed mechanism can be divided in five main steps. The following flowchart in Fig. 3 summarizes the *IQ-CS-DCA* mechanism, and each phase is briefly described.

- Initialization phase: During this phase, each SBS chooses randomly a channel from the pool of available channels and broadcasts a beacon signal on this channel.
- Measurement phase: Each SBS measures periodically the instantaneous beacon signal power on each of the available channels for a specific time duration. The received power can be computed considering both path loss and fading phenomena for a more complete analysis of the radio propagation environment.
- Creation of channel priority table: Each SBS creates the channel priority table based on average CCI power levels. In this step, the average CCI power can be computed either by using the first-ordering filtering similar to [9] or by using other learning/average mechanisms that use past CCI measurements and result in a stable assignment. The channel with the lowest CCI appears first in the priority table and the other channels with descending order of CCI follow.
- UEs-SBS association: During this phase, each UE associates with a SBS depending on various metrics (e.g., the highest receive signal strength indicator (RSSI) and the load due to other UEs associated with this SBS).
- Collection of requests: Each SBS collects the channel requests from the UEs.
- Creation of user QoS requirement priority tables: UEs are prioritized depending on their application priority, and the SBSs divide its priority table into multiple tables (depending on the number of UEs/applications). The first channel of each table is assigned to each UE depending on its application priority and the channel quality given by the CCI power level (better channels are given to UEs with stricter QoS requirements).
- Channel assignment: Each SBS assigns the channels to the users as follows based on the QoS priority tables.

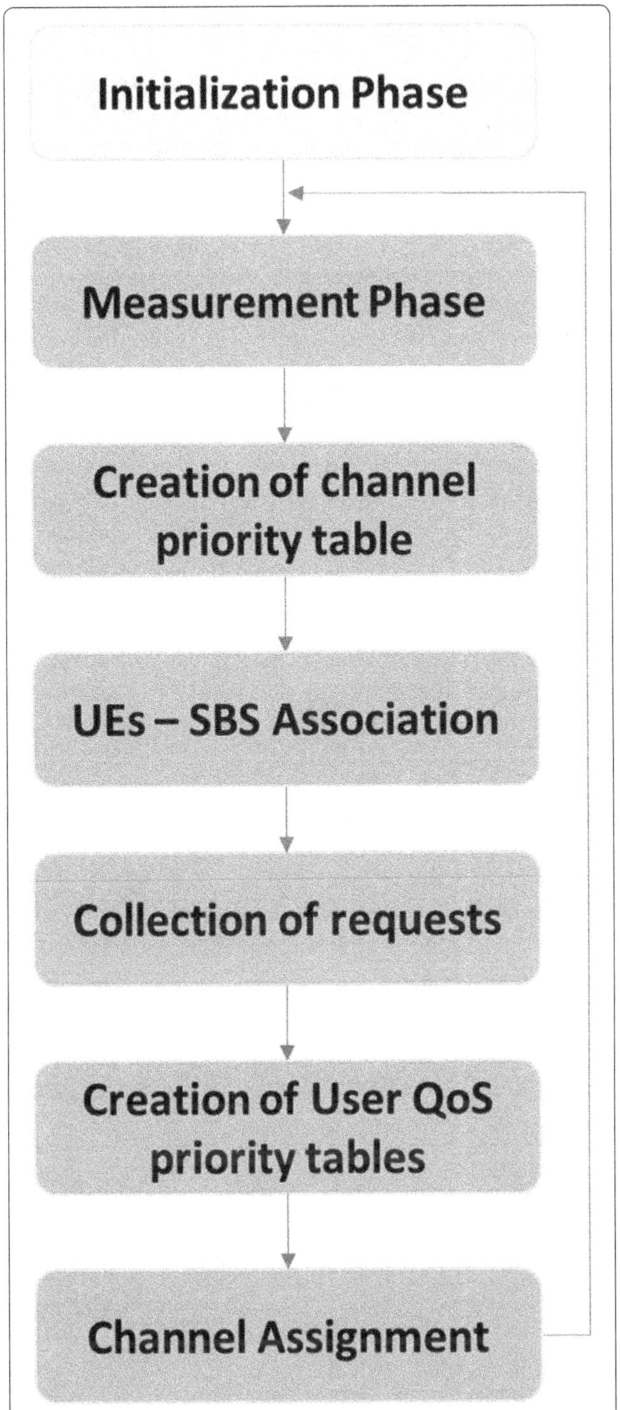

Fig. 3 Flowchart of IQ-CS-DCA mechanism

4.3 Implementation approach of the proposed algorithm

To evaluate our proposed solution, we have implemented two algorithms as a first stage. In order to provide results as a "proof of concept," we have introduce a state-of-the-art algorithm and our proposed algorithm which uses the interference levels acquired from the network as well as the QoS requirements from each UE.

The different QoS requirements are related to the 5G scenarios as explained at the SPEED-5G project. For example IoT users will have low priority in our algorithm. Ultra reliable communications will be at highest priority, meaning that they will be assigned to channels with better SINR value. Finally, the broadband communications where the users will either have a medium priority with respect to the quality of the channel assigned to or even a high priority given that could belong to a category of users that need to have low latencies and high throughputs.

In advance, those two solutions are compared with an algorithm from the presented state-of-the-art. The algorithm that was developed is the dynamic channel assignment scheme for distributed antenna networks found at [9] that our solution was based on. Figure 4 illustrates the procedure of the algorithm on the assignment of the channels for the multitude of the antennas as presented into the article.

Specifically, the first algorithm that has been investigated is called "random-based channel assignment (CA) algorithm" (Fig. 5). This solution does not have a certain logic for the assignment of channels that is why it is called 'random,' it arbitrarily allocates users to different channels without any knowledge of the channels current status or the whole system. As input, we have a set of UEs U that want to transmit, a set of macro BS M, a set of channels C, a set of available channels Ca ⊆ C. Then a certain procedure is followed in order to allocate certain channels to UEs as the flowchart illustrates.

The random-based channel assignment algorithm is used as a baseline in order to compare, evaluate, and

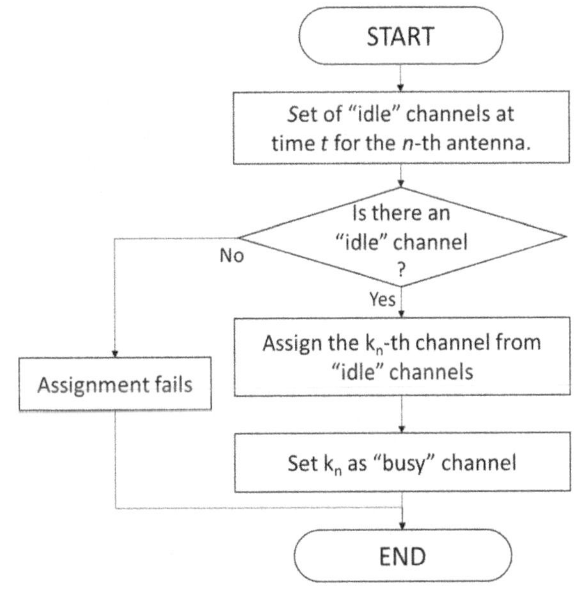

Fig. 4 Flowchart of channel assignment at the n-th antenna [9]

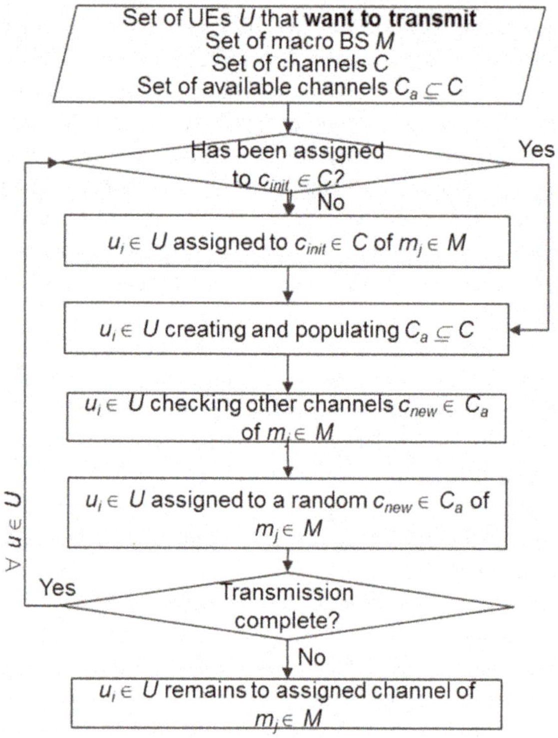

Fig. 5 Flowchart of the random-based CA algorithm

optimize the effectiveness of the next algorithm that we propose (Fig. 6). As input, we have again a set of UEs U that want to transmit, a set of macro BS M, a set of channels C, a set of available channels Ca ⊆ C. The selection procedure differs from the random channel assignment algorithm since here we introduce a control point for checking the best available channels in order to select these (if available). The best channel is identified according to the SINR and if the SINR of a new channel is better than the currently utilized one, then the UE will switch to the better channel. This algorithm enables context-aware RRM as each base station can collect the interference levels for each user that is connected to a specific channel in order to deduce the radio environment status and exploit it appropriately. In general, it is expected that through this algorithm, it will be possible to achieve better quality (e.g., higher throughput and less latency).

In order to calculate the SINR levels, every SBS at every cycle, creates the average interference that it calculates from the input of the UEs in the area for each other SBS. Instead of recalculating the signal strength at the location of the SBS, we are using the feedback from the SBS' served UE devices, to make the measurements more realistic and efficient. The UE devices store the signal strength per SBS in their physical layer variables. Therefore, we are averaging out this signal strength to calculate the average per rat interference. After this calculation, we then turn to the history of our RRM model transmission. Since it is not possible to know beforehand what collisions will occur on an air frame, we are using the previous transmission (history) to calculate these collisions and assume that statistically the impact will be the same. We create the resource utilization mask for each SBS, based on the RBs utilization data that were stored in the history variable.

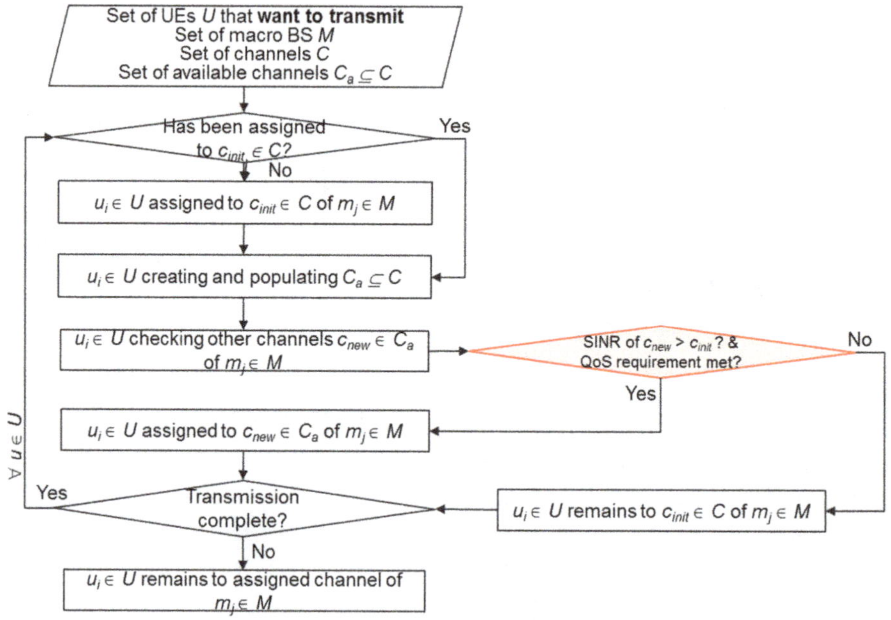

Fig. 6 Flowchart of our proposed algorithm

5 Evaluation aspects of the proposed algorithms

5.1 Simulation tool

For the evaluation of such concepts, extensive system level simulations are conducted. The implementation of our suggested solution was performed under a proprietary system-level simulation tool which is fully developed in Java with various capabilities and has been calibrated according to the 3GPP specifications. The simulator takes into account various parameters such as traffic level, available infrastructure elements, available channels and evaluates the various test cases. The calibration state of the proprietary simulator has been checked against the reference results of the 3GPP LTE calibration campaign [36.814] [15]. As a result, the cumulative distribution function (CDF) of coupling loss and downlink SINR have been checked in order to calibrate the tool with leading operators and vendors such as Nokia, Ericsson, Docomo, Huawei, and Telecom Italia.

The configuration is fully customizable so as to include various types of cells (i.e., macro and small cells).-Specifically, it is possible to customize the following: the size of playground; the area type (e.g., dense urban); the number and position of macro base stations and their inter-site distances (ISDs); the number and position of small cells per macro base station; the number and position of end-user devices in the playground; the number of available channels. In addition, the pathloss models for macrocells at 2 GHz band is set to $L = 128.1 + 37.6\log10(R)$, R in km, and for small cells is set to $L = 140.7 + 36.7\log10(R)$, R in km.

5.2 Simulation parameters

The parameters imported to the simulator are, 19 macro base stations (BS) each with three cells and also 9 small-cells per (BS) giving us a total of 228 cells (in which 171 are small cells uniformly distributed in the simulation playground) throughout the network. In addition, we have utilized 4 channels at 20 MHz bandwidth for every cell. The topology of the network created by the simulation tool is illustrated at (Fig. 7), and the differing shapes and symbols can be interpreted as the UEs and the wireless communication links. Specifically users are shown as small circles with four different colors (red, green, blue, and light blue) that represent the four channels that have been utilized for our scenarios. Furthermore, the green arrows illustrate the transmission process and the connection topology between UEs and BSs of each user to a specific cell of our network. The small cells are located close to the center of the macro cells and are working at the 3.5 Ghz band in contrary to the macro cells that are working on 2 Ghz, giving us a heterogeneous environment for our scenarios.

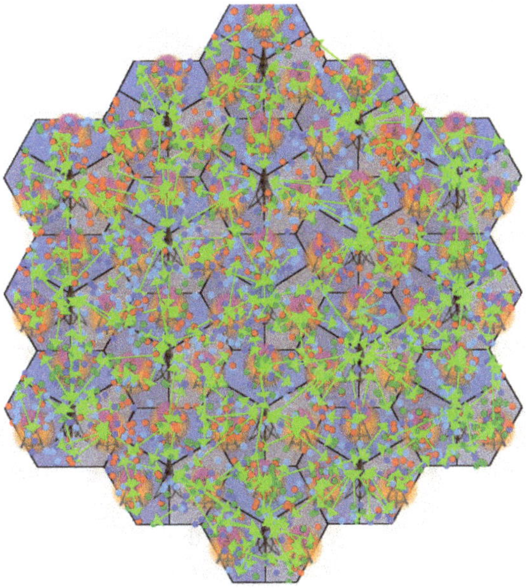

Fig. 7 Network topology created by the simulation tool

In addition, Table 1 presents the configuration of the base stations used in each case and their values that have been introduced to our simulator for the development and evaluation of our solution.

5.3 Experimentation scenarios and test cases

In order to analyze these two algorithms with the use of the simulation tool we had to introduce different scenarios and test cases that are summarized in our experiments Table 2. We have experimented with five values of sessions that every user requests per day ranging from 2880 up to 14,400 sessions. The file size requested from each user is 2 Mb, which means that a user can request from 4 up to 20 Mb/min. Those values could provide us with a broader knowledge of the algorithm capabilities for the specific network topology implemented to the simulator.

For each of the cases, three algorithms have been evaluated. Algorithm (A) is our proposed algorithm which builds on state-of-the-art and adds the notion of QoS prioritization. Algorithm (B) is a state-of-the-art algorithm which sorts available channels based on SINR values, and Algorithm (C) uses a random allocation of channels (not necessarily the best one).

Moreover, high, medium, and low priority services are considered where our proposed algorithm tries to

Table 1 Configuration of BSs

BS	MIMO mode	Bandwidth
Macro	2 × 2	20 Mhz
Small	Omni-antenna	20 Mhz

Table 2 Tested scenario cases

Test cases	Users	Sessions/day/user	Packet size
1	5000	14400	2MB
2	5000	11520	2MB
3	5000	8640	2MB
4	5000	5670	2MB
5	5000	2280	2MB

allocate the best possible channels firstly to the high priority services, then to the medium and finally to the low priority services.

5.4 Evaluation results

Figure 8 indicates the results from our experimentation. The barchart illustrates the cases of Table 1 and specifically the average air interface latency. On average, it is shown that our proposed algorithm outperforms the other two algorithms (up to 50%) especially in high and medium priority services by giving them a performance boost. On the contrary, low priority services seem that they do not benefit as much as the other two.

In Fig. 9, the results are sorted by priority levels, and we see in a clearer way the large benefit of our algorithm for high priority which is not the case for low priority services.

Furthermore, Fig. 10 illustrates the normalized throughput for each of the test cases and compared among each algorithm. It is evident that our algorithm performs better in almost every test case and especially in cases with higher loads (compared to less-loaded simulations).

Similarly, Fig. 11 illustrates the normalized throughput as of service priority levels and also here (as shown in latency charts), our solution seems to perform better especially in higher and medium priority services compared to low priority services. In this article, we investigated how a radio resource management algorithm utilizing contextual information acquired from the network and taking into account QoS requirement can cope on some specific scenarios. The test cases introduced here where designed in order to investigate the performance difference between the state-of-the-art and our proposed solution for an environment of almost one user per square meter at the case of 5000 users and files for 2 Mb size and variable number of requests per minute per user (ranging from 2 to 10). Our proposed solution was able to dynamically choose the best channel based on interference of the current position and thus allow each user to connect with higher speed and receive the file faster with less air interface latency.

On the contrary, the algorithms that used for comparison on average were making the less optimal selection of the channels (without giving priority based on QoS requirements), hence the users were not able to download at full speed and with higher loss packet ration, creating a continuously loop of poor selection of channels without being able to overcome this situation.

Furthermore, there are some differences between random allocation of channels and state-of-the-art algorithm when increased load is provided in the system. Specifically, Fig. 12 illustrates the differences between the algorithms for various priorities when 2× more sessions/day/user are tested in the system compared to test case 1. The random allocation has worst performance in high and medium priority services. The state-of-the-art algorithm performs better and our proposed algorithm has the best performance especially in high and medium priority services (which is the main aim of our study).

6 Challenges and future work

In the following, we discuss some challenges that we would like to address in the development of IQ-CS-DCA as well as some of the possible improvements left for future work.

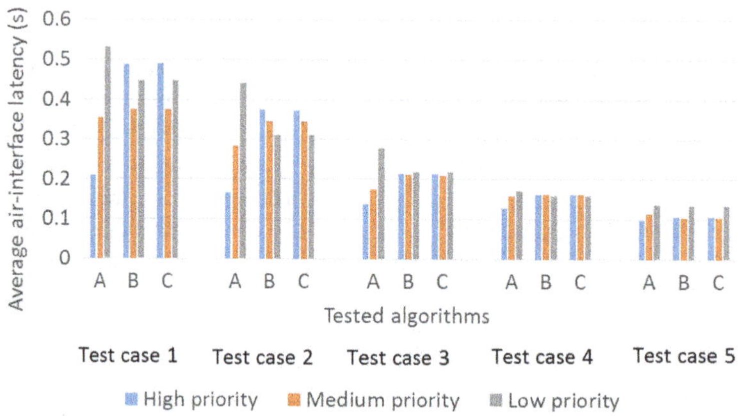

Fig. 8 Average air-interface latency for each test case(s)

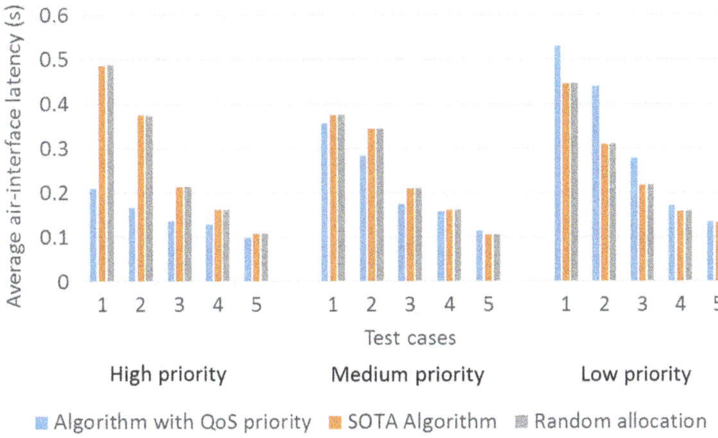

Fig. 9 Average air-interface latency for each service priority level

The ultra-densification of networks that is currently envisioned for 5G will bring new challenges to the radio access, especially related to interference management. In fact, ultra-dense networks are characterized by interference patterns that change quickly in time and that strongly depend on the realistic network deployment [16]. Therefore, interference mitigation techniques must be sufficiently dynamic and operate on a sufficiently small time-scale to ensure that the fluctuations in the interference are captured. Clearly, the same challenges also apply to other mechanisms that take into account interference measurements to operate, as in the case of the proposed IQ-CS-DCA.

Additionally, the aggregation of multiple radio access technologies (RATs), possibly operating on very diverse frequency bands with different characteristics, brings new challenges related to efficient ways to ensure the provision of the end-to-end QoS, as some RATs may not supply an explicit way to perform QoS management. This is currently a very active area of research in both academia and industry [17]. Examples of current technologies that will still play an important role in the decades to come are the Long Term Evolution (LTE) and the IEEE 802.11 (WiFi) family of standards. Table 3 gathers some of the key differences between the two, exemplifying the challenge due to the differences in the physical layer and the radio channel access.

As it has been referred in the Introduction section, a purely centralized approach to DCA might not be feasible in practice due to the large amount of feedback that must be exchanged between the BSs and the strict delay requirements that must be met by the network infrastructure. On the other hand, purely decentralized techniques might not be able to address all the aforementioned challenges. As a future work, we would like to investigate hybrid or semi-decentralized approaches, for instance cluster-based algorithms that take decisions with a minimum amount of feedback to be exchanged between the SBSs. Clearly, investigations should be performed to analyze the tradeoff between the complexity introduced due to the feedback channel required by these techniques versus the achieved benefits.

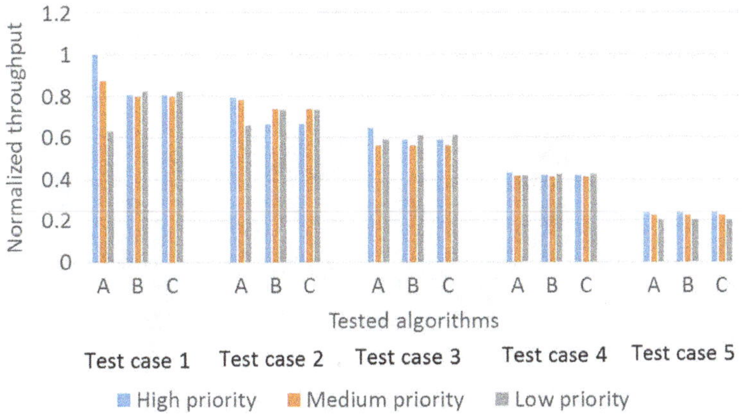

Fig. 10 Normalized throughput for each test case(s)

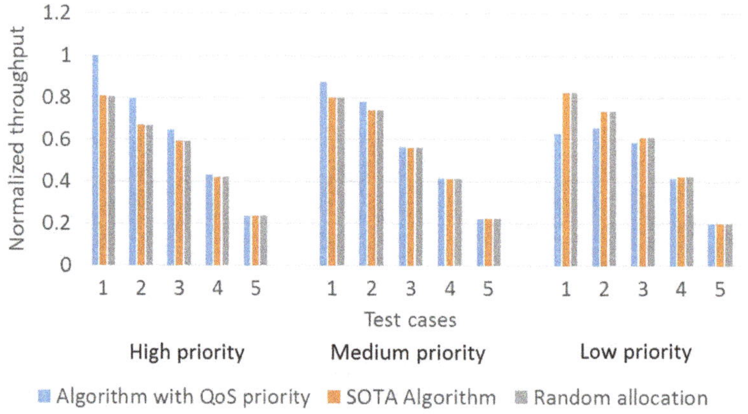

Fig. 11 Normalized throughput for each service priority level

Hybrid approaches to DCA can also bring benefits against the bursts of interference that can occur in heterogeneous networks, since they may avoid the strong interference factor generated by closely located BSs, during the initial measurement phase of the IQ-CS-DCA algorithm. It is worth saying that the assumption of a clustered architecture of SBSs is a viable hypothesis, since other functional entities of the network may already require this structure (e.g., to perform inter-cell interference coordination or soft handovers between SBSs). These approaches could leverage the X2 interface currently defined in LTE-Advanced, as well as the Xw interface newly introduced in LTE Release 13 between a 3GPP BS and a WiFi access point [18].

The approach presented in this paper is directly applicable to the case of heterogeneous RATs. Compared to an algorithm that deals with the underlying available channels agnostically, improvements are expected by providing additional information related to the different available technologies in the channel assignment step. As an example, a fast moving UE might better be assigned to a licensed band using LTE, rather than scheduled to an unlicensed band operating on WiFi. In other cases, though, it may be beneficial to off-load broadband static users to unlicensed WiFi in order to reduce the CCI generated in the licensed band. Additional information is needed by such evolved mechanisms, including the UE's capabilities (i.e., the supported bands and technologies), as well as a characterization of the user's mobility pattern, among others.

Finally, the growing importance of techniques for dynamic spectrum access in next-generation wireless systems should also be taken into consideration in the design of efficient DCA algorithms, together with the diverse characteristics of the underlying frequency bands. For instance, TV White Spaces (TVWS) are highly location dependent. TVWS frequencies might be available in a particular geographical area while being completely occupied at another location. Therefore, a centralized geolocation database might be necessary to implement a coexisting LTE-TVWS system. In such system, the information provided by the geolocation database is semi-

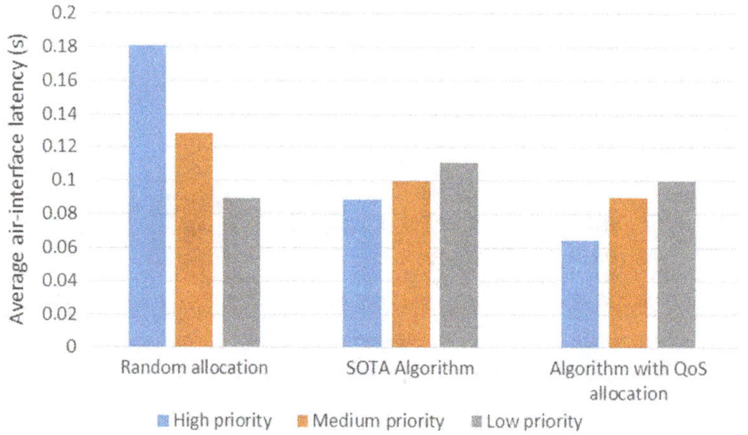

Fig. 12 Average air-interface latency for test case with 2× more sessions/day/user compared to test case 1 (s)

Table 3 Comparison between LTE and WiFi

	LTE	WiFi
Spectrum access	Licensed	Unlicensed
Channel bandwidths	1.4, 3 5, 10, 15, 20 MHz	5, 10, 20 MHz
Channel access method	Centralized	Contention-based
Physical layer	OFDMA/SC-FDMA	OFDM-CSMA
Optimized for mobility	Yes	No

static and does not change often in time. However, the inherent unpredictability of these frequencies might make them ideal only to best-effort applications with no QoS requirements, further highlighting the importance of intelligent RAT-aware algorithms, as pointed out in [17].

7 Conclusions

In this paper, we consider the case of a heterogeneous network scenario with macro-cell and small cells. Due to the inherent network architecture, one major challenge is the efficient management of CCI. In our work, numerous interference aware DCA mechanisms are discussed and the channel segregation approach is presented. A high level modified interference aware DCA mechanism that takes into account the differentiated QoS requirements of the users is proposed. Furthermore, the algorithmic approach of the mechanism is presented through some flow charts and two algorithmic approaches for the evaluation and the proof of concept are examined. The algorithm with the interference aware capability that uses the SINR measurements acquired from the network is able to provide better results. The advantage of our algorithm introduced here can further explored with the utilization of the small-cell at the macro cell edge in order to attract problematic traffic due to the great CCI presented in those areas. Finally, possible challenges and points for future work are recognized in order to have a more complete analysis of the resource allocation problem.

Acknowledgements
This work has been performed in the framework of the Horizon 2020 project SPEED-5G (ICT-671705) receiving funds from the European Union.

Funding
This work has received funding from European Commission, Horizon 2020, SPEED-5G project as mentioned also in the Acknowledgment section.

Authors' contributions
IPB contributed to the problem statement, algorithm implementation, simulation execution, and results. SV contributed to the problem formulation and related work search. AG contributed to the problem formulation, scenario definition, simulation execution, and result analysis. AM contributed to the implementation of state-of-the-art algorithm and simulation execution. FM contributed to the state-of-the-art with emphasis on the heterogeneous network scenario and relation to our proposed work. UH contributed to the definition of technical use cases and simulation scenarios in order to emphasize the benefits/advancements of our work with respect to the state-of-the-art. KT contributed to the algorithm implementation and simulation execution. PD contributed to the problem statement, formulation, and algorithm specification. All authors read and approved the final manuscript.

Competing interests
No competing interests are foreseen for this publication.

Author details
[1]WINGS ICT Solutions, Athens, Greece. [2]Sistelbanda, Valencia, Spain. [3]EURESCOM, Heidelberg, Germany. [4]University of Piraeus, Piraeus, Greece.

References
1. J.G. Andrews et al., "What will 5G be?" Selected areas in communications. IEEE Journal on **32**(6), 1065–1082 (2014)
2. P. Demestichas et al., 5G on the horizon: key challenges for the radio-access network. Vehicular Techn. Magazine, IEEE **8**(3), 47–53 (2013)
3. SPEED-5G EU project site: https://speed-5g.eu/. Accessed Sept 2017.
4. S. Vassaki, A. Georgakopoulos, F. Miatton, K. Tsagkaris, P. Demestichas, *Interference and QoS aware channel segregation for heterogeneous networks: a preliminary study* (2016 European Conference on Networks and Communications (EuCNC), Athens, 2016), pp. 195–199
5. A. Georgakopoulos, A. Margaris, K. Tsagkaris, P. Demestichas, Resource sharing in 5G contexts: achieving sustainability with energy and resource efficiency. IEEE Veh. Technol. Mag. **11**(1), 40–49 (2016)
6. S.-J. Kim, I. Cho, Y.-K. Kim, C.-H. Cho, A two-stage dynamic channel assignment scheme with graph approach for dense femtocell networks. IEICE Trans. Commun. **E97-B**(10), 2222–2229 (2014)
7. D. Goodman, S.A. Grandhi, and R. Vijayan, Distributed dynamic channel assignment schemes, Proc. IEEE 43rd Vehicular Technology Conference (VTC1993-Spring), 1993
8. Y. Furuya, Y. Akaiwa, Channel segregation, a distributed adaptive channel allocation scheme for mobile communication systems. IEICE Trans. Commun. **74**(6), 1531–1537 (1991)
9. R. Matsukawa, T. Obara, and F. Adachi. A dynamic channel assignment scheme for distributed antenna networks. Vehicular Technology Conference (VTC Spring), 75th. IEEE, 2012
10. S. Glisic, and B. Lorenzo. Advanced wireless networks: cognitive, cooperative & opportunistic 4G technology. London: Wiley ;2009.
11. Y. Matsumura et al., Interference-aware channel segregation based dynamic channel assignment for wireless networks. IEICE Trans. Commun. **98**(5), 854–860 (2015)
12. T. Katsuhiro, and F. Adachi. Multi-group interference-aware channel segregation based dynamic channel assignment. Network Infrastructure and Digital Content (IC-NIDC), 2014 4th IEEE International Conference on. IEEE, 2014
13. Y.Matsumura, et al. Interference-aware channel segregation based dynamic channel assignment using SNR-based transmit power control. Intelligent Signal Processing and Communications Systems (ISPACS), 2013 International Symposium on. IEEE, 2013
14. A. Mehbodniya, et al. Energy-efficient dynamic spectrum access in wireless heterogeneous networks. Communication Workshop (ICCW), 2015 IEEE International Conference on. IEEE, 2015
15. 3 GPP TR 36.814, Further advancements for E-UTRA physical layer aspects, March 2010.
16. B. Soret et al., Interference coordination for dense wireless networks. IEEE Commun. Mag. **53**(1), 102–109 (2015)
17. U. Herzog et al., Quality of service provision and capacity expansion through extended-DSA for 5G, European Conference on Networks and Communications (EuCNC 2016), 2016, in press
18. 3 GPP TR 37.870 V13.0.0, Study on multiple radio access technology (Multi-RAT) joint coordination, June 2015.

3

Uncoordinated pilot decontamination in massive MIMO systems

Jesper H. Sørensen* and Elisabeth de Carvalho

Abstract

This work concerns wireless cellular networks applying time division duplexing (TDD) massive multiple-input multiple-output (MIMO) technology. Such systems suffer from pilot contamination during channel estimation, due to the shortage of orthogonal pilot sequences. This paper presents a solution based on pilot sequence hopping, which provides a randomization of the pilot contamination. It is shown that such randomized contamination can be significantly suppressed through appropriate filtering. The resulting channel estimation scheme requires no inter-cell coordination, which is a strong advantage for practical implementations. Comparisons with conventional estimation methods show that the MSE can be lowered as much as an order of magnitude at low mobility. Achievable uplink and downlink rates are increased by 42 and 46%, respectively, in a system with 128 antennas at the base station.

Keywords: Massive MIMO, Pilot contamination, Kalman filter

1 Introduction

Muliple-input multiple-output (MIMO) technology [1] is finding its way into practical systems, like LTE and its successor LTE-Advanced. It is a key component for these systems' ability to improve the spectral efficiency. The success of MIMO technology has motivated research in extending the idea of MIMO to cases with hundreds, or even thousands of antennas, at transmitting and/or receiving side. This is often termed *massive MIMO*. In mobile communication systems, like LTE, the more realistic scenario is to have a massive amount of antennas only at the base station (BS), due to the physical limitations at the user equipment (UE). It has been shown that such a system [2], in theory, can eliminate entirely the effect of small-scale fading and thermal noise, when the number of BS antennas goes to infinity. The only remaining impairment is inter-cell interference, caused by imperfect channel state information (CSI), which is a result of non-orthogonality of training pilots used to gather the CSI. This is often referred to as *pilot contamination*. It is considered as one of the major challenges in massive MIMO systems [3].

Mitigation of pilot contamination has been the focus of several works recently. These fall into two categories: one with coordination among cells and one without. The first category includes [4], where it is utilized that the desired and interfering signals can be distinguished in the channel covariance matrices, as long as the angle-of-arrival spreads of desired and interfering signals do not overlap. A pilot coordination scheme is proposed to help satisfying this condition. The work in [5] utilizes coordination among BSs to share downlink messages. Each BS then performs linear combinations of messages intended for users applying the same pilot sequence. This is shown to eliminate interference when the number of BS antennas goes to infinity.

The category without coordination also includes notable contributions. A multi-cell precoding technique is used in [6] with the objective of not only minimizing the mean squared error of the signals of interest within the cell but also minimizing the interference imposed to other cells. In [7], it is shown that channel estimates can be found as eigenvectors of the covariance matrix of the received signal when the number of BS antennas grows large and the system has "favorable propagation." The work in [8–11] is based on examining the eigenvalue distribution of the received signal to identify an interference free subspace on which the signal is projected. It is

*Correspondence: jhs@es.aau.dk
Department of Electronic Systems, Aalborg University, Fredrik Bajers Vej 7, 9220 Aalborg, Denmark

shown that an interference free subspace can be identified when certain conditions are fulfilled concerning the number of BS antennas, user equipment antennas, channel coherence time, and the signal-to-interference ratio. Recently, in [12], a combination of the solutions in [4] and [8–11] was proposed. The resulting solution unites the strengths of these solutions leading to a more robust pilot decontamination.

In this paper, we propose pilot decontamination, which does not require inter-cell coordination and is able to exploit past pilot signals. It is based on pilot sequence hopping performed within each cell. Pilot sequence hopping means that every user chooses a new pilot sequence in each time slot. Thus, in every time slot, the pilot signal of a user is contaminated by a different set of interfering users, which means channel estimation is affected by a different set of interfering channels. If channel estimation is carried out based solely on the pilot sequence of the current slot, then pilot sequence hopping does not bring any gain. The key in our solution is a channel estimation that incorporates multiple time slots so that it can benefit from randomization of the pilot contamination. Recent work utilizing temporal correlation for channel estimation is found in [13], although not in combination with pilot hopping and not with the purpose of mitigating pilot contamination. Random selection of pilot sequences is also explored in [14] and [15]. Both works consider the random access problem in cellular networks. In [14], pilot contamination is avoided through a distributed collision detection algorithm, which enables users with weak channels to detect that they are contaminators of a user with a strong channel and as a result postpone their transmission. The work in [15] considers codeword transmissions that are spread across multiple time slots, each with a different contaminator. This decorrelates the contamination within a single codeword, which improves performance.

When the channel is time-variant and correlated across time slots, it is possible to exploit the information about the channel across time slots by an appropriate filtering and benefit from contamination randomization. In this paper, channel estimation across multiple time slots is performed using a modified version of the Kalman filter, which is capable of tracking the channel and the channel correlation. The level of contamination suppression depends on the channel correlation between slots of the UE of interest as well as the contaminators. In LTE, channel correlation between time slots is large even at medium-high speeds, making the proposed solution very efficient.

This work is an extension of the work in [16], where the concept of pilot sequence hopping in combination with a Kalman channel tracker is introduced. In this paper, the work is extended with more sophisticated mobility models and a generalization of the estimation algorithm, which allows higher order Kalman process models. Furthermore, the Bayesian Cramer-Rao lower bound is derived for the estimation problem at hand and applied as a benchmark in the numerical evaluations.

The remainder of this paper is organized as follows. Section 2 presents the applied channel and mobility models and the problem of pilot contamination. The proposed solution is described in Section 3 and analyzed in Section 4. Section 5 provides numerical results and a comparison to existing solutions. Finally, conclusions are drawn in Section 6.

2 System model

In this work, we denote scalars in lower case, vectors in bold lower case, and matrices in bold upper case. A superscript "T" denotes the transpose, and a superscript "H" denotes the conjugate transpose.

2.1 Channel model

This work treats a cellular system consisting of L cells with K users in each cell, see Fig. 1. A time division duplexing (TDD) massive MIMO scenario is considered, where the BS has M antennas and the UE has a single antenna. We restrict our attention to the channel estimation performed in a single cell, which we term "the cell of interest" and assign the index "0." The channel between the BS in the cell of interest and the kth user in the ℓth cell is denoted as $\boldsymbol{h}^{k\ell} = \left[h^{k\ell}(1)h^{k\ell}(2)\ldots h^{k\ell}(M)\right]$, where the individual channel coefficients are complex scalars. Note that for $\ell > 0$, $\boldsymbol{h}^{k\ell}$ refers to a channel between the BS of interest and a UE connected to a different BS. We furthermore restrict our attention to the estimation of a single channel coefficient; hence, a channel is denoted as the complex scalar $h^{k\ell}$. The work easily extends to vector estimations, in which case spatial correlation can be exploited for improved performance. A rich scattering environment is assumed, such that $h^{k\ell}$ can be modeled using Clarke's model [17]. Hence, in a time-slotted system, where a time slot has a length of t_s seconds, the channel coefficient in the nth time slot is

$$h_n^{k\ell} = \frac{1}{\sqrt{N_s}} \sum_{m=1}^{N_s} \exp\left(j2\pi f_d n t_s \cos\alpha_m + \phi_m\right), \qquad (1)$$

where N_s is the number of fixed scatterers associated with all BS/UE pairs, f_d is the maximum Doppler shift, α_m and ϕ_m are the angle of arrival and initial phase, respectively, of the wave from the mth scatterer. Both α_m and ϕ_m are independent and uniformly distributed in the interval $[-\pi, \pi)$, which results from random scatterer locations. Furthermore, $f_d = \frac{v}{c} f_c$, where v is the speed of the UE, c is the speed of light and f_c is the carrier frequency.

In a TDD massive MIMO system, collection of channel state information (CSI) is performed using uplink pilot

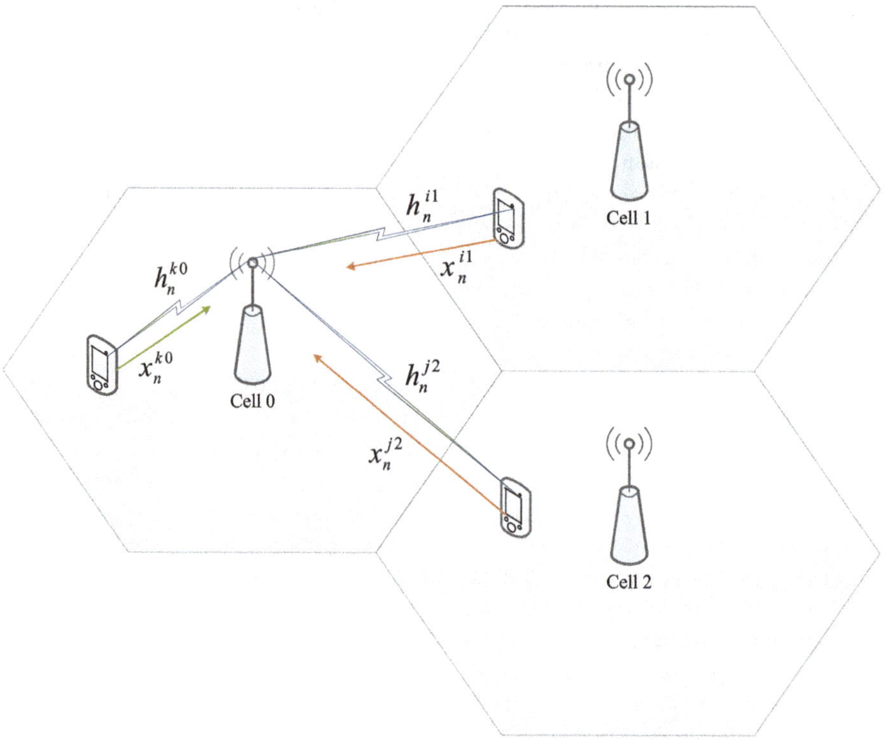

Fig. 1 A cellular system with three cells. Cell 0 is of interest and the neighboring cells will potentially cause interference (red arrows)

training. The CSI achieved this way is utilized in both downlink and uplink transmissions based on the channel reciprocity assumption. We define a pilot training period followed by an uplink and a downlink transmission period as a time slot. See Fig. 2 for an example of a transmission schedule with two time slots. We assume all transmissions in the system are synchronized, which represents a worst-case scenario from a pilot contamination perspective, as argued in [2]. During the nth pilot training period, the kth user in the ℓth cell transmits a pilot sequence $\boldsymbol{x}_n^{k\ell} = \left[x_n^{k\ell}(1)x_n^{k\ell}(2)\ldots x_n^{k\ell}(\tau)\right]^T$, where τ is the pilot sequence length. Ideally, all pilot sequences in the entire system are orthogonal, in order to avoid interference. However, this would require pilot sequences of at least length $L \times K$, which in most practical systems is not feasible. Instead, orthogonality within each cell only is ensured, i.e., $\tau = K$, thereby dealing with the potentially strongest sources of interference. As a result, all cells use the same set of pilots, potentially causing interference from neighboring cells. This is referred to as pilot contamination. We define the contaminating set, $\mathcal{C}_n^{k\ell}$, as the set of all pairs i, j, which identify all UEs applying the same pilot sequence in the nth time slot as the kth user in the ℓth cell. Hence, $\boldsymbol{x}_n^{ij} = \boldsymbol{x}_n^{k\ell} \forall i, j \in \mathcal{C}_n^{k\ell}$.

The pilot signal received by the BS of interest, concerning the kth user in the nth time slot can be expressed as

$$\boldsymbol{y}_n^{k0} = h_n^{k0}\boldsymbol{x}_n^{k0} + \sum_{i,j \in \mathcal{C}_n^{k0}} h_n^{ij}\boldsymbol{x}_n^{ij} + \boldsymbol{z}_n^{k0}, \tag{2}$$

where $\boldsymbol{z}_n^{k0} = \left[z_n^{k0}(1)z_n^{k0}(2)\ldots z_n^{k0}(\tau)\right]^T$ and $z_n^{k0}(j)$ are circularly symmetric Gaussian random variables with zero mean and unit variance for all j. Here, only signals leading to contamination are included in the sum term since any $h_n^{ij}\boldsymbol{x}_n^{ij}\ \forall\ i, j \notin \mathcal{C}_n^{k\ell}$ are removed when correlating with

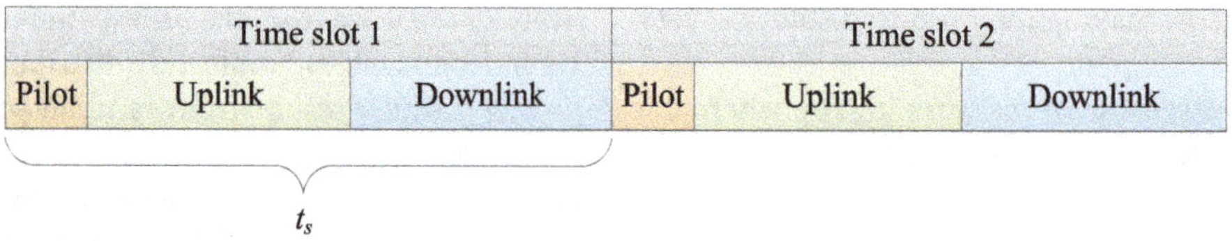

Fig. 2 Scheduling example

the applied pilot sequence. Hence, all contributions from the sum term are undesirable and will contaminate the CSI. Without loss of generality, we focus on the channel estimation for a single user in a single cell. Hence, in the remainder of the paper, we omit the superscript k for ease of notation.

2.2 Mobility model

In the employed mobility model, we restrict our attention to the consequences of mobility on the small-scale fading characteristics, i.e., f_d. Therefore, consequences like shadowing, varying path loss, and cell handover are disregarded. Since we employ pilot sequence hopping in this work, we can furthermore restrict our attention to the mobility of the UE of interest. This is explained in Section 3.1. We consider three different mobility models:

- *M1*: In this mobility model, the UE moves at a constant speed, v_1, for T_1 seconds.
- *M2*: (Train) This model emulates the mobility experienced in a train. Initially, the speed is zero for $T_{2,1}$ seconds. Then, the speed increases linearly, i.e., with constant acceleration, $\delta_{2,+}$, until a specified maximum speed, v_2. This speed is maintained for $T_{2,2}$ seconds, after which the speed is decreased linearly, with deceleration, $\delta_{2,-}$, until mobility has seized. Finally, the speed is kept at zero for $T_{2,3}$ seconds.
- *M3*: (Car) The third mobility model emulates the behavior of a car for T_3 seconds. The model operates with a vector of possible speeds, $\boldsymbol{v} = [v_0 v_1 \ldots v_{\max}]$, where the individual speeds are uniformly spaced between zero and $v_{\max}$. The initial speed is $v_0 = 0$. In every time slot, the speed is increased with probability p_+ and decreased with probability p_- and remains constant with probability $1 - p_+ - p_-$. Acceleration and deceleration are constant at $\delta_{3,+}$ and $\delta_{3,-}$, respectively. Speed changes always occur to the nearest speed in $\boldsymbol{v}$, and both acceleration from $v_{\max}$ and deceleration from v_0 result in no change.

Examples of all three mobility models are plotted in Fig. 3.

3 Pilot decontamination

The solution to pilot contamination proposed in this work consists of two components:

1. *Pilot sequence hopping*: This component refers to random shuffling of the pilots applied within a cell. This shuffle occurs between every time slot. The purpose of this component is to *decorrelate* the contaminating signals. When pilots are shuffled, the set of contaminating users will be replaced by a new set, whose channel coefficients are uncorrelated with those of the previous set.
2. *Kalman filtering:* The autocorrelation of the channel coefficient of the user of interest is high at low mobility. This means that information about the value of the current channel coefficient exists not only in the most recent pilot signal but also in past pilot signals. This can be extracted using a filter. Since the channel coefficients are time-varying, we are dealing with a tracking problem. For this purpose, a Kalman filter is attractive due to its excellent tracking capability and recursive structure, which provides good performance at low complexity. Since the contaminating signals have been decorrelated, the Kalman filter will suppress the impact of these signals, leading to pilot *decontamination*.

3.1 Pilot sequence hopping

Pilot sequence hopping is a technique where the UEs randomly switch to a new pilot sequence in between time slots. This must be coordinated with the BS, which in practice can be realized by letting the BS send a seed for a pseudorandom number generator to each UE. This ensures that the coordination overhead is limited to the initial connection phase, whereby it can be neglected. Random pilot sequence hopping is illustrated in Fig. 4 in the case of $\tau = K = 5$. Note how the identity of the contaminator changes between time slots, as opposed to a fixed pilot sequence schedule, where the contaminator remains the same UE. Consequently, the undesirable part of the pilot signal, i.e., the sum term in (2), varies rapidly between time slots compared to the variation caused by the mobility of a single contaminator in a fixed schedule. In fact, the impact of pilot sequence hopping, from a contamination perspective, can be viewed as a dramatic increase of the mobility of the contaminator. This in turn leads to a lowered autocorrelation, or decorrelation, in the contaminating signal, which is the motivation behind performing pilot sequence hopping.

The level of decorrelation is related to the time between two instances, where the same user acts as a contaminator. We refer to this as the collision distance, and we denote it as t_c^k for the kth user, see Fig. 4. Note that in the case of a fixed pilot schedule, $t_c^k = 1$. The goal of pilot sequence hopping is to maximize t_c^k, either in an expected sense or maxmin sense, i.e., maximization of the minimum value. The latter can be pursued through a minimal level of coordination of pilot sequence schedules among neighboring cells. However, this work is strictly restricted to a framework with no inter-cell coordination; hence, we focus on the expected value of t_c^k. If pilot sequence hopping is performed at random and $\tau = K$, then t_c^k follows a geometric distribution, such that

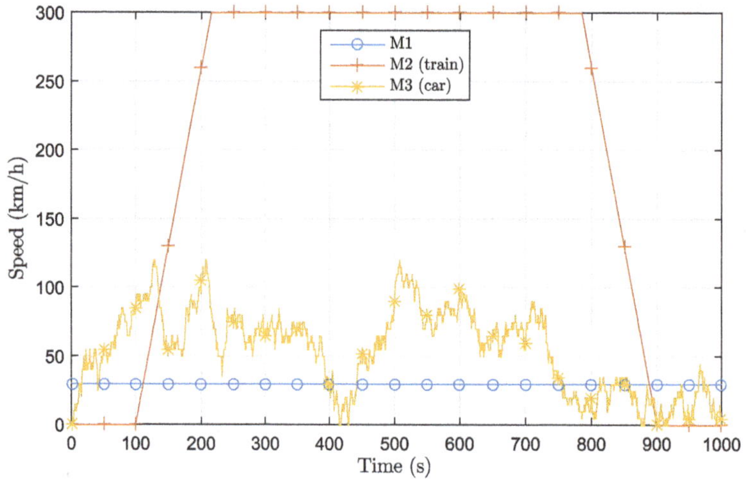

Fig. 3 Examples of mobility models

$$P\left(t_c^k = w\right) = (1-p)^{w-1}p, \qquad w = 1, 2, \ldots, \infty$$
$$p = \frac{1}{K}, \tag{3}$$

where $P\left(t_c^k = w\right)$ is the probability that the collision distance is w and p is the probability of a given UE being the next contaminator. We then have $\mathbb{E}\left[t_c^k\right] = K$, i.e., the expected collision distance increases with the number of users/pilots per cell, which follows intuition. Note that the collision distance is a user-specific measure, which holds for all potential contaminators in the system. Hence, the analysis still holds when considering systems with more than one neighboring cell.

The maximization of $\mathbb{E}\left[t_c^k\right]$ leads to a decorrelation of the contaminating signals. The benefit of this is reaped using appropriate filtering techniques. For this purpose, we have chosen a modified version of the Kalman filter, which is described next.

3.2 Modified Kalman filter

The problem of estimating a time-varying channel based on pilot signals, also termed channel tracking, can be solved using the Kalman filter [18]. The Kalman filtering

	Cell 0	Cell 1				
Time slot	UE of interest	UE 1	UE 2	UE 3	UE 4	UE 5
1	X_3	X_3	X_5	X_1	X_2	X_4
2	X_2	X_1	X_4	X_5	X_3	X_2
3	X_5	X_1	X_2	X_4	X_5	X_3
4	X_2	X_2	X_1	X_5	X_4	X_3
5	X_1	X_5	X_3	X_4	X_2	X_1
6	X_4	X_2	X_4	X_5	X_3	X_1
7	X_2	X_5	X_3	X_2	X_1	X_4
8	X_2	X_5	X_2	X_1	X_3	X_4
9	X_3	X_2	X_1	X_4	X_3	X_5
10	X_1	X_5	X_4	X_1	X_3	X_2

t_s t_c^5

Fig. 4 An example of a random pilot schedule for the UE of interest and potential contaminators in a neighboring cell. Green boxes represent pilots, which are orthogonal to the pilot from the UE of interest. Red boxes represent contamination and $\boldsymbol{x}_i$ denotes a pilot sequence

framework consists of a *process equation* and a *measurement equation*. The process equation expresses how the variable under estimation develops over time. We already chose such a model in (1); however, Clarke's model does not fit the structure of the Kalman filter. As a result, we choose an autoregressive (AR) model as an approximation of the model in (1). An AR model is said to have order $d+1$ if it expresses the current and d previous values as a function of the $d+1$ previous values. As d increases, the approximation of Clarke's model is increasingly valid. In our context, if we define $\boldsymbol{h}_n = \left[h_n \ldots h_{n-d}\right]^T$, the process equation for the Kalman filter is expressed as

$$\boldsymbol{h}_n = \boldsymbol{A}_n \boldsymbol{h}_{n-1} + \boldsymbol{v}_n^p, \tag{4}$$

$$\boldsymbol{A}_n = \begin{bmatrix} a_n^1 \ \ldots \ a_n^{d+1} \\ \boldsymbol{I}_d \quad \boldsymbol{0}_{d\times 1} \end{bmatrix}, \tag{5}$$

where $\boldsymbol{I}_d$ is the $d \times d$ identity matrix, $\boldsymbol{0}_{d\times 1}$ is a $d \times 1$ vector of zeros, and $\boldsymbol{v}_n^p = \left[v_n^p(1) \ldots v_n^p(d+1)\right]^T$ is the *process noise*, which is zero mean circularly symmetric Gaussian with covariance matrix $Q_n \boldsymbol{I}_{d+1}$, where

$$Q_n = \gamma_n^0 - \sum_{j=1}^{d+1} a_n^j \gamma_n^j. \tag{6}$$

Here, γ_n^j is the autocovariance of the channel coefficient at a lag of j time slots, and γ_n^0, i.e., channel power, is assumed known. Since $\gamma_n^j = \gamma_n^{-j}$, we can find γ_n^j for $j > 0$ by solving the Yule-Walker equations

$$\begin{bmatrix} \gamma_n^1 \\ \gamma_n^2 \\ \vdots \\ \gamma_n^{d+1} \end{bmatrix} = \begin{bmatrix} \gamma_n^0 & \gamma_n^{-1} & \cdots & \gamma_n^{-d} \\ \gamma_n^1 & \gamma_n^0 & \ddots & \vdots \\ \vdots & \ddots & \ddots & \gamma_n^{-1} \\ \gamma_n^d & \cdots & \gamma_n^1 & \gamma_n^0 \end{bmatrix} \times \begin{bmatrix} a_n^1 \\ a_n^2 \\ \vdots \\ a_n^{d+1} \end{bmatrix}. \tag{7}$$

The corresponding measurement equation for the Kalman filter is expressed based on (2) as follows:

$$\boldsymbol{y}_n = \boldsymbol{X}_n \boldsymbol{h}_n + \boldsymbol{v}_n^m, \tag{8}$$

$$\boldsymbol{X}_n = \begin{bmatrix} x_n(1) & 0 & \ldots & 0 \\ \vdots & \vdots & \tau \times d & \vdots \\ x_n(\tau) & 0 & \ldots & 0 \end{bmatrix}, \tag{9}$$

where $\boldsymbol{v}_n^m$ is the *measurement noise*, which is zero mean circularly symmetric Gaussian with covariance matrix $\sigma_o^2 \boldsymbol{I}_\tau + \sigma_c^2 \boldsymbol{X}_n \boldsymbol{X}_n^H$. Here, σ_o^2 and σ_c^2 are noise power and total contamination power (average over time), respectively, which are both assumed known.

In a conventional Kalman filter, $\boldsymbol{A}_n$ is assumed constant and known. However, this cannot be assumed in our case; thus, the varying elements, a_n^j, $j = 1, \ldots, d+1$, must be tracked along with the channel coefficients. For this purpose, we must modify the conventional Kalman filter to include an AR model tracker. First, we state the conventional Kalman filter [18] in our context, where the AR coefficients are assumed known.

For all n :

$$\boldsymbol{e}_n = \boldsymbol{y}_n - \boldsymbol{X}_n \boldsymbol{A}_{n-1} \hat{\boldsymbol{h}}_{n-1}, \tag{10}$$

$$\boldsymbol{R}_n = \boldsymbol{X}_n \boldsymbol{P}_n \boldsymbol{X}_n^H + \sigma_o^2 \boldsymbol{I}_\tau + \sigma_c^2 \boldsymbol{X}_n \boldsymbol{X}_n^H, \tag{11}$$

$$\boldsymbol{K}_n = \boldsymbol{P}_n \boldsymbol{X}_n^H \boldsymbol{R}_n^{-1}, \tag{12}$$

$$\hat{\boldsymbol{h}}_n = \boldsymbol{A}_n \hat{\boldsymbol{h}}_{n-1} + \boldsymbol{K}_n \boldsymbol{e}_n, \tag{13}$$

$$\boldsymbol{F}_n = \boldsymbol{I}_{d+1} - \boldsymbol{K}_n \boldsymbol{X}_n, \tag{14}$$

$$\boldsymbol{P}_{n+1} = \boldsymbol{A}_n \boldsymbol{F}_n \boldsymbol{P}_n \boldsymbol{A}_n^H + Q \boldsymbol{I}_{d+1}, \tag{15}$$

where $\boldsymbol{I}_\tau$ is the $\tau \times \tau$ identity matrix and $\hat{\boldsymbol{h}}_n$ is the estimate of $\boldsymbol{h}_n$.

For the tracking of the AR coefficients, an approach similar to the one in [19] is taken. In [19], the inclusion of a first-order AR coefficient tracker is presented for a Kalman predictor, i.e., a filter with the purpose of predicting the channel, h_n, based on all observations until $\boldsymbol{y}_{n-1}$. In this work, we extend this approach to higher order AR models taking all observations until $\boldsymbol{y}_n$ into account.

The approach is based on the principle of gradient descent. The gradient, ∇_n, with respect to $\boldsymbol{A}_n$ of the cost function, the mean squared error (MSE), is derived and used to adjust $\boldsymbol{A}_n$ in the direction of decreasing MSE. Note that this iterative numerical method is attractive since an analytical minimization of the cost function is complex to perform at every iteration. Furthermore, the minimization may only be feasible under certain conditions. Gradient descent is therefore a more robust and computationally simple solution. The gradient of the MSE is

$$\begin{aligned} \nabla_n(i,j) &= \frac{\partial}{\partial A_{n-1}(i,j)} \mathbb{E}\left[|\boldsymbol{e}_n|^2\right] \\ &= -\Re\Big[\left(\boldsymbol{q}_{n-1,i,j}^H \boldsymbol{A}_{n-1}^H \boldsymbol{X}_n^H + \hat{\boldsymbol{h}}_{n-1}^H \boldsymbol{\Gamma}_{i,j}^H \boldsymbol{X}_n^H\right) \boldsymbol{e}_n \\ &\quad + \boldsymbol{e}_n^H \left(\boldsymbol{X}_n \boldsymbol{A}_{n-1} \boldsymbol{q}_{n-1,i,j} + \boldsymbol{X}_n \boldsymbol{\Gamma}_{i,j} \hat{\boldsymbol{h}}_{n-1}\right)\Big], \\ \boldsymbol{\Gamma}_{i,j} &= \frac{\partial \boldsymbol{A}_n}{\partial A_n(i,j)}, \\ \boldsymbol{\Gamma}_{i,j}(k,\ell) &= \begin{cases} 1, & \text{if } i = k \text{ and } j = \ell, \\ 0, & \text{elsewhere.} \end{cases} \end{aligned} \tag{16}$$

Here, $\boldsymbol{q}_{n,i,j} = \frac{\partial \hat{\boldsymbol{h}}_n}{\partial A_n(i,j)}$ and is found by differentiating (13) with respect to $A_n(i,j)$, such that

$$\boldsymbol{q}_{n,i,j} = \boldsymbol{F}_n \left(\boldsymbol{A}_n \boldsymbol{q}_{n-1,i,j} + \boldsymbol{\Gamma}_{i,j} \hat{\boldsymbol{h}}_{n-1}\right) + \boldsymbol{M}_{n,i,j} \boldsymbol{e}_n. \tag{17}$$

where $\boldsymbol{M}_{n,i,j} = \frac{\partial \boldsymbol{K}_n}{\partial A_n(i,j)}$, which is found by differentiating (12) with respect to $A_n(i,j)$; hence,

$$\boldsymbol{M}_{n,i,j} = \boldsymbol{F}_n \boldsymbol{S}_{n,i,j} \boldsymbol{X}_n^H \boldsymbol{R}_n^{-1}. \tag{18}$$

We introduced $S_{n,i,j} = \frac{\partial P_n}{\partial A_n(i,j)}$, which is a differentiation of (15) with respect to $A_n(i,j)$, giving us

$$\begin{aligned} S_{n+1,i,j} &= \Gamma_{i,j} F_n P_n A_n^H + A_n F_n S_{n,i,j} F_n^H A_n^H \\ &\quad + A_n F_n P_n \Gamma_{i,j}^H + U_n I_{d+1} \end{aligned} \tag{19}$$

Finally, we have $U_n = \frac{\partial Q_n}{\partial A_n(i,j)}$, which can be found analytically by solving for Q_n in (6) and (7) and differentiating with respect to $A_n(i,j)$. Using ∇_n, we can then adjust A_n as follows:

$$A_n = A_{n-1} - \mu[\nabla_n]_{-\nu}^{+\nu}, \tag{20}$$

where μ is a parameter adjusting the convergence speed and the brackets denote truncations. The truncation to ν is for avoiding dramatic adjustments in situations with a high slope. The need for this will be explained in Section 5. In addition to the truncation, we enforce $|z_j| < 0.999$, where z_j are the roots of the polynomial $z^{d+1} - \sum_{j=1}^{d+1} a_n^j z^{d+1-j}$. This ensures a stationary AR process.

We can now state the modified Kalman filtering algorithm including an AR coefficient tracker:

For all n :

$$\begin{aligned} e_n &= y_n - X_n A_{n-1} \hat{h}_{n-1}, \\ R_n &= X_n P_n X_n^H + \sigma_o^2 I_\tau + \sigma_c^2 X_n X_n^H, \\ \nabla_n(i,j) &= -\Re\Big[\Big(q_{n-1,i,j}^H A_{n-1}^H X_n^H + \hat{h}_{n-1}^H \Gamma_{i,j}^H X_n^H\Big) e_n \\ &\quad + e_n^H \Big(X_n A_{n-1} q_{n-1,i,j} + X_n \Gamma_{i,j} \hat{h}_{n-1}\Big)\Big], \\ A_n &= A_{n-1} - \mu[\nabla_n]_{-\nu}^{+\nu}, \\ K_n &= P_n X_n^H R_n^{-1}, \\ \hat{h}_n &= A_n \hat{h}_{n-1} + K_n e_n, \\ M_{n,i,j} &= F_n S_{n,i,j} X_n^H R_n^{-1}, \\ q_{n,i,j} &= F_n \Big(A_n q_{n-1,i,j} + \Gamma_{i,j} \hat{h}_{n-1}\Big) + M_{n,i,j} e_n, \\ P_{n+1} &= A_n F_n P_n A_n^H + Q_n I_{d+1}, \\ S_{n+1,i,j} &= \Gamma_{i,j} F_n P_n A_n^H + A_n F_n S_{n,i,j} F_n^H A_n^H \\ &\quad + A_n F_n P_n \Gamma_{i,j}^H + U_n I_{d+1}. \end{aligned} \tag{21}$$

In the following subsection, we derive the lower bound on the MSE of an estimate of the channel coefficients. It serves as a benchmark in the numerical evaluations in Section 5.

4 Analysis

Initially, we present a simplified analysis of a toy example, in order to help the understanding of the benefit from pilot sequence hopping. Consider the ideal case of a constant channel between BS and UE of interest and a single contaminating neighboring cell. Noise is disregarded since attention is on decontamination. Moreover, for this toy example only, we assume an infinite amount of orthogonal pilot sequences and an infinite amount of users per cell, such that $\tau = K = \infty$ and $\mathbb{E}\left[t_c^k\right] = \infty$, which means contaminating signals in all time slots are independent. For simplicity, we assume $x_n^H x_n = 1$, such that the estimate in time slot n is

$$\hat{h}_n = h + h_n', \tag{22}$$

where h_n' is the channel of the contaminator in time slot n. We define the MSE of this estimate as $\mathbb{E}\left[\left(h - \hat{h}_n\right)^2\right]$.

Now, consider a new estimator, $\bar{\hat{h}}_n$, which is the average of all estimates until time slot n. Hence, we have

$$\bar{\hat{h}}_n = h + \frac{1}{n}\sum_{i=1}^{n} h_i'. \tag{23}$$

In this case, the error in the estimate is solely composed of the average of the contaminating signals, which are independent and have variance σ_c^2. Hence, $\mathbb{E}\left[\left(h - \bar{\hat{h}}_n\right)^2\right] = \frac{\sigma_c^2}{n+\sigma_c^2}$, if prior knowledge on h is a standard Gaussian. If pilot sequence hopping had not been performed, the MSE had remained $\frac{\sigma_c^2}{1+\sigma_c^2}$ since h_n' would be constant. Note that the MSE goes towards zero for $n \to \infty$, when pilot sequence hopping is performed. This is a result of the fact that a pilot signal in the infinite past carries as much information about the current channel as the most recent pilot signal, in the ideal example of a constant channel. Note also that for finite τ (and K) and thereby finite $\mathbb{E}\left[t_c^k\right]$, the MSE is lower bounded by $\frac{\sigma_c^2}{K+\sigma_c^2}$ since only a maximum of K independent estimates can be achieved. In a more practical example with a time-varying channel, the amount of information carried in a pilot signal decays over time. We can account for this in a more elaborate Bayesian analysis, which is described next.

4.1 Bayesian analysis

Given a set of observations, $Y_n = [y_1, \ldots, y_n]$, we are interested in deriving the distribution, in particular the variance, of the resulting estimate, $\hat{h}_n$, of the most recent channel coefficient, h_n. Here, $y_k = x_k h_k + v_k^m$, which through least squares estimation gives us a scalar observation $y_k^{ls} = \left(x_k^H x_k\right)^{-1} x_k^H y_k = h_k + v_k^m$, where v_k^m is the residual scalar noise, which is zero mean Gaussian with variance $\frac{\sigma_o^2}{\tau} + \sigma_c^2$. We define $y_n^{ls} = \left[y_1^{ls}, \ldots, y_n^{ls}\right]$ and can then express the conditional probability density function of the vector of channel coefficients $h_n = [h_1, \ldots, h_n]^T$ using Bayes' theorem as

$$f\left(h_n | y_n^{ls}\right) = \frac{f\left(y_n^{ls} | h_n\right) f\left(h_n\right)}{f\left(y_n^{ls}\right)}, \tag{24}$$

where $f(\boldsymbol{h}_n)$ is a multivariate Gaussian with mean vector $\mathbf{0}$ and covariance matrix $\boldsymbol{\Sigma}$, where $\boldsymbol{\Sigma}_{i,j} = \gamma_n^{i-j} = B_0\left(2\pi f_d t_s (i-j)\right)$ and B_0 is the zeroth-order Bessel function of the first kind. Furthermore, $f\left(\boldsymbol{y}_n^{ls}|\boldsymbol{h}_n\right)$ is a multivariate Gaussian with mean vector $\boldsymbol{h}_n$ and covariance matrix $\boldsymbol{C} = \left(\frac{\sigma_o^2}{\tau} + \sigma_c^2\right)\boldsymbol{I}_n$. It is well known that combining a Gaussian prior and a Gaussian likelihood provides a Gaussian posterior. This Gaussian posterior can be expressed as

$$f\left(\boldsymbol{h}_n|\boldsymbol{y}_n^{ls}\right) = \frac{1}{\sqrt{(2\pi)^n|\boldsymbol{V}|}} e^{-\frac{(\boldsymbol{h}_n-\boldsymbol{\mu}_h)^{\mathrm{H}}\boldsymbol{V}^{-1}(\boldsymbol{h}_n-\boldsymbol{\mu}_h)}{2}}, \quad (25)$$

$$\boldsymbol{V}^{-1} = \boldsymbol{C}^{-1} + \boldsymbol{\Sigma}^{-1}, \quad (26)$$

$$\boldsymbol{\mu}_h = \boldsymbol{V}\boldsymbol{\Sigma}^{-1}\boldsymbol{y}_n^{ls}. \quad (27)$$

Equations (26) and (27) provide the optimal coefficients of a Bayesian filter and the corresponding covariance. The lower right corner element of $\boldsymbol{V}$ is then the variance of a causal filter estimating the most recent channel coefficient, h_n.

With Eqs. (25) to (27) as a starting point, we can analyze filters based on different assumptions on the underlying model of the channel.

The proposed Kalman filter with a first-order AR model is based on a Markovian assumption, where a channel coefficient only depends on the previous channel coefficient, see Fig. 5. Under this assumption, the posterior of h_n simplifies to the following recursive expression:

$$f\left(h_n|\boldsymbol{y}_n^{ls}\right) = \frac{f\left(y_n^{ls}|h_n\right) f\left(h_n|\boldsymbol{y}_{n-1}^{ls}\right)}{f\left(y_n^{ls}|\boldsymbol{y}_{n-1}^{ls}\right)}, \quad (28)$$

$$f\left(h_n|\boldsymbol{y}_{n-1}^{ls}\right) = \int_{h_{n-1}} f(h_n|h_{n-1}) f\left(h_{n-1}|\boldsymbol{y}_{n-1}^{ls}\right) dh_{n-1}, \quad (29)$$

where $f\left(h_o|y_o^{ls}\right) = f(h_o)$ since no observation is made at time zero, and $f(h_o)$ is a standard Gaussian, such that the recursion is terminated. From the applied model, we have 0

$$f\left(y_n^{ls}|h_n\right) = \frac{1}{\sqrt{2\pi\left(\frac{\sigma_o^2}{\tau} + \sigma_c^2\right)}} e^{-\frac{\left(y_n^{ls}-h_n\right)^2}{2\left(\frac{\sigma_o^2}{\tau}+\sigma_c^2\right)}}, \quad (30)$$

$$f(h_n|h_{n-1}) = \frac{1}{\sqrt{2\pi(1-a^2)}} e^{-\frac{(h_n-ah_{n-1})^2}{2(1-a^2)}}, \quad (31)$$

$$f\left(h_{n-1}|\boldsymbol{y}_{n-1}^{ls}\right) = \frac{1}{\sqrt{2\pi\sigma_{h_{n-1}}^2}} e^{-\frac{\left(h_{n-1}-\mu_{h_{n-1}}\right)^2}{2\sigma_{h_{n-1}}^2}}, \quad (32)$$

where $a = B_0\left(2\pi f_d t_s\right)$. The integral in (29) can be viewed as the scaling factor in a product of the involved Gaussian distributions, such that

$$f\left(h_n|\boldsymbol{y}_{n-1}^{ls}\right) = \frac{1}{\sqrt{2\pi\left(a^2\sigma_{h_{n-1}}^2 + 1 - a^2\right)}} e^{-\frac{\left(h_n-a\mu_{h_{n-1}}\right)^2}{2\left(a^2\sigma_{h_{n-1}}^2+1-a^2\right)}}, \quad (33)$$

where a change of variable has been performed in (32) from h_{n-1} to ah_{n-1} in order to conform to its representation in (31). We can then express the scaled product of Gaussian distributions in (28) as follows:

$$f\left(h_n|\boldsymbol{y}_n^{ls}\right) = \frac{1}{\sqrt{2\pi\sigma_{h_n}^2}} e^{-\frac{\left(h_n-\mu_{h_n}\right)^2}{2\sigma_{h_n}^2}}, \quad (34)$$

$$\mu_{h_n} = \frac{\left(a^2\sigma_{h_{n-1}}^2 + 1 - a^2\right) y_n^{ls} + \left(\frac{\sigma_o^2}{\tau} + \sigma_c^2\right) a\mu_{h_{n-1}}}{a^2\sigma_{h_{n-1}}^2 + 1 - a^2 + \frac{\sigma_o^2}{\tau} + \sigma_c^2}, \quad (35)$$

$$\sigma_{h_n}^2 = \frac{\left(a^2\sigma_{h_{n-1}}^2 + 1 - a^2\right)\left(\frac{\sigma_o^2}{\tau} + \sigma_c^2\right)}{a^2\sigma_{h_{n-1}}^2 + 1 - a^2 + \frac{\sigma_o^2}{\tau} + \sigma_c^2}. \quad (36)$$

We are primarily interested in the evolution of the variance and thereby the MSE of the estimator. Figure 6 shows evaluations of (36) for different levels of mobility, $t_s = 0.5$ ms, $\sigma_c^2 = 0.6$, $\sigma_o^2 = 0.2$, and $\tau = 96$. It shows how the

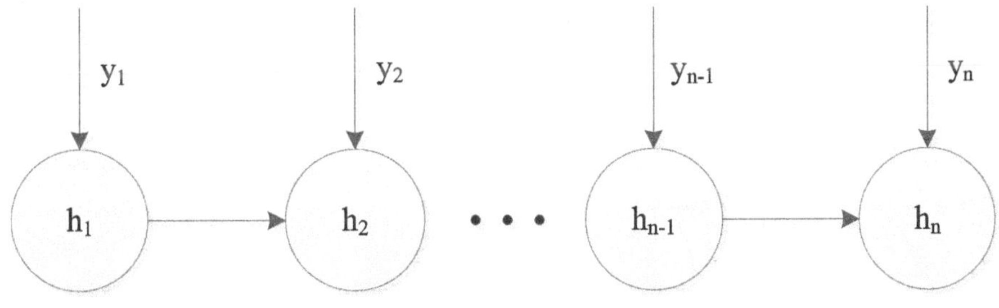

Fig. 5 Bayesian network representing the Markovian assumption applied in the Kalman filter

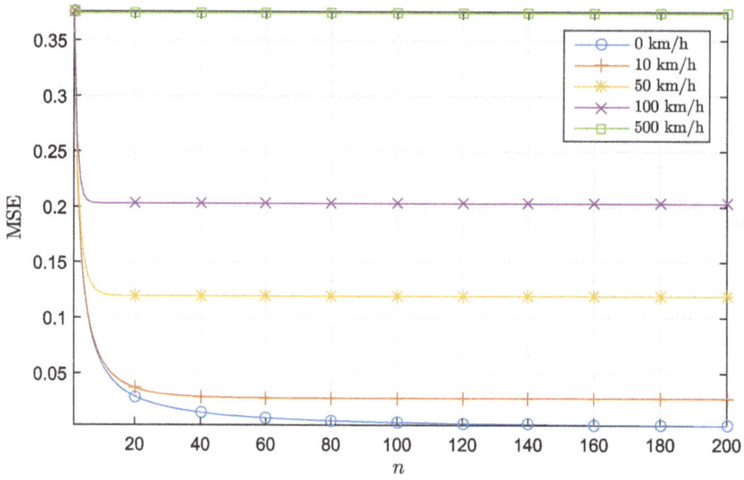

Fig. 6 MSE as a function of the number of pilot signals included in the filter. The results hold for the assumption of a first-order AR model and an optimal filter

MSE converges faster for higher mobilities since the information in past pilot signals decays faster. Note that these results act as a lower bound for the modified Kalman filter since we in (36) assume perfect knowledge of the AR coefficient, a. Any estimation error in the AR coefficient tracker will lead to increased MSE.

In the following subsection, we derive the lower bound on the MSE for arbitrary AR model order. Along with the bound for first-order AR models in (36), it serves as a benchmark in the numerical evaluations in Section 5.

4.2 Cramer-Rao lower bound

The Cramer-Rao lower bound (CRLB) [20] expresses the lower bound on the error covariance of any unbiased estimator based on a set of observations. In our context, the observations are $Y = [y_1, \ldots, y_n]$, and the estimation error covariance is $\mathbb{E}\left[\left(\hat{h}-h\right)\left(\hat{h}-h\right)^H\right]$, where $h = [h_1, \ldots, h_n]^T$. We follow a Bayesian framework, well suited to the tracking of time-varying parameters; thus, we employ the Bayesian CRLB. Having chosen Clarke's model as the channel model, it follows that the a priori distribution of the parameter h is well approximated as a Gaussian distribution. Furthermore, we adapt a compact formulation for the case of complex parameters (see [20, p. 529]). The complex Bayesian CRLB is expressed as

$$\mathbb{E}\left[\left(\hat{h}-h\right)\left(\hat{h}-h\right)^H\right] \geq J^{-1}, \tag{37}$$

where J is the Fisher information matrix. The matrix inequality means that $\mathbb{E}\left[\left(\hat{h}-h\right)\left(\hat{h}-h\right)^H\right] - J^{-1}$ is positive semidefinite. The Fisher information matrix is given by

$$J_{ij} = \mathbb{E}\left[-\frac{\partial^2 \log f_{Y,h}(Y,h)}{\partial h_i \partial h_j}\right], \qquad i,j = 1,\ldots,n. \tag{38}$$

Here, $f_{Y,h}(Y,h)$ is the joint probability density function of the observations and the channel coefficients. This can be expressed as

$$f_{Y,h}(Y,h) = f_{Y|h}(Y|h) f_h(h), \tag{39}$$

$$f_h(h) = c_1 \exp\left(-h^H \Sigma^{-1} h\right), \tag{40}$$

$$f_{Y|h}(Y|h) = \prod_{i=1}^{n} c_2 \exp\left(-\left(y_i - h_i x_i\right)^H C^{-1}\left(y_i - h_i x_i\right)\right), \tag{41}$$

where c_1 and c_2 are constants with independence from Y and h, Σ^{-1} is the inverse of the $n \times n$ covariance matrix of h, and C^{-1} is the inverse of the $\tau \times \tau$ observation error

Table 1 Simulation parameters

Parameter	Value	Description
σ_0^2	0.2	Noise variance
L	7	Number of cells
K	96	Users per cell
τ	96	Pilot length
μ	10^{-5}	Convergence speed
ν	100	Derivative cap
f_c	1.8 GHz	Carrier frequency
N_s	20	Number of scatterers
t_s	0.5 ms	Time between pilots
$\hat{h}_0$	$\mathbf{0}$	Initial estimate
q_0	$\mathbf{0}$	Initial differentiated estimate
P_1	$\mathbf{0}$	Initial error covariance
S_1	$\mathbf{0}$	Initial differentiated error covariance

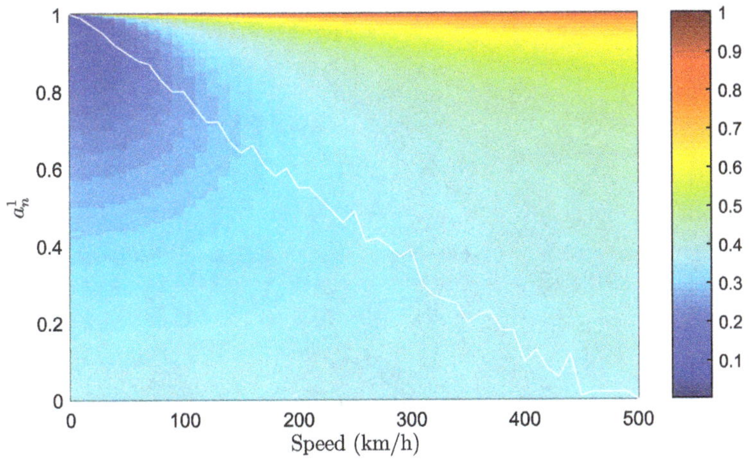

Fig. 7 MSE as a function of the autoregressive model coefficient and the speed of the UE. The coefficient with minimum MSE is marked with a white curve

covariance matrix. The CRLB at time n is the corresponding submatrix of J^{-1}; it gives a lower bound on channel estimation at time n accounting for the past observations.

By introducing the $n \times n$ matrix $\boldsymbol{\Omega}$, where $\boldsymbol{\Omega}_{i,j} = \mathbb{E}\left[-\frac{\partial^2 \log f_{Y|h}(Y|h)}{\partial h_i \partial h_j}\right]$ and combining (38) through (41), we get

$$
\begin{aligned}
J &= \boldsymbol{\Omega} + \boldsymbol{\Sigma}^{-1}, \\
\boldsymbol{\Omega}_{i,j} &= \begin{cases} \boldsymbol{x}_i^H \boldsymbol{C}^{-1} \boldsymbol{x}_i, & \text{if } i = j, \\ 0, & \text{if } i \neq j, \end{cases} \\
\boldsymbol{C} &= \sigma_o^2 \boldsymbol{I}_\tau + \sigma_c^2 \boldsymbol{x}_n \boldsymbol{x}_n^H, \\
\boldsymbol{\Sigma}_{i,j} &= B_0\left(2\pi f_d t_s (i-j)\right),
\end{aligned}
\tag{42}
$$

Working with the inverse of $\boldsymbol{\Sigma}$ may cause numerical problems at low mobility, where $\boldsymbol{\Sigma}$ is near-singular. This can be avoided by utilizing the matrix inversion lemma. We thus have

$$
\begin{aligned}
J^{-1} = \boldsymbol{\Sigma} - \boldsymbol{\Sigma}\left(\boldsymbol{I}_n \otimes \boldsymbol{x}_i\right)^H \Big(\left(\boldsymbol{I}_n \otimes \boldsymbol{C}^{-1}\right)^{-1} \\
+ \left(\boldsymbol{I}_n \otimes \boldsymbol{x}_i\right) \boldsymbol{\Sigma} \left(\boldsymbol{I}_n \otimes \boldsymbol{x}_i\right)^H \Big)^{-1} \left(\boldsymbol{I}_n \otimes \boldsymbol{x}_i\right) \boldsymbol{\Sigma},
\end{aligned}
\tag{43}
$$

where $\otimes$ denotes the Kronecker product. Furthermore, expression (43) allows a continuity with the case of no mobility for which the channel estimate of time n is the result of an average (see Eq. (23)).

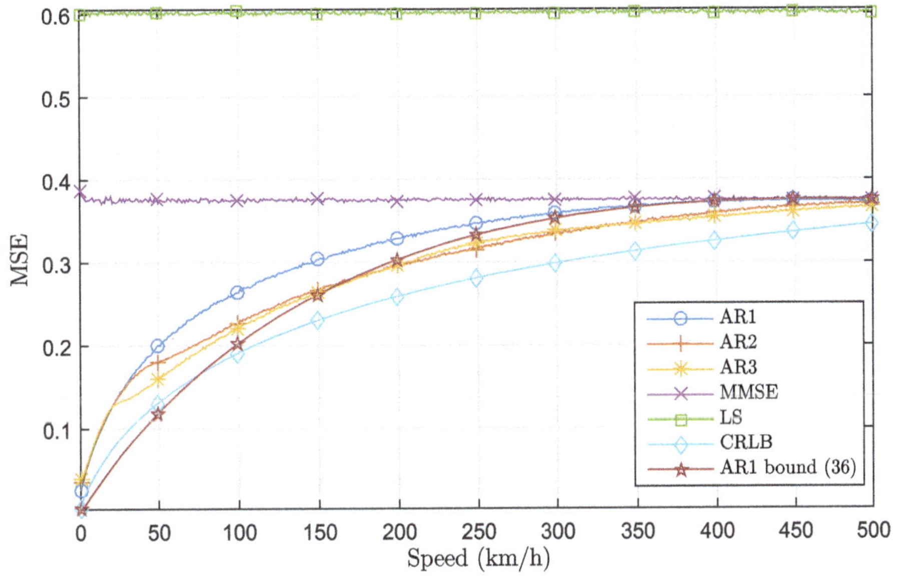

Fig. 8 Comparison between the proposed scheme and conventional solutions with respect to mean squared error as a function of mobility. The M1 mobility model is applied, and SIR is 2.2 dB

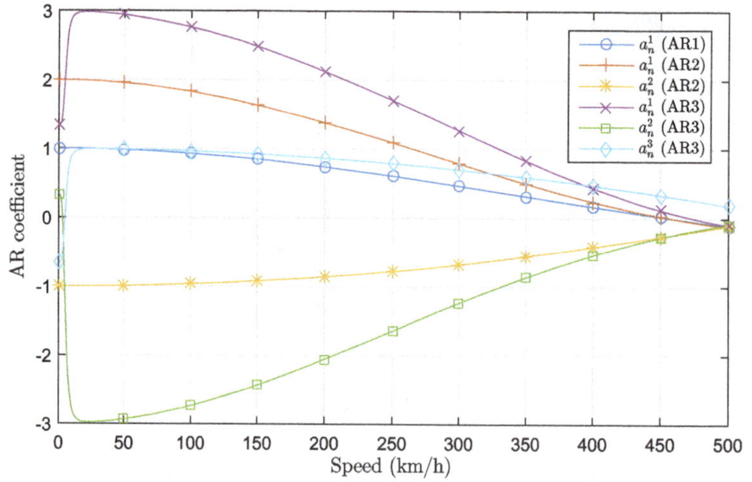

Fig. 9 AR model coefficients, found by solving (7), as a function of the speed of the UE

5 Numerical results

The proposed scheme has been simulated and compared to the conventional solutions of least squares (LS) estimation and minimum mean squared error (MMSE) estimation based on a single time slot. The expressions for the LS and MMSE estimators are given in (44) and (45), respectively. An overview of the parameters, which are common for all simulations, is given in Table 1. The choice of μ is based on experiments showing that this is a good compromise between convergence speed and limitation of noise-induced variance in the estimate. Throughout all simulations, we assume that contaminating signals have zero autocorrelation between time slots, which is justified by the choice of $K = 96$, such that $\mathbb{E}\left[t_c^k\right] = 96$. All simulation results are averages of 100 iterations of a scenario as specified by the mobility models in Section 2.2.

$$\hat{h}_n^{ls} = \left(\boldsymbol{x}_n^H \boldsymbol{x}_n\right)^{-1} \boldsymbol{x}_n^H \boldsymbol{y}_n, \tag{44}$$

$$\hat{h}_n^{mmse} = \boldsymbol{x}_n^H \left(\boldsymbol{x}_n \boldsymbol{x}_n^H + \sigma_n^2 \boldsymbol{I}_\tau + \sigma_c^2 \boldsymbol{x}_n \boldsymbol{x}_n^H\right)^{-1} \boldsymbol{y}_n. \tag{45}$$

Initially, results are shown for the conventional Kalman filter expressed in Eqs. (10) through (15), using a first-order AR model as the process equation. MSE, defined as $\mathbb{E}\left[\left(h_n - \hat{h}_n\right)^2\right]$, as a function of the user mobility, v, and the AR coefficient, a_n^1, is shown in Fig. 7. From this figure, it is evident how important it is to have an accurate AR model, which suits the current mobility of the UE of interest. This stresses the need for the modification of the Kalman filter, as proposed in Section 3.2. Moreover, it is seen that the derivative of the MSE with respect to a_n^1 may

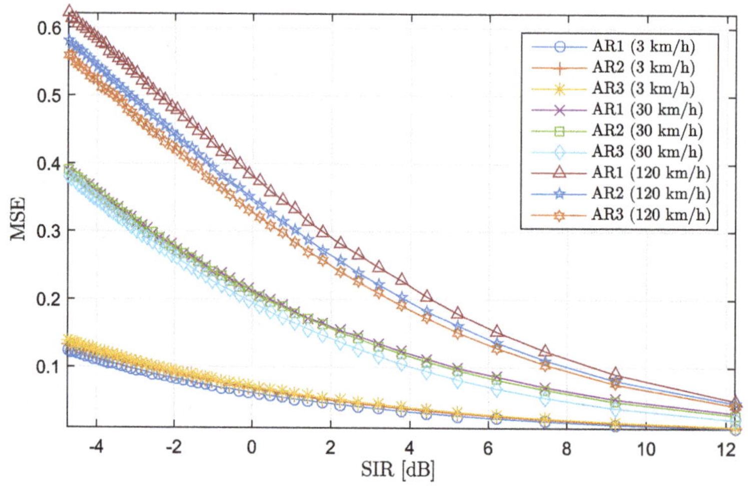

Fig. 10 Comparison between AR models with different orders with respect to mean squared error as a function of the signal-to-interference ratio

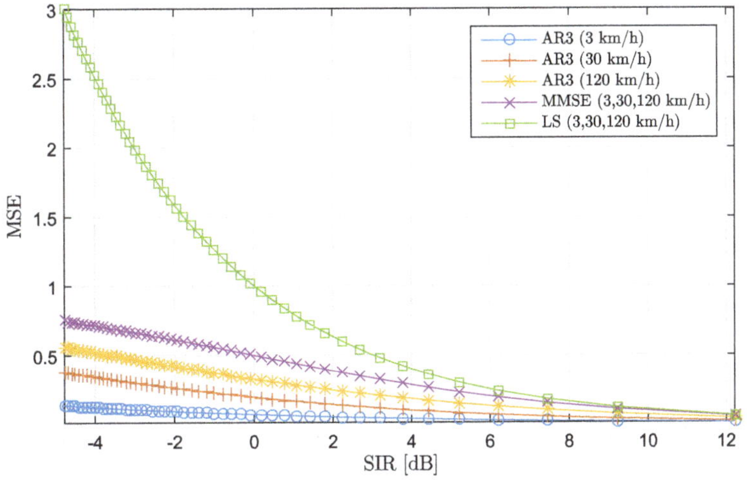

Fig. 11 Comparison between the proposed scheme and conventional solutions with respect to mean squared error as a function of the signal-to-interference ratio

attain very high values at high a_n^1. This can cause undesirably high variance in the estimate of the optimal a_n^1, which motivates the use of a derivative cap, ν.

Figure 8 shows a comparison of the proposed estimator, the LS estimator, and the MMSE estimator, with respect to MSE as a function of the speed of the UE using mobility model M1. The simulations apply $T_1 = 100$ s and $\sigma_c^2 = 0.6$. For the proposed estimator, AR models of first, second, and third order are included. In all three cases, the initial AR model coefficients are numerically optimized through a parameter sweep. The results were $a_0^1 = 0.3$ for the first-order model, $a_0^1 = 0.8$ and $a_0^2 = 0.2$ for the second-order model, and $a_0^1 = 0.7$, $a_0^2 = -0.3$, and $a_0^3 = -0.2$ for the third-order model. The results show that a significant performance improvement is achieved at medium and high levels of mobility when increasing the AR model order from one to two. Further increasing to a third-order model yields a more mixed result. At medium mobility, a significant gain is achieved, whereas at higher mobility, the performances of the second- and third-order models are quite close and take turns in being the better. At low mobility, no gain is achieved when increasing the order of the AR model. At first glance, this may be surprising, but when looking at the coefficients found from solving the Yule-Walker equations, see Fig. 9, it is evident that low mobility presents a particularly challenging tracking task for a third-order AR model, due to large dynamics. Compared to the conventional methods of LS and MMSE, the proposed scheme offers a performance in an entirely different league. The

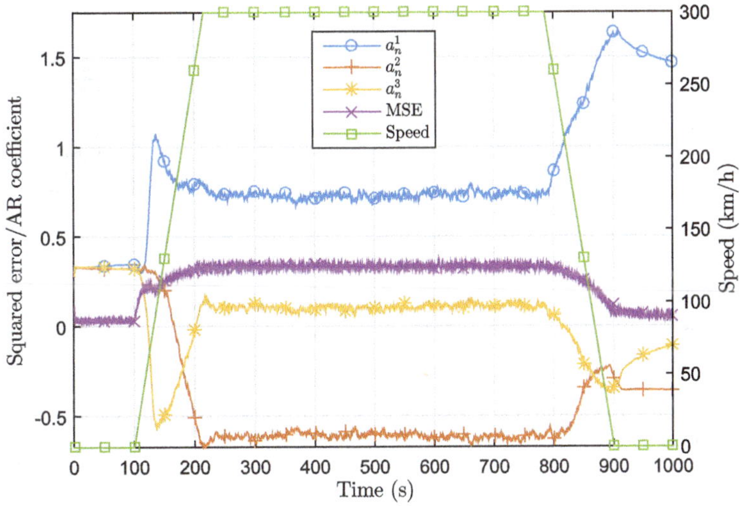

Fig. 12 Example of a simulation with a third-order AR model and the M2 mobility model. SIR is 2.2 dB

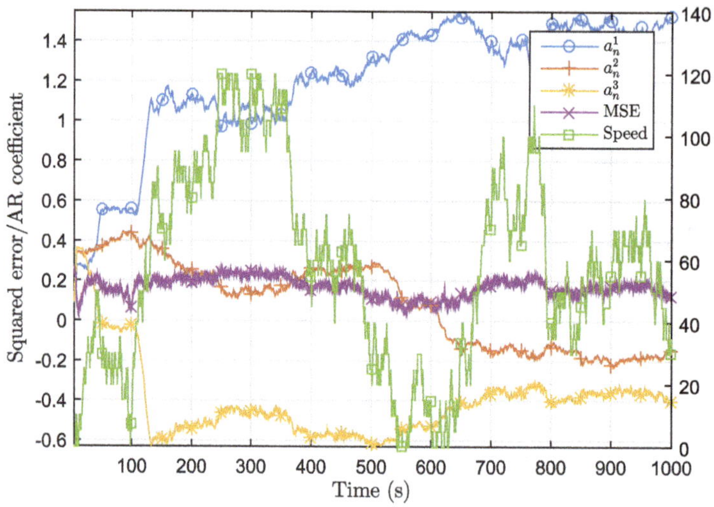

Fig. 13 Example of a simulation with a third-order AR model and the M3 mobility model. SIR is 2.2 dB

gain decreases as speed increases, but only at unusually high speeds is the gain insignificant. There is still a gap to the Cramer-Rao lower bound, although much of it has been closed when comparing to the conventional methods. When comparing to the bound associated with a first-order AR model, as expressed in (36), it is seen that the lack of performance gain at high mobility is largely explained by the choice of process model for the Kalman filter.

A different perspective is given in Fig. 10. Here, the MSE is plotted as a function of the signal-to-interference ratio (SIR), at typical mobility levels as defined by 3GPP [21]. Again, it is seen how increasing the order of the AR model is an advantage at medium and high mobility, but not at lower speeds. Figure 10 shows that this holds in a wide range of SIR.

From the same perspective, Fig. 11 shows a comparison of the conventional methods and the proposed scheme with a third-order model. Here, it is seen that the performance improvement is achieved in the entire SIR range. Decreasing the SIR is particularly penalizing the LS method, while MMSE and the proposed method are better able to cope with the increased interference.

Next, we look at the more sophisticated mobility models, M2 and M3. For M2, we use parameters $T_{2,1} = T_{2,3} = 100$ s, $T_{2,2} = 800$ s, $\delta_{2,+} = 2.6$ m/s^2, $\delta_{2,-} = -2.6$ m/s^2, and $v_2 = 300$ km/h. An example of a simulation with a single sequence of channel realizations is shown in Fig. 12. For M3, we use parameters $T_3 = 1000$ s, $v = [05 \ldots 120]$ km/h, $p_+ = p_- = 0.00025$, $\delta_{3,+} = 2.5$ m/s^2, and $\delta_{3,-} = -4$ m/s^2. Figure 13 shows a simulation example with this mobility model. It is seen that with

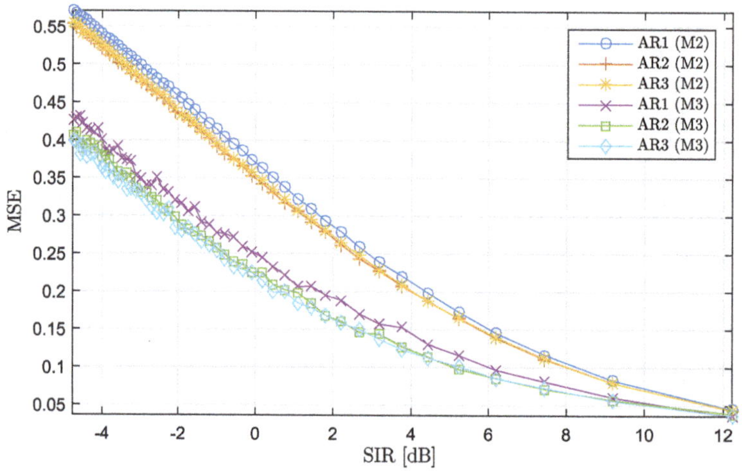

Fig. 14 Comparison of different AR model orders when using M2 and M3 mobility models

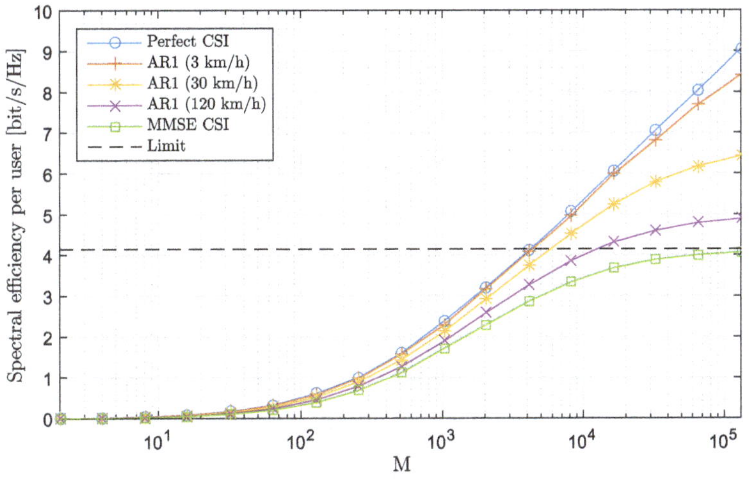

Fig. 15 The achievable downlink rate as a function of the number of antennas, *M*, at the BS

both mobility models, the AR parameter tracker is able to adjust to the varying speed and thereby adapt to the varying amount of information available in past pilot signals. In general, we expect to see a low MSE when the speed is low and vice versa. The figures confirm that this is in fact achieved. Figure 14 shows a comparison of the different AR model orders. It shows that increasing the order also provides performance enhancements in a wide range of SIR when considering M2 and M3.

Finally, we evaluate how the improvements of the channel estimates translate into increased achievable uplink and dowlink rates in a TDD massive MIMO system with M antennas at the BS. The achievable rates of such a system were derived in [22, p. 4]. We apply the proposed scheme for each individual channel coefficient in the system and evaluate the resulting achievable uplink and downlink rates. Furthermore, we evaluate the achievable rates in a system with perfect CSI at the BS and a system with CSI achieved with a conventional MMSE estimator at the BS. These act as upper and lower bounds, respectively, to the proposed scheme. The results are shown in Figs. 15 (downlink) and 16 (uplink). It is evident that the proposed scheme provides a significant improvement in achievable rates compared to the system with MMSE estimation. At low mobility, it even comes fairly close to a system with perfect CSI. Although the improvements are more visible at very high and impractical values of M, significant relative improvements are also achieved at lower values of M. As an example, the achievable uplink and downlink rates are increased by 42 and 46%, respectively, for $M = 128$, compared to MMSE estimation. Another important observation is that the system with MMSE

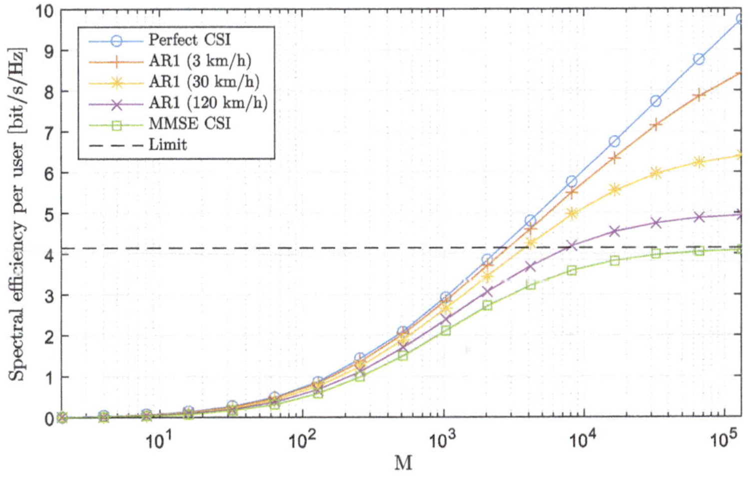

Fig. 16 The achievable uplink rate as a function of the number of antennas, *M*, at the BS

estimation converges to a limit as M goes to infinity. The limit is found as $\log_2\left(1+\frac{1}{\sigma_c^2}\right)$ and is the well-known limitation caused by pilot contamination. As in the case of perfect CSI, the proposed scheme is able to break this limit, which demonstrates its ability to mitigate pilot contamination.

6 Conclusions

We have presented a solution to pilot contamination in channel estimation, which is a major challenge in TDD massive MIMO systems. It is based on a combination of a pilot sequence hopping scheme and a modified Kalman filter. The pilot sequence hopping scheme involves random shuffling of the assigned pilot sequences within a cell, which ensures decorrelation in the time dimension of the contaminating signals. This is essential since it enables subsequent filtering to suppress the contamination. For this filtering, the Kalman filter has been chosen, due to its ability to track a time-varying state. However, a conventional Kalman filter is not able to adapt to changes in the underlying model, which is necessary when users have unknown and varying levels of mobility. For this problem, we have presented a modified Kalman filter, which can adapt the underlying model based on a minimization of the mean squared error.

Numerical evaluations show that the proposed solution can suppress a significant portion of the contamination at low and moderate levels of mobility. Even at high mobility, i.e., car speeds of 100 to 130 km/h, the proposed solution can provide a noticeable gain over conventional estimation methods.

Acknowledgements

The research presented in this paper was supported by the Danish Council for Independent Research (Det Frie Forskningsråd) DFF - 1335 – 00273.

Authors' contributions

Both authors have contributed to the design and analyis of the proposed methods as well as the writing of this manuscript. Both authors read and approved the final manuscript.

Competing interests

The authors declare that they have no competing interests.

References

1. T Brown, E De Carvalho, P Kyritsi, *Practical Guide to the MIMO Radio Channel*. (John Wiley & Sons, Hoboken, 2012)
2. TL Marzetta, Noncooperative cellular wireless with unlimited numbers of base station antennas. Wirel. Commun. IEEE Trans. **9**(11), 3590–3600 (2010). doi:10.1109/TWC.2010.092810.091092
3. F Rusek, D Persson, BK Lau, EG Larsson, TL Marzetta, O Edfors, F Tufvesson, Scaling up MIMO: opportunities and challenges with very large arrays. Signal Proc. Mag. IEEE. **30**(1), 40–60 (2013). doi:10.1109/MSP.2011.2178495
4. H Yin, D Gesbert, M Filippou, Y Liu, A coordinated approach to channel estimation in large-scale multiple-antenna systems. Sel. Commun. IEEE J. **31**(2), 264–273 (2013). doi:10.1109/JSAC.2013.130214
5. A Ashikhmin, T Marzetta, in *Information Theory Proceedings (ISIT), 2012 IEEE International Symposium On*. Pilot contamination precoding in multi-cell large scale antenna systems, (2012), pp. 1137–1141. doi:10.1109/ISIT.2012.6283031
6. J Jose, A Ashikhmin, TL Marzetta, S Vishwanath, Pilot contamination and precoding in multi-cell TDD systems. Wirel. Commun. IEEE Trans. **10**(8), 2640–2651 (2011). doi:10.1109/TWC.2011.060711.101155
7. HQ Ngo, EG Larsson, in *Acoustics, Speech and Signal Processing (ICASSP), 2012 IEEE International Conference On*. EVD-based channel estimation in multicell multiuser MIMO systems with very large antenna arrays, (2012), pp. 3249–3252. doi:10.1109/ICASSP.2012.6288608
8. RR Muller, L Cottatellucci, M Vehkapera, Blind pilot decontamination. Sel. Topics Signal Proc. IEEE J. **8**(5), 773–786 (2014). doi:10.1109/JSTSP.2014.2310053
9. RR Mueller, M Vehkaperae, L Cottatellucci, in *Smart Antennas (WSA), 2013 17th International ITG Workshop On*. Blind pilot decontamination (VDE Verlag, Offenbach, 2013), pp. 1–6
10. RR Muller, M Vehkapera, L Cottatellucci, in *Signals, Systems and Computers, 2013 Asilomar Conference On*. Analysis of blind pilot decontamination, (2013), pp. 1016–1020. doi:10.1109/ACSSC.2013.6810444
11. L Cottatellucci, RR Müller, M Vehkaperä, in *Vehicular Technology Conference (VTC Spring), 2013 IEEE 77th*. Analysis of pilot decontamination based on power control, (2013), pp. 1–5. doi:10.1109/VTCSpring.2013.6691891
12. H Yin, L Cottatellucci, D Gesbert, RR Müller, G He, Robust pilot decontamination based on joint angle and power domain discrimination. IEEE Trans. Signal Process. **64**(11), 2990–3003 (2016). doi:10.1109/TSP.2016.2535204
13. J Choi, DJ Love, P Bidigare, Downlink training techniques for FDD massive mimo systems: open-loop and closed-loop training with memory. Sel. Topics Signal Proc. IEEE J. **8**(5), 802–814 (2014). doi:10.1109/JSTSP.2014.2313020
14. E Björnson, E de Carvalho, JH Sørensen, EG Larsson, P Popovski, A random access protocol for pilot allocation in crowded massive mimo systems. IEEE Trans. Wireless Commun. **16**(4), 2220–2234 (2017). doi:10.1109/TWC.2017.2660489
15. E de Carvalho, E Björnson, EG Larsson, P Popovski, in *2016 IEEE International Conference on Acoustics, Speech and Signal Processing (ICASSP)*. Random access for massive mimo systems with intra-cell pilot contamination, (2016), pp. 3361–3365. doi:10.1109/ICASSP.2016.7472300
16. JH Sørensen, E De Carvalho, in *IEEE Global Telecommunications Conference, GLOBECOM*. Pilot decontamination through pilot sequence hopping in massive mimo systems, (2014)
17. R Clarke, A statistical theory of mobile-radio reception. Bell Syst. Tech. J. **47**(6), 957–1000 (1968)
18. S Haykin, *Adaptive Filter Theory*. (Pearson Education, Limited, Harlow, 2013)
19. K-Y Han, S-W Lee, J-S Lim, K-M Sung, Channel estimation for OFDM with fast fading channels by modified Kalman filter. Consum. Electron. IEEE Trans. **50**(2), 443–449 (2004). doi:10.1109/TCE.2004.1309406
20. SM Kay, *Fundamentals of Statistical Signal Processing: Estimation Theory*. (Prentice-Hall, Inc., Upper Saddle River, 1993)
21. 3GPP: Spatial channel model for multiple input multiple output (MIMO) simulations. TR 25.996, 3rd Generation Partnership Project (3GPP) (September 2012). http://www.3gpp.org/ftp/Specs/html-info/25996.htm. Accessed 13 Sept 2017
22. J Hoydis, S ten Brink, M Debbah, Massive MIMO in the UL/DL of cellular networks: how many antennas do we need? Sel. Areas Commun. IEEE J. **31**(2), 160–171 (2013). doi:10.1109/JSAC.2013.130205

4

SHAM: Scalable Homogeneous Addressing Mechanism for structured P2P networks

Manaf Zghaibeh* and Najam Ul Hassan

Abstract

In designing structured P2P networks, scalability, resilience, and load balancing are features that are needed to be handled meticulously. The P2P overlay has to handle large scale of nodes while maintaining minimized path lengths in performing lookups. It has also to be resilient to nodes' failure and be able to distribute the load uniformly over its participant. In this paper, we introduce SHAM: a *S*calable, *H*omogenous, *A*ddressing *M*echanism for structured P2P networks. SHAM is a multi-dimensional overlay that places nodes in the network based on geometric addressing and maps keys onto values using consistent hashing.

Our simulation results show that SHAM locates keys in the network efficiently, is highly resilient to major nodes' failure, and has an effective load balancing property. Furthermore, unlike other DHTs and due to its distinguished naming scheme, SHAM deploys homogenous addressing which drastically reduces latency in the underlying network.

Keywords: P2P, Structured, Performance, DHT, Overlay, Latency

1 Introduction

Locating keys efficiently in structured P2P networks, commonly referred to as *distributed hash tables* (DHTs), is always associated with the variant of maintaining limited routing tables at nodes. Some research adopted and worked on the principle that in order to lower the cost of search, the node has to acquire the most information available from the overlay [1, 2]. Thus, the more routing information the node gathers, the less the lookup will take. From another angle to this argument, other research insisted on keeping the size of the routing tables as minimum as possible while sacrificing some of the lookup performance [3–6]. The point here is that maintaining large routing tables is not viable and, thus, giving up few additional hops to reach the destination is more satisfying. Proposed DHTs differ in this regard and are classified based on the number of hops they require in order to land a lookup at its destination. We categorize the classes in this respect into single-hop overlays, constant degree overlays, multi-dimension overlays, and logarithmic overlays.

In single-hop overlays, a destination can be reached within one hop from the source node while maintaining a global knowledge about other nodes in the overlay [1, 2, 7–11]. However, although with the assertion that space requirements are cheap and bandwidth is abundant, still, keeping up with the continuous change in churn-intensive overlays is unfeasible.

Constant degree overlays on the other hand aim to reduce the global knowledge to a certain limit while increasing the latency from one single hop to a *fixed* number of hops regardless of the number of nodes in the system. Nevertheless, such DHTs are still not suitable for large networks where maintaining the routing table remains a challenge [12–14].

In multi-dimension overlays [4] and logarithmic overlays [3, 5, 6], neither the size of the routing table nor the latency is fixed. These systems work on balancing the two constraints based on the number of participating nodes: They tend to optimize the size of the routing table in order to reach a reduced latency. Such systems are known for their flexibility, scalability, and resilience to nodes' failure.

In similar direction, our motivation behind the work in this paper is to propose an addressing algorithm for

*Correspondence: mzghaibeh@du.edu.om
Department of Electrical and Computer Engineering, Dhofar University, P. O. Box 2509, 211, Salalah, Oman

structured P2P networks that can balance between the latency and the size of a routing table. In addition to that, our goal for the mechanism is to be robust and resilient against nodes' failures, to be able to scale without sacrificing performance, and to have an effective load balancing property. We propose *SHAM*: a Scalable Homogeneous Addressing Mechanism for structured P2P networks. SHAM handles three preeminent procedures. First, it inserts and situates nodes into the overlay. Second, it maps keys onto available nodes in the overlay. And third, it retrieves the keys. Similar to other structured P2P systems [3, 4, 15], SHAM employs consistent hashing to generate a naming space for keys. The use of consistent hashing is known to balance the load since each node in the system will theoretically have the same number of keys [3]. However, unlike other systems in the same class, SHAM does not require hashing the identifier of the node (*IP/port*) in order to place it in the system. Yet, it deterministically places the node in the system and assigns it a unique identifier based on homogeneous geometric pattern.

We will show in this paper by matching SHAM up with the two prominent DHTs, CAN and Chord, the following:

1. SHAM has better routing capabilities than CAN in all dimensions. Furthermore, by adjusting its dimensions, SHAM has a shorter path length than Chord up to a certain network sizes with minimized routing tables.
2. Unlike Chord, load balancing is always maintained in SHAM without additional cost. This is an issue that has always been neglected in discussing Chord: Although Chord maintains entries for $O(\log N)$ nodes, however, the use of its load balancing property is associated with each node holding $\log^2$ N entries.
3. The addressing scheme adopted in SHAM enables the overlay to be used to reduce latency in the underlying network. This is not applicable in Chord since placing nodes in the overlay is the responsibility of the consistent hashing. In SHAM, however, nodes can be placed in the overlay based on their actual geographical location. Thus, similar to load balancing, latency in underlying network is another limitation in Chord that SHAM overcomes.

Lookups in SHAM are bounded to $O(N^{1/d})$ where d is the number of dimensions in the system. While node insertion affects $\left(3^d - 1\right)$ other nodes, each node in SHAM maintains $2\left(3^d - 1\right)$ routing entries.

The paper is organized in the following manner: Section 2 presents the design of SHAM. Results and discussion are presented in Section 3. The review of related work is in Section 4. And we conclude our work in Section 5.

2 Design

SHAM is created and maintained using three procedures. The node joining, the key inserting, and the key locating procedures. In this section, we describe its design and clarify the above mentioned procedures in detail.

In its core, SHAM is a self-managing structured P2P network built on a d-dimensional *hexadecimal* coordinate space to host nodes; see Figure 1. It consists of nodes that organize themselves into an overlay. Each node in SHAM is able to (1) receive and forward routing requests, (2) host and locate keys, (3) situate newcomers in the network.

Nodes in a structured P2P overlay must be assigned unique identifiers to indicate their locations in the network: *nodeIds*. This is achieved in many systems by hashing the IP address of the node using a consistent hashing function to generate its identifier. Nodes are then ordered in the network based on their nodeIds. For example, in Chord, the nodes are placed orderly on an identifier circle modulo 2^m, where m is a system parameter [3]. In

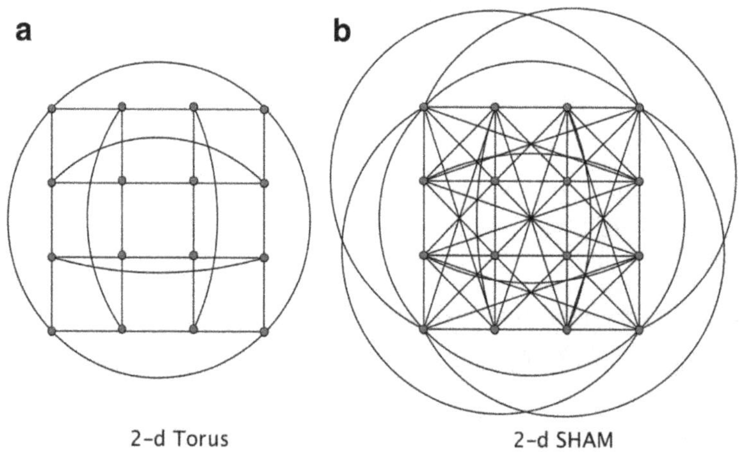

Fig. 1 a 2-d torus. **b** 2-d SHAM

like manner, each node in SHAM is assigned a unique identifier. However, a node's identifier in SHAM is completely independent of its IP address and is deterministically selected from a 2^m space. The design eliminates the consistent hashing phase by directly assigning an identifier to each node joining the system based on the *pre-constructed* geometry of the overlay. Thus, SHAM situates the node first in the overlay based on geometric addressing, then it concludes its nodeId from the address of its position. To accomplish that, we devise that the location of a node in our system is a d-tuple $(X^1, .., X^d)$, such that $X^d = \left(x^d_{r_d-1}..x^d_1x^d_0\right)$ is the d coordinate in the system and r_d is a system distribution parameter concerned with the topology of the overlay, where for a system with 2^m nodes, we have $m = \sum_{i=1}^{d} r_i$. Moreover, for any given node, $X^1, .., X^d$ bear hexadecimal representation of the location of the node in the system. Accordingly, the nodeId for node u in SHAM is simply the concatenation of its coordinates from X^1 to X^d to form the string $(X^1X^2..X^d)$. Throughout this paper, we shall use the term *nodeId* to refer to both the node's location and identifier.

Since addressing in SHAM is sequential, nodes that share the same coordinate are labeled in sequence in the other coordinates. The identifier space suggests addressing wraps around on each coordinate. Thus, we perceive the addressing as rings on regular coordinates $X^1, X^2, ..X^d$ and diagonal coordinates $X^1X^2, ..X^{d-1}X^d$.

Figure 2 illustrates the hexadecimal addressing in a 16-node 2-d system using two rules. The first rule is when $r_1 = r_2 = 2$, $X^1 = \left(x^1_1x^1_0\right)$, and $X^2 = \left(x^2_1x^2_0\right)$, where $x^1_0 = 0$, x^1_1 range is $\{1 - 4\}$, $x^2_1 = 1$, and the range of x^2_0 is $\{A - D\}$. The second rule is when $r_1 = 2$, $r_2 = 3$, $X^1 = \left(x^1_1x^1_0\right)$, and $X^2 = \left(x^2_2x^2_1x^2_0\right)$, where $x^1_1 = 0$, the range of x^1_0 is $\{1 - 4\}$, $x^2_2 = 1$, $x^2_1 = 1$, and the range of x^2_0 is $\{A - D\}$.

Figure 3 presents an example of addressing in a 3-d SHAM system with $r_1 = r_2 = r_3 = 2$ with a node *13F9BC* having its complete set of direct neighbors. The address of the node will be translated onto coordinates: $X^1 = 13$, $X^2 = $ F9, and $X^3 = $ BC.

2.1 Routing tables

In a d-dimensional fully occupied SHAM system, every node has $(3^d - 1)$ direct neighbors it must be aware of. Strictly speaking, a direct neighbor of node u is sequential to u in addressing in a specific direction. Thus, the routing table of any node has $\left(3^d - 1\right)$ permanent entries for those neighbors. Each entry includes the nodeId and the IP address of the neighbor, the nodeId and IP address of the successor of that neighbor in the same direction, a *timeout* counter, and the weights of the paths of the node in every direction. The timeout counter is used to measure the connectivity and the availability of a neighbor. While the path weight of a node in a specific direction, or simply the *weight*, represents the number of consecutive adjacent successors of the node in that direction. The path weights are maintained by the node to realize the growth of the overlay on all direction. If a direct neighbor has not been placed yet, or it has left the system, a *Null* value is entered in its relative position in the routing table. Figure 4 shows an example of path weight measurement.

The weight counters are also used in gap detection. The *gap* notion is used in the remaining of this text to indicate an unfilled zone in the overlay caused by the departure of one or more nodes. This is the only way a gap is created in SHAM. Nodes monitor the change of the weights of their

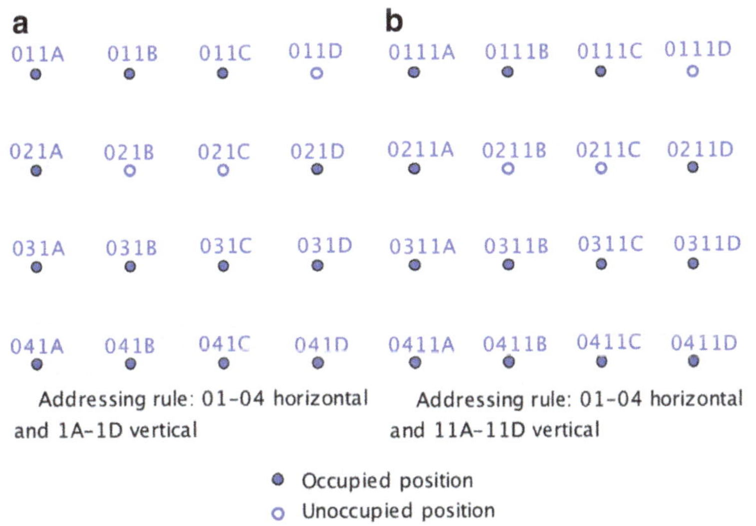

Fig. 2 A 16-node 2-d system with two different addressing schemes. **a** Addressing rule: 01-04 horizontal and 1A-1D vertical. **b** Addressing rule: 01-04 horizontal and 11A-11D vertical

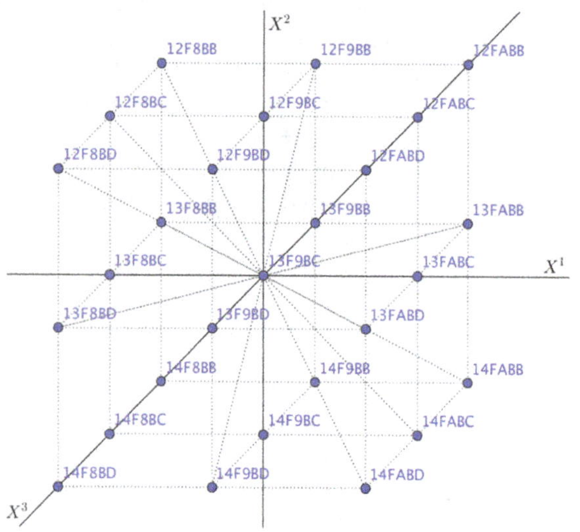

Fig. 3 Hexadecimal addressing in a 3-d SHAM system. Node 13F9BC has 26 direct neighbors organized in sequential order on each coordinate

neighbors continuously to detect the formation of gaps. For instance, say that node *b* observes that the weight of its direct neighbor *c* has decreased after sometime; *b* then deduces that one of the nodes down the direction of *c* has left the system.

To entrench routing in SHAM, the node's routing table includes temporary entries for *remote* neighbors. A node's remote neighbor in a direction *d* is not sequential in addressing to that node, yet it is the first available node in that direction. The sole purpose of remote neighbors is to facilitate routing. Their entries are created in case a direct neighbor does not exist. The essence here is that any node *must* have $(3^d - 1)$ routing entries in total, direct and remote. Table 1 shows routing entries for node 031C shown in Fig. 2a.

2.2 Node join

In P2P systems, peer dynamic refers to the ability of nodes to join, leave, or even fail without any delay or restriction imposed [16]. Thus, it is important for the DHT to have a simplified joining procedure in order to maintain its resilience and scalability when multiple nodes attempt to join. Here, we mention two methods of admitting newcomers to the overlay.

In Chord, after hashing the newcomer's IP address and generating its nodeld, the system relies on some external mechanism to introduce the newcomer to an existing node in the overlay. Afterwards, the existing node will generate a lookup message to find the successor of the newcomer based on its generated nodeld. The newcomer will then connect to its successor and join the overlay. CAN on the other hand assumes the existence of a domain name server that maintains the addresses of bootstrapping nodes which hold partial lists of available nodes in the overlay. In order to join the overlay, the newcomer has first to obtain the address of one of the bootstrapping nodes. Later, the newcomer contacts this node in order to retrieve a list of randomly selected available nodes. Finally, the newcomer sends a join request to one of those nodes which accordingly splits its zone in half and assigns it to the newcomer.

The joining procedure in SHAM is distinctively simple as well. Adding a nascent node to the system entails the following steps.

2.2.1 *Bootstrap*

Similar to Chord, to join SHAM, node *u* contacts a bootstrapping server that holds the IP addresses of nodes already situated in the system. The bootstrapping server responds with a list containing the IP addresses of nodes that had recently joined the system. Discovering and contacting a node in the system engages procedures outside of the overlay, i.e., contacting a web-server that is known to provide the addresses of nodes in SHAM. This server is simply a *Rendezvous* host that is not attached in anyway to overlay [17]. Hence, it does not impinge on the functionality of the mechanism such as node and key handling since its role is limited to providing addresses of available nodes.

2.2.2 *Establish a connection to an existing node*

After receiving a list from the bootstrapping server, the nascent node *u* selects a node *p* from the list and sends

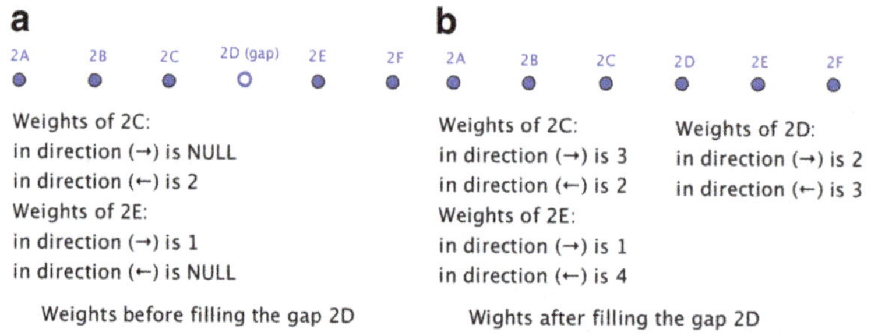

Fig. 4 Weight measurement in a 2-d SHAM. **a** Weights before filling the gap 2D. **b** Weights after filling the gap 2D

Table 1 Routing table entries for node 031C in Fig. 2a

Direction	Neighbor				Successor		Remote	
	NodeId	IP	Timer	Weight	NodeId	IP	NodeId	IP
→	031D	92.110.XXX.XXX	20	4	031A	67.209.XXX.XXX	Null	Null
↘	041D	87.89.XXX.XXX	611	2	011A	167.80.XXX.XXX	Null	Null
↓	041C	212.23.XXX.XXX	221	2	011C	67.115.XXX.XXX	Null	Null
↙	041B	89.213.XXX.XXX	342	4	011A	167.80.XXX.XXX	Null	Null
←	031B	55.10.XXX.XXX	512	4	031A	67.209.XXX.XXX	Null	Null
↖	Null	Null	0	0	Null	Null	011A	167.80.XXX.XXX
↑	Null	Null	0	0	Null	Null	011C	67.115.XXX.XXX
↗	021D	114.23.XXX.XXX	343	4	011A	167.80.XXX.XXX	Null	Null

it a join request which contains its IP address. If no reply is received, u chooses another node and sends a new join request.

2.2.3 *The existing node positions the nascent in the network*

Once the join request is received, p refers to its routing table to find a position for the newcomer. If p has a more than one direct spot[1], it will situate n in one of them directly such that the newcomer will get a maximum number of neighbors upon joining; see Fig. 5a.

However, if p does not have a direct spot, it forwards the join request to another node in the overlay. This forwarding is not arbitrarily done. p consults its routing table to determine on which direction the join request has to traverse. The node elects the direction that reflects the minimum weight entry in the routing table assuming that either a gap exists or that the network lacks growth in that direction. This also helps limit forwarding the request through the overlay. Figure 5b illustrates an example of forwarding the join request based on the weight entry. In this 36-node 2-d system, the host 2C does not have a direct empty spot. Thus, based on its routing table entries, the weight of its link is 1 in the direction of 3B, while the weight in directions 2B and 3C is 2. As a result of that, the join request will be sent to 3B.

After deciding on the position of u, p assigns the newcomer a nodeId tuple and embraces it in the system's routing information:

1. p and u update their routing tables with each other's information.
2. p provides u with its successor information on the direction u was inserted.
3. p sends u's information to their common direct neighbors. Accordingly, each one of them repeats the previous two steps with u.

All of the preceding steps engage adjusting the weight entries of related nodes.

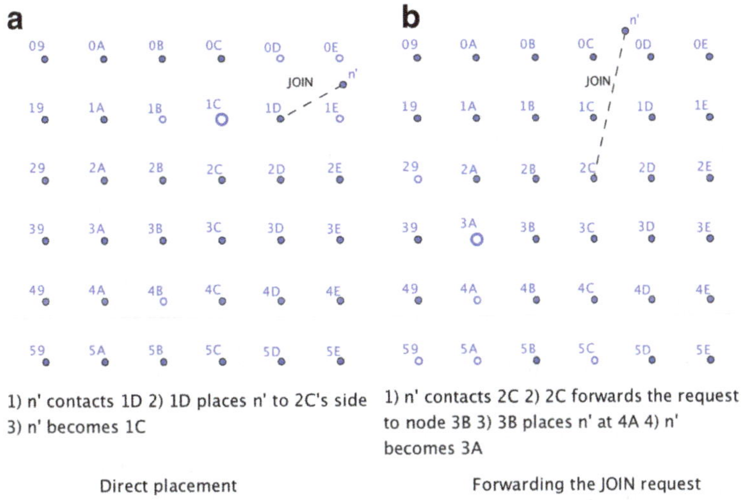

Fig. 5 Two cases of node n′ joining the system. **a** Direct placement: n′ will be placed by 1D based on a greedy manner at position 1C. **b** Forwarding the join request: 2C does not have an empty spot. The join request will be sent to node 3B which will place n′ at 3A

2.2.4 *Graceful departure*

Before a node leaves the system gracefully, it chooses a direct neighbor to become the *heir* to its keys. As a consequence, the heir receives the keys from the departing node and assumes their ownership. Furthermore, the new owner duplicates the keys on its direct neighbors. The selection of the heir is based on the weights of the direct neighbors. The departing node aims to choose a neighbor which reflects the maximum value of weight. This is a safeguard to store the keys at nodes which have as many neighbors as possible, thus, duplicating them more efficiently. Finally, the departing node informs the neighbors which are "uncommon" with the heir about its departure to delete its keys.

2.2.5 *Ungraceful departure*

In the case of a node's departure, the system needs to be informed and updated as well. However, due to the dynamism of P2P networks, nodes tend to leave ungracefully without informing the system. Nodes in SHAM rely on system maintenance messages to keep their routing tables updated. However, in the absence of such messages, SHAM nodes depend on the timeout counters and self-initiated *heartbeat* messages to check the availability of their neighbors. Each neighbor receives the heartbeat message responds with an *alive* message and resets the timeout counter of the neighbor that sent it.

The heartbeat messages used in SHAM are the same as alive messages that are communicated to detect node's failure in other systems, i.e., PASTRY [15]. That is, a heartbeat or alive message is a simple ping command from one peer to another peer in the overlay. Thus, such "soft" messages are not considered as an overhead for the traffic of the overlay as they might traverse through nodes that are not in the overlay.

2.3 Maintenance and recovery

Chord uses periodic stabilization protocol at each node to learn about newly joined nodes, update successor and predecessor, and fix finger tables. Nodes in SHAM announce the changes in their routing tables by piggybacking these changes on traffic messages such as node and key placements and key search [8]. Moreover, in addition to the heartbeats that are invoked when timeout counters expire, SHAM relies on two mechanisms to keep the routing tables up to date: the Join and the Routing Restoration mechanisms.

The join mechanism is the first recovery tool in the system. Whenever a new node joins the system, any other node that either handles its placement or forwards its join request in the overlay will update its routing table accordingly. In the routing restoration mechanism, the system recovers stale entries that are caused by the ungraceful departure of nodes. In this procedure, if a node's direct neighbor and its successor go offline ungracefully, it will send a *hunt* message looking for a node that succeeds the departed successor on the same direction. The hunt message will be forwarded until it reaches a node that is "currently" a neighbor of the first available successor of the departed node. Consequently, the neighbor responds by sending the routing information of the successor directly to the requester. The requesting node in turn updates its routing information to reflect the successor as a link in that direction.

Successors also enforce the stability of the system. Fundamentally, each node should be in a relation with $(3^d - 1)$ other nodes. However, in a partially occupied system and due to nodes' departure, there exist gaps and

---

Algorithm 1 The *update* procedure between nodes p and u

$p.update(u)$
$R_p^{d^u} \leftarrow u$ {$R_p^{d^u}$: Add u to p's routing table in direction of u}
$R_u^{d^p} \leftarrow p$ {$R_u^{d^p}$: Add p to u's routing table in direction of p}
if $(p.remote)^{d^u} == u^{d^u} + 1$ **then**
 $R_u^{d^{u+1}} \leftarrow u^{d^u} + 1$
 $R_{u+1}^{d^u} \leftarrow u$
 $p.successor(u^{d^u}) == u^{d^u} + 1$
 if $(p - 1) \neq NULL$ **then**
  $(p - 1).successor(p^{d^u}) = u$
  $(u).successor(p^{d^p}) = p - 1$
 end if
else
 $(u.remote)^{d^u} = (p.remote)^{d^u}$
 $p.successor(u^{d^u}) = (p.remote)^{d^u}$
end if
$(p.remote)^{d^u} = NULL$
$u.publish()$
$u.weight()$

Algorithm 2 The *publish* procedure between nodes p and u

$u.publish()$
Require: ψ_u: set of available direct neighbors of u
for all $j \in \psi_u$ **do**
 $u \leftarrow \xi_j = \{\Re_u \cap \Re_j\}$ {ξ_j: common available neighbors between u and j}
 for all $v \in \xi_j$ **do**
  if $v \notin \psi_u$ **then**
   $v.update(u)$
  end if
 end for
end for

Algorithm 3 The *weight* procedure for node u

```
u.weight()
ξ_k = {ℜ_u ∩ ℜ_p}
for all members of ξ_k do
    if u^{d^j} + 1 ≠ NULL then
        while u^{d^j}.link < maximum diameter do
            u^{d^j}.link = (u^{d^j} + 1).link + 1
        end while
    else
        u^{d^j}.link = NULL
    end if
end for
```

unoccupied positions in the system. SHAM requires each node to hold entries for the successors of its direct neighbors as a contingency plan. As a result of that, each node maintains $2(3^d - 1)$ entries of peers' addresses which fortifies the system and reduces the average path length as we will discuss later. Also, this is important to enhance connectivity and routing in the system. In case a direct neighbor fails or leaves the system, the node will move the successor nodeId and IP address from the routing table to the relative remote neighbor entry. Therefore, the node in SHAM has to maintain between $(3^d - 1)$ and $2(3^d - 1)$ entries in its routing table.

2.4 Key handling

Values are data items needed to be uniformly distributed over the P2P network [18–20]. In structured P2P networks this is usually carried out by using DHTs. Each value in the system will be paired with a key that is generated by a known hashing function. Each node identifier (usually the IP address of the node accompanied with a port number) will be hashed using the same hashing function to generate the nodeId. As a result of that, both of the generated nodeId and the key will fall within the same identifier space. Having created the namespace for nodes and values, every (*key, value*) pair will be stored at a node that has an identifier matches the generated key. If a match is not found, the key will be stored at a node with the closest identifier. This eventually makes every node accountable for a group of values. Our system as mentioned before does not require hashing the node's IP address to generate its nodeId, yet, it is derived directly from the node's position in the overlay. Choosing a hexadecimal representation in SHAM congregates the nodeIds and keys within the same namespace. We relax our guard when choosing the address space since no two nodes will have the same nodeIds. The criterion in choosing the address space becomes related to the load on each node in particular: The more nodes in the system, the less load on each node. On the other hand, the size of the namespace must be large enough to neglect the probability of having two values hash to the same identifier [3].

2.4.1 Key mapping

Key mapping in SHAM is similar to that in other DHT mechanisms. In SHAM, CAN, and Chord, the key will be cached at a node that has a nodeId that is closest to the key. Each time a new node joins with closer nodeId to that of they key, the key will be forwarded to it. Specifically, in SHAM, keys are distributed to nodes based on the coordinate system $\left(X^1, .., X^d\right)$. The address of the key will be resolved to get its X^1, X^2, .., and X^d coordinates. For an address space with a tuple $\left(X^1, .., X^d\right)$, any key in the system will be regarded as a concatenation of $\left(x^1_{r_1-1}..x^1_1x^1_0, .., x^d_{r_d-1}..x^d_1x^d_0, z\right)$ bytes, where r_d is the number of hexadecimal bytes along the X^d coordinate and z is a suffix of hexadecimal bytes in which $z = 0$ if the address space is the same size of the key space. After extracting the coordinates of the key, it will be forwarded through the system until it reaches a node that matches its X^1, X^2, .., and X^d coordinates. If no node has been placed in the system with that address yet, the key will be stored at a node that has the closest identifier to that of the key. Later on, whenever a node whose identifier is closer to the key joins the system, the key will be reallocated to that newcomer. The reallocation of the key may continue until it is stored at node whose nodeId matches the key[2].

Figure 6 shows an example of key mapping in a 2-d SHAM system, (X^1, X^2). In the figure, node 3E receives key 1B23F which is resolved to $X^1 = 1$, $X^2 = B$, and $z = 23F$. Thus, the node deduces that the key should be stored at node 1B. 3E forwards the key to another node that has a closer identifier to that of the key. In this particular case, 3E forwards the key to node 1C. After receiving the key, 1C applies the same procedure to decide on which direction the key should be forwarded. Since 1B is a direct neighbor to 1C, and since it has not been placed in the system yet, 1C caches the key locally. After 1B joins the system, the key 1B23F will forwarded to 1B.

2.4.2 Key duplicate

Replication techniques play a major role in reducing latency, improving load balancing, and enhancing availability [21–24]. Some known techniques are synchronous, asynchronous, dynamic, full, and neighborhood replication [25–27]. Such schemes differ based on their complexity. For example, synchronous, asynchronous, and dynamic replication techniques require frequent messaging in order to keep the system updated. Whereas the full replication technique is much simpler in principal, yet it is associated with high cost in maintaining the replicas of all keys in the system. In SHAM, we employ the neighborhood replication: Once the key is stored at a node, it will be directly replicated at the node's neighbors only as a

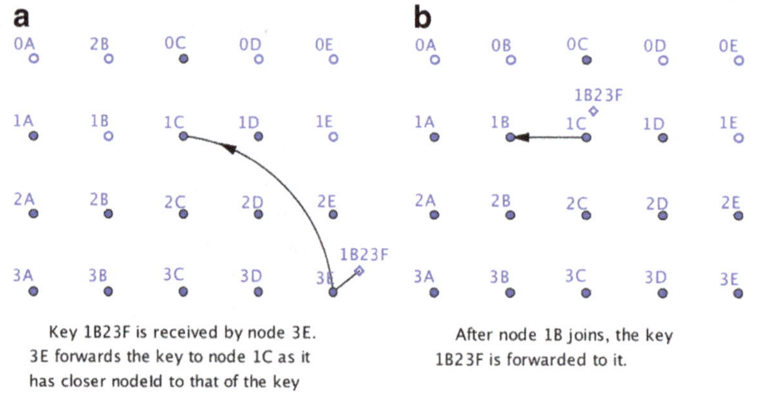

Fig. 6 Key mapping in a 2-d SHAM. **a** Key 1B23F is received by node 3E. 3E forwards the key to node 1C as it has closer nodeId to that of the key. **b** As node 1B joins, the key 1B23F is forwarded to it

safeguard that the key will still be an accessible contingent upon the failure of its host. In this scheme, no messaging is required to update the system, i.e., there is no need to know who has what. Moreover, the cost of maintaining the direct neighbors' keys is considerably a small compromise for enhancing the availability of keys.

We require the node to discriminate between its own keys and its neighbors' keys. The discrimination is crucial to prevent the neighbors' keys from being replicated over and over again in the network. Strictly speaking, the node will only replicate its original keys. This rule holds until the neighbor which replicated its keys departs. In that case, as an heir, the node assumes the neighbor's keys to be original and replicate them at its neighbors except those which are common with the departed.

2.4.3 Key search

The presence of the weight entries in the routing tables fortifies the routing efficiency of the system. Those entries give the node sufficient knowledge of the approximate depth of the network from all $(3^d - 1)$ directions. Searching for a key in our system comprises two steps. First, resolve the direction of which the search will traverse. Then, estimate the distance (*depth*) of the key from the requesting node. Using simple mathematical interpretation, the node can resolve the direction on which the query should traverse. Consequently, the node will approximate its distance from the requested key based on the weight of the neighbor which is in that direction. The mathematical interpretation is an inherited characteristic from the topology of the system. The property of sequential addressing places the nodes within specific intervals from each other. Table 2 shows the distance between a node and its direct neighbors in a 2-dimensional system X^1X^2.

Figure 7 illustrates an example on key search in a 2-d system that uses an addressing rule with $r_1 = r_2 = 1$ and $X^1 = (x_0^1)$, $X^2 = (x_0^2)$. In the figure, when node 5C searches for key 1731*A*, it first determines the location of node 17 in the system, which is $5 - 1 = 4$ steps away on the upward direction of X^2 coordinate and $C - 7 = 5$ steps away on the leftward direction of X^1 coordinate. Accordingly, 5C forwards the request to node 3A which is in the upward and leftward direction in its routing table. The forwarding continues by 3A following the same procedures.

3 Discussion

We compare SHAM with two major DHT systems: CAN and Chord. Each one of them represents a family of DHT routing protocols. There are more recent DHT routing protocols that have been recently developed [28–30]. However, the majority of these systems adopt the paradigms of CAN or Chord or they do not fall into the class of systems that balance between the routing performance and the size of the routing table.

We choose three areas of comparison in this paper: routing performance, robustness against failures, and load balancing. In addition to that, we explore a distinctive

Table 2 Address resolution in 2-d SHAM

	With direct neighbors	
$X^1 - 1X^2 - 1$	$X^1 - 1X^2$	$X^1 - 1X^2 + 1$
$X^1X^2 - 1$	X^1X^2	$X^1X^2 + 1$
$X^1 + 1X^2 - 1$	$X^1 + 1X^2$	$X^1 + 1X^2 + 1$
	With other neighbors	
$X^1 - jX^2 - j$	$X^1 - jX^2$	$X^1 - jX^2 + j$
$X^1X^2 - j$	X^1X^2	$X^1X^2 + j$
$X^1 + jX^2 - j$	$X^1 + jX^2$	$X^1 + jX^2 + j$

Note: $j = 2, 3, .., E, F$
$1 + F = 0$ and $0 - 1 = F$

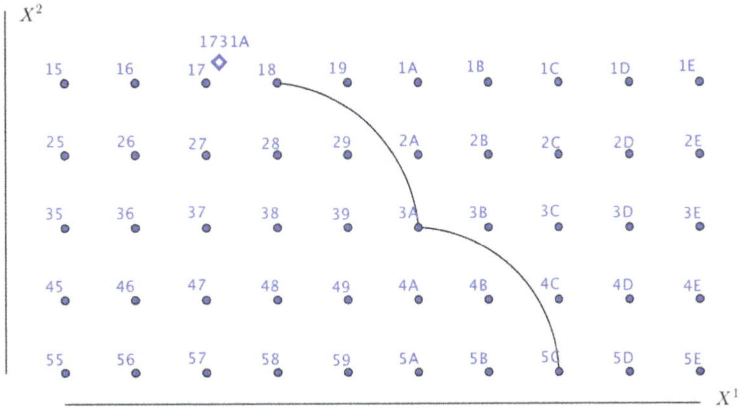

Fig. 7 Key search in a 2-d SHAM system. Following the address resolution procedure, node 5C forwards the query to node 3A from its routing table. Subsequently, 3A forwards the query to node 18. Based on key replication technique used in SHAM, node 18 will answer the query

characteristic of SHAM, homogeneous addressing, and show its effectiveness in reducing latency in the underlying network.

3.1 Average path length in CAN vs. SHAM

In CAN, the average path length for a d-dimension system is $\frac{d}{4}N^{\frac{1}{d}}$, where each dimension has an average path length of $\frac{1}{4}N^{\frac{1}{d}}$. In SHAM, however, the average path length for each dimension is $\frac{1}{4}N^{\frac{1}{d}}$, whereas for a d-dimension system the average path length becomes $\frac{1}{8}N^{\frac{1}{d}}$. This is because the search in SHAM takes diagonal paths between coordinates while in CAN the search traverses on the coordinates as Fig. 8 illustrates [4, 31]. Furthermore, each hop in SHAM is being performed from a node to its direct neighbor's successor which reduces the path length by half.

3.2 Performance against CAN and Chord

In Fig. 9, we present the results of the first experiment in matching SHAM up with CAN and Chord. The basis in this part of the simulation is to saturate the overlays with nodes by setting Poisson arrival rate $\lambda = 1$ and the departure rate $\mu = 0$. The overlay size varied from 2^{10} nodes with a query rate of 0.1 and an update rate of 0.001 to 2^{23} nodes with a query rate 0.001 and an update rate of 0.0001. Table 3 summarizes these parameters.

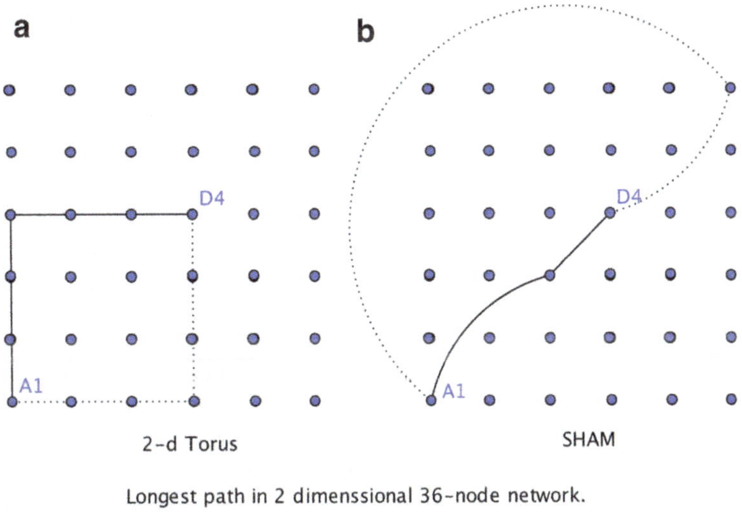

Fig. 8 Depiction of routing in SHAM vs. Torus. **a** In 2-d torus: the longest path is between nodes A1 and D4. It requires the lookup six hops to reach D4 from A1 (solid line). Another possible same length route is the dashed line. **b** In case of SHAM: the longest path is also between A1 and D4. It requires the lookup two hops to reach D4 (solid line). Another same length route is the dashed line

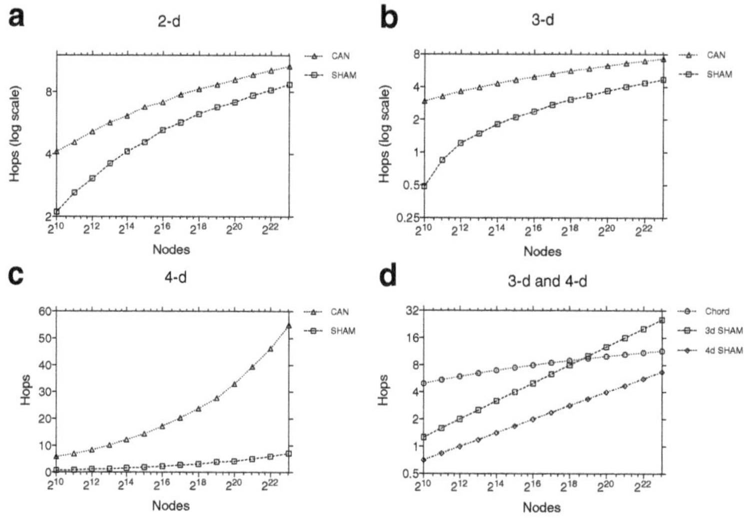

Fig. 9 a–d Average path length in SHAM, CAN, and Chord

This experiment is comprised of two steps: First, we compare the performance of SHAM with CAN in 2-d, 3-d, and 4-d systems. Second, we compare the performance of SHAM in 3-d and 4-d systems with Chord.

Regarding CAN, the results seen in the figure support our analysis that SHAM performs better than CAN and achieves a shorter path length. The advantage in performance is clearly seen in Fig. 9a–c. Distinctly, the performance of SHAM in 3-d is even better than CAN in 4-d. Part of this improvement in performance is credited to the use of diagonal paths between coordinates which minimizes the average path length by a factor of *d*. The other part is ascribed to the double hops that are being performed from the node to the successor of its direct neighbor which further reduces the average path length by a factor of 2.

On the other hand, the results presented in Fig. 9d show that up to a certain network size, SHAM performs better than Chord in 3-d and 4-d. This is consistent with expectation: in 3-d and 4-d, SHAM has shorter path length than Chord in networks of sizes up to 2^{19} and 2^{27} nodes, respectively. Thus, increasing the number of dimensions plays a significant role in reducing the average path length in SHAM comparing to Chord.[3]

Table 3 Simulation parameters 1

Operating system	Red hat
Simulator	C++
Processor	Xeon E7 @ 3.20–3.50 GHz
Cache	45 MB
RAM	16 GB
Network size	$2^{10} - 2^{23}$
Number of keys	$2^{10} - 2^{23}$
Arrival rate λ	1
Departure rate μ	0
Query rate per node	{0.001–0.1} per time interval
Update rate per node	{0.0001–0.01} per time interval

3.3 Failed lookups

Another part of the analysis is failed lookups, i.e., queries that do not result in keys. Although Chord has a better performance after specific network sizes in 4-d, however, when it comes to failed lookups, SHAM noticeably outperforms Chord.

To examine this venue, we run the simulation on fully occupied 2^{16}-node 3-d SHAM and Chord overlays and measure the number of failed queries in each system. The scenario of this second experiment is to fail nodes at different capacities in both systems by increasing the departure rate, then monitor the percentage of queries that encounter failed nodes through their paths to destination; see Table 4. From Fig. 10a, we notice that in SHAM and Chord, as the percentage of failed nodes increases, the probability that a query will face at least a failed node also increases. However, Fig. 10b, which represents the percentage of failed lookups, signifies the difference in performance between the two systems. In SHAM, when the percentage of failed nodes is 20% for instance, the probability that a query will encounter at least a failed node is around 65%, yet, the percentage of failed queries is less than 10%. This is not the case for Chord: For the same percentage of failed nodes, the probability of facing at least a failed node is around 80%, while the percentage of failed lookups reaches more than 35%.

Table 4 Simulation parameters 2

Network size	65,536
Number of keys	262,144
Graceful departure %	30%
Departure rate μ	{0–0.2}
Query rate	{0.2–0.5} per time interval
Update rate	{0.1–1} per time interval

This reduction in failed lookups in SHAM comparing to Chord is related to the following. First, it is related to the connectivity of the system in SHAM where each node maintains entries to $2(3^d - 1)$ nodes in total, thus, having different possible routes to the destination. Second, each key in SHAM is being replicated at the direct neighbors of the host. As a result of that, even if the host failed, once the query reaches one of its direct neighbors, it would be considered as a successful hit. In Chord, however, a failed lookup is attributed to one of these two reasons: either the node that is hosting the key has failed which means a failed lookup or the finger table of some of the predecessors are inconsistent which hinders forwarding the query [3].

The last part of the comparison is to examine the additional number of hops visited when lookups face failed nodes. In this experiment, we use the previous network configuration in SHAM and Chord with two failure settings: 5 and 20%. Figure 11 illustrates that SHAM outperforms Chord in this part of the performance as well. For a 5% failure, in Fig. 11a, around 84% of lookups encountering at most two additional hops is observed in SHAM, while in Chord, 86% of lookups encounter at most three additional hops. On the other hand, Fig. 11b shows the results of the same experiment for a 20% failure rate. The figure shows that both systems start sluggishly with 11 and 6% of at most two additional hops visited for SHAM and Chord respectively. However, as seen in the figure, 61 and 56% of lookups encountered at most five and seven additional hops in SHAM and Chord respectively.

3.4 Load balancing

Load balancing is an important aspect of structured P2P networks [18, 32–35]. A robust and resilient system should be able to fairly distribute the keys in the network over the participating nodes [32, 34, 35]. This will avert inundating nodes with keys and hence improving resources accessibility in the network. Consistent hashing uniformly distribute keys over the namespace as discussed earlier. However, in structured P2P system where both of the nodeId and the key are generated using the consistent hashing, there exist situations where the node identifiers do not uniformly cover the entire namespace system [3]. Thus, some nodes will be overloaded with keys while other nodes do not cache any keys. Chord maintains uniformity in this regard by requiring each node to cache its keys at an additional $O(\log N)$ *virtual* nodes.

SHAM emulates the property of consistent hashing by deterministically assigning nodeIds based on the predefined addressing scheme. Moreover, if a node that is supposed to store a key has not been placed yet, the key will be cached at a node that has a nodeId which is closest to the key's identifier temporarily. Whenever a node with closer nodeId is placed in the system, the key will be forwarded to it. This ensures that each node will roughly receive the expected load. However, most importantly, SHAM benefits from its key replicating scheme at direct neighbors in enhancing its load balancing. This is nearly similar to Chord with two differences. First, in Chord, the node will need to discover its virtual nodes and saves their routing information whereas in SHAM, those nodes are already in the node's routing table. Second, the cost of adding the virtual nodes in Chord is increased [3]. For example, in an overlay of one million nodes, a Chord's node has to maintain entries for $\log^2 N \leq 400$ nodes. In 4-d SHAM, however, for the same network, each node maintains entries for $2\left(3^4 - 1\right) = 160$ other nodes in total.

Figure 12 signifies the load balancing capability in SHAM. The figure presents the results obtained by simulating a 2-d SHAM overlay with a capacity of 2^{12} nodes and 2^{14} keys. Intuitively, if the system is fully occupied, we expect theoretically each node to hold $2^{14}/2^{12} = 4$ keys. However, the deficiency in load balancing in DHTs arises when the overlay is not fully occupied or the namespace is not entirely covered. Therefore, to show the efficiency of SHAM in load balancing, we run the experiment on

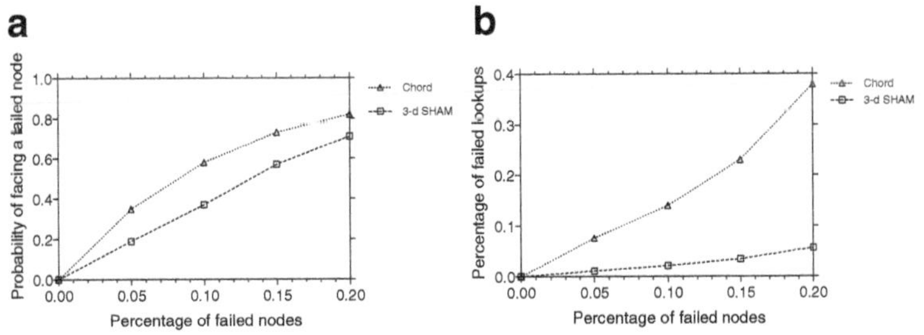

Fig. 10 Performance comparison between SHAM and Chord. **a** Probability that a query will encounter a failed node. **b** Percentage of failed lookups

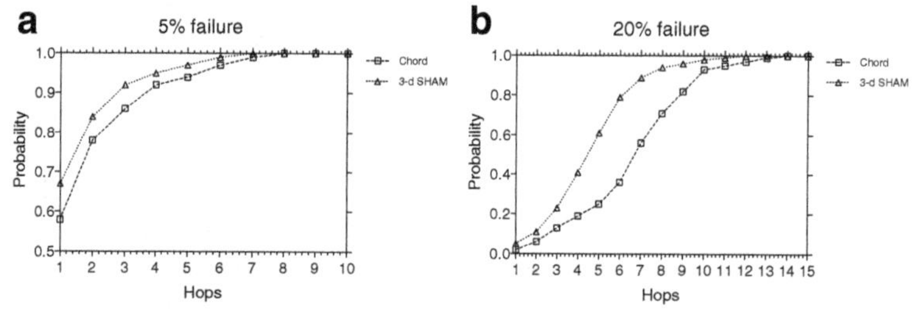

Fig. 11 Performance comparison between SHAM and Chord. CDF of number of additional hops visited in case a lookup encounters failed nodes in a network of 65,000 nodes. **a** Percentage of failed nodes is 5%. **b** Percentage of failed nodes is 20%

a partially occupied overlay. The first figure, Fig. 12a, shows how SHAM distributes 2^{14} keys over nodes without invoking key replication procedure. The results indicate that when 2^{11} nodes are present, load balancing is maintained in the overlay with around 60% of nodes holding between 10 to 16 keys each. Additionally, Fig. 12b presents the results of running the same experiment when the key replicating procedure is invoked. The results indicate that replicating keys at neighbor nodes adds more balance in distributing keys: around 23% of nodes are storing 160 keys each. The tradeoff here is in the increase in the number of keys the node holds; however, we believe that this increase is viable and practical since not only it enhances load balancing but also it increases availability in the network.

3.5 Homogeneous addressing

In addition to load balancing, nodes dynamics, and self-organizing capabilities, reduced latency is a crucial property in P2P networks [23, 24, 36, 37]. Latency in P2P network is related to the distributed nature of nodes. Nodes with adjacent addresses in the overlay might be located in different geographical areas. This mismatch between the overlay and the IP network is the major factor for increasing latency. For example, in SHAM, a direct neighbor is within one hop, yet, this same neighbor could be many hops away in the IP network. A remedy to this problem is to enable nodes to connect to physically nearby neighbors. Other method is utilizing clustering to impose physical proximity of overlay neighbors or even using geographical routing, i.e., geographic awareness [38, 39].

In SHAM, we propose positioning nodes with adjacent physical addresses at close spots in the overlay as a solution to this problem. SHAM can function as to arrange nodes in classes, where each class holds a range of nodes with close physical locations. Thus, once a newcomer arrives to the bootstrap server, a traceroute command is sufficient to indicate to which class the newcomer belongs. In that case, the bootstrap server constructs a list of nodes from that class to handle the positioning of the newcomer. Figure 13 shows the results of testing the latency in SHAM under two conditions. First, when nodes are placed with no regard to their geographical positions; see Fig. 13a. And second, when we apply homogenous IP/location addressing scheme; see Fig. 13b.

In this experiment, we simulate a 2-d SHAM system having 2^{11} nodes distributed over five IP classes with each class representing a geographical area. The latency between nodes within the same class is drawn uniformly at random from $[5, 15]$ intervals. Similarly, the interclass latency is selected from the space $[16, 45]$. The scenario here is to send queries from source nodes to their most distant nodes in the overlay and measure the latency these queries accumulate in a heterogeneous and homogeneous addressing schemes. Noticeably, the fluctuation in delay in first condition is due to the heterogeneity of nodes locations in the overlay. On the other hand, when we apply homogenous addressing in SHAM, fluctuation has been

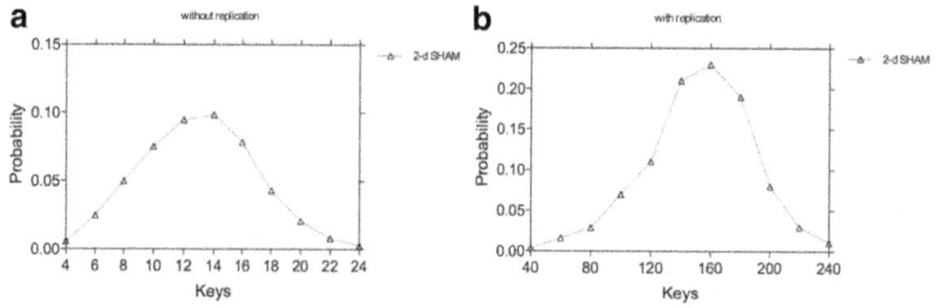

Fig. 12 Distribution of 2^{14} keys over 2^{12} nodes. **a** Without replication. **b** With replication

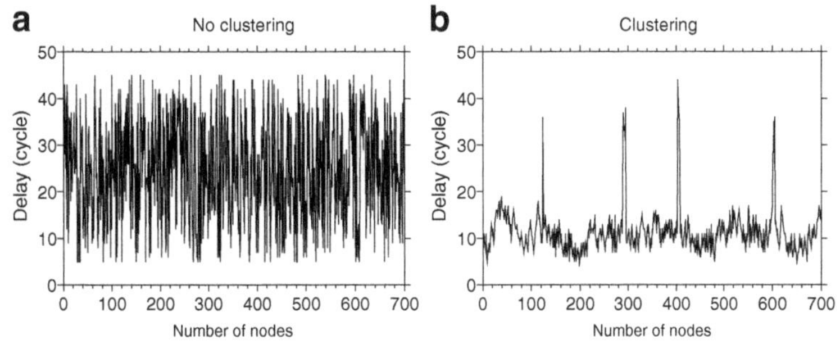

Fig. 13 Latency when no clustering is imposed. **a** No clustering. **b** Clustering

reduced drastically with occasional overshoots. The overshoots are affiliated with forwarding the queries from one class of addresses to another.

4 Related work

In classifying structured P2P overlays based on their routing performance, we put them in four major groups: single-hop overlays, constant degree overlays, multi-dimension overlays, and logarithmic degree overlays. In single-hop overlays and as the name implies, routing can be resolved within one hop. Some of the known $O(1)$ overlays are D1HT [1], OneHop [2], Kelips [7], and EpiChord [8]. As mentioned before, in order to resolve lookups within one step from source to destination in such overlays, nodes must maintain global knowledge about the status of the overlay. Thus, a major drawback to the $O(1)$s is the high cost associated with maintaining the global routing tables and their inability to handle churn-intensive workloads.

Constant degree overlays take fixed number of hops in resolving lookups regardless of the number of nodes in the system. One of the well-known systems in this class is Cycloid [12]. The system is a d-dimensional cube-connected cycles. Every vertex in the cube is replaced by a *cycle* of 3° d nodes each maintaining a routing table of seven entries. Although routing is achieved within $O(d)$, however, stabilizing and maintaining valid routes degrade the system's performance.

The third category in DHTs is the multi-dimension overlays. Our system falls within this category. The most prominent system in this category and one of the earliest DHT systems that have been proposed is CAN [4, 31]. Nodes self-organize themselves in CAN in a virtual d-dimensional Cartesian coordinate space built on a d-torus. To join the system, a newcomer must acquire the information of an available node in the overlay. This information is obtained usually from a bootstrapping server. A join request is then sent by the newcomer destined to a point P in the overlay. The message is routed using CAN routing algorithm through the available nodes until it reaches a node which is the owner of the zone where P exists. The node then divides its own zone by half and assigns one half to the newcomer. The split process is performed in such a way that the zones will be re-merged upon the departure of nodes. Each node in CAN holds entries for a number of neighbors based on the dimension of the coordinate system. Thus, insertion process in CAN affects $O(d)$ existing nodes. Average path length in CAN is bounded to $O(\mathrm{N}^{1/d})$. If d is chosen such that $d = (\log N)/2$, routing performance converges to $O(\log N)$. Although our mechanism seems similar to CAN, however, in addition to lowering the cost of search, there are also major points that distinguish SHAM. First, unlike CAN, the overlay in SHAM is already organized into fixed addresses, or spots, that will be occupied when nodes join, whereas in CAN, the *spot* will be created for the node upon its arrival, which may include zone splitting. Second, situating a newcomer in SHAM is much simpler and requires minimum effort comparing to CAN, which involves forwarding the join request to the selected point P as mentioned earlier. Third, SHAM employs homogeneous addressing which is not possible in the CAN structure.

The last category in our discussion is logarithmic degree overlays. In this category of DHTs, routing is reduced by half in each step the query takes towards the target. Systems such as Chord, Ulysses, Pastry, Kademilia, Tpastry, and P-grid belong to this class [3, 5, 15, 40–43].

Chord is the most renowned DHT in this category. It utilizes consistent hashing to generate an $m - bit$ identifier for nodes and keys. Nodes are then organized in an identifier circle modulo 2^m. The node identifier will be generated from the node's IP address while the identifier of the key will be generated by hashing the key itself. This arrangement makes both of the nodes and keys have the same naming space. Chord assigns keys to nodes based on the naming space as well. A key k will be stored at its *successor* node whose nodeId matches the identifier of the key in the space. If no match is found in the system, i.e., if the node whose nodeId matches the key has not been placed in the system or has left the system, the key will be cached at the next node whose nodeId follows the identifier of they key in the space in a clockwise manner. Also, Chord recognizes a *predecessor* to a node (or a key) as a node whose nodeId precedes that node in the circle in a counterclockwise manner. Each node in Chord maintains a routing table of $O(\log N)$ entries in N-node system. Lookups are resolved in $O(\log N)$ messages to other nodes, while insertion or deletion of a nodes affect $O(\log^2 N)$. However, in order for Chord to be load balanced, each node has to store its keys at additional $O(\log N)$ virtual nodes. Thus, increasing the routing table for the purposes of routing and storing to $O(\log^2 N)$. For reader's reference, other DHTs that are derived from Chord are [28–30].

Ulysses is a P2P structured system that adopts the *static butterfly* topology and attempts to reduce the $O(\log N)$ latency by a factor of log log N [5]. Naming space in Ulysses is based on a row-level convention. In a network with l levels and N nodes, each DHT node is depicted by a tuple of row and level identifiers (P, l), where P is a binary string signifying the row to which the node belongs and l denotes the level in that row. The row identifier can be mapped into a concatenation of bits between 0 and $k-1$, where k is the size of the dimension in the static butterfly topology. The space in Ulysses is divided into disjoint zones in which each DHT node is responsible for a specific zone. A node with an identifier (P, l) caches all keys (α, l) where P is a prefix to α. Routing in Ulysses traverses the network through levels. If a node (P, l) searches for a key (α, i), the query will be forwarded to the next level $(P', l+1)$. α in this case will match a range of the binary string P' in level $l+1$. The forwarding continues in Ulysses through levels where at each step, the search for α narrows down until the target level is reached. Ulysses is optimized such that its nodes maintain routing tables of an average size of size log N in order to achieve a minimum diameter of $\lceil \frac{\log N}{\log \log N} \rceil + 1$. Although Ulysses reduces the path length while maintaining log N routing entries, however, this comes at a cost of uniformity. Thus, two major drawback exist in Ulysses: node congestion and load balancing.

In addition to routing performance in DHTs, we also list some of the overlays that discuss the geographical awareness in structured P2P networks [39, 44, 45].

Jedda and Mouftah [39] proposes the Geographic-Aware Content Addressable Network (GCAN) to solve the issue of geographic awareness in object naming service architecture, ONS, by placing it on top of Chord-like systems. According to the paper, GCAN preserves the complexity of Chord to $O(\log N)$ and the routing table size to $O(\log N)$ as well. GCAN is built on a grid that could cover a certain geographical area. This area is divided into cell using vertical and horizontal chords. Later, using a discretization procedure, cells will be filled with nodes that represent nameservers.

SpatialP2P is a decentralized mechanism that provides node indexing and key storing for multi-dimensional data [44]. According to the paper, the proposed framework upholds the major requirements of DHTs such as indexing, retrieval, and load balancing. It is built on grid that is divided into *cells* that hold the spatial data and are identified by the grid's coordinates. Spatial data can be stored at one single cell or they can span over a group of cells. Moreover, nodes in the system are also identified by the same coordinates making cells and nodes sharing the same coordinates eventually map to each other. Nodes in SpatialP2P maintain lists of successors and indexed nodes as well. Successors are needed for connectivity and routing, while indexed nodes are essential for enhancing the lookups. The cost of search in SpatialP2P when mapped to a one-dimensional Chord-like system is bounded to 3log N. Although the paper discusses load balancing, however, the experiment tested the scenario of having fully occupied network whereas the issue of load balancing is when the network is not fully occupied.

Geodemlia is yet another P2P overlay that is based on Kademlia [40, 45]. It is a location-based search algorithm that allows nodes to locate keys around specific geographical area. Geodemlia is to be built on static nodes that provide storage and search functionalities for mobile devices. Nodes in Geodemlia are positioned on a sphere inspired by the shape of the earth. Accordingly, each node can be located using longitude and latitude angles using some location services, i.e., GPS or IP locators. Nodes in the system split the geographical area into predefined directions, where for each direction, the area will be divided into distance buckets that store fixed number of the system's nodes. Geodemlia, as a geographical awareness mechanism and similar to SpatialP2P, preserves the locality and directionality of data in the overlay. However, the assumptions of the static nature of nodes and load balancing are a matter of question.

5 Conclusions

In this paper, we presented SHAM, an addressing mechanism for structured P2P overlays. SHAM is robust, highly scalable, and decentralized. SHAM is simple, it uses a hashing function to generate keys' identifiers, and a primitive hexadecimal scheme to address nodes in the overlay. In performance, given a key, SHAM can route queries to that key in $1/8\left(N^{1/d}\right)$ steps, with each node maintaining $2\left(3^d-1\right)$ routing entries, where d is the number of dimensions and N is the number of nodes in the steady state.

With a limited increase in the size of the routing tables and the use of diagonal paths, we have shown that SHAM significantly outperforms CAN. Moreover, SHAM even performs better than the dominant Chord system. Our results demonstrated that failed lookups were noticeably reduced in SHAM comparing to Chord. We attributed this reduction to the presence of multiple routes to the destination and to the duplication of keys in the system.

The rigidness of SHAM stems from the use of the gap filling mechanism in which the priority is to situate newcomers in vacant positions formed by departed nodes. Thus, entrenching routing as nodes perform double hopping to the successors of their direct neighbors.

Finally, with the use of homogeneous addressing scheme, nodes which have close geographical locations can be positioned adjacent to each others in the overlay. Thus, giving SHAM a major advantage in reducing latency in the network.

Endnotes

[1] The term *spot* in this paper refers to an empty position in the overlay.

[2] Key refers to the (key, value) pair.

[3] Choosing the number of dimensions in SHAM such that $d = \log(N)$ reduces SHAM to be from the same family as Chord. Thus, reducing the average path length to $\frac{1}{2}\log(N)$.

Authors' contributions
MZ proposed and designed the mechanism and the simulator. NUH assisted in the analysis of the output data. Both authors read and approved the final manuscript.

Competing interests
The authors declare that they have no competing interests.

References

1. LR Monnerat, CL Amorim, in *Proceedings of the 20th IEEE International Parallel and Distributed Processing Symposium*. D1HT: a Distributed OneHop Hash Table, (Rhodes Island, 2006), p. 10. doi:10.1109/IPDPS.2006.1639278
2. A Gupta, B Liskov, R Rodrigues, in *Proceedings of 1st Symposium on Networked Systems Design and Implementation*. Efficient Routing for Peer-to-peer Overlays (USENIX Association Berkeley, San Francisco, 2004), pp. 113–116
3. I Stoica, R Morris, D Karger, MF Kaashoek, H Balakrishnan, in *Proceedings of the 2001 conference on Applications, technologies, architectures, and protocols for computer communications (SIGCOMM '01)*. Chord: A Scalable P2P Lookup Service for Internet Applications (ACM, New York, 2001), pp. 149–160. doi:http://dx.doi.org/10.1145/383059.383071
4. S Ratnasamy, P Francis, M Handley, R Karp, S Schenker, in *Proceedings of the 2001 conference on Applications, technologies, architectures, and protocols for computer communications (SIGCOMM '01)*. A Scalable Content Addressable Network (ACM, New York, 2001), pp. 161–172. doi:http://dx.doi.org/10.1145/383059.383072
5. A Kumar, S Merugu, J Xu, X Yu, in *Proceedings of the 11th IEEE International Conference on Network Protocols*. Ulysses: a Robust, Low-Diameter, Low-Latency Peer-to-peer Network, (2003), pp. 258–267. doi:10.1109/ICNP.2003.1249776
6. F Kaasheok, D Karger, in *Proceedings of the 2nd International Workshop, IPTPS*. Koorde: a Simple Degree-optimal Distributed Hash, (Berkeley, 2003). doi:10.1007/978-3-540-45172-3_9
7. I Gupta, K Birman, P Linga, A Demers, R van Renesse, in *Proceedings of the 2nd International Workshop, IPTPS*. Kelips: Building an Efficient and Stable P2P DHT Through Increased Memory and Background Overhead, (Berkeley, 2003). doi:10.1007/978-3-540-45172-3_15
8. B Leong, B Liskov, E Demaine, in *Proceedings of the 12th IEEE International Conference on Networks, ICON*. EpiChord: Parallelizing the Chord Lookup Algorithm with Reactive Routing State Management, vol. 1, (2004), pp. 270–276. doi:10.1109/ICON.2004.1409145
9. L Monnerat, CL Amorim, in *Proceedings of the 2009 IEEE Global Telecommunications Conference*. Peer-to-Peer Single Hop Distributed Hash Tables, (Honolulu, 2009), pp. 1–8. doi:10.1109/GLOCOM.2009.5425764
10. L Monnerat, CL Amorim, An Effective Single-Hop Distributed Hash Table with High Lookup Performance and Low Traffic Overhead. Concurr. Comput. Pract. Experience, 1767–1788 (2015). doi:http://dx.doi.org/10.1002/cpe.3342
11. J Risson, A Harwood, T Moors, in *IEEE Transactions on Parallel and Distributed Systems*. Topology Dissemination for Reliable One-Hop Distributed Hash Tables, (2009), pp. 680–694. doi:10.1109/TPDS.2008.145
12. H Shen, C-Z Xu, G Chen, in *Proceedings of the 18th International Parallel and Distributed Processing Symposium*. Cycloid: a Scalable Constant-degree Lookup-Efficient P2P Overlay Network, (Santa Fe, 2004), p. 26. doi:10.1109/IPDPS.2004.1302935
13. MI Yousuf, S Kim, in *Proceedings of the 21st IEEE International Conference on Network Protocols (ICNP)*. Kistree: A Reliable Constant Degree DHT, (Goettingen, 2013), pp. 1–10. doi:10.1109/ICNP.2013.6733613
14. D Li, X Lu, J Wu, in *Proceedings of IEEE 24th Annual Joint Conference of the IEEE Computer and Communications Societies*. FISSIONE: a Scalable Constant Degree and Low Congestion DHT Scheme Based on Kautz Graphs, (2005), pp. 1677–1688. doi:10.1109/INFCOM.2005.1498449
15. A Rowstron, P Druschel, in *Proceedings of the IFIP/ACM International Conference on Distributed Systems Platforms Heidelberg*. Pastry: Scalable, Decentralized Object Location, and Routing for Large-Scale Peer-to-Peer Systems, (2001), pp. 329–350. doi:10.1007/3-540-45518-3_18
16. J Augustine, G Pandurangan, P Robinson, S Roche, E Upfal, in *Proceedings of IEEE 56th Annual Symposium on Foundations of Computer Science*. Enabling Robust and Efficient Distributed Computation in Dynamic Peer-to-Peer Networks (Foundations of Computer Science (FOCS), Berkeley, 2015), pp. 350–369. doi:10.1109/FOCS.2015.29
17. V Venkataraman, K Yoshida, P Francis, in *Proceedings of the 2006 IEEE International Conference on Network Protocols*. Chunkyspread: Heterogeneous Unstructured Tree-Based Peer-to-Peer Multicast, (Santa Barbara, 2006), pp. 2–11. doi:10.1109/ICNP.2006.320193
18. E Balaji, G Gunasekaran, in *2016 International Conference on Information Communication and Embedded Systems (ICICES)*. Efficient Range Query Processing, Load Balancing and Fault Tolerance with Popular Web Cache in DHT, (Chennai, 2016), pp. 1–4. doi:10.1109/ICICES.2016.7518891
19. D Thatmann, A Butyrtschik, A Kupper, in *Proceedings of the 9th International Conference on Signal Processing and Communication Systems (ICSPCS)*. A Secure DHT-Based Key Distribution System for Attribute-Based Encryption and Decryption, (Cairns, 2015), pp. 1–9. doi:10.1109/ICSPCS.2015.7391732
20. Y Zhang, L Liu, *Distributed Line Graphs: a Universal Technique for Designing DHTs Based on Arbitrary Regular Graphs*, vol. 24, (Beijing, 2008), pp. 152–159. doi:10.1109/ICDCS.2008.35
21. Z Qiu, J Pérez, P Harrison, in *Proceedings of the 7th ACM/SPEC on International Conference on Performance Engineering, ICPE '16*. Tackling Latency via Replication in Distributed Systems (ACM, New York, 2016), pp. 197–208. doi:10.1145/2851553.2851562
22. D Wang, G Joshi, G Wornell, Using straggler replication to reduce latency in large-scale parallel computing. SIGMETRICS Perform. Eval. Rev. **43**(3), 7–11 (2015)
23. A Vulimiri, B Godfrey, R Mittal, J Sherry, S Ratnasamy, S Shenker, in *Proceedings of the ninth ACM conference on Emerging networking experiments and technologies. Low latency via redundancy*. Low latency via redundancy (ACM, New York, 2013), pp. 283–294. doi:10.1145/2535372.2535392
24. A Vulimiri, O Michel, PB Godfrey, S Shenker, in *11th ACM Workshop on Hot Topics in Networks (HotNets-XI)*. More Is Less: Reducing Latency via Redundancy (ACM, New York, 2012), pp. 13–18. doi:10.1145/2390231.2390234

25. E Cohen, S Shenker, in *Proceedings of ACM SIGCOMM'02*. Replication Strategies in Unstructured Peer-to-Peer Networks (ACM, New York, 2002), pp. 177–190. doi:10.1145/633025.633043
26. Q Lv, P Cao, E Cohen, K Li, S Shenker, in *Proceedings of ICS '02*. Search and Replication in Unstructured P2P Networks (ACM, New York, 2002), pp. 84–95. doi:10.1145/514191.514206
27. X Shen, et al., *Handbook of Peer-to-Peer Networking, 3*. (Springer Science+Business Media, LLC 2010. doi:10.1007/978-0-387-09751-0 1
28. W Xiong, DQ Xie, LX Peng, J Liu, in *Proceedings of 2011 International Conference on Electronic & Mechanical Engineering and Information Technology*. PrChord: A Probability Routing Structured P2P Protocol, Harbin, Heilongjiang, 2011), pp. 3142–3145. doi:10.1109/EMEIT.2011.6023753
29. S Wang, S Yang, L Guo, in *Proceedings of the Third International Conference on Communications and Mobile Computing*. LiChord: a Linear Code Based Structured P2P for Approximate Match, (Qingdao, 2011), pp. 118–121. doi:10.1109/CMC.2011.31
30. Y Wang, X Li, Q Jin, J Ma, in *Proceedings of the 9th International Conference on Ubiquitous Intelligence and Computing and 9th International Conference on Autonomic and Trusted Computing*. AB-Chord: an Efficient Approach for Resource Location in Structured P2P Networks, (Fukuoka, 2012), pp. 278–284. doi:10.1109/UIC-ATC.2012.158
31. S Ratnasamy, A Scalable Content-Addressable Network. Technical report, University of California at Berkley. www.icir.org/sylvia/thesis.ps. Accessed Sept 2017
32. A Takeda, T Oide, A Takahashi, T Suganuma, in *Proceedings of the 18th International Conference on Network-Based Information Systems*. Efficient Dynamic Load Balancing for Structured P2P Network, (Taipei, 2015), pp. 432–437. doi:10.1109/NBiS.2015.66
33. D Liu, Z Yu, in *Proceedings of the 5th International Conference on Electronics, Communications and Networks (CECNet 2015)*. Towards Load Balance and Maintenance in a Structured P2P Network for Locality Sensitive Hashing, (2015). doi:10.1007/978-981-10-0740-8_46
34. B Godfrey, K Lakshminarayanan, S Surana, R Karp, I Stoica, in *Twenty-third Annual Joint Conference of the IEEE Computer and Communications Societies*. Load Balancing in Dynamic Structured P2P Systems, INFOCOM 2004, (2004), pp. 2253–2262. doi:10.1109/INFCOM.2004.1354648
35. Q Vu, B Ooi, M Rinard, K Tan, Histogram-Based Global Load Balancing in Structured Peer-to-Peer Systems. IEEE Trans. Knowl. Data Eng, 595–608 (2009). doi:10.1109/TKDE.2008.182
36. J Ghimire, M Mani, N Crespi, T Sanguankotchakorn, Delay and Capacity Analysis of Structured P2P Overlay for Lookup Service. Telecommun. Syst. **58**, 33–54 (2015). doi:10.1007/s11235-014-9872-9
37. N Varyani, S Nikhil, VS Shekhawat, in *2016 30th International Conference on Advanced Information Networking and Applications Workshops (WAINA)*. Latency and Routing Efficiency Based Metric for Performance Comparison of DHT Overlay Networks, (Crans-Montana, 2016), pp. 337–342. doi:10.1109/WAINA.2016.16
38. S Ratnasamy, B Karp, S Shenker, D Estrin, R Govindan, L Yin, F Yu, Data-centric storage in sensor nets with GHT, a geographic hash table. Mob. Netw. Appl. **8**(4), 427–442 (2003). doi:10.1023/A:1024591915518
39. A Jedda, HT Mouftah, in *2015 6th International Conference on the Network of the Future (NOF)*. Enhancing DHT-based Object Naming Service Architectures with Geographic-awareness, (Montreal, 2015), pp. 1–6. doi:10.1109/NOF.2015.7333309
40. P Maymounkov, D Mazieres, in *IPTPS f01: Revised Papers from the First International Workshop on Peer-to-Peer Systems*. Kademlia: a Peer-to-Peer Information System Based on the XOR Metric, (2002), pp. 53–65. doi:10.1007/3-540-45748-8_5
41. BY Zhao, L Huang, J Stribling, SC Rhea, AD Joseph, J Kubiatowicz, in *IEEE Journal on Selected Areas in Communications*. Tapestry: a Resilient Global-Scale Overlay for Service Deployment, (2004), pp. 41–53. doi:10.1109/JSAC.2003.818784
42. K Aberer, M Hauswirth, M Punceva, R Schmidt, in *IEEE Internet Computing*. Improving data access in P2P systems, (2002), pp. 58–67. doi:10.1109/4236.978370
43. K Aberer, A Datta, M Hauswirth, *Peer-to-Peer Systems and Applications P-Grid: Dynamics of Self Organization Processes in Structured P2P Systems*. (Springer Verlag, 2005), pp. 137–153. doi:10.1007/11530657_10
44. V Kantere, S Skiadopoulos, T Sellis, Storing and Indexing Spatial Data in P2P Systems. IEEE Trans. Knowl. Data Eng. **21**(2), 287–300 (2009). doi:10.1109/TKDE.2008.139
45. C Gross, B Richerzhagen, D Stingl, C Munker, D Hausheer, R Steinmetz, in *Proceedings of IEEE P2P 2013*. Geodemlia: Persistent Storage and Reliable Search for P2P Location-based Services, (Trento, 2013), pp. 1–2. doi:10.1109/P2P.2013.6688730

5

Fuzzy logic-based call admission control in 5G cloud radio access networks with preemption

Tshiamo Sigwele[1*†], Prashant Pillai[2†], Atm S. Alam[3] and Yim F. Hu[1]

Abstract

Fifth generation (5G) cellular networks will be comprised of millions of connected devices like wearable devices, Androids, iPhones, tablets, and the Internet of Things (IoT) with a plethora of applications generating requests to the network. The 5G cellular networks need to cope with such sky-rocketing traffic requests from these devices to avoid network congestion. As such, cloud radio access networks (C-RAN) has been considered as a paradigm shift for 5G in which requests from mobile devices are processed in the cloud with shared baseband processing. Despite call admission control (CAC) being one of radio resource management techniques to avoid the network congestion, it has recently been overlooked by the community. The CAC technique in 5G C-RAN has a direct impact on the quality of service (QoS) for individual connections and overall system efficiency. In this paper, a novel fuzzy logic-based CAC scheme with preemption in C-RAN is proposed. In this scheme, cloud bursting technique is proposed to be used during congestion, where some delay tolerant low-priority connections are preempted and outsourced to a public cloud with a penalty charge. Simulation results show that the proposed scheme has low blocking probability below 5%, high throughput, low energy consumption, and up to 95% of return on revenue.

Keywords: Call admission control (CAC), Cloud radio access network (C-RAN), Preemption, Fuzzy logic, 5G

1 Introduction

In recent years, a large number of mobile devices and multimedia services in recent years has resulted in gigantic demands for larger system capacities and higher data rates over large coverage areas in high mobility environments. As a result, radio access networks (RAN) have tremendously grown so complex and are becoming so difficult to manage and control. Maintaining quality of service (QoS) for real-time (RT) and non-real time (NRT) services while optimizing resource utilization is a major challenge for next generation systems like fifth generation (5G). The 5G cellular networks will be comprised of millions of devices like wearable devices, Androids, iPhones, tablets, and Internet of Things (IoT) connected to the network with a plethora of applications. The 5G cellular networks will need to cope with the explosive increase of traffic requests from these devices to avoid network overload and traffic congestion in the core network. The 5G will comprise of cloud-based architecture called cloud-RAN (C-RAN) which was introduced as a way of solving the drawbacks of conventional RAN by pooling BS resources to a centralized cloud. Virtualization concept is used on general purpose processors (GPPs) to dynamically allocate BS processing resources to different virtual baseband units (vBBU) in the BBU pool.

Call admission control (CAC) is a scheme that offers an effective way of avoiding network congestion and can play a key role in the provision of guaranteed QoS and avoid traffic congestion in 5G. The basic function of a CAC algorithm is to accurately decide whether a connection can be accepted into a resource-constrained network without violating the service commitments made to the already admitted connections. On the other hand, However, traditional CAC schemes are not suitable for 5G C-RAN while an efficient CAC scheme aims to optimize call blocking probability (CBP), call dropping probability (CDP), and system utilization.

*Correspondence: t.sigwele@bradford.ac.uk
†Equal contributors
[1]Faculty of Engineering and Informatics, University of Bradford, BD7 1DP Bradford, UK
Full list of author information is available at the end of the article

There are many reasons as to why conventional CAC schemes are not suitable for 5G. First, conventional CAC approaches in cellular networks suffer uncertainties due to real-time processing of radio signals and the time varying nature of parameters such as speed, location, direction, channel conditions, available power, etc. Many of these traditional CAC schemes are ineffective leading to incorrect request admission when the network is actually incapable of servicing the request or incorrect rejection when there are actually enough resources to service the request. Some of these CAC schemes tend to assume network state information is static [1]. However, in practice, the network is dynamic and values measured keep changing. Second, as stated in our previous work [1], traditional CAC schemes are based on stand-alone RAN base station (BS) architectures while 5G will be based on centralized cloud BSs. These BSs are preconfigured for peak loads and have unshared processing and computation resources located in the BS cell areas. These BS resources cannot be shared to address varied traffic needs on other cell areas, causing poor resource utilization, high CBP, and CDP. As such, there is a need for efficient CAC schemes suitable for 5G. Intelligent CAC schemes based on intelligent decision-making techniques like fuzzy logic are a promising solution and solve the problem of imprecision and uncertainties in cellular networks [2]. The schemes mimic the cognitive behavior of human mind without the need for complex mathematical modeling making them adaptive, less complex, flexible, and suitable to cope with the rapidly changing network conditions of cellular networks in 5G.

This paper presents a fuzzy logic-based CAC scheme using preemption in 5G C-RAN. During congestion, some delay tolerant NRT low priority connections are preempted and outsourced to a public cloud with a pricing penalty to accommodate the RT connections, a technique called cloud bursting [3]. This work is the continuation of our published works in [1] and [2] where the former proposed CAC in 5G C-RAN without fuzzy logic while the latter proposed CAC in 5GC-RAN using fuzzy logic without preemption. The work in this paper will add preemption and cloud bursting technique. Below are the contributions of this paper:

(i) A CAC scheme based on fuzzy logic with preemption in 5G C-RAN is proposed. The fuzzy logic avoids uncertainties caused by traditional CAC schemes in distributed RAN systems.
(ii) A cloud bursting technique is proposed where during congestion, low priority delay tolerant NRT connections are preempted and outsourced to a public cloud at a certain price penalty to accommodate the RT connections. It is assumed that the public cloud is of infinite processing capacity as such it cannot get congested, as such it will not be captured in the simulation.
(iii) A rigorous simulation study is conducted for validating the proposed scheme, which shows a significant performance improvement.

CAC with preemption technique have been previously studies in the past, but in this work, preemption in CAC have been implemented using fuzzy logic technique which significantly improves blocking probability, also, CAC on its own have not been studied in C-RAN and this is the first work to study CAC in C-RAN. Also, cloud bursting have not been implemented in CAC before and is introduced in our work.

The rest of this paper is organized as follows: Section 2 presents the related works on CAC schemes. The proposed fuzzy logic CAC scheme in 5G C-RAN is presented in Section 3. Section 4 presents the simulation model and the obtained performance results. Finally, Conclusions and further works are presented in Section 5.

2 Related work

There are many ways of categorizing CAC schemes such as parameter based, measurement based, utility based, centralized/distributed, static/dynamic, etc. Comprehensive surveys can be found here [4–7]. This paper concentrate on intelligent CAC schemes which are based on intelligent decision-making techniques for solving the problem of error and uncertainties in conventional CAC schemes [8]. They are adaptive and flexible, thus making them suitable to cope with the rapidly changing network conditions and bursty traffic that can occur in 5G networks to give an efficient network management scheme. A fuzzy logic CAC scheme for stand alone BSs for high-speed networks was proposed in [9]. Even though the author used fuzzy logic to better estimate equivalent capacity, he does not show how the schemes performs in terms of CBP. In [10], the author proposed a fuzzy logic CAC approach scheme for long-term evolution (LTE). Even though the proposed scheme shows better call rejection than the quality index-based approach, the CAC scheme is based on standalone BS architecture with low BS utilisation not suitable for 5G. A method of fuzzy admission control for multimedia applications scheme is proposed in [11]. In this method, for multimedia applications, two fuzzy controllers have been introduced allowing better estimation of QoS. The drawbacks of this scheme is that it has many fuzzy controllers that can magnify CAC complexity and computation latency. In [12], a CAC scheme using genetic algorithm (GA) has been proposed for roaming mobile users with low handoff latency in next generation wireless systems. The scheme provides high network utilization, minimum cost, but it is not suitable for real-time applications since GA is very slow and cannot be used for real-time decision-making. A neural network approach for CAC with QoS guarantee in multimedia high-speed networks is proposed in [13]. It is an integrated method that combines linguistic control capabilities and the learning abilities of a neural network. Even though the scheme provides higher system utilization, it requires large computational resources working in parallel. A novel learning approach to solve the CAC in multimedia cellular networks with multiple classes of traffic is presented in [14]. The near optimal CAC policy is obtained through a form of neuro-evolution algorithm. This method guarantees that the specified CDP remains under a pre -defined upper bound while retaining acceptable CBP. This scheme is black box learning approach since the knowledge of its internal working of the scheme is never known.

3 Proposed CAC scheme

3.1 C-RAN architecture

C-RAN is a paradigm shift for next-generation RANs like 5G. C-RAN is described using four C's which stand for; clean, centralized processing, collaborative radio, and real-time cloud computing [1]. The C-RAN architecture adopted in this paper is shown in Fig. 1. The C-RAN concept separates the radio and antenna parts from the digital baseband parts and pools multiple baseband units (BBUs) in a central office called the BBU pool. These digital only BSs, called vBBUs, are linked via high bandwidth, low latency fiber to remote radio heads (RRHs). GPPs like X86 and ARM processors are used to house the BBUs and using cloud computing virtualization concept, multiple vBBU virtual machines (VMs) are dynamically provisioned in accordance to traffic demands.

3.2 Problem formulation

The main problem is that next-generation cellular networks like 5G C-RAN will have to process many requests from billions of devices, as such there will be traffic congestion in this 5G network. The question to be answered is how can efficient CAC schemes be deviced and then be incorporated in 5G C-RAN to improve CBP and resource utilization while maintaining the required QoS.

3.3 Fuzzy logic-based CAC scheme

In this paper, fuzzy logic scheme is used for performing CAC in 5G C-RAN because of its simplicity and robustness [6]. Fuzzy logic techniques resembles the human decision-making with an ability to generate precise solutions from certain or approximate information. Fuzzy logic avoids uncertainties and computational complexities brought by many CAC schemes and does not require precise inputs, and can process any number of inputs. Fuzzy logic incorporates a simple, rule-based approach based on natural language to solve control problem rather than attempting to model a system mathematically. In the proposed scheme, baseband signals from multiple cells are no longer processed on their stand-alone BBUs but processed on GPPs in the cloud using the concept of cloud computing. The GPPs are software defined enabling multiple radio signal from different cells to be processed in one computer platform. This is made possible through virtualization technology where hardware components are abstracted from software components. The vBBUs

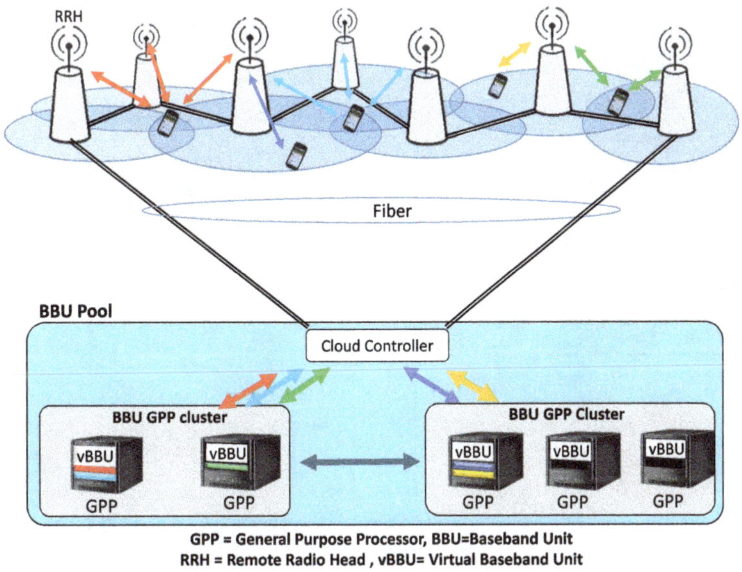

Fig. 1 C-RAN architecture. A figure showing the C-RAN architecture

are dynamically provisioned to service traffic requests from cells. The vBBU performs baseband signal processing of specific cell traffic. The traffic demand from cells is mapped into baseband processing resource such that every RRH traffic is serviced by its own vBBU.

Figure 2 shows the proposed fuzzy CAC system model diagram for 5G C-RAN which is located in the BBU pool inside the cloud controller. The model consists of various modules comprising of the operator's C-RAN infrastructure for normal processing of requests when the congestion is low and a third party public C-RAN infrastructure for handling requests for the operator's C-RAN during congestion. Connection requests that are processed in the public infrastructure are charged a certain price by the charging manager depending on the type of service and the size of the connection request. The resource estimator estimates the available capacity in the operator's C-RAN infrastructure and indicate whether the cloud is congested or not. The model also comprise of the fuzzy controller which performs the CAC decisions for incoming requests from users. The fuzzy controller takes as inputs three variables which are effective capacity, Ec, in Kbps, service type St and normalized available capacity, Ac and the output is the admittance decision, Ad. The admittance decision is either accept a request, reject a request or preempt some low priority requests and outsource them to a public cloud. The traffic requests are divided into two groups, namely, RT and NRT traffic as shown below.

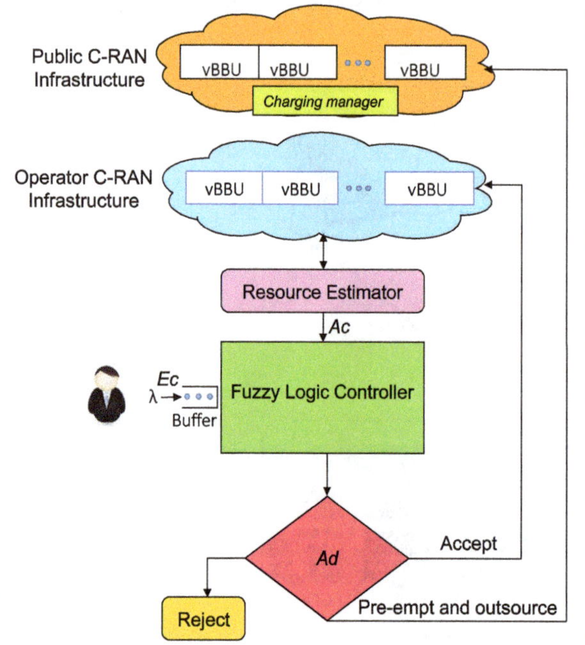

Fig. 2 Proposed system model. A figure showing the proposed system model

- RT classes. These are called guaranteed bit rate (GBR) which include VoIP, live streaming, video call, and real-time gaming. This type of services are delay sensitive.
- NRT classes. This are called non-GBR which include buffered streaming and transmission control protocol (TCP)-based services like web browsing, email, file transfer protocol (ftp), and point to point (p2p). These types of services are delay tolerant.

3.3.1 *Cloud bursting technique for preempted connections*

The cloud bursting technique allows the operators to dynamically extend their infrastructure by renting third-party resources [15]. During congestion of the operator's C-RAN infrastructure, when a high priority RT connections arrives as illustrated in Fig. 3 and the cloud is congested, two things happen, either the low priority NRT connections are preempted from the operator's C-RAN and then bursted into the public C-RAN infrastructure to accommodate the high priority RT connections or the RT connection is dropped if there are no NRT connections to preempt in the operator's C-RAN. RT connections are never outsourced to the public cloud because they are delay sensitive. Only NRT connections are outsourced to the public cloud. An agreement is made between the operator and the public cloud operator, and a certain price is charged for outsourcing some NRT connections. When a NRT connection arrives and the operator's cloud is congested, the NRT connection is forwarded to the public cloud as shown in Fig. 3 with a certain price penalty where the request will be charged by the charging manager.

3.3.2 *Structure of fuzzy logic controller*

The fuzzy controller of the proposed scheme takes three inputs: (i) effective capacity, Ec; (ii) available capacity, Ac; and (iii) network congestion factor, Nc and output the

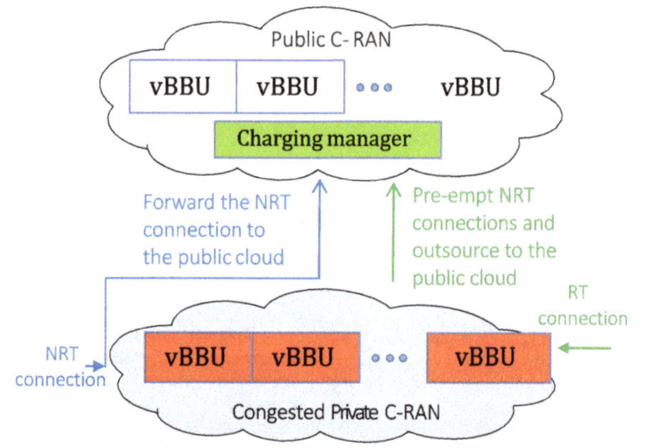

Fig. 3 Cloud bursting model. A figure showing the cloud bursting technique when RT and NRT connections arrive into the congested BBU pool

admittance decision, Ad. Below is the description of the structure of the proposed fuzzy logic controller.

Membership functions: Trapezoidal and triangular membership functions are chosen for simplicity. The membership functions for input and output linguistic parameters are shown in Fig. 4. The values of the membership functions have been chosen based on commonly used values of membership functions in various literature. For the fuzzy controller, the term sets for Ec, St, Ac, Nc, and Ad are defined as follows:

i) $T(\text{Ec}) = \{\text{Low, Medium, High}\}$
ii) $T(\text{St}) = \{\text{NRT, RT}\}$
iii) $T(\text{Ac}) = \{\text{NotEnough, Enough}\}$
iv) $T(\text{Ad}) = \{\text{Accept, Reject, Preempt}\}$

Fuzzy rule base: The fuzzy rule base consists of a series of fuzzy rules shown in Table 1. These control rules are of the following form: IF 'condition', THEN 'action'. Example, if St is 'RT' and 'St' is 'Not Enough' and 'Ec' is 'High' then 'Reject'.

Defuzzification method: The center of gravity (COG) [1] method is used for defuzzification to convert the degrees of membership of output linguistic variables into crisp/numerical values. The COG method is adopted since the membership functions used are simple triangular and trapezoidal shapes with low computational complexity and can be expressed as [1]:

$$Z_{\text{COG}} = \frac{\int_z \mu(z) z dz}{\int_z \mu(z) dz} \tag{1}$$

3.3.3 Queueing system for preempted connections

The connections in the cloud follows the $M/M/c/K$ queueing model or Erlang B model [16]. In the $M/M/c/K$ model, the request arrival is governed by a Poisson process at arrival rate λ and the service times are exponentially distributed with parameter μ and there are c servers in the cloud processing the requests from the front of the queue. The variable K denotes the capacity of the system. The buffer is considered to be of a finite size, and connection requests greater that the queue length are dropped. The model can be described as a continuous time Markov chain which is a type of a birth–death process. The server utilization, ρ, is written as [16]:

$$\rho = \frac{\lambda}{c\mu}, \rho < 1 \tag{2}$$

The variable ρ should be less than one for the queue to be stable otherwise the queue will grow without bound.

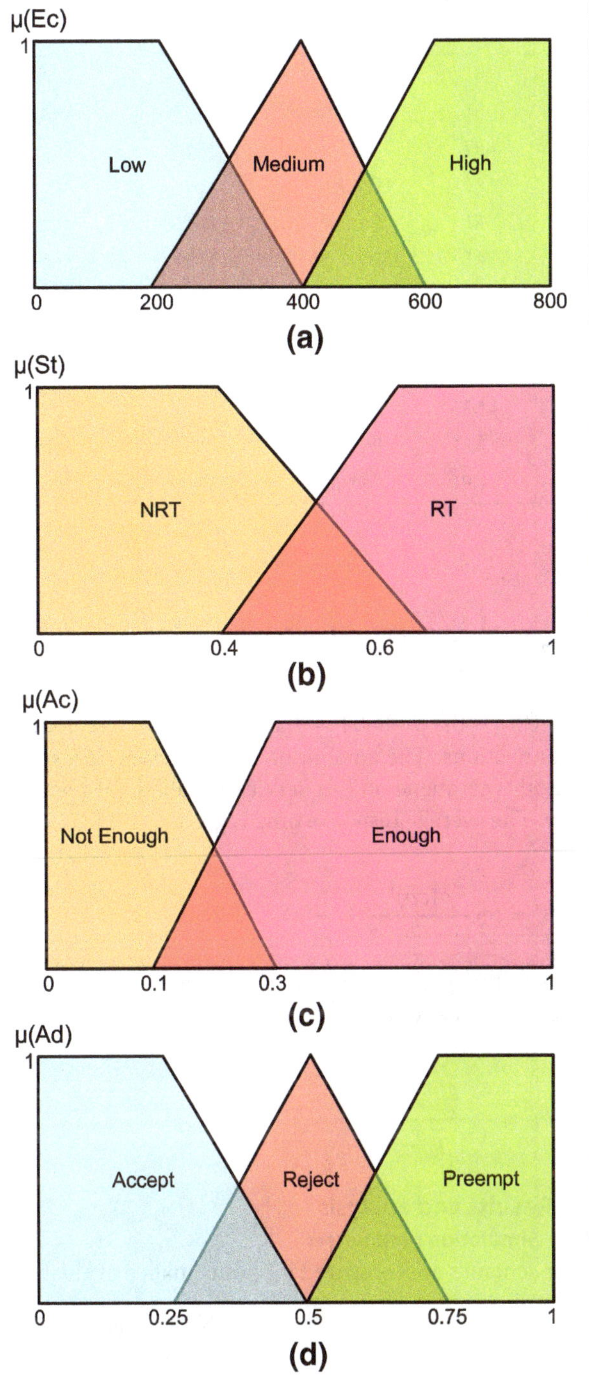

Fig. 4 Membership functions for (**a**) effective capacity, Ec (**b**) service type, St (**c**) available capacity, Ac (**d**) admittance, Ad. A figure showing the the membership functions of various fuzzy logic terms

The probability that the system contain n connections can be written as [16]:

$$\pi_0 = \left[\sum_{n=0}^{c} \frac{\lambda^n}{\mu^n n!} + \frac{\lambda^c}{\mu^c c!} \sum_{n=c+1}^{K} \frac{\lambda^{n-c}}{\mu^{n-c} c^{n-c}} \right]^{-1} \tag{3}$$

Table 1 Fuzzy rule base for fuzzy controller

Rule	St	Ac	Ec	Ad
1	RT	Not Enough	Low	Outsource
2	RT	Not Enough	Medium	Reject
3	RT	Not Enough	High	Reject
4	RT	Enough	Low	Accept
5	RT	Enough	Medium	Accept
6	RT	Enough	High	Accept
7	NRT	Not Enough	Low	Outsource
8	NRT	Not Enough	Medium	Outsource
9	NRT	Not Enough	High	Outsource
10	NRT	Enough	Low	Accept
11	NRT	Enough	Medium	Accept
12	NRT	Enough	High	Accept

$$\pi_n = \begin{cases} \frac{(\lambda/\mu)^n}{n!}\pi_0, & for\ n = 1, 2, \ldots, c \\ \frac{(\lambda/\mu)^k n}{c^{n-c}c!}\pi_0, & for\ n = c+1, \ldots, K. \end{cases} \tag{4}$$

where π_n is the probability that the cloud system contains n connections. The amount of time a connection spends in both the queue and in service is called the response time. The average response time is given as [16]:

$$T = \pi_0 \frac{\rho(c\rho)^c}{(1-\rho)^2 c!} + \frac{1}{\mu} \tag{5}$$

Then the probability that an arriving connection is blocked can be written using Erlang B formula as [16]:

$$P_b = \frac{\frac{\rho^c}{c!}}{\sum_{i=0}^{c} \frac{\rho^i}{i!}} \tag{6}$$

4 Results and analysis

4.1 Simulation parameters

Four schemes are compared for performance evaluation;

1) CAC scheme on distributed RAN systems with stand alone BBUs serving individual BSs from our previous work in [1],
2) CAC on C-RAN without fuzzy logic applied from our previous work in [1],
3) CAC with fuzzy-logic on C-RAN without preemption from our previous work in [2], and
4) the proposed CAC with fuzzy logic on C-RAN with preemption in this paper.

The Matrix Laboratory (MATLAB) was used to simulate the proposed framework. For simulation and performance evaluation, the following four traffic classes or service types were considered as shown in Table 2 from [17]:

- VoIP as RT service
- Conversational video (live streaming) as RT service
- ftp as NRT service
- web browsing or www as NRT service

The MBR values are taken as the values for E_C. Four traffic classes are evaluated for simplicity, but the proposed framework applies to multiple traffic classes. The value of λ was varied with every simulation, and 100 calls were generated for each traffic class. The simulation time was kept at 500 s. The membership function for the inputs and output of the fuzzy controller are shown in Fig. 4. It is assumed that the network operator operating the private cloud enters into an agreement with the public cloud operator which involves the service level agreement (SLA) which involves the cost. The cost of accepting a connection request in the public cloud is assumed to be 10% of what the private C-RAN operator will make when processing the request. It should be noted that the request size, duration, and QoS can form the basis of how much the request can be charged but this will be considered in the future, but in this paper, only 10% is deducted by the public cloud.

4.2 Simulation results

Figures 5, 6, and 7 shows a comparison of the combination of the input terms Ec, Ac, St, and the output term Ad when the fuzzy rules in Table 1 are applied. The figure shows that as the value of Ec increases, the admittance (Ad) decreases meaning that when the value Ec of a particular service is Low, the admittance becomes Accept and as the value of Ec increases, the admittance becomes Preempt. Also, for available capacity (Ac), the figures show that when Ac is NotEnough, the admittance value becomes higher meaning that there is Preemption of NRT connections. As the value of Ac increases (to Enough), the admittance tend to Accept. Finally for service type (St), when the value of St increases from NRT to RT, the admittance decreases from Preemption to Accept since the NRT requests are preempted and the RT connections are accepted.

Table 2 Simulation parameters [17]

QCI	Service	Type	Priority	Delay	PER	MBR/E_C
1	VoIP	GBR	2	100 ms	10^{-2}	12 Kbps
3	Conversational video	GBR	4	150 ms	10^{-3}	240 Kbps
8	ftp	Non-GBR	8	300 ms	10^{-6}	512 Kbps
9	www	Non-GBR	9	300 ms	10^{-6}	512 Kbps

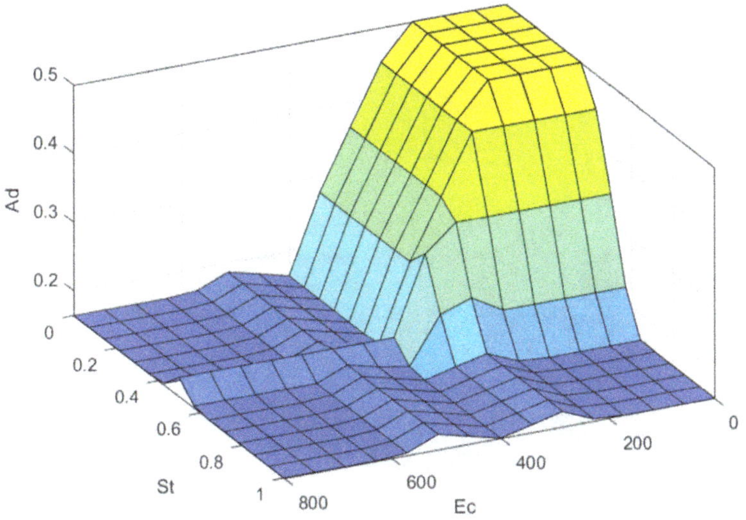

Fig. 5 Comparison for inputs St, Ec, and the output Ad. A figure showing the comparison of fuzzy inputs and output

Figure 8 shows the blocking probability versus offered traffic load. The figure shows that for all the schemes, as the offered traffic increases, the blocking probability also increases. The CBP of the CAC distributed RAN is higher than all the other schemes because the baseband computing power is limited as each cell is covered by a single BBU with limited capacity. The blocking probability of CAC C-RAN with no fuzzy scheme also performs poorly with blocking probability greater than threshold at 40% offered traffic load due to improper and uncertain decision-making of the admission control scheme without fuzzy logic. The fuzzy C-RAN without preemption performs well up to 90% traffic load compared to the previous two schemes because fuzzy logic avoids imprecisions and uncertainties when performing admission control. The fuzzy C-RAN with preemption scheme performs better than all the rest with 100% traffic below blocking probability threshold of 5% because, instead of connection requests being blocked, they are forwarded to a public cloud as such more connections are accepted in the system.

Figure 9 shows the resource utilization in the private C-RAN cloud for different traffic arrival rates. The figure shows that as the arrival rate increases, the resource utilization in the cloud also increases because more requests are being processed and occupies the available capacity. The fuzzy C-RAN with preemption and the fuzzy C-RAN without preemption scheme have the same but higher

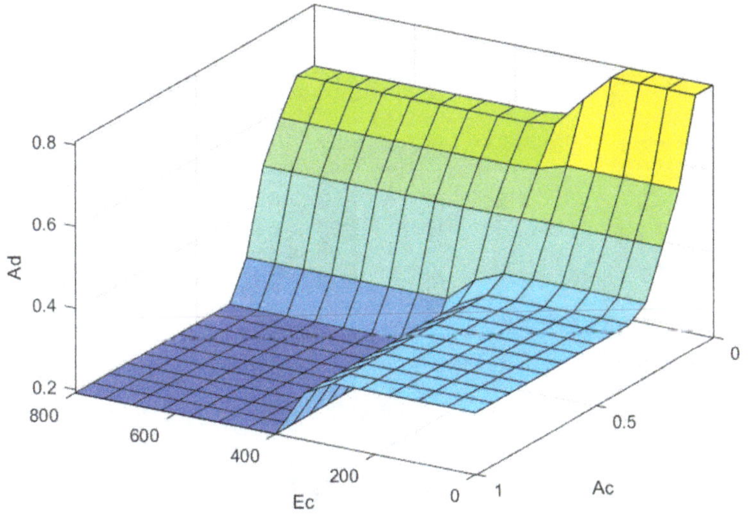

Fig. 6 Comparison for inputs Ec, Ac, and the output Ad. A figure showing the comparison of fuzzy inputs and output

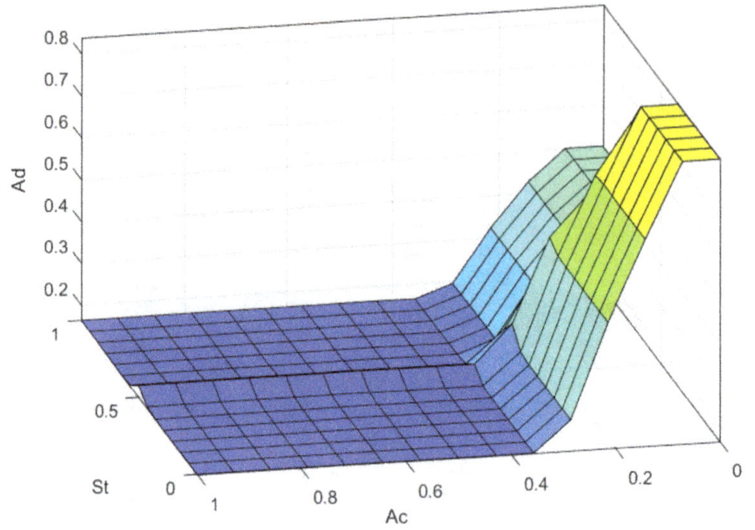

Fig. 7 Comparison for the inputs St, Ac, and the output Ad. A figure showing the comparison of fuzzy inputs and output

resource utilization than all the other schemes because the BBUs in the cloud are shared and a single BBU can process requests from multiple cells. It can be noticed that preemption have no impact on resource utilization. The CAC C-RAN with no fuzzy scheme has high utilization than the CAC distributed RAN scheme because in the latter, BBUs are stand alone and BBU processing resources are not shared to address varied traffic needs in the cell area. Figure 10 shows the response time versus offered traffic load for C-RAN system. The figure shows that as the offered traffic increases, the response time increases because more requests take more time to be processed. The figure shows that the preempted NRT connections take more time to be processed because they are forwarded to the public cloud which incurs more delays, but this does not affect the NRT preempted connections because they are delay tolerant. The new RT connections are delay sensitive, and they have small response time because they are processed in the private cloud and not in the public cloud.

Figure 11 shows the operators revenue for peak traffic periods. At peak traffic periods, the CAC distributed RAN scheme has a blocking probability of 0.5 which means the revenue is 50%, where the lower revenue is due to higher blocking probability. The CAC C-RAN with no fuzzy scheme has a blocking probability of 20% at peak traffic leading to a revenue of 80%. The fuzzy C-RAN without preemption scheme has a blocking probability of 10% at peak traffic leading to 90% revenue for the operator while the fuzzy C-RAN with preemption scheme has the

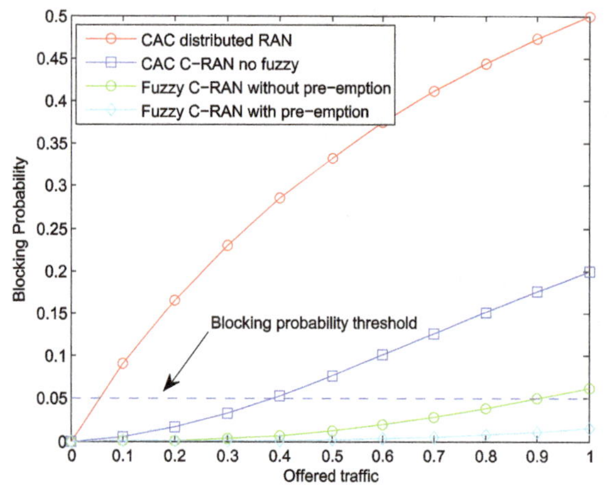

Fig. 8 Blocking probability versus offered traffic load

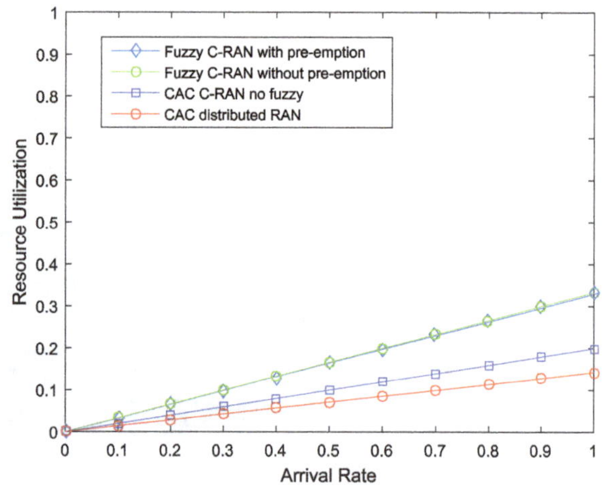

Fig. 9 System utilization versus arrival rate. A figure showing how the system utilization in the BBU pool varies with the change in arrival rate of requests

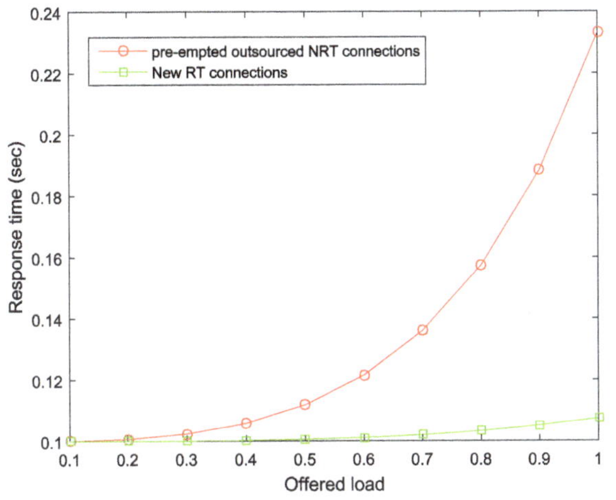

Fig. 10 Response time versus offered traffic load. A figure showing how the service response time/delay varies with the increase in the offered traffic load

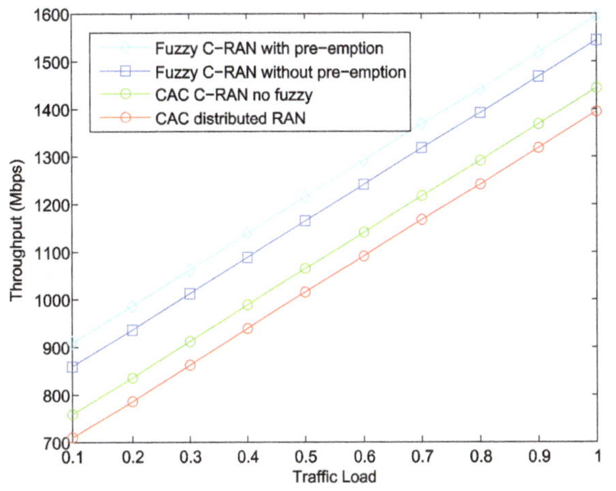

Fig. 12 System throughput. A figure showing the total throughput of the C-RAN system

highest revenue than all the schemes which is 95% since more requests are accepted in both the private and public cloud. Figure 12 shows the the total network throughput for different traffic loads, and it can be shown that for both schemes, as the traffic load increases, the network throughput also increases. The throughput is for the entire network, is calculated at the BBU pool, and is expected to be larger. The fuzzy C-RAN with preemption scheme has a higher throughput than all the other schemes with 900 and 1600 Mbps during low and peak traffic, respectively, and the scheme is 28.6% effective compared to CAC distributed RAN scheme. This is because more connections are being accepted as more computing resources are being provided by the public cloud using the cloud bursting technique. The fuzzy C-RAN without preemption has a throughput of 880 and 1550 Mbps during low and peak traffic, respectively, and outperforms the CAC-distributed RAN scheme by 25.7%. The CAC C-RAN with no fuzzy performs better than the CAC-distributed scheme by 8.6%. The CAC-distributed RAN performs poorly that the rest of the schemes with 700 and 1400 Mbps during low and peak traffic, respectively, because it has high blocking probability due to limited baseband computing resources.

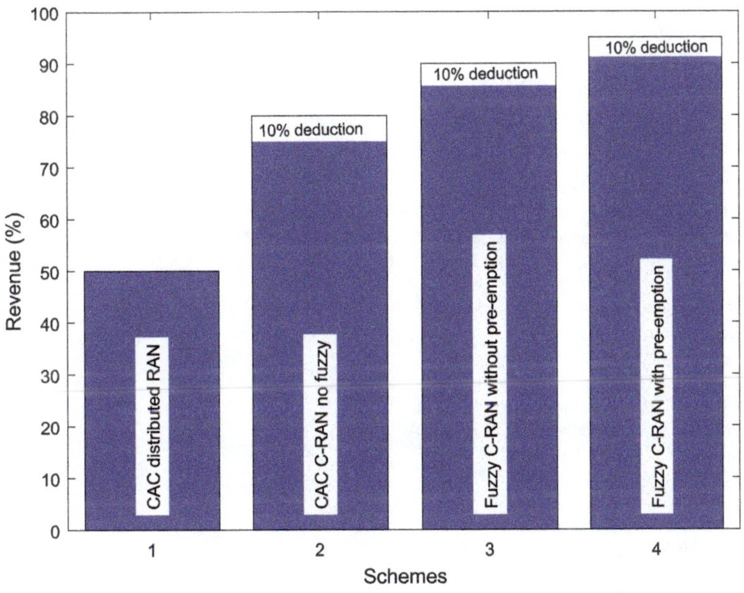

Fig. 11 Operator revenue. A figure showing the operator's income from hosting connection requests from users

5 Conclusions

In this paper, a fuzzy logic-based call admission control (CAC) scheme is proposed in fifth generation (5G) cloud radio access networks (C-RAN). The fuzzy logic avoids uncertainties caused by traditional CAC schemes in distributed RAN systems. A cloud bursting technique used proposed where during congestion, low priority delay tolerant non real-time (NRT) connections are preempted and outsourced to a public cloud at a certain price penalty. The simulation results shows that the proposed scheme has low blocking probability which is within blocking probability threshold limit of 5%. The proposed scheme has a return revenue of 95%.

Acknowledgements
The authors would like to thank MoESDP-DTEF (Ministry of Education Skills and Development Planning, Department of Tertiary Education and Financing, Botswana, Africa) for financing this research.

Authors' contributions
All the authors have contributed significantly to this research articles. Below are the author's contributions: Mr TS have contributed significantly to this paper on the coming up with mathematical models, running simulations, and writing up the paper. Dr PP has contributed significantly on the aspects of supervision, the organization of the paper, reviewing, and the validation of the proposed framework. Dr ASA has also contributed significantly in the related work section and also helped in drawing the diagrams in this paper. Prof YFH has also contributed significantly in this work by proof reading the work and validating the proposed framework and also restructuring the paper. All authors read and approved the final manuscript.

Competing interests
The authors declare that they have no competing interests.

Author details
[1]Faculty of Engineering and Informatics, University of Bradford, BD7 1DP Bradford, UK. [2]Faculty of Technology, Design and Environment, Oxford Brookes University, Oxford, UK. [3]Institute of Communication Systems, 5G Innovation Centre University of Surrey, Düsternbrooker Weg 20, GU2 7XH Guildford, UK.

References

1. T Sigwele, P Pillai, Y Hu, in *IEEE Future Internet of Things and Cloud*. Call admission control in cloud radio access networks, (2014)
2. T Sigwele, P Pillai, Y Hu, in *International Conference on Wireless and Satellite Systems*. Elastic call admission control using fuzzy logic in virtualized cloud radio base stations, (2015)
3. M Farahabady, YC Lee, AY Zomaya, Pareto-optimal cloud bursting. IEEE Trans. Parallel Distrib. Syst. **25**, 2670–2682 (2014)
4. MH Ahmed, Call admission control in wireless networks: a comprehensive survey. IEEE Commun. Surv. Tutorials. **7**(1-4), 50–69 (2005)
5. D Niyato, E Hossain, Call admission control for QoS provisioning in 4G wireless networks: issues and approaches. IEEE Netw. **19**(5), 5–11 (2005)
6. Q Liang, NN Karnik, JM Mendel, Connection admission control in ATM networks using survey-based type-2 fuzzy logic systems. IEEE Trans. Syst. **30**(3), 329–39 (2000)
7. Y Liu, M Meng, in *Future Computer and Communication*. Survey of admission control algorithms in IEEE 802.11e wireless LANs, (2009)
8. V Kolici, T Inaba, A Lala, G Mino, S Sakamoto, L Barolli, in *Network-Based Information Systems*. A fuzzy-based cac scheme for cellular networks considering security, (2014)
9. L Barolli, A Koyama, T Yamada, S Yokoyama, T Suganuma, N Shiratori, in *12th International Workshop on Database and Expert Systems Applications*. A fuzzy admission control scheme for high-speed networks, (2001)
10. CT Ovengalt, K Djouani, A Kurien, A fuzzy approach for call admission control in lte networks. Procedia Comput. Sci. **32**, 237–244 (2014)
11. L Barolli, M Durresi, K Sugita, A Durresi, A Koyama, in *19th International Conference on Advanced Information Networking and Applications*. A cac scheme for multimedia applications based on fuzzy logic, (2005)
12. D Karabudak, C Hung, B Bing, in *Proceedings of the 2004 ACM Symposium on Applied Computing*. A call admission control scheme using genetic algorithms, (2004)
13. RG Cheng, CJ Chang, LF Lin, A QoS-provisioning neural fuzzy connection admission controller for multimedia high-speed networks. IEEE/ACM Trans. Networking. **7**(1), 111–121 (1999)
14. X Yang, J Bigham, in *International Joint Conference on Artificial Intelligence*. A call admission control scheme using neuroevolution algorithm in cellular networks, (2007)
15. S Acs, M Kozlovszky, P Kacsuk, in *Applied Computational Intelligence and Informatics*. A novel cloud bursting technique, (2014)
16. AO Allen, *Probability statistics and queueing theory, 1edn*. (Academic Press, London, 2014)
17. K Leonhard, How to dimension user traffic in 4G networks. (Slideshare, 2014). https://www.slideshare.net/mobile/althafhussain1023/how-to-dimension-user-traffic-in-lte. Accessed 1 June 2017

6

Optimal power allocation and throughput performance of full-duplex DF relaying networks with wireless power transfer-aware channel

Xuan-Xinh Nguyen and Dinh-Thuan Do*

Abstract

In terms of modern applications of wireless sensor networks in smart cities, relay terminals can be employed to simultaneously deliver both information and energy to a designated receiver by harvesting power via radio frequency (RF). In this paper, we propose time switching aware channel (TSAC) protocol and consider a dual-hop full-duplex (FD) relaying system, where the energy constrained relay node is powered by RF signals from the source using decode-and-forward (DF) relaying protocols. In order to evaluate system performance, we provide an analytical expression of the achievable throughput of two different communication modes, including instantaneous transmission and delay-constrained transmission. In addition, the optimal harvested power allocation policies are studied for these transmission modes. Most importantly, we propose a novel energy harvesting (EH) policy based on FD relaying which can substantially boost the system throughput compared to the conventional half-duplex (HD) relaying architecture in other transmission modes. Numerical results illustrate that our proposed protocol outperforms the conventional protocol under the optimal received power for energy harvesting at relay. Our numerical findings verify the correctness of our derivations and also prove the importance of FD transmission mode.

Keywords: Energy harvesting, Full-duplex, Decode-and-forward, Outage capacity, Delay-constrained

1 Introduction

It is undoubted that wireless communication systems have attracted much research interest in recent years. In particular, energy-aware radio access solutions can be implemented to deal with the massive increase in the consumption of energy in telecommunication networks and the efficient use of power is important for energy optimization. In addition, applications based on internet of things networks have become increasingly popular, so they require novel approaches for energy saving applied in low-power devices. Energy harvesting is the amount of energy available at the transceiver node powered by surrounding energy sources such as solar, wave, vibration, and radio frequency.

*Correspondence: dodinhthuan@tdt.edu.vn
Wireless Communications Research Group, Faculty of Electrical and Electronics Engineering, Ton Duc Thang University, 19 Nguyen Huu Tho Street, Ho Chi Minh City, Vietnam

Since energy harvesting plays an important role in relaying regardless of power optimization at relays which are assumed to be powered by ideal power sources. In sensor networks and cellular networks, wireless devices using rechargeable or replaceable batteries are often out of order in a short period of time, as the battery powered devices in such wireless networks usually suffer from limited operating times. Unlike portable devices, the maintenance cost of sensor nodes is often higher in case they are replaced or recharged. Additionally, it is noted that it may be dangerous to replace batteries in toxic environments and powering medical sensors implanted inside human bodies is also challenging. To supply a perpetual power in such networks, energy harvesting is considered as a potential method to prolong lifetime of wireless devices [1, 2]. To take advantage of information transfer over wireless channels, receivers can scavenge power from the transmitted signal. Since ambient radio signals can carry energy and information, energy harvesting brings more major advantages compared to the conventional grid power supply [3].

Since radio frequency (RF) signals are capable of carrying both information and energy, a new concept in green wireless communications was put forward, namely simultaneous wireless information and power transfer (SWIPT). To take advantage of SWIPT, more practical receiver architectures have been developed with two separated circuits to carry out energy harvesting and information decoding [4, 5]. There are two major schemes in the receiver, including time switching and power splitting. The authors in [4] presented performance of system under capability of energy harvesting applied in a simple single-input single-output scenario while multiple-input multiple-output (MIMO) broadcasting scenarios was introduced in [6].

In order to obtain practical insights in term of optimal time and power allocation. The work in [7] focused on time allocation policy for the two transmitters in case the efficiency of energy transfer is maximized by an energy beamformer under the impact of Channel State Information (CSI) received in the uplink by the energy transmitter. Furthermore, in [8], the time fraction in TSR impacts on the optimal throughput and such parameter can be found in a numerical method.

We next consider several system models regarding existing cooperative networks with capability of energy harvesting. Firstly, the employed relay in SWIPT networks [9, 10] or the source terminal [11] can harvest energy from the radiated signal of the source terminal or the employed relay. Secondly, in multi-hop networks, energy is transferred to remote terminals via multi-hop [12, 13]. In multi-hop systems, the high path loss of the energy-bearing signal can be eliminated [12]. Unlike [12], the authors in [13] investigated a multi-antenna relay adopting two separate terminals with capability of information processing and power transfer, respectively, and the expressions of the transmission rate and outage probability were presented under the impact of remote energy transfer. In addition, relay selection is considered as solution to determine a tradeoff between the efficiency for the information transmission and the amount of energy forwarded to the energy receivers [14–16].

Furthermore, full-duplex (FD) mode was evaluated, in which it allows transmitting and receiving signals at the same frequency band at the same time slot. Various theoretical analysis and practical designs have been conducted in terms of FD networks like in [17–21]. Thanks to the use of FD mode, the resources are utilized more efficiently and it can double the spectral efficiency compared to half-duplex (HD) mode. However, due to practical constraints, the performance of FD communication can be affected by the self-interference (SI) stemming from FD node transmission.

In addition, energy harvesting along with throughput optimization has been mentioned in previous works. In [22] and [23], throughput optimization with constraints was studied under a static channel condition for obtaining best efficiency of energy harvesting transmitters.

Moreover, energy harvesting based on power control policies for wireless powered transmission over fading channels suffer from several problems, i.e., the randomness of RF energy source, wireless fading channels and the maximum power constraints. To address this situation, several existing works have considered offline optimal power control designs for fading channels [24, 25]. The work in [25] considered that the offline optimization in an efficient optimal solution was presented to achieve optimal energy efficiency. Although the authors in [26] investigated the offline scheduling and formulated the corresponding performance optimization problems of two-way relay networks, the statistics of energy and fading channels are assumed to be parameters at the transmitter, and optimal function can be derived in a numerical manner under high computation.

The authors in [27] considered that the optimal time splitting coefficients for the full-duplex dual-hop relaying lead to enhance the system throughput in comparison with the traditional half-duplex relaying scheme for all kinds of modes including instantaneous transmission, delay-constrained transmission, and delay-tolerant transmission. The energy-constrained FD relay node can be applied in the multiple-input single-output (MISO) system; the optimal power allocation and beamforming design are investigated in [28]. In another line of research, the FD decode-and-forward system using the time-switching protocol is embedded in the multiple antenna-assisted relay to obtain more energy from the source and transfer signal to the destination as in [29, 30]. Interestingly, in order to achieve the maximal throughput performance, the optimal time switching coefficient is adaptively selected based on channel state information (CSI), accumulated energy, and threshold signal-to-noise ratio (SNR) [31, 32] and [33].

Motivated from the previous works [27], we focus on optimal throughput of a two-hop case, where RF energy harvesting powers the relay. In this paper, the impact of FD transmission is investigated in terms of the system throughput to determine the performance of RF energy harvesting relaying system. The two-antenna configuration is proposed in the FD mode, where the relay is equipped with two antennas, one for transmitting signal and the other one for receiving signal simultaneously. In this paper, two different transmission modes are investigated, namely instantaneous transmission and delay-constrained transmission. Furthermore, we examine the throughput of DF relaying protocols and characterize the fundamental trade-off between energy harvesting and system throughput. In order to compare with the effect of the FD relaying architecture, the HD relaying architecture is also investigated. The main aim of this paper is that FD relaying is an attractive and promising solution to enhance the throughput of RF energy harvesting-based relaying systems.

Comparing to [27] and [28], although this work consider a same system model with those works, our investigation also design a novel wireless power transfer strategy to improve the system performance. The authors in [27] consider a conventional time splitting protocol, in which relay switches from energy harvesting to information transfer with fixed EH fraction time allocated. While [28] design a self-energy recycling strategy for energy harvesting at relay, this EH manner can collect transmitted signal itself and scavenge to energy. Contrarily, our work redesigns the conventional time switching protocol as in [27] to a novel EH schedule. The proposed EH will acquire the channel state information (CSI) to allocate amount of EH time. In particular, the relay transmission power can be preset level. The magnitude of self interference at relay thus can be controlled based on relay transmitted power, which can improve the system performance.

The main contributions of our paper are summarized as follows:

- A new protocol for wireless power transfer called time switching aware channel protocol (TSAC) is proposed in FD DF relaying networks and further analysis is presented as well.
- We provide analytical expressions in instantaneous transmission mode for both cases, i.e., one antenna and dual antennas at relay node in EH-based networks.
- The outage probability and average throughput for delay-constrained transmission mode are derived in closed-form expressions for tractable computation. The optimal values can be achieved in various simulation results.
- The advantages of the proposed protocol are also compared with the previous works. The most critical performance metric (i.e., optimal throughput efficiency) is thoroughly analyzed and systematically validated via comparative simulations.
- It can be seen that optimal power allocation leads to enhance throughput and resulting in system performance in comparison with non-power allocation solution in the literature.

The remainder of the paper is organized as follows. In Section 2, the energy harvesting cooperative scenario with one source node equipped two antennas in FD mode is considered and two different strategies for single antenna or dual antennas for energy harvesting are investigated. In Section 3, the power allocation in instantaneous transmission is evaluated while outage probability and optimal throughput in delay-constraint are given in Section 4 for performance evaluation. Section 5 examines HD mode for performance comparison with FD relay. Numerical results and useful insights are provided in Section 6. Eventually, Section 7 draws a conclusion for our paper.

2 System model and energy harvesting protocol

2.1 System model

As shown in Fig. 1a, we consider a two-hop relaying network consisting of one source denoted by $\mathcal{S}$, one destination denoted by $\mathcal{D}$, and one intermediate node denoted as $\mathcal{R}$. It is assumed that the source and destination node are equipped with single antenna, i.e., $\mathcal{S}$ and $\mathcal{D}$ are provided with single antenna, while $\mathcal{R}$ is equipped with two antennas, and operated in FD mode. In addition, the relay node is assumed to have no extra embedded energy sources so it requires to harvest energy from the received RF signal from the source node [2, 5, 8, 14, 20, 27–29, 32]. The TSAC protocol is shown in Fig. 1b, in which the relay adjusts EH time to meet the installed transmit power every block time. Therefore, at time slot ith, the relay node will energy harvesting with $\alpha_i T$ seconds.

Figure 2 shows basic block diagram for the SWIPT system, where TSAC is deployed. The system has one switching unit and one combining unit compared to conventional systems as Fig. 2a or as system models introduced in [5, 27]. As illustrated in Fig. 2b, we design an architecture exploiting dual antennas to harvest energy at relay during the energy harvesting phase as [27]. In this scheme, we consider WPT phase in which the switching units change the role of both transmitting and receiving antenna to receive energy via RF signal as [27, 30]. Then, those RF signals are combined and used to feed the energy harvester. It is worth noting that case 2 is a natural choice, since it fully exploits the available hardware resources (i.e., antenna components in multiple-input multiple-output (MIMO) systems) to collect more energy [30]. Nevertheless, due to its straightforward implementation, case 1 could be more practical in certain applications.

2.2 Channel model

We assume that the $\mathcal{S} \rightarrow \mathcal{R}$ and $\mathcal{R} \rightarrow \mathcal{D}$ channel links include both large-scale path loss and statistically independent small-scale Rayleigh fading. We denote d_1 is distance between source and relay node and d_2 is distance between relay and destination node. We also denote the main channels such as h and k are the links from the source to first antenna and second antenna at relay, respectively, and g is the channel from relay to destination node. It is also assumed that the main channels experience Rayleigh fading and remain constant over the block time

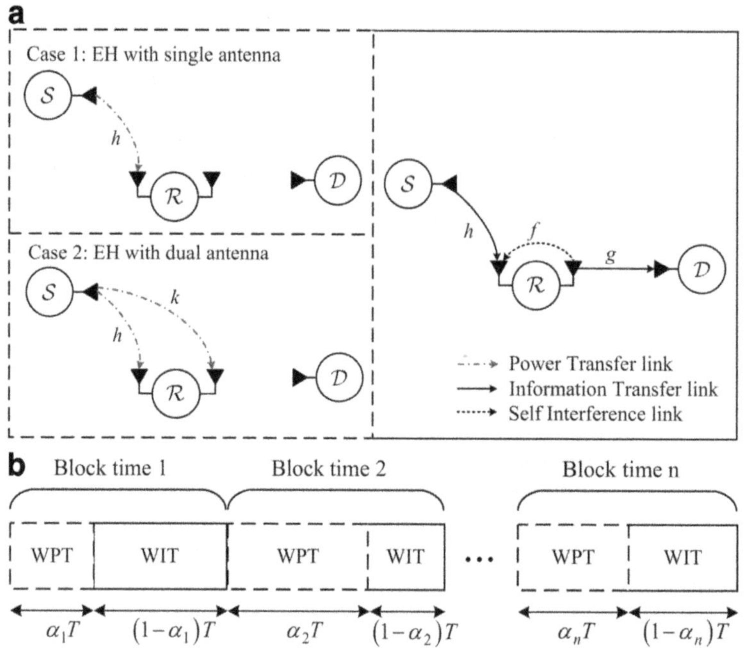

Fig. 1 Two-hop relaying networks in full-duplex communication with SWIPT. **a** Describing two phases of cooperative relay energy harvesting communications with two scenarios of energy harvesting at relay node. **b** TSAC protocol for energy harvesting full-duplex relaying network

T and varies independently and identically from one block to the other.

In this paper, we also use character f to represent the SI link at relay node as normal manner. Due to the short distance of this link, the line-of-sight (LoS) path is likely to represent to SI channel; hence, it can be shown that the Rician distribution can handle such SI channel as [18]. However, due to the complicated Rician fading probability density function (PDF), the analytical expressions become extremely difficult. Fortunately, the alternative model of the Nakagami-m fading distribution provides a very good approximation to the Rician distribution. Motivated by this, and to simplify in the analysis, we adopt the Nakagami-m fading with fading severity factor m_f and mean λ_f to model the loop interference channel in this paper.

So that $|h|^2$ and $|k|^2$ are independent and identically distributed (i.i.d.), exponential random variables with mean λ_h and λ_k, where $\lambda_h = \lambda_k$, $|g|^2$ is exponentially distributed with mean λ_g. The self-interference power link at relay, i.e., $|f|^2$, is a Gamma random variable distributed as $\Gamma\left(m_f, \lambda_f/m_f\right)$, in this paper we also assume m_f is an integer number.

The transmit power of source and relay are represented by $P_{\mathcal{S}}$ and $P_{\mathcal{R}}$, respectively. Due to the shadowing effect, the direct transmission between source node and destination node does not exist [2, 5, 8, 20, 27, 29, 32]. The FD mode causes self-interference, which can be addressed

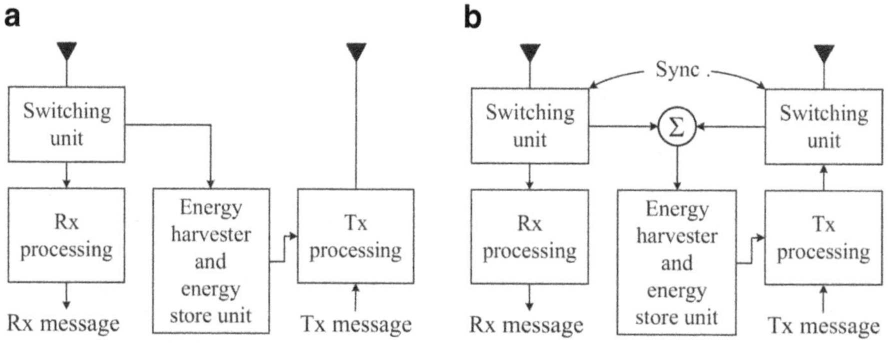

Fig. 2 Receiver architecture of relay with energy harvesting. **a** Relay node with single antenna EH. **b** Relay node with dual antennas EH

by the novel methods in the literature as [17, 19] and these algorithms are beyond the scope of our paper. Unfortunately, the residual self-interference still exists after interference suppression in the practical receiver architecture and it also impairs the performance of FD relay networks as [17–21, 27]. In this paper, we mainly focus on the impact of the residual self-interference on system performance in terms of the harvested power.

2.3 TSAC protocol description

In this subsection, the energy harvesting protocol is presented. In spirit of [31–33] suggesting adaptive time switching strategies for SWIPT system, we redesigned TS protocol related to CSI for FD relay transmission mode, named TSAC protocol. The detail of modified TSAC protocol is described as below. The harvested energy is stored in a rechargeable battery and then totally used to feed power circuits and transmit information to the destination node. Particularly in TSAC policy, each communication block time is slit into two slots, including wireless power transfer (WPT) slot and wireless information transfer (WIT) slot as mentioned in [4–7, 11, 12, 27, 31–33]. In each block time, WPT slot represents the first $\alpha_i T$ of block time while the WIT slot stands for the rest of $(1-\alpha_i)\,T$ of block time. During WIT phase, the source transmits its symbol toward the intermediate relay simultaneously, the cooperative relay retransmits its decoded symbol to destination at the same time and bandwidth. The relay thus suffers from loop interference (LI).

Before further description, some important symbols are listed and defined in Table 1. Moreover, regarding the proposed WPT policy, $\mathcal{R}$ will operate with the preset transmit power, $P_{\mathcal{R}}$, and thus $\mathcal{R}$ node needs to exactly determine the time duration for WPT to harvest a sufficient amount of preset energy, $\mathcal{E}_i = P_{\mathcal{R}}\,(1-\alpha_i)\,T$.

This process can be done as steps below. Because relay only harvests sufficient amount of energy, $\mathcal{E}_i^{EH} = f(\alpha_i)$ (where $f(\alpha_i)$ means that function of α_i), one can exactly determine the EH time by equaling amount of harvested energy and that of preset energy, i.e., $\mathcal{E}_i^{EH} = \mathcal{E}_i$. Finally, the EH time duration, α_i, is derived. It is noted that the suggested TSAC protocol does not require more additional time slot since the total frame for communication is the same as [27] (see Remark 1), and the preset relay transmit power is a constant value ($P_{\mathcal{R}}$) while the EH time is a function of the random variable WPT channel gain(s) as [31–33].

Table 1 List of important symbols

Symbol	Definition
$P_{\mathcal{S}}$	The fixed transmission power, preset at source node.
$P_{\mathcal{R}}$	The fixed transmission power, preset at relay node.
$\mathcal{E}_i$	The relay transmission energy during time slot ith corresponding to preset power, $P_{\mathcal{R}}$.
$P_{\mathcal{R},i}^{EH}$	The harvested power from EH at relay node at time slot ith.
$\mathcal{E}_i^{EH}$	The relay transmission energy during time slot ith corresponding to harvested power $P_{\mathcal{R},i}^{EH}$.
α_i	The duration time allocated to EH in time slot ith.

As aforementioned, to exactly determine this duration of energy harvesting time, channel gain(s) of WPT link is an important parameter(s). Therefore, we assume that the channel state information (CSI) during the first hop is available at source and relay node, which can be obtained by using novel estimation algorithm. Interestingly, the TSAC protocol adjusts the WPT time duration, α_i, in each time slot to satisfy amount of installed power $P_{\mathcal{R}}$. In contrast, the fixed-time allocation protocol in [5, 27], the harvested power, $P_{\mathcal{R}}$, vary in each block based on channel gain.

2.4 Signal model

In wireless information transfer phase (WIT) of SWIPT systems, the received signal at $\mathcal{R}$, $y_{\mathcal{R},i}$ and $\mathcal{D}$, $y_{\mathcal{D},i}$ are given respectively as follows:

$$y_{\mathcal{R},i} = \frac{h_i}{\sqrt{d_1^m}}\sqrt{P_{\mathcal{S}}}x_i + f_i\sqrt{P_{\mathcal{R}}}\hat{x}_i + n_{\mathcal{R},i}, \tag{1}$$

and

$$y_{\mathcal{D},i} = \frac{g_i}{\sqrt{d_2^m}}\sqrt{P_{\mathcal{R}}}\hat{x}_i + n_{\mathcal{D},i}. \tag{2}$$

where i is the block time index, x_i and $\hat{x}_i$ are message symbol at $\mathcal{S}$ and decoded symbol at $\mathcal{R}$ with unit power and zero average, respectively. We also assume that the relay decodes when $\mathcal{S}-\mathcal{R}$ link does not suffer from outage. It is assumed that $n_{\mathcal{R},i}$ and $n_{\mathcal{D},i}$ are additive white Gaussian noise (AWGN) at $\mathcal{S}$ and $\mathcal{D}$ in block time ith, respectively.

From (1) and (2), the signal interference noise ratio (SINR) at $\mathcal{R}$ and $\mathcal{D}$ in the time slot, ith are determined by

$$\gamma_i^{\mathcal{R}} = \frac{P_{\mathcal{S}}\,|h_i|^2}{P_{\mathcal{R}}d_1^m\,|f_i|^2 + d_1^m\sigma_{\mathcal{R},i}^2}, \tag{3}$$

and

$$\gamma_i^{\mathcal{D}} = \frac{P_{\mathcal{R}}\,|g_i|^2}{d_2^m\sigma_{\mathcal{D},i}^2}, \tag{4}$$

where noise terms at $\mathcal{R}$ and $\mathcal{D}$ are zero mean and variances of $\sigma_{\mathcal{R},i}^2$, $\sigma_{\mathcal{D},i}^2$, respectively.

Based on the DF relaying scheme, the end-to-end SINR at block time ith, is expressed as follow

$$\gamma_i^{e2e} = \min\left(\gamma_i^{\mathcal{R}}, \gamma_i^{\mathcal{D}}\right). \tag{5}$$

In WPT time slot of SWIPT, we investigate the performance of two schemes based on the number of antennas in EH phase at relay as [27], where (i) only one antenna is responsible for receiving signals and harvesting RF signals in EH phase or (*ii*) either two antennas equipped for FD communication can be used during the EH stage.

2.4.1 Single antenna for energy harvesting

In this case, only one antenna receives signals and it scavenges the RF signal to convert to DC signal. Therefore, the received signal at the relay can be expressed by

$$y_{\mathcal{R},i}^{EH} = \frac{h_i}{\sqrt{d_1^m}}\sqrt{P_S}x_i + n_{\mathcal{R},i}. \quad (6)$$

In principle, the harvested energy can be stored in an inexpensive capacitor and then entirely use for information transmission. Thus, the harvested energy at $\mathcal{R}$, $\mathcal{E}_i^{EH}$, is given by

$$\mathcal{E}_i^{EH} = \eta\alpha_i T\frac{P_S |h_i|^2}{d_1^m}, \quad (7)$$

where $0 \leq \eta \leq 1$ is the energy conversion efficiency that depends on the rectification process and the energy harvesting circuitry. It is noted that we ignore the harvested energy from noise. So that the harvested power can be determined during $(1-\alpha_i)T$ as $P_{\mathcal{P},i}^{EH} = \mathcal{E}_i^{EH}/(1-\alpha_i)T$.

We consider the fixed time scheme assigned for power transfer duration in [5], in which power collected at relay depends on CSI. Contrarily, in our proposed TSAC protocol, the power transmission at relay denoted by $P_{\mathcal{R}}$ corresponding with optimal throughput is preset before the transmission energy at relay thus can be calculated by

$$\mathcal{E}_i = P_{\mathcal{R}}(1-\alpha_i)T. \quad (8)$$

Finally, in order to calculate the fraction EH time, we setting (7) equal to (8), i.e., $\mathcal{E}_i^{EH} = \mathcal{E}_i$, the allocated fraction of time for power transfer in case of single antenna at any block time is computed as

$$\alpha_i = \frac{P_{\mathcal{R}}d_1^m}{\eta P_S|h_i|^2 + P_{\mathcal{R}}d_1^m}. \quad (9)$$

2.4.2 Dual antennas for energy harvesting

In case of FD communication, the expected relay node is provided with two antennas to receive and transmit signals and such operations are utilized thanks to the collection of RF signal in EH duration. Thus, the received signal at relay node in this stage can be expressed by

$$y_{\mathcal{R},i}^{EH} = \frac{(h_i + k_i)}{\sqrt{d_1^m}}\sqrt{P_S}x_i + n_{\mathcal{R},i}. \quad (10)$$

Similarly, the harvested and transmitted energy regardless of the harvested energy from noise at $\mathcal{R}$ is given by

$$\mathcal{E}_i^{EH} = \eta\alpha_i T\frac{P_S\left(|h_i|^2 + |k_i|^2\right)}{d_1^m}. \quad (11)$$

In order to determined EH time, one setting (8) equal to (11), the fraction of time for power transfer in dual antennas mode for energy harvesting at any block times is determined as

$$\alpha_i = \frac{P_{\mathcal{R}}d_1^m}{\eta P_S\left(|h_i|^2 + |k_i|^2\right) + P_{\mathcal{R}}d_1^m}. \quad (12)$$

Remarks 1 *In* (9) *and* (12), *the duration of time allocated in energy harvesting phase is a function of some variables, including channel gains, optimal power at relay node , the distance between* $\mathcal{S}$ *and* $\mathcal{R}$, *i.e.* d_1, *energy harvesting coefficient and power transmission at source, i.e.* $P_{\mathcal{S}}$. *Specifically, this time fraction is always less than one, i.e.,* $\alpha_i < 1$, *which implies the allocated time which is required for simultaneous wireless information and power transfer.*

Remarks 2 *Consider energy harvesting time in* (9) *and* (12), *to convenience in represent we use two characters* α_i^{single} *and* α_i^{dual} *to denote* α_i *in* (9) *and* (12), *respectively. We can observe that, since* $|k_i|^2 \geq 0$ *and other parameters are invariable then* $\alpha_i^{single} \geq \alpha_i^{dual}$. *This implies that the time EH with dual antennas is less than that with single antenna. In other words, the amount of WIT time with dual antennas EH case is greater than that with single antenna EH case.*

Remarks 3 *The proposed modifying TSAC protocol for FD cooperative relaying system is not more require additional time slot since the total frame for communication is as same as* [27] *(see Remark* 1*), and the prior installed relay transmit power is a constant value* ($P_{\mathcal{R}}$) *while the EH time is a function of the random variable WPT channel gain(s) as* [31–33].

It is worth noting that the Tables 2 and 3 confirmed that our study aims to find power allocation to obtain optimal performance in energy harvesting networks. Such scheme relies on calculation of CSI feedback signal and so-called offline power allocation due to requiring low complexity in design of real applications.

3 Optimal power allocation for maximized instantaneous throughput

Energy harvesting is an effective method to enhance the performance of the power-constrained relay with available RF signals, in which the transmission power at relay is allocated to maximize the instantaneous throughput. In this paper, the optimal transmitted power in energy harvesting scheme is proposed for FD relay networks.

The instantaneous throughput of FD relay can be expressed by

$$C = (1-\alpha)\log_2\left(1 + \gamma^{e2e}\right). \quad (13)$$

Table 2 The previous relevant works and their contributions

Reference	EH protocol	Transmission mode	Relay mode	Works area
Nasir et al. [5]	TS, PS	Half duplex	AF	Throughput analysis
Nasir et al. [33]	Continues and discrete TS	Half duplex	AF, DF	Throughput analysis
Chen et al. [17]	Non EH	Full duplex	DF	Optimal power allocation
Zeng and Zhang [28]	TS with energy recycling	Full duplex	AF	Optimizing relay transmission power
Zhong et al. [27]	TS	Full duplex	AF, DF	Optimizing throughput
Ding et al. [32]	Adaptive TS	Half duplex	AF, DF	Optimizing throughput
Ding et al. [15]	PS with storage	Half duplex	DF	Outage performance analysis
Krikidis [16]	PS with storage	Half duplex	DF	Outage performance analysis
Our work	TS aware channel	Full duplex	DF	Optimizing transmit power at relay Throughput analysis.

The optimal transmit power allocated for relay node is formulated as

$$P_{\mathcal{R}}^{opt} = \arg\max_{P_{\mathcal{R}}} \{C(P_{\mathcal{R}})\}, \quad \text{s.t.} 0 \leq P_{\mathcal{R}}. \tag{14}$$

3.1 Energy harvested by single antenna

Substituting (3), (4), (5), and (9) into (13), the instantaneous throughput is the function of relay transmit power which can be written as follow

$$C(P_{\mathcal{R}}) = \left(1 - \frac{P_{\mathcal{R}} d_1^m}{\eta P_{\mathcal{S}} |h|^2 + P_{\mathcal{R}} d_1^m}\right) \times log_2 \left(1 + \min\left(\frac{P_{\mathcal{S}}|h|^2}{P_{\mathcal{R}} d_1^m |f|^2 + d_1^m \sigma_{\mathcal{R}}^2}, \frac{P_{\mathcal{R}}|g|^2}{d_2^m \sigma_{\mathcal{D}}^2}\right)\right). \tag{15}$$

Theorem 1 *The optimum transmitted power at the energy-constrained relay for maximizing instantaneous throughput is calculated by*

$$P_{\mathcal{R}}^{opt} = \begin{cases} \frac{e^{\mathcal{W}\left(\frac{\pi_2\pi_3-1}{e}\right)+1}-1}{\pi_2}, & e^{\mathcal{W}\left(\frac{\pi_2\pi_3-1}{e}\right)+1} < \pi_2\frac{\sqrt{\Delta}-\pi_0}{2\pi_1}+1, \\ \frac{\sqrt{\Delta}-\pi_0}{2\pi_1}, & \text{otherwise.} \end{cases} \tag{16}$$

where $\pi_0 = \frac{d_1^m\sigma_{\mathcal{R}}^2}{P_{\mathcal{S}}|h|^2}$, $\pi_1 = \frac{d_1^m|f|^2}{P_{\mathcal{S}}|h|^2}$, $\pi_2 = \frac{|g|^2}{d_2^m\sigma_{\mathcal{D}}^2}$, $\pi_3 = \frac{\eta P_{\mathcal{S}}|h|^2}{d_1^m}$, $\Delta = \pi_0^2 + \frac{4\pi_1}{\pi_2}$, *and* $\mathcal{W}(x)$ *is the Lambert function in* [34], *where* $\mathcal{W}(x)$ *can be found due to the problem solving of* $\mathcal{W}\exp(\mathcal{W}) = x$.

Proof Please see in Appendix A. □

3.2 Energy harvested by dual antennas

Similarly, substituting (3), (4), (5), and (12) into (13), the instantaneous throughput can be formulated as below

$$C(P_{\mathcal{R}}) = \left(1 - \frac{P_{\mathcal{R}} d_1^m}{\eta P_{\mathcal{S}}\left(|h|^2+|k|^2\right) + P_{\mathcal{R}} d_1^m}\right) \times \log_2\left(1 + \min\left(\frac{P_{\mathcal{S}}|h|^2}{P_{\mathcal{R}} d_1^m |f|^2 + d_1^m\sigma_{\mathcal{R}}^2}, \frac{P_{\mathcal{R}}|g|^2}{d_2^m\sigma_{\mathcal{D}}^2}\right)\right). \tag{17}$$

The optimal relay transmission power is derived in the same way as (14).

Theorem 2 *The optimal allocation of transmit power at relay in case of two antennas equipped for EH for maximizing instantaneous mode is given by*

$$P_{\mathcal{R}}^{opt} = \begin{cases} \frac{e^{\mathcal{W}\left(\frac{\pi_2\pi_4-1}{e}\right)+1}-1}{\pi_3}, & e^{\mathcal{W}\left(\frac{\pi_2\pi_4-1}{e}\right)+1} < \pi_2\frac{\sqrt{\Delta}-\pi_0}{2\pi_1}+1, \\ \frac{\sqrt{\Delta}-\pi_0}{2\pi_1}, & \text{otherwise.} \end{cases} \tag{18}$$

where $\pi_4 = \frac{\eta P_{\mathcal{S}}\left(|h|^2+|k|^2\right)}{d_1^m}$ *and other related parameters are defined as in Theorem* 1.

Table 3 CSI requirement for proposed TSAC protocol schemes

EH type	Transmission mode	CSI requirements	Channel information
Single antennas	Instantaneous	High	$\|h_i\|^2$, $\|g_i\|^2$, $\|f_i\|^2$
Single antennas	Delay-constrained	Medium	$\|h_i\|^2$, and distributions of $\|g_i\|^2$, $\|f_i\|^2$
Dual antennas	Instantaneous	High	$\|h_i\|^2$, $\|k_i\|^2$, $\|g_i\|^2$, $\|f_i\|^2$
Dual antennas	Delay-constrained	Medium	$\|h_i\|^2$, $\|k_i\|^2$, and distributions of $\|g_i\|^2$, $\|f_i\|^2$

Proof The Theorem can be explained by using in a similar way to Theorem 1. □

Remarks 4 *In practical wireless system, only causal channel information and harvested energy are useful in calculation of power allocation. Unfortunately, the offline power allocation policy is not readily applicable in reality. Considering at a given time slot, the CSI in the second hop of the relaying network and the upcoming harvested energy are not known in advance and hence power allocation is evaluated in stochastic circumstance. However, such solution only requires low complexity when comparing online power allocation which is high computational complexity and may not be implementable in practice.*

4 Analysis of outage probability and throughput in delay-constrained transmission mode

In this section, we consider the optimal throughput and outage probability of two-hop relaying networks in FD transmission mode. In this scenario, the system is designed with the fixed transmission rate or the preset SNR threshold γ_0. In particular, the system suffers from outage when the SNR performance, i.e., γ_i^{e2e} drops below the threshold value. Therefore, the expression of outage probability can be obtained by

$$OP_i = \Pr\left\{\gamma_i^{e2e} < \gamma_0\right\} = \Pr\left\{\min\left(\gamma_i^{\mathcal{R}}, \gamma_i^{\mathcal{D}}\right) < \gamma_0\right\}, \tag{19}$$

and the expected value of outage probability can be obtained by substituting (3) and (4) into (19).

As aforementioned in protocol description subsection 2.3, since the relay node will harvest a sufficient amount of energy, $\mathcal{E}_i$, before operating WIT phase, it is noted that the relay be certainly performed because the harvesting energy time is always less than 1, see Remark 1. Therefore, the relay transmit power $P_{\mathcal{R}}$ is a constant value during the transmission duration, see Remark 3. This quantity of power is preset before (such as by technicians). It is worth noting that the $\gamma_i^{\mathcal{R}}, \gamma_i^{\mathcal{D}}$ are independent in term of $P_{\mathcal{R}}$ as in [31–33].

With the fact that such channel is independent to each other, it can be shown that [17, 20]

$$\begin{aligned} E\{OP_i\} &= 1 - E_{|h_i|^2,|f_i|^2}\left\{\Pr\left(\gamma_i^{\mathcal{R}} > \gamma_0\right)\right\} \times E_{|g_i|^2}\left\{\Pr\left(\gamma_i^{\mathcal{D}} > \gamma_0\right)\right\} \\ &= 1 - E_{|h_i|^2,|f_i|^2}\left\{\Pr\left(\frac{P_{\mathcal{S}}|h_i|^2}{P_{\mathcal{R}} d_1^m |f_i|^2 + d_1^m \sigma_{\mathcal{R},i}^2} > \gamma_0\right)\right\} \\ &\quad \times E_{|g_i|^2}\left\{\Pr\left(\frac{P_{\mathcal{R}}|g_i|^2}{d_2^m \sigma_{\mathcal{D},i}^2} > \gamma_0\right)\right\}. \end{aligned} \tag{20}$$

where $E\{.\}$ is the expectation function.

In the delay-constrained transmission mode, the throughput efficient in the time slot ith, which is the function of outage probability and the EH duration time, can be formulated by

$$\tau_i^{DC} = (1 - OP_i)(1 - \alpha_i). \tag{21}$$

Therefore, the average throughput efficient of system is written by

$$\tau^{DC} = E\left\{\tau_i^{DC}\right\} = E\{1 - OP_i\} - E\{(1 - OP_i)\alpha_i\}. \tag{22}$$

The optimal value of transmit power for throughput optimization can be presented as below

$$\begin{aligned} P_{\mathcal{R}}^{opt} &= \arg\max_{P_{\mathcal{R}}} \left\{\tau^{DC}(P_{\mathcal{R}})\right\} \\ \text{s.t} \quad & 0 \le P_{\mathcal{R}}. \end{aligned} \tag{23}$$

4.1 Energy harvested by single antenna

The system throughput in this case can be determined as (22) and the next theorem is proposed.

Theorem 3 *When the relay uses single antenna to harvest energy, then the average throughput efficient of the system is given by*

$$\begin{aligned} \tau^{DC} &= \exp\left(-\frac{\gamma_0 d_2^m \sigma_{\mathcal{D}}^2}{\lambda_g P_{\mathcal{R}}}\right) \times \left\{\exp\left(-\frac{\vartheta}{\lambda_h}\right)\left(\frac{\omega\lambda_f}{m_f\lambda_h} + 1\right)^{-m_f}\right. \\ &\quad - \frac{\chi}{\lambda_h}\exp\left(\frac{\chi}{\lambda_h}\right) \times \left[\exp\left(\frac{m_f(\chi+\vartheta)}{\omega\lambda_f}\right)\right. \\ &\quad \left.\times \mathrm{Ei}\left(-(\chi+\vartheta)\left(\frac{m_f}{\omega\lambda_f} + \frac{1}{\lambda_h}\right)\right) - \mathrm{Ei}\left(-\frac{\chi+\vartheta}{\lambda_h}\right)\right] \\ &\quad \left.- \frac{\chi}{\lambda_h}\exp\left(\frac{m_f\vartheta}{\omega\lambda_f}\right)\sum_{i=1}^{m_f-1}\frac{1}{i!}\left(\frac{m_f}{\omega\lambda_f}\right)^i \times \tau_{31}\right\}, \end{aligned} \tag{24}$$

where $\chi = \frac{P_{\mathcal{R}} d_1^m}{\eta P_{\mathcal{S}}}$, $\omega = \frac{\gamma_0 P_{\mathcal{R}} d_1^m}{P_{\mathcal{S}}}$, $\vartheta = \frac{\gamma_0 d_1^m \sigma_{\mathcal{R}}^2}{P_{\mathcal{S}}}$, *Ei is the exponential integral function as (Eq. 8.211) in* [35], *and*

$$\begin{aligned} \tau_{31} &\triangleq (-1)^{i+1}\vartheta^i \exp\left(\frac{\chi}{\lambda_h}\right)\mathrm{Ei}\left(-(\chi+\vartheta)\left(\frac{1}{\lambda_h} + \frac{m_f}{\omega\lambda_f}\right)\right) \\ &\quad + \sum_{k=1}^{i}\binom{i}{k}(-1)^i\vartheta^{i-k}\int_{\vartheta}^{\infty}\exp\left(-x\left(\frac{1}{\lambda_h} + \frac{m_f}{\omega\lambda_f}\right)\right)\frac{x^k}{x+\chi}dx. \end{aligned}$$

Proof The detailed explanation of the Theorem 3 is provided in Appendix B. □

Alternatively, we can use the following closed-form when the SI channel modeled by Rayleigh fading as following corollary.

Corollary 1 *When* $m_f = 1$ *(or the SI link at relay experience Rayleigh fading model) the throughput performance in case of single antenna for EH at relay node is given as*

$$\tau^{DC} = \exp\left(-\frac{\gamma_0 d_2^m \sigma_{\mathcal{D}}^2}{\lambda_g P_{\mathcal{R}}}\right) \times \left\{\exp\left(-\frac{\vartheta}{\lambda_h}\right)\frac{\lambda_h}{\lambda_h + \omega\lambda_f} + \frac{\chi}{\lambda_h}\exp\left(\frac{\chi}{\lambda_h}\right)\mathrm{Ei}\left(-\frac{\chi+\vartheta}{\lambda_h}\right) - \frac{\chi}{\lambda_h}\exp\left(\frac{\chi+\vartheta}{\omega\lambda_f}+\frac{\chi}{\lambda_h}\right)\mathrm{Ei}\left(-\frac{\chi+\vartheta}{\omega\lambda_f}-\frac{\chi+\vartheta}{\lambda_h}\right)\right\}. \tag{25}$$

Proof The Corollary 1 can be obtained easily by substituting $m_f = 1$ into the expression in Theorem 3. □

Extracting (23) by using (24), the optimal transfer power at relay with the aim of optimizing the system throughput can be obtained. Since $\tau_{DC}(P_{\mathcal{R}})$ is a concave function of $P_{\mathcal{R}}$, the optimal value $P_{\mathcal{R}}^{opt}$ can be obtained by solving the equation $\partial\tau_{DC}(P_{\mathcal{R}})/\partial P_{\mathcal{R}} = 0$. Although the derivation of closed-form optimal expressions is challenging, the optimal value can be obtained by using numerical simulations which are presented in the next section.

4.2 Energy harvested by dual antennas

In this case, we also study the throughput in the delay-constrained mode with the following theorem.

Theorem 4 *We consider a cooperative scenario with two-hop source-relay-destination pair and two antennas provided at energy harvesting relays, the throughput can be determined as*

$$\tau^{DC} = \exp\left(-\frac{\gamma_0 d_2^m \sigma_{\mathcal{D}}^2}{\lambda_g P_{\mathcal{R}}}\right) \times \left\{\exp\left(-\frac{\vartheta}{\lambda_h}\right)\left(\frac{\omega\lambda_f}{m_f\lambda_h}+1\right)^{-m_f} - \frac{\chi}{\lambda_h}\exp\left(\frac{\chi}{\lambda_k}\right) \times \left\{\lambda_k\exp\left(-\frac{\vartheta+\chi}{\lambda_k}\right) + (\vartheta+\chi)\,\mathrm{Ei}\left(-\frac{\vartheta+\chi}{\lambda_k}\right) \times \frac{\omega\lambda_f}{m_f}\mathrm{Ei}\left(-\frac{\vartheta+\chi}{\lambda_k}\right) - \frac{\omega\lambda_f}{m_f}\exp\left(\frac{m_f(\vartheta+\chi)}{\omega\lambda_f}\right) \times \mathrm{Ei}\left(-(\vartheta+\chi)\left(\frac{1}{\lambda_k}+\frac{m_f}{\omega\lambda_f}\right)\right)\right\} + \exp\left(\frac{m_f(\vartheta+\chi)}{\omega\lambda_f}\right) \times \tau_{43}\right\}, \tag{26}$$

where χ, ω, ϑ, Ei is defined below Theorem 3, and

$$\tau_{43} \triangleq \sum_{i=1}^{m_f-1}\frac{1}{i!}\left(\frac{m_f}{\omega\lambda_f}\right)^i\left\{(-1)^i(\vartheta+\chi)^i\frac{\omega\lambda_f\lambda_h}{\omega\lambda_f+m_f\lambda_h} \times \left[\exp\left(-\left(\frac{1}{\lambda_h}+\frac{m_f}{\omega\lambda_f}\right)(\vartheta+\chi)\right)\mathrm{E}_1\left(\frac{\vartheta+\chi}{\lambda_k}\right) - \mathrm{E}_1\left((\vartheta+\chi)\left(\frac{1}{\lambda_k}+\frac{1}{\lambda_h}+\frac{m_f}{\omega\lambda_f}\right)\right)\right] + \sum_{k=1}^{i}\binom{i}{k}(-(\vartheta+\chi))^{i-k}\int_{\vartheta+\chi}^{\infty} t^k\exp\left(-x\left(\frac{1}{\lambda_h}+\frac{m_f}{\omega\lambda_f}\right)\right) \times \mathrm{E}_1\left(\frac{t}{\lambda_k}\right)dx\right\}.$$

Proof Please see in Appendix C. □

When the SI channel is modeled by Rayleigh fading connection, the throughput performance of system is given as corollary below.

Corollary 2 *When $m_f = 1$ (or the SI link at relay experience Rayleigh fading model) the throughput performance in case of dual antenna for EH at relay node is given as*

$$\tau^{DC} = \exp\left(-\frac{\gamma_0 d_2^m \sigma_{\mathcal{D}}^2}{\lambda_g P_{\mathcal{R}}}\right) \times \left\{\exp\left(-\frac{\vartheta}{\lambda_h}\right)\frac{\lambda_h}{\lambda_h+\omega\lambda_f} - \frac{\chi}{\lambda_h}\exp\left(\frac{\chi}{\lambda_k}\right) \times \left[\lambda_k\exp\left(-\frac{\vartheta+\chi}{\lambda_k}\right) + (\vartheta+\chi)\,\mathrm{Ei}\left(-\frac{\vartheta+\chi}{\lambda_k}\right) - \omega\lambda_f\left(\exp\left(\frac{\vartheta+\chi}{\omega\lambda_f}\right)\mathrm{Ei}\left(-\frac{\vartheta+\chi}{\lambda_k}-\frac{\vartheta+\chi}{\omega\lambda_f}\right) - \mathrm{Ei}\left(-\frac{\vartheta+\chi}{\lambda_k}\right)\right)\right]\right\}. \tag{27}$$

Proof It is easy to obtain the Corollary 2 by substituting $m_f = 1$ into the expression in Theorem 4. □

Remarks 5 *Regarding impact of self-interference (SI) cancellation, the active SI suppression methods were shown experimentally to be capable of facilitating FD at short range communication. To apply energy harvesting in a real FD system, it is critical to accurately measure and suppress the SI in such FD networks. In this paper, we investigate SI in system performance through numerical simulation results.*

5 Half-duplex relaying networks

In HD transmission mode, the relay node uses single antenna for energy harvesting and information processing. In order to compare with the FD relaying systems, we consider different transmission modes. Some of the results have been derived in [31–33]; however, we present here to make our work self-contained.

In this scenario, the received signal at relay in HD mode is presented as

$$y_{\mathcal{R},i} = \frac{h_i}{\sqrt{d_1^m}}\sqrt{P_{\mathcal{S}}}x_i + n_{\mathcal{R},i}, \tag{28}$$

And the SNR at relay node given by

$$\gamma_i^{\mathcal{R}} = \frac{P_{\mathcal{S}}\left|h_i\right|^2}{d_1^m\sigma_{\mathcal{R},i}^2}. \tag{29}$$

5.1 Instantaneous transmission mode

In this scenario, the instantaneous throughput can be illustrated as

$$C_{HD} = \frac{1}{2}(1-\alpha)\log_2\left(1+\gamma^{e2e}\right). \quad (30)$$

The optimal allocation for power transmission at relay is similar to (14)

Theorem 5 *The optimal power allocation in HD relaying networks for instantaneous transmission mode can be calculated as*

$$P_{\mathcal{R}}^{opt} = \begin{cases} \frac{e^{\mathcal{W}\left(\frac{2\pi_2\theta-1}{e}\right)+1}-1}{\pi_2}, & e^{\mathcal{W}\left(\frac{2\pi_2\theta-1}{e}\right)+1} < \frac{1}{\pi_0}+1, \\ \frac{1}{\pi_0\pi_2}, & \text{otherwise.} \end{cases} \quad (31)$$

where $\theta = \pi_3$ in case of single antenna, $\theta = \pi_4$ in case of dual antennas. And $\pi_0, \pi_2, \pi_3, \pi_4$ are defined as below Theorem 1.

Proof The proof is similar to that of Theorem 1. □

5.2 Delay-constrained transmission mode

The throughput of HD relaying can be obtained as

$$\tau_i^{DC} = \frac{1}{2}(1-OP_i)(1-\alpha_i). \quad (32)$$

In comparison with the calculation of throughput in FD mode, in (21), the factor 2 can be seen by the denominator denoting as transmission efficiency. In following calculation, we can obtain two theorem which are solved similarly as that in Theorem 4.

Theorem 6 *In the HD mode relaying, the throughput in case of single antenna given by*

$$\tau^{HD} = \frac{1}{2}\exp\left(-\frac{\gamma_0^{HD}d_2^m\sigma_{\mathcal{D}}^2}{\lambda_g P_{\mathcal{R}}}\right)\times\left\{\exp\left(-\frac{\gamma_0^{HD}d_1^m\sigma_{\mathcal{R}}^2}{\lambda_h P_{\mathcal{S}}}\right) + \frac{\chi_1}{\lambda_h}\exp\left(\frac{\chi_1}{\lambda_h}\right)\mathrm{Ei}\left(-\frac{(\chi_1+\vartheta_1)}{\lambda_h}\right)\right\}. \quad (33)$$

Theorem 7 *In case that the energy is harvested by dual antennas, the throughput of HD relaying networks can be determined as*

$$\tau^{HD} = \frac{1}{2}\exp\left(-\frac{\gamma_0^{HD}d_2^m\sigma_{\mathcal{D}}^2}{\lambda_g P_{\mathcal{R}}}\right)\left\{\exp\left(-\frac{\gamma_0^{HD}d_1^m\sigma_{\mathcal{R}}^2}{\lambda_h P_{\mathcal{S}}}\right) - \frac{\chi_1}{\lambda_h\lambda_k}\exp\left(\frac{\chi_1}{\lambda_k}\right)\times\left[\lambda_k\exp\left(-\frac{\vartheta_1+\chi_1}{\lambda_k}\right) + (\vartheta_1+\chi_1)\,\mathrm{Ei}\left(-\frac{\vartheta_1+\chi_1}{\lambda_k}\right)\right]\right\}, \quad (34)$$

where $\chi_1 = \frac{P_{\mathcal{R}}d_1^m}{2\eta P_{\mathcal{S}}}, \vartheta_1 = \frac{\gamma_0^{HD}d_1^m\sigma_{\mathcal{R}}^2}{P_{\mathcal{S}}}$.

6 Numerical results

In this section, numerical simulation results are demonstrated to validate analytical expressions as concerns in the previous section, and the impact of key system parameters is investigated in detail in terms of system throughput. The energy harvesting efficiency is set $\eta = 0.8$, while the path loss exponent is set $m = 3$. The distances, d_1 and d_2 are normalized values which are set $d_1 = 3$ and $d_2 = 1$, respectively. The simulation environments are associated with the detailed parameters as $\gamma_0 = 10$ dB, $\lambda_f = 0.01, \lambda_h = 1, \lambda_k = 1, \lambda_g = 1, P_{\mathcal{S}} = 26$ dB and noise terms as $\sigma_{\mathcal{D}}^2 = -5$ dB, $\sigma_{\mathcal{R}}^2 = -10$ dB. In this subsection, we present simulation results to verify the analytical results. The throughput performance is calculated by averaging the throughput values over a 100,000 blocks, while fading channels for each block is perfectly independent.

Figure 3 illustrates the impact of received power at relay node on the instantaneous throughput. To compare the performance of FD and HD relay, we set $|f|^2 = 0.05, |h|^2 = 1.2, |g|^2 = 1.4, |k|^2 = 1$. The figure proves that the analytical results based on curves match well with the simulation results. It is noted that the optimal received power at relay for single antenna in HD case and FD case is smaller than those schemes with dual antennas, and the highest instantaneous throughput of FD case outperforms the HD case. When the received power is small, more energy can be collected to facilitate the information transmission while more received power or strong loop-back interference is produced such as the excessive amount of harvested energy, which is the primary cause of worse system performance.

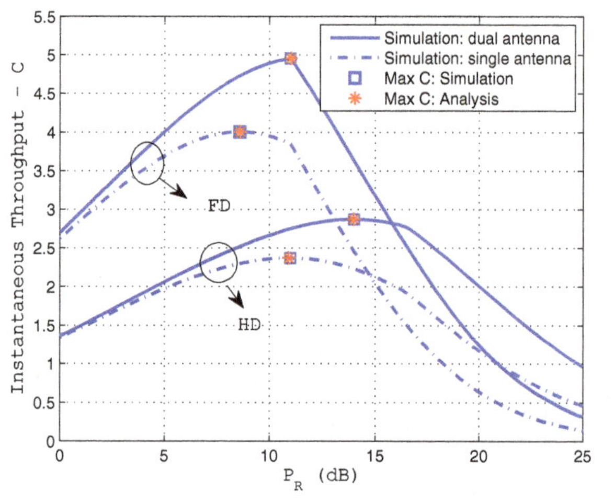

Fig. 3 Instantaneous throughput versus the received power at the relay

The average throughput in delay constraint mode is presented in Fig. 4. It can be observed that the dual antenna case outperforms the single antenna case in terms of the received power at relay. In fact, the dual antennas case can harvest more energy with less transmit power. However, when the transmit power is larger than the optimal harvested power, it results in lower average throughput caused by loop-back interference. On the other hand, the gap between throughput of single antenna and dual antennas is still trivial due to short time for energy harvesting. In general, there should be a balance between energy harvesting capability and noise processing. One can increase $P_{\mathcal{R}}$ to the optimal value, at approximately 10 dB to achieve optimal throughput. As the previous illustration, it is confirmed that FD relay is better than HD relay when the received power at relay is small.

Next, it can be observed that throughput performance improves as increasing transmit power of the source node as in Fig. 5. To evaluate the impacts of the factor m_f in Nakagami channel of SI on system performance, in this experience we set $\lambda_f = 0.1$ and $P_{\mathcal{R}} = P_{\mathcal{S}} - 10$ (dB). The smallest of m_f brings the highest throughput performance. A general observation is that the dual antenna model significantly outperforms the single antenna counterpart at any values of transmit power of the source. Interestingly, when the transmit power overcome about 40 dB, the throughput still remain the peak value in stable floor.

To illustrate the impact of noise variance at relay and destination node in case of optimal throughput, Figs. 6 and 7 also show that dual antennas in FD mode outperform FD single antenna. This can be explained as follows: A lower noise variance decreases the outage probability and and hence we obtain more throughput. Conversely, as for higher noise variance, the outage probability increases while the throughput falls dramatically. Thus, the system

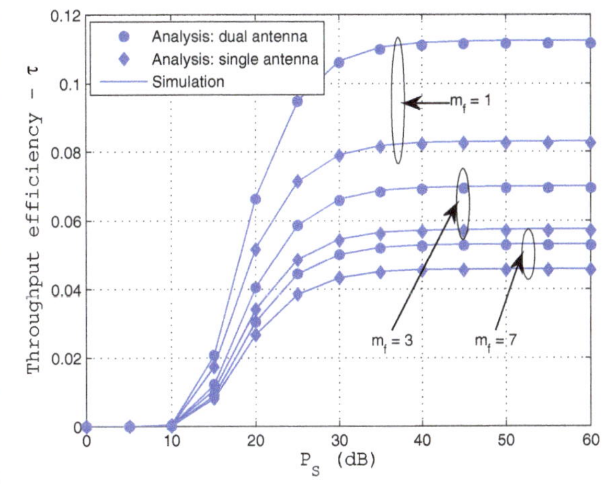

Fig. 5 Throughput efficiency versus $P_{\mathcal{S}}$ with difference case of m_f, $\lambda_f = 0.1$ and $P_{\mathcal{R}} = P_{\mathcal{S}} - 10$ (dB)

performance is affected more by erroneous information processing, leading to a waste of resources during the whole block time. In addition, it can be seen that the impact of noise term at relay affects the noise term at destination more. As a result, throughput performance in Fig. 7 declines sharply when noise varies from − 5 to 5 dB.

In Fig. 8, the simulation results present the throughput efficiency versus γ_0 and compare with existing protocols in [27] and [28]. The main observation is the effect of threshold SNR on throughput in delay constraint mode using TSAC, which is investigated with different quantities of antennas, especially in comparison with traditional energy harvesting protocols. The simulation parameters are installed as $\alpha = 0.2, P_{\mathcal{R}} = 10$ dB. The throughput in case of TSAC is better than the traditional case due to optimal time for energy harvesting which is solved in

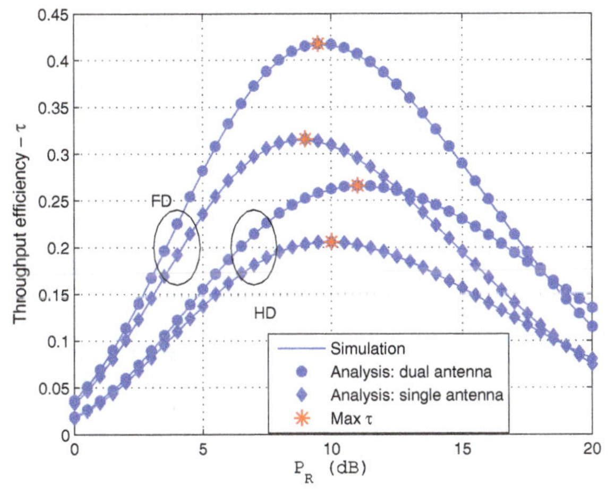

Fig. 4 Throughput efficiency versus $P_{\mathcal{R}}$ in FD relay with $m_f = 1$

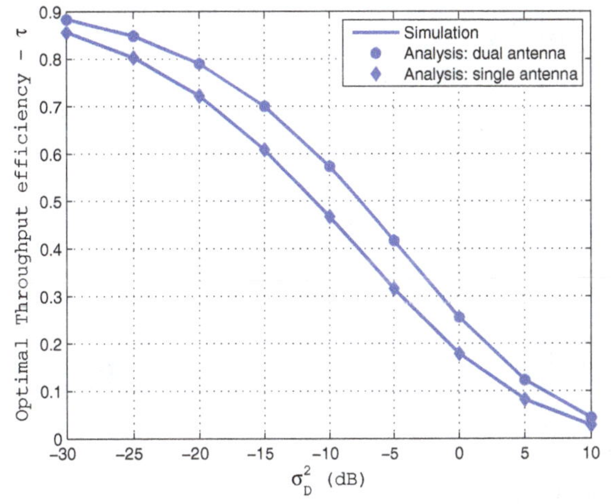

Fig. 6 Throughput efficiency versus $\sigma_{\mathcal{D}}^2$ with $m_f = 1$

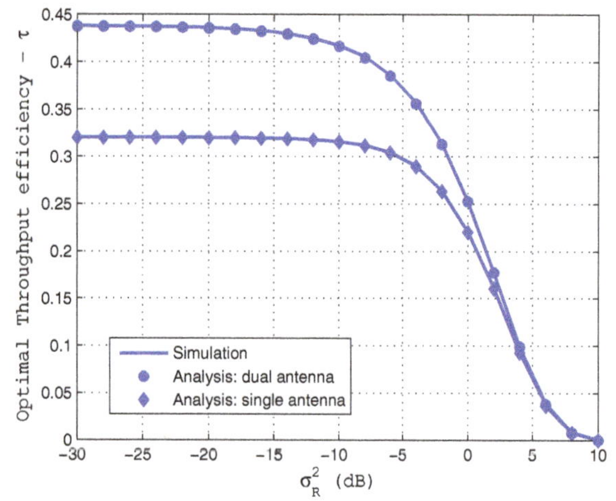

Fig. 7 Throughput efficiency versus $\sigma_{\mathcal{R}}^2$ with $m_f = 1$

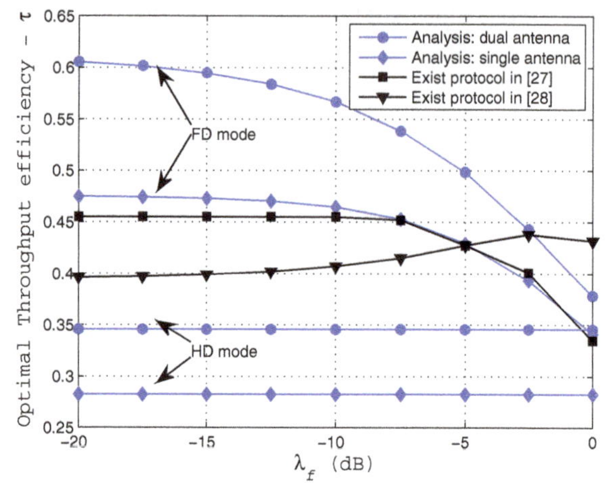

Fig. 9 The impact of residual self-interference on the throughput and in comparison with existing protocols in [27, 28] with $\gamma_0 = 5$ dB and $m_f = 1$

the proposed TSAC. In case of γ_0 described below 14 dB, the proposed protocol related to the optimal throughput efficiency outperforms the existing protocols. However, in high threshold regime, the self-energy recycling protocol proposed in [28] is slightly better than those schemes. In general, this figure reveals that the FD relay has worse performance at higher threshold SNR, which means the system may require higher bit per channel.

As depicted in Fig. 9, the dual antenna FD relay based on TSAC outperforms the single antenna in terms of residual self-interference. When the residual self-interference is greater than −5 dB, throughput decreases significantly, since this loop-back noise impairs the overall performance. Therefore, the self-interference should be remained lower than acceptable value to satisfy system performance and the FD transmission is only beneficial, if the loop-back noise is eliminated. When SI factor ranges from −20 to −5 dB, the performance of TSAC scheme is superior to the existing protocols, but when SI parameters are greater than −5 dB, the optimal throughput of the proposed protocol in [28] is a prime candidate. Thanks to the EH protocol proposed in [28], the loop interference at relay is reused for self-energy recycling. As a result, part of the energy (loop energy) is used for information transmission by the relay and it improves the scheme's performance.

According to Fig. 10, the throughput changes dramatically when we move the relay node in several positions between the source and destination node. The results are then obtained by setting $d_1 = 0.2 \rightarrow 3.8$, $d_2 = 4 - d_1$.

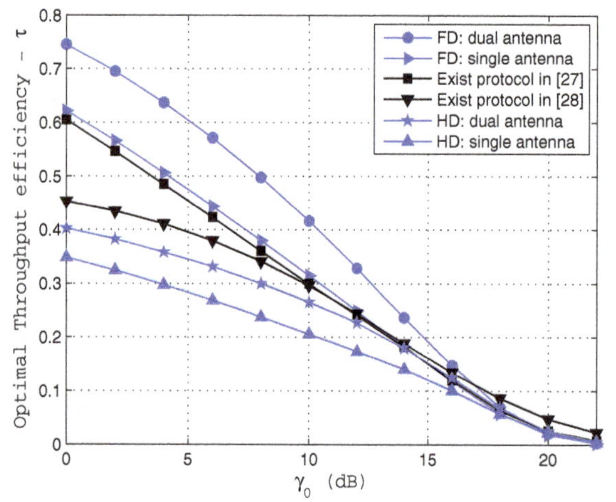

Fig. 8 Throughput efficiency versus γ_0 and the comparison with existing protocols in [27, 28] with $\lambda_f = -20dB$ and $m_f = 1$

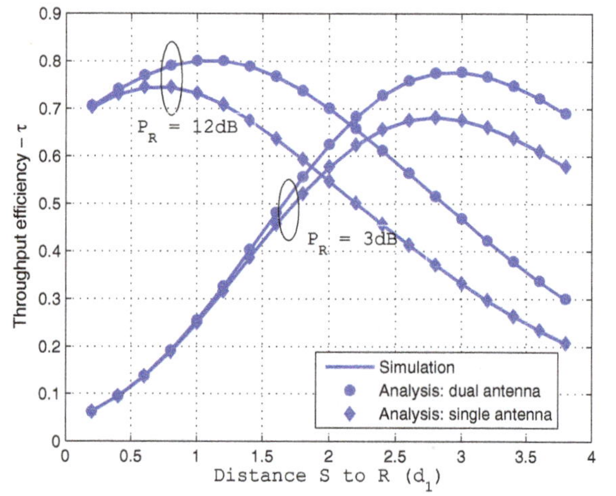

Fig. 10 Throughput efficiency versus the distance of the first hop in FD relay with $m_f = 1$

The distance between relay and source has a capability of energy harvesting which can illustrate how much energy can be obtained via wireless channel. It can be seen clearly that opposite trends for optimal throughput with the received power at relay, $P_{\mathcal{R}} = 3$ dB and $P_{\mathcal{R}} = 12$ dB.

7 Conclusion

In this paper, the throughput of FD and HD relaying in RF energy harvesting systems is investigated. Interestingly, the number of antennas equipped at each relay has significant influence to throughput performance due to the harvested energy at the relay node. Regarding optimal throughput, analytical expressions for the outage probability and throughput capacity of the system were derived. Therefore, the optimal time switching of energy harvesting was comprehensively evaluated. It is confirmed that by employing dual antennas at relay for energy harvesting is beneficial, and the throughput gain is significant when transmit power at source and the received power at relay are carefully calculated. In addition, in comparison with HD relaying networks, our results indicate that FD relaying can substantially boost the system throughput with optimal power allocation policy at energy harvesting-enabled relay. Via mathematical and numerical analysis, the optimal throughput in both instantaneous transmission mode and delay constraint transmission mode can be obtained. More importantly, in order to compute optimal time switching fractions in energy harvesting protocol so-called TSAC, it solely relies on the channel statistics without the need of instantaneous CSI, and it has become an attractive solution to implement in future RF energy harvesting cooperative systems. Finally, for enhancing energy harvesting efficiency, future works should take into account the extra energy with multiple antenna system model (MIMO) to transmit wireless power.

Appendix

A Proof of Theorem 1

Setting $\pi_0 = \frac{d_1^m \sigma_{\mathcal{R}}^2}{P_{\mathcal{S}}|h|^2}$ and $\pi_1 = \frac{d_1^m |f|^2}{P_{\mathcal{S}}|h|^2}$, $\pi_2 = \frac{|g|^2}{d_2^m \sigma_{\mathcal{D}}^2}$, $\pi_3 = \frac{\eta P_{\mathcal{S}}|h|^2}{d_1^m}$, the formula (15) can be re-expressed as

$$C(P_{\mathcal{R}}) = \left(1 - \frac{P_{\mathcal{R}}}{\pi_3 + P_{\mathcal{R}}}\right) \times \log_2\left(1 + \min\left(\frac{1}{\pi_0 + \pi_1 P_{\mathcal{R}}}, \pi_2 P_{\mathcal{R}}\right)\right). \tag{35}$$

In (35), two cases can be calculated as below

$$C(P_{\mathcal{R}}) = \begin{cases} \frac{\pi_3}{\pi_3+P_{\mathcal{R}}} \log_2(1+\pi_2 P_{\mathcal{R}}), & \text{if } \pi_2 P_R < \frac{1}{\pi_0+\pi_1 P_{\mathcal{R}}}, \\ \frac{\pi_3}{\pi_3+P_{\mathcal{R}}} \log_2\left(1+\frac{1}{\pi_0+\pi_1 P_{\mathcal{R}}}\right), & \text{otherwise.} \end{cases} \tag{36}$$

In case $\pi_2 P_R < \frac{1}{\pi_0+\pi_1 P_{\mathcal{R}}}$ and combine the power at relay, e.g $P_{\mathcal{R}}$, is the non-negative number, we achieve $0 \le P_{\mathcal{R}} < \frac{\sqrt{\Delta}-\pi_0}{2\pi_1}$.

Taking the derivative of $C(P_R)$ with respect to $P_{\mathcal{R}}$ and setting equal zero, we have

$$\frac{\pi_2}{1+\pi_2 P_{\mathcal{R}}} = \frac{\ln(1+\pi_2 P_{\mathcal{R}})}{\pi_3 + P_{\mathcal{R}}}. \tag{37}$$

After some algebraic manipulations, we have

$$\frac{\pi_2\pi_3 - 1}{e} = \exp\left\{\ln\left(\frac{1+\pi_2 P_{\mathcal{R}}}{e}\right)\right\} \ln\left(\frac{1+\pi_2 P_{\mathcal{R}}}{e}\right). \tag{38}$$

Using the form of Lambert function, (38) can be written as

$$\ln\left(\frac{1+\pi_2 P_{\mathcal{R}}}{e}\right) = \mathcal{W}\left(\frac{\pi_2\pi_3 - 1}{e}\right). \tag{39}$$

With (39), after some algebraic manipulations, we get

$$P_{\mathcal{R}}^{opt} = \frac{e^{\mathcal{W}\left(\frac{\pi_2\pi_3-1}{e}\right)+1} - 1}{\pi_2}. \tag{40}$$

In case 2, where $\frac{1}{\pi_0+\pi_1 P_{\mathcal{R}}} \le \pi_2 P_R$ or $P_{\mathcal{R}} \ge \frac{\sqrt{\Delta}-\pi_0}{2\pi_1}$ the instantaneous throughput is $C(P_{\mathcal{R}}) = \log_2\left(1+\frac{1}{\pi_0+\pi_1 P_{\mathcal{R}}}\right)\frac{\pi_3}{\pi_3+P_{\mathcal{R}}}$. Getting the first coefficient of throughput, $C(P_{\mathcal{R}})$ with respect of $P_{\mathcal{R}}$ variable, we have

$$\frac{\partial\{C(P_{\mathcal{R}})\}}{\partial P_{\mathcal{R}}} = -\frac{\pi_0}{(\pi_3+P_{\mathcal{R}})^2}\log_2\left(1+\frac{1}{\pi_0+\pi_1 P_{\mathcal{R}}}\right) - \frac{1}{\ln(2)}\frac{\pi_1}{(\pi_0+\pi_1 P_{\mathcal{R}})(\pi_0+\pi_1 P_{\mathcal{R}}+1)}\frac{\pi_3}{\pi_3+P_{\mathcal{R}}}. \tag{41}$$

In (41), the derivative expression involves a negative value. Hence, $C(P_{\mathcal{R}})$ is the restrictive function with the increase in variable, $P_{\mathcal{R}}$. Therefore, the optimal power can be obtained as

$$P_{\mathcal{R}}^{opt} = \frac{\sqrt{\Delta} - \pi_0}{2\pi_1}. \tag{42}$$

This is end of proof.

B Proof of Theorem 3

Substituting (9), (20) into (22), then putting $\chi = \frac{P_{\mathcal{R}} d_1^m}{\eta P_{\mathcal{S}}}$, $\omega = \frac{\gamma_0 P_{\mathcal{R}} d_1^m}{P_{\mathcal{S}}}$ and $\vartheta = \frac{\gamma_0 d_1^m \sigma_{\mathcal{R}}^2}{P_{\mathcal{S}}}$, as thus the average throughput of system can be expressed

$$\tau^{DC} = \underbrace{E_{|g_i|^2}\left\{\Pr\left(\frac{P_{\mathcal{R}}|g_i|^2}{d_2^m \sigma_{\mathcal{D},i}^2} > \gamma_0\right)\right\}}_{\tau_1^{DC}} \times \left\{\underbrace{E_{|h_i|^2,|f_i|^2}\left\{\Pr\left(|h_i|^2 > \omega|f_i|^2 + \vartheta\right)\right\}}_{\tau_2^{DC}} - \underbrace{E_{|h_i|^2,|f_i|^2}\left\{\Pr\left(|h_i|^2 > \omega|f_i|^2 + \vartheta\right)\frac{\chi}{|h_i|^2 + \chi}\right\}}_{\tau_3^{DC}}\right\}. \quad (43)$$

The first item could be evaluated with the fact that the $|g_i|^2$ channel experience exponential distribution, yields

$$\begin{aligned}\tau_1^{DC} &= E_{|g_i|^2}\left\{\Pr\left(P_{\mathcal{R}}|g_i|^2 > \gamma_0 d_2^m \sigma_{\mathcal{D}}^2\right)\right\} \\ &= F_{|g_i|^2}\left(\frac{\gamma_0 d_2^m \sigma_{\mathcal{D}}^2}{\lambda_g P_{\mathcal{R}}}\right) = \exp\left(-\frac{\gamma_0 d_2^m \sigma_{\mathcal{D}}^2}{\lambda_g P_{\mathcal{R}}}\right),\end{aligned} \quad (44)$$

where $F_X(x)$ is the cumulative distribution function (cdf) function of random variable X.

And the second item can be obtained as

$$\begin{aligned}\tau_2^{DC} &\triangleq E_{|h_i|^2,|f_i|^2}\left\{\Pr\left(|h_i|^2 > \omega|f_i|^2 + \vartheta\right)\right\} \\ &= \int_0^\infty F_{|h_i|^2}(\omega x + \vartheta) f_{|f_i|^2}(x)\,dx \\ &\overset{(a)}{=} \exp\left(-\frac{\vartheta}{\lambda_h}\right)\frac{m_f^{m_f}}{\Gamma(m_f)\lambda_f^{m_f}}\int_0^\infty x^{m_f-1}\exp\left(-x\left(\frac{\omega}{\lambda_h} + \frac{m_f}{\lambda_f}\right)\right)dx \\ &\overset{(b)}{=} \exp\left(-\frac{\vartheta}{\lambda_h}\right)\left(\frac{\omega\lambda_f}{m_f\lambda_h} + 1\right)^{-m_f},\end{aligned} \quad (45)$$

where $f_X(x)$ is the pdf of random variable X, step (a) is done by $|f_i|^2$ following the gamma distribution and $|h_i|^2$ experience exponential distribution, stage (b) is revealed by using (eq. 3.381.4) in [35].

Finally, the third element is determined as

$$\begin{aligned}\tau_3 &\triangleq E_{|h_i|^2,|f_i|^2}\left\{\Pr\left(\omega|f_i|^2 < |h_i|^2 - \vartheta\right)\frac{\chi}{|h_i|^2 + \chi}\right\} \\ &= \int_\vartheta^\infty \frac{\chi}{x + \chi} F_{|f_i|^2}\left(\frac{x - \vartheta}{\omega}\right) f_{|h_i|^2}(x)\,dx \\ &= \frac{\chi}{\lambda_h}\int_\vartheta^\infty \exp\left(-\frac{x}{\lambda_h}\right)\left(1 - \exp\left(-\frac{m_f(x - \vartheta)}{\omega\lambda_f}\right) \times \sum_{i=0}^{m_f-1}\frac{1}{i!}\left(\frac{m_f}{\omega\lambda_f}\right)^i (x - \vartheta)^i\right)\frac{1}{x + \chi}dx \\ &\overset{(a)}{=} \frac{\chi}{\lambda_h}\left\{-\exp\left(\frac{\chi}{\lambda_h}\right)\mathrm{Ei}\left(-\frac{\chi + \vartheta}{\lambda_h}\right) + \exp\left(\frac{\chi}{\lambda_h}\right)\exp\left(\frac{m_f(\chi + \vartheta)}{\omega\lambda_f}\right) \times \mathrm{Ei}\left(-(\chi + \vartheta)\left(\frac{m_f}{\omega\lambda_f} + \frac{1}{\lambda_h}\right)\right) - \exp\left(\frac{m_f\vartheta}{\omega\lambda_f}\right)\sum_{i=1}^{m_f-1}\frac{1}{i!}\left(\frac{m_f}{\omega\lambda_f}\right)^i \underbrace{\times \int_\vartheta^\infty \exp\left(-x\left(\frac{1}{\lambda_h} + \frac{m_f}{\omega\lambda_f}\right)\right)\frac{(x - \vartheta)^i}{x + \chi}dx}_{\tau_{31}}\right\},\end{aligned} \quad (46)$$

where the step (a) can be derived by applying given in ([35], Eq. 3.352.2). The last item in (46) can be reduced as

$$\begin{aligned}\tau_{31} &\triangleq (-1)^i\vartheta^i\int_\vartheta^\infty \exp\left(-x\left(\frac{1}{\lambda_h} + \frac{m_f}{\omega\lambda_f}\right)\right)\frac{1}{x + \chi}dx \\ &+ \sum_{k=1}^{i}\binom{i}{k}(-1)^i\vartheta^{i-k}\int_\vartheta^\infty \exp\left(-x\left(\frac{1}{\lambda_h} + \frac{m_f}{\omega\lambda_f}\right)\right)\frac{x^k}{x + \chi}dx \\ &= (-1)^{i+1}\vartheta^i\exp\left(\frac{\chi}{\lambda_h}\right)\mathrm{Ei}\left(-(\chi + \vartheta)\left(\frac{1}{\lambda_h} + \frac{m_f}{\omega\lambda_f}\right)\right) \\ &+ \sum_{k=1}^{i}\binom{i}{k}(-1)^i\vartheta^{i-k}\int_\vartheta^\infty \exp\left(-x\left(\frac{1}{\lambda_h} + \frac{m_f}{\omega\lambda_f}\right)\right)\frac{x^k}{x + \chi}dx.\end{aligned} \quad (47)$$

Substituting (44), (45), (46), and (47) into (43), Theorem 3 is simply derived. This ends the proof.

C Proof of Theorem 4

Substituting (12), (20), into (22) the average throughput of system can be expressed

$$\tau^{DC} = \underbrace{E_{|g_i|^2}\left\{\Pr\left(\frac{P_{\mathcal{R}}|g_i|^2}{d_2^m\sigma_{\mathcal{D},i}^2} > \gamma_0\right)\right\}}_{\tau_1^{DC}} \times \left\{\underbrace{E_{|h_i|^2,|f_i|^2}\left\{\Pr\left(|h_i|^2 > \omega|f_i|^2 + \vartheta\right)\right\}}_{\tau_2^{DC}} - \underbrace{E_{|h_i|^2,|f_i|^2,|k_i|^2}\left\{\Pr\left(|h_i|^2 > \omega|f_i|^2 + \vartheta\right)\frac{\chi}{|h_i|^2+|k_i|^2+\chi}\right\}}_{\tau_4^{DC}}\right\}. \tag{48}$$

The first and second elements (i.e. τ_1^{DC}, τ_2^{DC}) can be obtained as (44), (45) in Appendix 7, respectively. And the third item can be expressed as

$$\begin{aligned}
\tau_4^{DC} &\triangleq {}_{|h_i|^2,|f_i|^2,|k_i|^2}\left\{\Pr\left(\omega|f_i|^2 < |h_i|^2 - \vartheta\right)\frac{\chi}{|k_i|^2+|h_i|^2+\chi}\right\}\\
&= \int_{\vartheta}^{\infty} F_{|f_i|^2}\left(\frac{x-\vartheta}{\omega}\right) f_{|h_i|^2}(x) \int_0^{\infty} \frac{\chi}{t+x+\chi} f_{|k_i|^2}(t)\, dt dx\\
&= \frac{\chi}{\lambda_h}\int_{\vartheta}^{\infty} F_{|f_i|^2}\left(\frac{x-\vartheta}{\omega}\right)\exp\left(\frac{-x}{\lambda_h}\right)\exp\left(\frac{x+\chi}{\lambda_k}\right)\mathrm{E}_1\left(\frac{x+\chi}{\lambda_k}\right)dx\\
&= \frac{\chi}{\lambda_h}\exp\left(\frac{\chi}{\lambda_k}\right)\times\left\{\tau_{41} - \exp\left(\frac{m_f(\vartheta+\chi)}{\omega\lambda_f}\right)\times(\tau_{42}+\tau_{43})\right\},
\end{aligned} \tag{49}$$

in which the terms τ_{41} and τ_{42} can be derived as

$$\begin{aligned}
\tau_{41} &\triangleq \int_{\vartheta+\chi}^{\infty} \mathrm{E}_1\left(\frac{t}{\lambda_k}\right)dt = \int_0^{\infty}\mathrm{E}_1\left(\frac{t}{\lambda_k}\right)dt - \int_0^{\vartheta+\chi}\mathrm{E}_1\left(\frac{t}{\lambda_k}\right)dt\\
&= \lambda_k\exp\left(-\frac{\vartheta+\chi}{\lambda_k}\right) + (\vartheta+\chi)\,\mathrm{E}_1\left(\frac{\vartheta+\chi}{\lambda_k}\right).
\end{aligned} \tag{50}$$

The integral can be derived with help of (Eq. 5.221.8) and (Eq. 6.224.1) given in [35] and

$$\begin{aligned}
\tau_{42} &\triangleq \int_{\vartheta+\chi}^{\infty}\exp\left(-\frac{m_f t}{\omega\lambda_f}\right)\mathrm{E}_1\left(\frac{t}{\lambda_k}\right)dt\\
&= \int_0^{\infty}\exp\left(-\frac{m_f t}{\omega\lambda_f}\right)\mathrm{E}_1\left(\frac{t}{\lambda_k}\right)dt - \int_0^{\vartheta+\chi}\exp\left(-\frac{m_f t}{\omega\lambda_f}\right)\mathrm{E}_1\left(\frac{t}{\lambda_k}\right)dt\\
&= \frac{\omega\lambda_f}{m_f}\left\{\exp\left(-\frac{m_f(\vartheta+\chi)}{\omega\lambda_f}\right)\mathrm{E}_1\left(\frac{\vartheta+\chi}{\lambda_k}\right)\right.\\
&\quad\left. -\mathrm{E}_1\left((\vartheta+\chi)\left(\frac{1}{\lambda_k}+\frac{m_f}{\omega\lambda_f}\right)\right)\right\},
\end{aligned} \tag{51}$$

and

$$\begin{aligned}
\tau_{43} &\triangleq \int_{\vartheta+\chi}^{\infty}\exp\left(-\frac{m_f t}{\omega\lambda_{f_r}}\right)\sum_{i=1}^{m_f-1}\frac{1}{i!}\left(\frac{m_f}{\omega\lambda_f}\right)^i (t-(\vartheta+\chi))^i\,\mathrm{E}_1\left(\frac{t}{\lambda_k}\right)dt\\
&= \sum_{i=1}^{m_f-1}\frac{1}{i!}\left(\frac{m_f}{\omega\lambda_f}\right)^i\left\{(-1)^i(\vartheta+\chi)^i\underbrace{\int_{\vartheta+\chi}^{\infty}\exp\left(-x\left(\frac{1}{\lambda_h}+\frac{m_f}{\omega\lambda_f}\right)\right)\mathrm{E}_1\left(\frac{t}{\lambda_k}\right)dx}_{\tau_{431}}\right.\\
&\quad\left.+\sum_{k=1}^{i}\binom{i}{k}(-(\vartheta+\chi))^{i-k}\int_{\vartheta+\chi}^{\infty}t^k\exp\left(-x\left(\frac{1}{\lambda_h}+\frac{m_f}{\omega\lambda_f}\right)\right)\mathrm{E}_1\left(\frac{t}{\lambda_k}\right)dx\right\},
\end{aligned} \tag{52}$$

and

$$\begin{aligned}
\tau_{431} &\triangleq \int_{\vartheta+\chi}^{\infty}\exp\left(-x\left(\frac{1}{\lambda_h}+\frac{m_f}{\omega\lambda_f}\right)\right)\mathrm{E}_1\left(\frac{t}{\lambda_k}\right)dx\\
&= \frac{\omega\lambda_f\lambda_h}{\omega\lambda_f+m_f\lambda_h}\left\{\exp\left(-\left(\frac{1}{\lambda_h}+\frac{m_f}{\omega\lambda_f}\right)(\vartheta+\chi)\right)\right.\\
&\quad\left.\times\mathrm{E}_1\left(\frac{\vartheta+\chi}{\lambda_k}\right) - \mathrm{E}_1\left((\vartheta+\chi)\left(\frac{1}{\lambda_k}+\frac{1}{\lambda_h}+\frac{m_f}{\omega\lambda_f}\right)\right)\right\},
\end{aligned} \tag{53}$$

where the last integral can be derived by applying (Eq. 3.352.2) given in [35] To this end, pulling everything together and after some simple mathematical manipulations, Theorem 4 is derived. This is the end of the proof.

Funding
The author declares that there is no fund of this work.

Authors' contributions
The individual contributions of each authors to the manuscript are the same. Both authors read and approved the final manuscript.

Competing interests
The author declares that there is no competing interests.

References

1. H Chingoska, Z Hadzi-Velkov, I Nikoloska, N Zlatanov, Resource Allocation in Wireless Powered Communication Networks With Non-Orthogonal Multiple Access. IEEE Wirel. Commun. Lett. **5**(6), 684–687 (2016)
2. D-T Do, Energy-Aware Two-Way Relaying Networks under Imperfect Hardware: Optimal Throughput Design and Analysis. Telecommun. Syst. (Springer). **62**(2), 449–459 (2015)
3. LR Varshney, in *Proc. of IEEE International Symposium in Information Theory.* Transporting information and energy simultaneously, (Toronto, 2008), pp. 1612–1616

4. X Zhou, R Zhang, CK Ho, Wireless information and power transfer: Architecture design and rate-energy trade-off. IEEE Trans. Wirel. Commun. **61**(11), 4754–4767 (2013)
5. AA Nasir, X Zhou, S Durrani, RA Kennedy, Relaying protocols for wireless energy harvesting and information processing. IEEE Trans. Wirel. Commun. **12**(7), 3622–3636 (2013)
6. R Zhang, CK Ho, MIMO broadcasting for simultaneous wireless information and power transfer. IEEE Trans. Wirel. Commun. **12**(5), 1989–2001 (2013)
7. X Chen, C Yuen, Z Zhang, Wireless energy and information transfer tradeoff for limited feedback multi-antenna systems with energy beamforming. IEEE Trans. Veh. Technol. **63**(1), 407–412 (2014)
8. KT Nguyen, DT Do, XX Nguyen, NT Nguyen, DH Ha, in *Proc. of AETA 2015: Recent Advances in Electrical Engineering and Related Sciences*. Wireless information and power transfer for full duplex relaying networks: performance analysis, (Ho Chi Minh, 2015), pp. 53–62
9. HS Nguyen, DT Do, M Voznak, Two-Way Relaying Networks in Green Communications for 5G: Optimal Throughput and Trade-off between Relay Distance on Power Splitting-based and Time Switching-based Relaying SWIPT. AEU Int. J. Electron. Commun. **70**(3), 1637–1644 (2016)
10. DT Do, HS Nguyen, A Tractable Approach to Analyze the Energy-Aware Two-way Relaying Networks in Presence of Co-channel Interference. EURASIP J. Wirel. Commun. Netw. **271** (2016)
11. DWK Ng, ES Lo, R Schober, Wireless information and power transfer: energy efficiency optimization in OFDMA systems. IEEE Trans. Wirel. Commun. **12**(12), 6352–6370 (2013)
12. X Chen, C Yuen, Z Zhang, Wireless energy and information transfer tradeoff for limited feedback multi-antenna systems with energy beamforming. IEEE Trans.Veh. Technol. **63**(1), 407–412 (2014)
13. G Yang, CK Ho, YL Guan, Dynamic resource allocation for multiple-antenna wireless power transfer. IEEE Trans. Signal Process. **62**(14), 3565–3577 (2014)
14. DS Michalopoulos, HA Suraweera, R Schober, Relay selection for simultaneous information transmission and wireless energy transfer: A tradeoff perspective. IEEE J. Sel. Areas Commun. **33**(8), 1578–1594 (2015)
15. Z Ding, I Krikidis, B Sharif, HV Poor, Wireless information and power transfer in cooperative networks with spatially random relays. IEEE Trans. Wirel. Commun. **13**(8), 4440–4453 (2014)
16. I Krikidis, Relay selection in wireless powered cooperative networks with energy storage. IEEE J. Sel. Areas Commun. **33**(12), 2596–2610 (2015)
17. L Chen, S Han, W Meng, C Li, Optimal Power Allocation for Dual-Hop Full-Duplex Decode-and-Forward Relay. IEEE Commun. Lett. **19**(3), 471–474 (2015)
18. M Duarte, C Dick, A Sabharwal, Experiment-driven characterization of full-duplex wireless systems. IEEE Trans. Wirel. Commun. **11**(12), 4296–4307 (2012)
19. Y Hua, P Liang, Y Ma, AC Cirik, Q Gao, A method for broadband full-duplex MIMO radio. IEEE Signal Process. Lett. **19**(12), 793–796 (2012)
20. H Liu, KJ Kim, KS Kwak, HV Poor, Power Splitting-Based SWIPT With Decode-and-Forward Full-Duplex Relaying. IEEE Trans. Wirel. Commun. **15**(11), 7561–7577 (2016)
21. I Krikidis, HA Suraweera, PJ Smith, C Yuen, Full-duplex relay selection for amplify-and-forward cooperative networks. IEEE Trans. Wirel. Commun. **11**(12), 4381–4393 (2012)
22. K Tutuncuoglu, A Yener, Optimum transmission policies for battery limited energy harvesting nodes. IEEE Trans. Wirel. Commun. **11**(3), 1180–1189 (2012)
23. J Yang, S Ulukus, Optimal packet scheduling in an energy harvesting communication system. IEEE Trans. Commun. **60**(1), 220–230 (2012)
24. F Yuan, Q Zhang, S Jin, H Zhu, Optimal harvest-use-store strategy for energy harvesting wireless systems. IEEE Trans. Wirel. Commun. **14**(2), 698–710 (2015)
25. J Xu, R Zhang, Throughput optimal policies for energy harvesting wireless transmitters with non-ideal circuit power. IEEE J. Select. Areas Commun. **32**(2), 322–332 (2014)
26. Z Chen, Y Dong, P Fan, KB Letaief, Optimal Throughput for Two-Way Relaying: Energy Harvesting and Energy Co-Operation. IEEE J. Sel. Areas Commun. **34**(5), 1448–1462 (2016)
27. C Zhong, HA Suraweera, G Zheng, I Krikidis, Z Zhang, Wireless information and power transfer with full duplex relaying. IEEE Trans. Commun. **62**(10), 3447–3461 (2014)
28. Y Zeng, R Zhang, Full-duplex wireless-powered relay with self-energy recycling. IEEE Wirel. Commun. Lett. **4**(2), 201–204 (2015)
29. M Mohammadi, BK Chalise, HA Suraweera, C Zhong, G Zheng, I Krikidis, Throughput analysis and optimization of wireless-powered multiple antenna full-duplex relay systems. IEEE Trans. Commun. **64**(4), 1769–1785 (2016)
30. Z Ding, C Zhong, DWK Ng, M Peng, HA Suraweera, R Schober, HV Poor, Application of smart antenna technologies in simultaneous wireless information and power transfer. IEEE Commun. Mag. **53**(4), 86–93 (2015)
31. X Ji, J Xu, YL Che, Z Fei, R Zhang, Adaptive Mode Switching for Cognitive Wireless Powered Communication Systems. IEEE Wirel. Commun. Letters. **6**(3), 386–389 (2017)
32. H Ding, X Wang, DB da Costa, Y Chen, F Gong, Adaptive Time-Switching Based Energy Harvesting Relaying Protocols. IEEE Trans. Commun. **65**(7), 2821–2837 (2017)
33. AA Nasir, X Zhou, S Durrani, RA Kennedy, Wireless-powered relays in cooperative communications: Time-switching relaying protocols and throughput analysis. IEEE Trans. Commun. **63**(5), 1607–1622 (2015)
34. RM Corless, GH Gonnet, DE Hare, DJ Jeffrey, DE Knuth, On the Lambert W function. Adv. Comput. Math. **5**(1), 329–359 (1996)
35. A Jeffrey, D Zwillinger, *Table of integrals, series, and products*, 7th edn. (Academic Press, New York, 2007)

7

Individual secrecy of fading homogeneous multiple access wiretap channel with successive interference cancellation

Kaiwei Jiang[1,2*], Tao Jing[1], Zhen Li[1], Yan Huo[1] and Fan Zhang[1]

Abstract

We investigate individual secrecy performance in a K-user quasi-static Rayleigh fading homogeneous multiple access wiretap channel (MAC-WT), where a legitimate receiver employs successive interference cancellation (SIC) decoding. We first evaluate individual secrecy performance under an arbitrary SIC order by deriving closed-form expressions with respect to secrecy outage probability and effective secrecy throughput (EST) as main metrics. The resulting closed-form expressions disclose a significant impact on the secrecy performance from the order of SIC decoding. Therefore, we propose three SIC decoding order scheduling schemes: (1) round-robin scheme, absolutely fair and served as a benchmark; (2) suboptimal scheme, based on each user's main channel condition; and (3) optimal scheme, based on each user's achievable secrecy rate. Comparison results show that the last two schemes outperform the first one with regard to both the EST and the multi-user diversity gain, whereas the performance of the suboptimal scheme is highly close to that of the optimal scheme which is usually impractical due to a requirement for the eavesdropper's channel state information (CSI).

Keywords: Multiple access wiretap channel, Successive interference cancellation, Secrecy performance, Scheduling scheme

1 Introduction

Since Wyner introduced the notion of the *wiretap channel* in his seminal work [1], the studies of keyless physical-layer secrecy transmission have made tremendous progress. He initially characterized a rate-equivocation region for a degraded wiretap channel, which was extended to a more general non-degraded version in [2]. The Gaussian wiretap channel was investigated in [3], which confirmed that the secrecy capacity is the difference between the capacities of main and eavesdropper channels. Thus, a more favorable main channel means the existence of a positive secrecy capacity. However, during wireless transmission, fading channels rather than stationary channels with Gaussian noise are a natural situation, where a more favorable main channel is not always guaranteed. Consequently, researchers characterized another two secrecy metrics, *secrecy outage probability* and *ergodic secrecy capacity* [4, 5]. Besides, one interesting measure, *effective secrecy throughput* (EST), was introduced in [6] by Nan Yang et al. evaluating an average secrecy rate at which messages are transmitted to a legitimate receiver confidentially.

A great deal of research work addressed how to improve secrecy performance, some from the perspective of precoding and beamforming [7–10], some via the utilization of artificial noise or jamming [11–14], and some jointly applying these approaches [15–18]. Overall, the purpose of exploiting these techniques is to enlarge the capacity difference between main and eavesdropper channels. One can further refer to a recent survey of physical-layer security which has been investigated from information theory to security engineering in [19].

Inspired by a capacity improvement on an individual user in a multi-user system with a successive interference cancellation (SIC) receiver [20], in this work, we consider a secrecy scenario as depicted in Fig. 1, where K transmitters send confidential messages to a legitimate receiver

*Correspondence: kwjiang@bjtu.edu.cn
[1]School of Electronics and Information Engineering, Beijing Jiaotong University, 100044 Beijing, China
[2]School of Electric and Information Engineering, Taizhou Vocational and Technical College, 318000 Taizhou, China

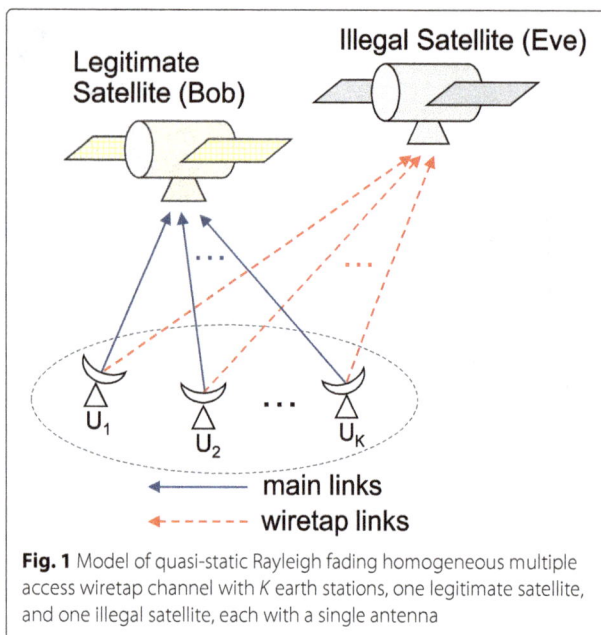

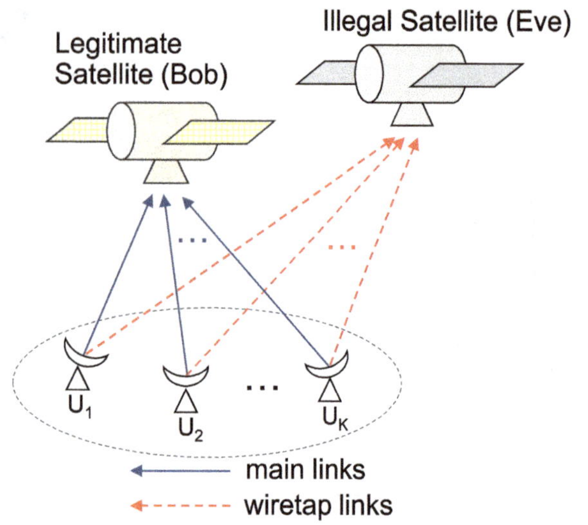

Fig. 1 Model of quasi-static Rayleigh fading homogeneous multiple access wiretap channel with *K* earth stations, one legitimate satellite, and one illegal satellite, each with a single antenna

who employs SIC decoding with the existence of an eavesdropper. We investigate the impacts of the SIC order and multi-user diversity on individual secrecy performance and propose three SIC order scheduling schemes.

The system model in this work is a typical multiple access wiretap channel (MAC-WT)[1], which has been intensively studied recently. However, all of these references [21–34] considered problems from an information-theoretical perspective and characterized the setting as a whole. Ender Tekin et al. characterized secrecy rate regions and an upper bound for the secrecy sum-rate for a Gaussian MAC-WT via the Gaussian encoding in [21]. They also addressed a fading MAC-WT in [22], providing achievable ergodic secrecy rate regions as well as its outer bounds and giving optimal power allocations to maximize the sum-rate. Hassan Zivari-Fard et al. [23] investigated a 2-user MAC-WT with a common message which can be decoded by both the legitimate receiver and the eavesdropper, and derived general inner and outer bounds on a secrecy capacity region for both discrete memoryless and Gaussian cases. The authors further addressed this type of MAC-WT (with a common message) in [24], where only one transmitter's private messages should be kept secret from the adversary. They derived general inner and outer bounds for both imperfect and perfect secrecy conditions for the adversary under discrete memoryless, less noisy, and Gaussian versions of the MAC-WT. Yet, the secure degrees of freedom (DoF) of such Gaussian signaling-based schemes are zero, which means the schemes are suboptimal and leads to further work for new encoding techniques. One commonly used encoding technique is the real interference alignment [25–27]. Other techniques, e.g., uniform distributed source coding and polar coding, are also developed and applied for the MAC-WT [28–30]. Moreover, two alignment techniques, *scaling-based alignment* (SBA), and *ergodic secret alignment* (ESA), were proposed in [31] to achieve an ergodic secrecy rate region for a 2-user fading MAC-WT. The SBA performs repetition coding at two consecutive time instances while the ESA performs repetition coding at two carefully chosen time instances. Another research direction of interest on the MAC-WT is secure DoF, which was also fully studied in recent years [25, 32–34]. Xie and Ulukus [25] derived the sum secure DoF of a *K*-user Gaussian MAC-WT. The entire secure DoF region of the *K*-user Gaussian MAC-WT was determined in [32]. Mukherjee and Ulukus [33] considered the secure DoF of the case with no eavesdropper's channel state information (CSI). The results showed the sum secure DoF is less than that of the case with a subset or all of the transmitters knowing the eavesdropper's CSI. The secure DoF of a MIMO MAC-WT was studied in [34], showing that the optimal sum secure DoF is affected by the number of eavesdropper antennas.

The problems in this work are also different from the existing literature regarding multi-user uplink wiretap channels [35–37]. Jin et al. [35] and Zou et al. [36] ignored cochannel interference deliberately in the system models. Although cochannel interference was considered in [37], where uplink secure communications in a cellular network are studied, such cochannel interference is not from the users in a cell but from neighboring base stations.

Secrecy problems related to SIC were studied in [38], where the authors only considered the case of two transmitter-receiver pairs and tried to find the coordinated beamforming vectors at the transmitters to reach an ergodic secrecy rate balancing point. In [39], we investigated individual secrecy performance for a *K*-user MAC-WT, where the legitimate receiver possesses multiple antennas and employs two specific decoding methods, zero-forcing (ZF), and minimum mean-square error (MMSE), jointly with SIC. For simplicity, we ignored the cochannel interference to the eavesdropper during modeling. We later complicated the eavesdropper setting from a single-antenna to multi-antenna scenario in [40], where, besides the derivations of individual secrecy performance, we further proposed one SIC order scheduling scheme based on each user's relative distance to the eavesdropper over the legitimate receiver, and gave a solution to the problem of uplink optimal power allocation.

In this paper, we continue to study individual secrecy performance for the *K*-user MAC-WT with SIC. We take into account the cochannel inference to the eavesdropper for evaluating the secrecy performance. Moreover, we propose three new SIC decoding order scheduling schemes. Both the main and eavesdropper channels are assumed to experience quasi-static Rayleigh fading, just as in [39, 40]. We further assume the users are homogeneous, i.e., all the users experience the same received power on average at each of the receivers (the legitimate receiver and the eavesdropper). We call such a network as the homogeneous MAC-WT, which is assumed throughout this paper. Similar to [39, 40], secrecy outage probability and effective secrecy throughput are utilized as our key measures in evaluating the secrecy performance.

It is worthwhile to mention that an assumption of homogeneous (also call "symmetric" in some literature) networks is widely used to address wireless communication issues. In [41], homogeneous users were assumed in the system model to address a secrecy issue of multi-user downlinks. Reid et al. [42], Nitinawarat [43], Chou et al. [44] studied related wireless communication problems based on symmetric multiple access channels. A 2-user symmetric MAC-WT was investigated in [45], while reference [46] modeled a K-user symmetric MIMO MAC-WT for analyzing secure DoF.

To demonstrate how the SIC order impacts the individual secrecy performance, we derive the closed-form expressions of the secrecy outage probability and the effective secrecy throughput with a specified SIC order. Observing these closed-from expressions, we can assess the impact of the SIC order.

To compare these three SIC order scheduling schemes, we first make succinct comparisons of the EST of the *best* case (i.e., no cochannel interference) for each scheme, since the performance of the best case predominates the other cases, which will be seen clearly later in this paper. Then, with the aid of simulations, we make full performance comparisons of these schemes regarding the maximum sum-EST and the multi-user diversity gain.

The main contributions of the paper are summarized as follows:

1) Model the fading homogeneous MAC-WT and derive closed-form expressions of the cumulative distribution function (CDF) of signal-to-interference-plus-noise ratios (SINRs) for any individual transmitter at the legitimate receiver and the eavesdropper, respectively.
2) Evaluate the impacts of the SIC order and multi-user diversity on the secrecy performance in terms of secrecy outage probability, effective secrecy throughput, and other secrecy performance metrics by deriving the corresponding closed-form expressions.
3) Propose and evaluate three SIC decoding order scheduling schemes, namely, round-robin scheme, suboptimal scheme, and optimal scheme.

The rest of the paper is organized as follows: in Section 2, we model the quasi-static Rayleigh fading homogeneous MAC-WT. Section 3 investigates the individual secrecy performance. Three SIC order scheduling schemes are proposed in Section 4. Numerical results and further discussions are presented in Section 5. Finally, Section 6 concludes the work.

Notation: $\mathcal{A}\backslash\mathcal{B}$ denotes set $\mathcal{A}$ minus set $\mathcal{B}$. $\mathcal{CN}$, Exp, and χ^2_m specify the circular symmetric complex Gaussian distribution, the exponential distribution and the chi-squared distribution with m degrees of freedom, respectively. log denotes the base 2 logarithm. $[\cdot]^+ = \max(0, \cdot)$. $\mathbf{E}(\cdot)$ specifies the expectation operator.

2 Fading homogeneous MAC-WT modeling

As illustrated in Fig. 1, K earth stations ($U_1, \ldots, U_K$) are intended to transmit their confidential messages to a legitimate satellite (Bob) through main multiple access links, while an illegal satellite (Eve) attempts to intercept data from a specific user U_k, $k \in \mathcal{K} \triangleq \{1, \ldots, K\}$, through a corresponding wiretap link with the existence of cochannel interference. Assuming that all of the nodes (K users, i.e., earth stations, Bob, and Eve) are equipped with a single antenna, the main channel coefficient between U_k and Bob is denoted as h_k, while g_k denotes the eavesdropper channel coefficient between U_k and Eve. We suppose Bob has instantaneous knowledge of the realization of h_k, but only has the statistic of g_k. The signal transmitted from U_k is denoted as x_k, which has an identical average power constraint of P for all $k \in \mathcal{K}$.

Therefore, the instantaneous composite signals received at Bob and Eve can be formulated as

$$y_b = \sum_{i=1}^{K} h_i x_i + w_b, \tag{1}$$

$$y_e = \sum_{i=1}^{K} g_i x_i + w_e, \tag{2}$$

where w_b and w_e are circular symmetric complex Gaussian random variables with the variances of N_b and N_e, respectively, i.e., $w_b \sim \mathcal{CN}(0, N_b)$ and $w_e \sim \mathcal{CN}(0, N_e)$.

We assume both the main and wiretap links experience quasi-static Rayleigh fading.[2] As such, $|h_k|^2$ and $|g_k|^2$ are exponentially distributed, the means of which are denoted as δ_k^2 and σ_k^2, respectively.

We further assume all the K users are situated in a small region (e.g., a diameter of several kilometers) relative to the distances between the users and both satellites (maybe several hundred or thousand kilometers). Such a scenario may exist in a military base or a television station where multiple earth stations need to transmit their signals to a relay satellite. The assumption implies all the users are approximately equidistant from each of the satellites. Thus, all these users can be considered to be homogeneous, i.e., the average power received at Bob/Eve from each user is the same, as long as they have identical average transmit power.

Interestingly, in this case, the K single-antenna users can also correspond to one K-antenna transmitter, where each antenna transmits one independent data stream. Therefore, the system can also be regarded as a multiple-input-single-output (MISO) wiretap channel, where the eavesdropper attempts to intercept messages from one specific data stream. Different from issues of the MISO wiretap channel in [6, 47–49], we focus on the impact of the SIC order on the individual secrecy performance (i.e., secrecy performance of each data stream in this equivalent MISO wiretap channel model).

2.1 Achievable individual secrecy rate with SIC

In traditional multi-user decoding, a receiver decodes each user's data by treating all other users' data as noise. Such a method is not capacity-achieving. In this network model, Bob employs SIC decoding by subtracting already-decoded data from the composite signal, reducing the amount of cochannel interference for the next user's decoding. Therefore, the main capacity of an individual user is determined not only by its signal-to-noise ratio (SNR) but also by its own order in SIC decoding. When a user is decoded at first, it is just the same as the traditional way of decoding for that user, achieving the *worst* case. In contrast, it achieves the *best* case, i.e., no cochannel interference, if it is decoded at last. In general, we can express the main capacity of an individual user (i.e., U_k) with SIC as

$$C_{b,k}^{(\Im)} = \log\left(1 + \frac{\xi_k}{1+\sum_{j\in\Im}\xi_j}\right) = \log\left(1+\gamma_k^{(\Im)}\right), \quad (3)$$

where $\xi_k = \frac{|h_k|^2 P}{N_b}$ ($\forall k \in \mathcal{K}$) is the instantaneous SNR of U_k at Bob with the mean of $\overline{\xi_k} = \frac{\delta_k^2 P}{N_b} = 1/\lambda_k$; $\Im \subseteq \mathcal{K}\backslash\{k\}$, implying U_k is decoded just before the users in the set $\Im$ (index) during SIC decoding; and $\gamma_k^{(\Im)}$ denotes the SINR of U_k at Bob, the superscript of which indicates where the cochannel interference comes from.

It is reasonable to assume that Eve cannot do any SIC decoding before it is able to intercept a message successfully from any user. Consequently, the wiretap capacity for U_k can be expressed as

$$C_{e,k} = \log\left(1 + \frac{\eta_k}{1+\sum_{j\neq k}\eta_j}\right) = \log\left(1+\rho_k\right), \quad (4)$$

where $\eta_k = \frac{|g_k|^2 P}{N_e}$ ($\forall k \in \mathcal{K}$) is the instantaneous SNR of U_k at Eve with the mean of $\overline{\eta_k} = \frac{\sigma_k^2 P}{N_e} = 1/\mu_k$, and ρ_k denotes the SINR of U_k at Eve.

Thereby, in this MAC-WT channel, the instantaneous achievable individual secrecy rate with SIC can be formulated as [21, 50]

$$C_{s,k}^{(\Im)} = \left[C_{b,k}^{(\Im)} - C_{e,k}\right]^{+}. \quad (5)$$

2.2 Statistics of SINRs

Since the derivation of the secrecy outage probability requires the statistics of $\gamma_k^{(\Im)}$ and ρ_k, we decide to derive them in advance in this subsection.

Before deriving the statistic of $\gamma_k^{(\Im)}$, we first investigate the random variables $\xi_1, \xi_2, \ldots, \xi_K$. As $\xi_k \propto |h_k|^2$ ($\forall k \in \mathcal{K}$), the random variables $(\xi_1, \xi_2, \ldots, \xi_K)$ are mutually independent and exponentially distributed, i.e., $\xi_k \sim \text{Exp}(\lambda_k)$ ($\forall k \in \mathcal{K}$). The CDF of $\gamma_k^{(\Im)}$ can be obtained,

$$F_{\gamma_k^{(\Im)}}(\gamma_k) = \begin{cases} 1 - \frac{\prod_{i\in\Im}\lambda_i}{\prod_{i\in\Im}(\lambda_i+\lambda_k\gamma_k)}e^{-\lambda_k\gamma_k}, & \gamma_k \geq 0 \\ 0, & \gamma_k < 0. \end{cases} \quad (6)$$

Similarly, the random variables $\eta_1, \eta_2, \ldots, \eta_K$ are exponentially and independently distributed, i.e., $\eta_k \sim \text{Exp}(\mu_k)$ ($\forall k \in \mathcal{K}$). The CDF of ρ_k can be formulated in the same fashion,

$$F_{\rho_k}(\rho_k) = \begin{cases} 1 - \frac{\prod_{i\neq k}\mu_i}{\prod_{i\neq k}(\mu_i+\mu_k\rho_k)}e^{-\mu_k\rho_k}, & \rho_k \geq 0 \\ 0, & \rho_k < 0. \end{cases} \quad (7)$$

The detailed derivations of (6) and (7) can be found in Appendix A.

Due to the homogeneousness, $\lambda_1 = \lambda_2 = \ldots = \lambda_K$ and $\mu_1 = \mu_2 = \ldots = \mu_K$. Thus, (6) and (7) can be further simplified to

$$F_{\gamma^{(n)}}(\gamma) = 1 - \frac{e^{-\lambda\gamma}}{(1+\gamma)^n}, \quad \gamma \geq 0 \quad (8)$$

$$F_{\rho}(\rho) = 1 - \frac{e^{-\mu\rho}}{(1+\rho)^{K-1}}, \quad \rho \geq 0. \quad (9)$$

Here, the subscript k is omitted, as each user has the same attributes, and the superscript $\Im$ is changed into its cardinality n, i.e., the number of cochannel interferers.

Their probability density functions (PDFs) can be yielded by differentiating the related CDFs in (8) and (9), respectively.

$$f_{\gamma^{(n)}}(\gamma) = \frac{(1+\gamma)\lambda + n}{(1+\gamma)^{n+1}}e^{-\lambda\gamma}, \quad \gamma > 0 \quad (10)$$

$$f_{\rho}(\rho) = \frac{(1+\rho)\mu + K - 1}{(1+\rho)^{K}}e^{-\mu\rho}, \quad \rho > 0. \quad (11)$$

Notably, the subscript k of notations will be automatically omitted in the rest of the paper if necessary.

3 Individual secrecy performance with SIC

This section investigates the individual secrecy performance, termed secrecy performance for short hereinafter, under an arbitrary SIC order in terms of *secrecy outage probability*, *effective secrecy throughput*, and some other secrecy performance metrics. The closed-form expressions are derived and related performance analysis is explored.

3.1 Secrecy outage probability

The secrecy outage probability is the probability of the achievable secrecy rate that is less than a predefined secrecy rate R_s,

$$P_{\text{so}}^{(n)}(R_s) = \Pr\left(C_s^{(n)} < R_s\right). \tag{12}$$

Here, $C_s^{(n)}$ denotes the instantaneous achievable individual secrecy rate with n cochannel interferers, and its prototype definition can be found in (5).

Due to the independence between the main and wiretap links, the random variables $\gamma^{(n)}$ and ρ are also independent. Along with (8), (9), (10) and (11), the secrecy outage probability can be derived as follows:

$$\begin{aligned}
P_{\text{so}}^{(n)}(R_s) &= \Pr\left(\frac{1+\gamma^{(n)}}{1+\rho} < 2^{R_s}\right) \\
&= \int_0^\infty \int_0^{2^{R_s}\rho+2^{R_s}-1} f_\rho(\rho) f_{\gamma^{(n)}}(\gamma)\, d\gamma\, d\rho \\
&= \int_0^\infty f_\rho(\rho) F_{\gamma^{(n)}}\left(2^{R_s}\rho + 2^{R_s} - 1\right) d\rho \\
&= \int_0^\infty f_\rho(\rho)\, d\rho - \int_0^\infty f_\rho(\rho) \frac{e^{-\lambda(2^{R_s}\rho+2^{R_s}-1)}}{\left(2^{R_s}+2^{R_s}\rho\right)^n} d\rho \\
&= 1 - \int_0^\infty \frac{(1+\rho)\mu + K - 1}{(1+\rho)^K} e^{-\mu\rho} \frac{e^{-\lambda(2^{R_s}\rho+2^{R_s}-1)}}{\left(2^{R_s}+2^{R_s}\rho\right)^n} d\rho \\
&= 1 - \underbrace{\frac{e^{-\lambda(2^{R_s}-1)}}{2^{nR_s}}}_{I_1} \underbrace{\{\Theta_1(n) + \Theta_2(n)\}}_{I_2},
\end{aligned} \tag{13}$$

where

$$\begin{aligned}
\Theta_1(n) &= \mu \mathbf{U}\left(K+n-1, \mu + 2^{R_s}\lambda\right), && (14)\\
\Theta_2(n) &= (K-1)\, \mathbf{U}\left(K+n, \mu + 2^{R_s}\lambda\right). && (15)
\end{aligned}$$

The function $\mathbf{U}(k,\theta)$ has the following definition,

$$\mathbf{U}(k,\theta) = \int_0^\infty \frac{e^{-\theta x}}{(1+x)^k} dx. \tag{16}$$

The integral result of this function can be looked up from ([51], Eq. (3.353.2)),

$$\mathbf{U}(k,\theta) = \frac{\sum_{j=1}^{k-1} (j-1)!(-\theta)^{k-j-1} - (-\theta)^{k-1} e^\theta \mathrm{Ei}(-\theta)}{(k-1)!}. \tag{17}$$

Here, $\mathrm{Ei}(x)$ is the exponential integral function with the definition

$$\mathrm{Ei}(x) = \int_{-\infty}^x \frac{e^t}{t} dt. \tag{18}$$

By observing (16), it is easy to find that $\mathbf{U}(k,\theta)$ is a decreasing function of θ and/or k. Particularly, $\mathbf{U}(k,\theta) \to 0$ as k or $\theta \to \infty$. Since the term I_2 in (13) is a non-negative linear combination of the decreasing function $\mathbf{U}(k,\theta)$, it decreases with n or R_s increasing by fixing the other arguments. So does the term I_1. Moreover, the item I_1 has an exponential decay with either n or R_s increasing.

Thereby, we can attain that $P_{\text{so}}^{(n)}(R_s)$ is an increasing function of n and/or R_s.

In the path-loss model, we realize $\lambda \propto d_b^\alpha$ and $\mu \propto d_e^\alpha$ (here, suppose $N_b = N_e$), where d_b denotes the straight-line distance between an individual user and Bob, d_e is each user's distance to Eve, and α is the path-loss exponent. μ can then be transformed into the form of $(d_e/d_b)^\alpha \lambda$. Substituting it into the corresponding expressions in (14) and (15), we can obtain the result in (13) from the perspective of locations. On the other hand, such a form of expression makes it possible for Bob to reversely derive the insecure range of any individual user for a certain secrecy quality-of-service (QoS) (secrecy outage probability under a predefined secrecy rate). Concretely, for a specified SIC order as well as a certain secrecy QoS, the distance d_e is available via numerical root-finding, and it is exactly the radius of an insecure range centered on that user.

Accordingly, the *secrecy transmission probability* can be deduced,

$$\begin{aligned}
P_{\text{st}}^{(n)}(R_s) &= 1 - P_{\text{so}}^{(n)}(R_s) \\
&= \frac{e^{-\lambda(2^{R_s}-1)}}{2^{nR_s}} \{\Theta_1(n) + \Theta_2(n)\},
\end{aligned} \tag{19}$$

which is, obviously, a decreasing function of n and/or R_s.

3.2 Effective secrecy throughput

In accordance with [6], the EST is the product of a secrecy rate R_s and the corresponding secrecy transmission probability $P_{\text{st}}^{(n)}(R_s)$,

$$\begin{aligned}
T^{(n)}(R_s) &= P_{\text{st}}^{(n)}(R_s) R_s \\
&= \frac{e^{-\lambda(2^{R_s}-1)}}{2^{nR_s}} \{\Theta_1(n) + \Theta_2(n)\} R_s.
\end{aligned} \tag{20}$$

Since $P_{\text{st}}^{(n)}(R_s)$ is a decreasing function of n, $T^{(n)}(R_s)$ declines with an increase in n. Moreover, $P_{\text{st}}^{(n)}(R_s)$ decreases exponentially with R_s increasing. Thus, multiplying it with R_s makes the product increase at first and then decrease quickly, which denotes the existence of the maximum EST over R_s.

The maximum EST is expressed as

$$T_{\max}^{(n)}\left(R_s^{*(n)}\right) = P_{\mathrm{st}}^{(n)}\left(R_s^{*(n)}\right) R_s^{*(n)}. \quad (21)$$

Apparently, the *optimal secrecy rate* $R_s^{*(n)}$ to achieve the maximum EST is not identical for different numbers of cochannel interferers n. Interestingly, $R_s^{*(n)}$ with a small value of n is higher than that with a big one, because $P_{\mathrm{st}}^{(n)}(R_s)$ under a small value of n decays more slowly over R_s. We can view such a phenomenon more clearly in Fig. 2.

For the path-loss model, it is possible for each of the users to calculate its optimal secrecy rate with the locations of Bob and Eve. Yet, it is usually difficult for each user to obtain Eve's location. In practice, the estimation of the optimal secrecy rate for each user can be done by Bob instead, and each user tunes to its optimal secrecy rate to gain the maximum EST with an instruction from Bob.

More numerical details for the EST and the estimations of optimal secrecy rates will be examined later in Section 5.

3.3 Other secrecy performance metrics

One of the most important metrics is the *positive secrecy rate probability* (denoted as $P_{\mathrm{ps}}^{(n)}$) which is equivalent to the probability of $\gamma^{(n)} > \rho$. It can be easily obtained, $P_{\mathrm{ps}}^{(n)} = P_{\mathrm{st}}^{(n)}(0)$.

Another interesting performance measure is the ϵ-outage secrecy rate which is defined as the highest secrecy rate when the secrecy outage probability is not greater than ϵ. It can be formulated by

$$P_{\mathrm{so}}^{(n)}\left(C_s^{(n)}(\epsilon)\right) = \epsilon. \quad (22)$$

Although we cannot achieve the closed-form expression of the ϵ-outage secrecy rate due to the complexity of (13), it is possible to obtain the result by numerical root-finding.

3.4 Asymptotic behaviors

We proceed with the asymptotic behaviors of the secrecy performance for an extreme value of K or μ while limiting the other parameters.

Observing (17), an interesting equation can be obtained,

$$\lambda \mathbf{U}(k-1,\lambda) + (k-1)\,\mathbf{U}(k,\lambda) = 1. \quad (23)$$

By applying (23), we rewrite (13) as

$$P_{\mathrm{so}}^{(n)}(R_s) = 1 - \frac{e^{-\lambda(2^{R_s}-1)}}{2^{nR_s}}(1-\Phi), \quad (24)$$

where,

$$\Phi = 2^{R_s}\lambda \mathbf{U}\left(K+n-1, \mu+2^{R_s}\lambda\right) + n\mathbf{U}\left(K+n, \mu+2^{R_s}\lambda\right). \quad (25)$$

Obviously, (24) is a decreasing function of K and/or μ.

Note that $\Phi \to 0$, as K or $\mu \to \infty$. Thus, we can obtain the following asymptotic expression of the secrecy outage probability,

$$\lim_{K or \mu\to\infty} P_{\mathrm{so}}^{(n)}(R_s) = 1 - \frac{e^{-\lambda(2^{R_s}-1)}}{2^{nR_s}}, \quad (26)$$

which is completely reduced to the *transmission outage probability*.

Accordingly, the asymptotic expression of the EST can be formulated as

$$\lim_{K or \mu\to\infty} T^{(n)}(R_s) = \frac{e^{-\lambda(2^{R_s}-1)}}{2^{nR_s}} R_s. \quad (27)$$

Further, we get the asymptotic expression of the maximum EST,

$$\lim_{K or \mu\to\infty} T_{\max}^{(n)}\left(R_s^{*(n)}\right) = \frac{e^{-\lambda\left(2^{R_s^{*(n)}}-1\right)}}{2^{nR_s^{*(n)}}} R_s^{*(n)}, \quad (28)$$

where $R_s^{*(n)}$ is the optimal secrecy rate to achieve the maximum value of this asymptotic EST, and it should not be greater than the related main capacity.

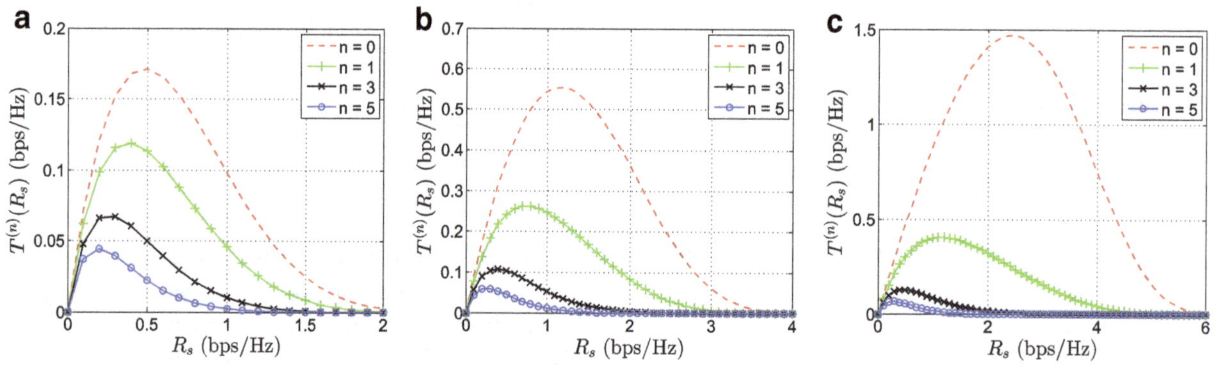

Fig. 2 Relationship between effective secrecy throughput $T^{(n)}(R_s)$ and R_s with different n on basis of $\beta = 1$. **a** SNR = − 3 dB. **b** SNR = 3 dB. **c** SNR = 10 dB

By observing (27), we note the EST declines dramatically with the weight of 2^{-nR_s}. For $n = 3$ and $R_s = 1$ bps/Hz, $\lim_{K or \mu\to\infty} T^{(3)}(1)$ is only about 1/8 of that for the best case ($n = 0$). In other words, the EST with $n = 0$ overwhelms the other cases.

Remark 1 *As the SIC order in this section is assumed to be random, the individual secrecy performance obtained above is an average result.*

Remark 2 *Although it is predictable that a decrease in n and/or an increase in K improve the individual secrecy performance in this SIC-based homogeneous MAC-WT network, the closed-form expressions clearly show how the SIC order and the multi-user diversity impact the secrecy performance. Moreover, it provides a foundation for understanding the SIC order scheduling schemes proposed in the upcoming section.*

4 SIC order scheduling schemes

As the order of SIC affects the secrecy performance significantly, we further explore SIC order scheduling policies from the perspective of Bob. Three SIC order scheduling schemes are studied in this section. First, we consider a conventional round-robin scheduling scheme, which is also used as a benchmark. Next, a suboptimal scheduling scheme is proposed, which schedules on the basis of users' main channel conditions. Finally, we study an optimal scheduling scheme, which is based on achievable individual secrecy rates. We refer to all these three schemes as *round-robin scheme, suboptimal scheme*, and *optimal scheme* for short, respectively. It is essential to clarify that the scheme name optimal scheme does not mean the best scheme of all (including not mentioned in this work) but only signifies superiority over the other two schemes.

4.1 Round-robin scheme

In this scheme, all the K users take the same chance for all possible SIC orders, which brings extreme fairness for each user. Specifically, there are exactly $K!$ corner points in the K-user capacity region with SIC decoding, each one corresponding to an SIC order among the users ([20], Chapter 6.1.4). Bob takes turns to access each of the $K!$ corner points for an equal period of time (i.e., $\frac{1}{K!}$ of scheduling duration), which makes every user experience each of the total $K!$ SIC orders for equal time. It seems impossible for Bob to take all the turns around within one time slot, especially with a large K. Such a problem can be solved by changing the scheduling duration from one time slot to an appropriate number of time slots, e.g., $K!$ time slots, one time slot for each corner point.

As is known to all, a secrecy outage event occurs, when an achievable secrecy rate falls below a predefined secrecy rate R_s. Nevertheless, the secrecy outage probability on each corner point is not necessarily the same. In fact, the secrecy outage probabilities on most of the corner points are different. Thus, an average secrecy outage probability is usually used as a performance metric in this case.

A straightforward way to get the average secrecy outage probability of an individual user for a predefined secrecy rate R_s is to calculate the secrecy outage probability on each corner point separately, add up all the results, and average over the total number of corner points $K!$. As the system in this work is homogeneous, there exactly exist $(K-1)!$ corner points for an individual user with n cochannel interferers (i.e., the total number of SIC orders in which the specified user is decoded at $(n+1)^{th}$ from the end). Therefore, the average secrecy outage probability of an individual user can be formulated as

$$\begin{aligned} P_{\mathrm{so}}^{\mathrm{av}}(R_s) &= \frac{1}{K!}\sum_{n=0}^{K-1}(K-1)!\,P_{\mathrm{so}}^{(n)}(R_s) \\ &= 1-\frac{1}{K}\sum_{n=0}^{K-1}\frac{e^{-\lambda(2^{R_s}-1)}}{2^{nR_s}}\{\Theta_1(n)+\Theta_2(n)\}. \end{aligned} \quad (29)$$

Subsequently, the average individual EST can be easily obtained, the maximum expression of which is denoted as

$$T_{\max}^{\mathrm{av}}(R_s) = \max_{R_s}\left(\left(1-P_{\mathrm{so}}^{\mathrm{av}}(R_s)\right)R_s\right). \quad (30)$$

However, the maximum value acquired by (30) is suboptimal, as the optimal secrecy rate R_s^* for the above maximum value is fixed for all possible n. The optimal maximum value can be obtained by taking n into account for each optimal secrecy rate, just as in Subsection 3.2,

$$T_{\max}^{\mathrm{av(opt)}}\left(R_s^{*(0)},\ldots,R_s^{*(K-1)}\right) = \frac{1}{K}\sum_{n=0}^{K-1}T_{\max}^{(n)}\left(R_s^{*(n)}\right), \quad (31)$$

where $R_s^{*(n)}$ and $T_{\max}^{(n)}\left(R_s^{*(n)}\right)$ is given in (21).

Accordingly, the maximum sum-EST in this scheme can be achieved by multiplying the maximum average EST with K,

$$\begin{aligned} T_{\max}^{\mathrm{sum}}\left(R_s^{*(0)},\ldots,R_s^{*(K-1)}\right) &= K T_{\max}^{\mathrm{av(opt)}}\left(R_s^{*(0)},\ldots,R_s^{*(K-1)}\right) \\ &= \sum_{n=0}^{K-1}T_{\max}^{(n)}\left(R_s^{*(n)}\right), \end{aligned} \quad (32)$$

which is equivalent to the sum of the maximum EST of an individual user over all possible n.

Although the round-robin scheme is a good way for fairness, it is not an optimal option for achieving the maximum sum-EST from Bob's point of view. We next introduce the other two schemes, i.e., the suboptimal scheme and the optimal scheme, which achieve more maximum sum-EST. The former needs no CSI of Eve for Bob, while the latter requires Eve's CSI. Although the optimal scheme is usually impractical, we still study it for the purpose of comparisons.

4.2 Suboptimal scheme

The suboptimal scheme is based on instantaneous main channel gains in a certain time slot. To be concrete, the SIC order is sorted by Bob according to each user's instantaneous channel gain from the lowest to the highest. That is, the user with the lowest instantaneous channel gain is decoded at first (no interference cancellation), while the user with the highest gain is decoded at last (no cochannel interference).

The reason why Bob sorts the SIC order that way (from the lowest gain to the highest) is that the EST of an individual user with the best case (decoded at last, i.e., $n = 0$, we call such EST as best-case-EST later) overwhelms the other cases for the same SNR. Meanwhile, the EST increases with an increase in SNR. Therefore, Bob allocates the best case to the user with the best main channel condition (i.e., achieving the highest SNR for the same transmit power), expecting to achieve more maximum sum-EST.

Although it is difficult to derive the closed-form expressions of the EST for all possible cases (from the best to the worst case) in this scheme, the closed-form expression of one special case, i.e., the best case for the user with the best channel condition, can be derived,[3]

$$T^{\dagger(0)}(R_s) = \left(1 - P_{so}^{\dagger(0)}(R_s)\right) R_s = -\sum_{i=1}^{K} \binom{K}{i} (-1)^i e^{-\lambda(2^{R_s}-1)i} \left\{\Theta_1'(i) + \Theta_2'(i)\right\} R_s, \tag{33}$$

where

$$\Theta_1'(i) = \mu \mathbf{U}\left(K-1, \mu + i2^{R_s}\lambda\right), \tag{34}$$

$$\Theta_2'(i) = (K-1)\, \mathbf{U}\left(K, \mu + i2^{R_s}\lambda\right). \tag{35}$$

The process of the derivation can be found in Appendix B.

Note that $\Theta_1'(1) = \Theta_1(0)$ and $\Theta_2'(1) = \Theta_2(0)$. We transform (33) into

$$T^{\dagger(0)}(R_s) = \mathrm{K}e^{-\lambda(2^{R_s}-1)}\{\Theta_1(0) + \Theta_2(0)\}R_s - \Delta = \mathrm{K}T^{(0)}(R_s) - \Delta, \tag{36}$$

where

$$\Delta = \sum_{i=2}^{K} \binom{K}{i} (-1)^i e^{-\lambda(2^{R_s}-1)i} \left\{\Theta_1'(i) + \Theta_2'(i)\right\} R_s. \tag{37}$$

We can acquire that the first term in (36) has a K-fold gain over $T^{(0)}(R_s)$, which means the multi-user diversity gain is achieved significantly in this scheme.

Since the maximum best-case-EST predominates the maximum sum-EST for both the round-robin and suboptimal schemes, it is illustrative to contrast the performance of these two schemes with respect to the maximum best-case-EST instead of the maximum sum-EST. According to (36), $T_{\max}^{\dagger(0)}\left(R_s^{*(0)}\right)$ is significantly greater than $T_{\max}^{(0)}\left(R_s^{*(0)}\right)$. As such, we can infer that the performance of the suboptimal scheme outperforms that of the round-robin scheme significantly with regard to the maximum sum-EST. More numerical analysis will be continued in Section 5.

Let us further explore the asymptotic expression of $T^{\dagger(0)}(R_s)$, as $K \to \infty$.

By applying (23) to (34) and (35), we obtain,

$$\Theta_1'(i) + \Theta_2'(i) = 1 - i2^{R_s}\lambda \mathbf{U}\left(K-1, \mu + i2^{R_s}\lambda\right), \tag{38}$$

which approaches to 1, as $K \to \infty$.

Then, the asymptotic expression of (33) can be derived as,

$$\lim_{K\to\infty} T^{\dagger(0)}(R_s) = \lim_{K\to\infty} -\sum_{i=1}^{K} \binom{K}{i} (-1)^i e^{-\lambda(2^{R_s}-1)i} R_s = \lim_{K\to\infty} \left\{1 - \left(1 - e^{-\lambda(2^{R_s}-1)}\right)^K\right\} R_s = R_s. \tag{39}$$

Subject to the constraint of the main capacities, the maximum value of $\lim_{K\to\infty} T^{\dagger(0)}(R_s)$ is given by,

$$\lim_{K\to\infty} T_{\max}^{\dagger(0)}(R_s^*) = \log\left(1 + \max_{i\in\mathcal{K}} \xi_i\right). \tag{40}$$

4.3 Optimal scheme

The optimal scheme schedules the SIC order based on achievable individual secrecy rates, which implies a need for Eve's CSI. Concretely, Bob sorts the SIC order according to each user's instantaneous achievable secrecy rate ($n = 0$) from the lowest to the highest, i.e., the user with the highest achievable secrecy rate is decoded at last. For the same reason that the best-case-EST predominates the sum-EST for this scheme, we compare the performance of this scheme with the above two schemes in terms of best-case-EST for simplicity. Before achieving the closed-form expression of the best-case-EST of the user with the highest achievable secrecy rate, we first derive the corresponding secrecy outage probability, which is formulated as[4]

$$P_{so}^{\ddagger(0)}(R_s) = Pr\left(\max_{i\in\mathcal{K}} C_{s,i}^{(0)} < R_s\right) = \prod_{i\in\mathcal{K}} \Pr\left(C_{s,i}^{(0)} < R_s\right) = \left[P_{so}^{(0)}(R_s)\right]^K, \tag{41}$$

where $C_{s,i}^{(0)}$ denotes the ith user's achievable secrecy rate with no cochannel interference.

Subsequently, the best-case-EST for this scheme can be expressed as

$$T^{\ddagger(0)}(R_s) = \left(1 - \left[P_{\text{so}}^{(0)}(R_s)\right]^K\right) R_s, \tag{42}$$

which is obviously bigger than $T^{(0)}(R_s)$ by observing (20). Next, we try to prove that $T^{\ddagger(0)}(R_s)$ is also not less than $T^{\dagger(0)}(R_s)$.

Proof We first rewrite the expression of $P_{\text{so}}^{\ddagger(0)}(R_s)$ as

$$\begin{aligned} P_{\text{so}}^{\ddagger(0)}(R_s) &= \left[P_{\text{so}}^{(0)}(R_s)\right]^K \\ &= \left[\int_0^\infty \left(1 - e^{-\lambda(2^{R_s}\rho + 2^{R_s} - 1)}\right) f_\rho(\rho) d\rho\right]^K \\ &= \left[\mathbf{E}\left(1 - e^{-\lambda(2^{R_s}\rho + 2^{R_s} - 1)}\right)\right]^K. \end{aligned} \tag{43}$$

Due to the Jensen's inequality and the convexity of $(\cdot)^K$, $\left[\mathbf{E}\left(1-e^{-\lambda(2^{R_s}\rho+2^{R_s}-1)}\right)\right]^K \leq \mathbf{E}\left(\left(1-e^{-\lambda(2^{R_s}\rho+2^{R_s}-1)}\right)^K\right) \stackrel{(a)}{=} P_{\text{so}}^{\dagger(0)}(R_s)$, where (a) can be obtained from (53) in Appendix A. We then get $T^{\ddagger(0)}(R_s) \geq T^{\dagger(0)}(R_s)$, which completes the proof. □

Nevertheless, $T^{\dagger(0)}(R_s)$ is very close to $T^{\ddagger(0)}(R_s)$ even with moderate values of SNR and K. In other words, the suboptimal scheme is close to the optimal scheme with respect to the EST. It is worth explaining why the performance (in terms of EST) gap between the practical suboptimal scheme and the ideal optimal scheme is not so much significant.

As mentioned previously, the suboptimal scheme sorts the SIC order by users' channel gains (equivalently, main capacities) from the lowest to the highest, while the optimal scheme schedules via users' achievable secrecy rates. In light of (5), the user with the highest achievable secrecy rate also has an appreciably high probability of being with the highest main capacity. Similarly, the user with the lowest achievable secrecy rate has an appreciably high probability of owning the lowest main capacity. It is applicable for the other cases between the highest achievable secrecy rate and the lowest. Therefore, we can achieve very similar performance statistically just by using users' channel gains instead of achievable secrecy rates for scheduling. In other words, we can obtain the conclusion that the suboptimal scheme is highly close to the optimal scheme with respect to the EST. Such a conclusion will be confirmed once more in Section 5.

5 Numerical results and discussions

In this section, numerical results are presented to further examine and verify the analytical results mentioned above. On account of the consistency of the secrecy outage probability and the effective secrecy throughput as well as the limited space, we only demonstrate the numerical results in terms of effective secrecy throughput. Additionally, the performance comparison of these three SIC order scheduling schemes regarding the maximum sum-EST is examined as well. We make an additional discussion to end this section.

By setting the average received SNR of an individual user at Bob, i.e., $\overline{\xi}$, as the benchmark, the average received SNR at Eve is specified as $\overline{\eta} = \beta\overline{\xi}$, i.e., $\lambda/\mu = \beta$. Hereinafter, the "SNR" refers in particular to $\overline{\xi}$. We assume the default value of K is 10.

5.1 Secrecy performance

5.1.1 *Effective secrecy throughput*

Figure 2 depicts EST curves versus R_s for different values of n and SNR. Here, $\beta = 1$ is supposed. Note that the curve increases first and then decreases for each value of n and SNR, which confirms the inference in Subsection 3.2 that there exists a maximum value for each EST curve, namely, $T_{\max}^{(n)}\left(R_s^{*(n)}\right)$. Another verification is also done that the optimal secrecy rate $R_s^{*(n)}$ to achieve the maximum EST for different numbers of co-channel interferers is different. The lesser the n is, the higher the $R_s^{*(n)}$ becomes. Observing the comparisons from Fig. 2a to c, it is easy to find $T^{(n)}(R_s)$ for each specific n increases when the SNR increases. Interestingly, the gap between $T_{\max}^{(0)}\left(R_s^{*(0)}\right)$ and any other $T_{\max}^{(n)}\left(R_s^{*(n)}\right)$ (where n is not equal to 0) becomes bigger and bigger with an increase in SNR. Also, it confirms that $T_{\max}^{(0)}\left(R_s^{*(0)}\right)$ overwhelms the maximum EST for other cases. This is why we make the performance comparisons of the scheduling schemes by mainly evaluating $T^{(0)}(R_s)$, $T^{\dagger(0)}(R_s)$, and $T^{\ddagger(0)}(R_s)$.

5.1.2 *Optimal secrecy rate estimation*

We here demonstrate how Bob estimates the optimal secrecy rate for each user from a viewpoint of locations. Since each user's location is available to Bob, the distance between any individual user to Bob can be acquired by Bob itself. Likewise, each user's distance to Eve can also be calculated by Bob, as long as Bob has the information of Eve's location. Obviously, it is much more feasible for Bob to detect Eve's location than knowing its CSI.

To be more specific, we take a simple example. Suppose Bob is 200 km right above the users and Eve is 132 km away from Bob in the horizontal direction. Hence, $d_b = 200$ km, $d_e \approx 240$ km and $d_e/d_b \approx 1.2$. We further assume the path-loss exponent $\alpha = 2.5$, and the SNR perceived at Bob is 3 dB. Then, the maximum EST

Table 1 Maximum EST and corresponding optimal secrecy rate for each n

(bps/Hz)	$n = 0$	$n = 1$	$n = 2$	$n = 3$	$n = 4$
$T^{(n)}_{max}$	0.5566	0.2655	0.1588	0.1078	0.0790
$R^{*(n)}_s$	1.16	0.76	0.52	0.38	0.30

and the corresponding optimal secrecy rate for each n can be estimated via numerical root-finding. The results are listed in Table 1. We can also obtain the results for other values of distance ratio. The complete relationships between distance ratios and EST curves are demonstrated in Fig. 3.

We note that the distance ratio does not affect the EST significantly. When Eve is extremely far away, with the limitation of the main capacities, the EST is still limited. Otherwise, when Eve is very close to the users, the EST is also not reduced too much. That is because the cochannel interference at Eve is increased as well when the SNR at Eve rises up.

5.2 Comparison of scheduling schemes

Figure 4 illustrates the best-case-EST curves for these three SIC order scheduling schemes, i.e., round-robin scheme, suboptimal scheme and optimal scheme, for three different cases of SNR, with an assumption of $\beta = 1$. The best-case-EST for the round-robin scheme ($T^{(0)}(R_s)$, green curves) is much lesser than that for the suboptimal (red curves) or optimal (blue curves) scheme. The gaps of the maximum best-case-EST between the round-robin scheme and the other two schemes enlarge when the SNR increases. In contrast, the best-case-EST for the suboptimal scheme ($T^{\dagger(0)}(R_s)$) is very close to that for the optimal scheme ($T^{\ddagger(0)}(R_s)$), which shows the more practical suboptimal scheme is a good option to replace the ideal but impractical optimal scheme.

In Fig. 5, it shows that $T^{\dagger(0)}(R_s)$ (blue line-marker curves) and $T^{\ddagger(0)}(R_s)$ (red line-marker curves) increases significantly with K increasing while $T^{(0)}(R_s)$ (black curves) does not increase much. Such a phenomenon shows that the multi-user diversity impacts the round-robin scheme very little, while the other two schemes achieve significant multi-user diversity gains. Similarly, the multi-user diversity gain for $T^{\dagger(0)}(R_s)$ is very close to that for $T^{\ddagger(0)}(R_s)$. Moreover, the gaps between $T^{\dagger(0)}(R_s)$ or $T^{\ddagger(0)}(R_s)$ and $T^{(0)}(R_s)$ expand drastically with K increasing.

In order to compare the performance of the scheduling schemes in terms of the maximum sum-EST numerically, we first calculate $T^{(n)}_{max}(R^{*(n)}_s)$ for all n with four different cases of SNR by numerical root-finding. The results are listed in Table 2. Here, $\beta = 1$. Observing the data, it confirms once more that the maximum sum-EST is dominated by the maximum best-case-EST. Similarly, the maximum values of best-case-EST for the other two schemes, $T^{\dagger(0)}_{max}$ and $T^{\ddagger(0)}_{max}$, are also calculated by numerical root-finding and listed in the right-hand columns of Table 3. From the table, we find $T^{\dagger(0)}_{max}$ or $T^{\ddagger(0)}_{max}$ is even larger than the maximum sum-EST for the round-robin scheme (listed in the second column of Table 3) except the case of SNR = − 3 dB for $T^{\dagger(0)}_{max}$, where $T^{\dagger(0)}_{max}$ is only a little less. Since the expressions of the maximum sum-EST for the suboptimal and optimal schemes are $\sum_{n=0}^{K-1} T^{\dagger(n)}_{max}$ and $\sum_{n=0}^{K-1} T^{\ddagger(n)}_{max}$, respectively, it obviously verify the maximum sum-EST for the suboptimal or optimal scheme is much larger than that for the round-robin scheme. In other words, the suboptimal or optimal scheme is much better than the round-robin scheme from the perspective of the maximum sum-EST in this model of MAC-WT.

To further evaluate the values of $T^{\dagger(n)}_{max}$ and $T^{\ddagger(n)}_{max}$ for the other cases (i.e., $n = 1, \ldots, K$), we perform simulations consisting of 2000 independent trials to obtain the average results of the EST curves for both schemes. Figure 6 demonstrates the simulation results (the cases for $n > 2$ are ignored, due to their low values for the curves) on basis of SNR = 3 dB. From Fig. 6, we observe the simulation curves for $T^{\dagger(0)}(R_s)$ (black dash-dot lines) and $T^{\ddagger(0)}(R_s)$ (red dashed lines) almost converge to those from the analytical results (blue circles for the suboptimal scheme and magenta diamonds for the optimal scheme), which shows

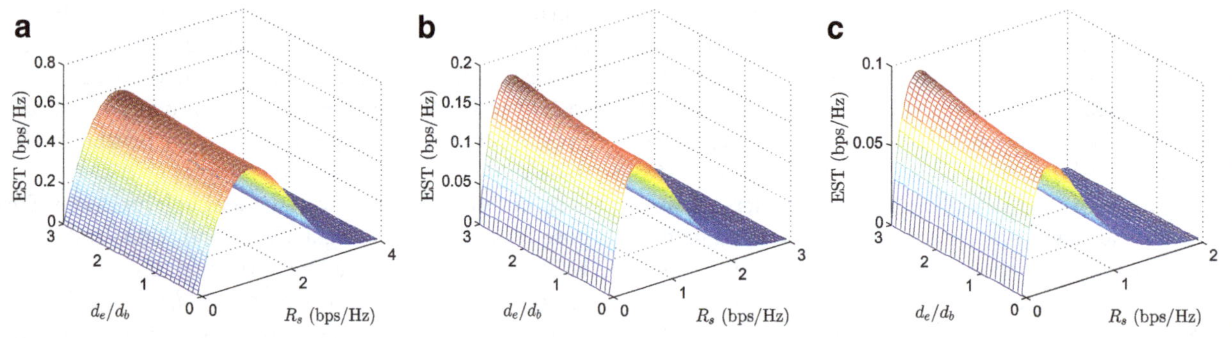

Fig. 3 Relationship between EST and distance ratio (d_e/d_b) on basis of SNR = 3 dB and $\alpha = 2.5$. **a** $n = 0$. **b** $n = 2$. **c** $n = 4$

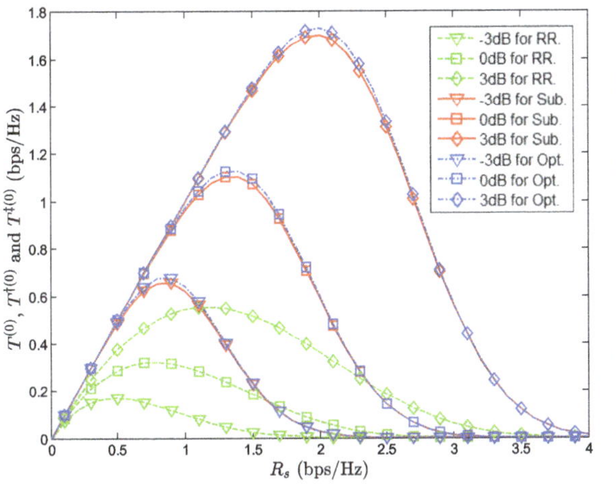

Fig. 4 Comparison of best-case-EST for round-robin scheme (RR.), suboptimal scheme (Sub.), and optimal scheme (Opt.) with different cases of SNR

the simulation results are reliable. Meanwhile, the simulation curves for both the suboptimal (black dash-dot lines) and optimal (red dashed lines) schemes are highly close to each other for $n > 0$, just like the case of $n = 0$. Although, for a big value of n, the maximum values of EST for these two schemes are poorer than the round-robin scheme, by observing the simulation results, $T^{\dagger(1)}_{\max}$ and $T^{\ddagger(1)}_{\max}$ are still superior to $T^{(1)}_{\max}$, and the maximum EST of either scheme for $n = 2$ is not too much less than that for the round-robin scheme.

To demonstrate the impacts of the multi-user diversity on $T^{\dagger(n)}$ and $T^{\ddagger(n)}$ for the other cases (i.e., $n = 1, \ldots, K$), we perform the simulation once more and achieve the results as plotted in Fig. 7 (partly, black curves for the suboptimal scheme and red curves for the optimal scheme),

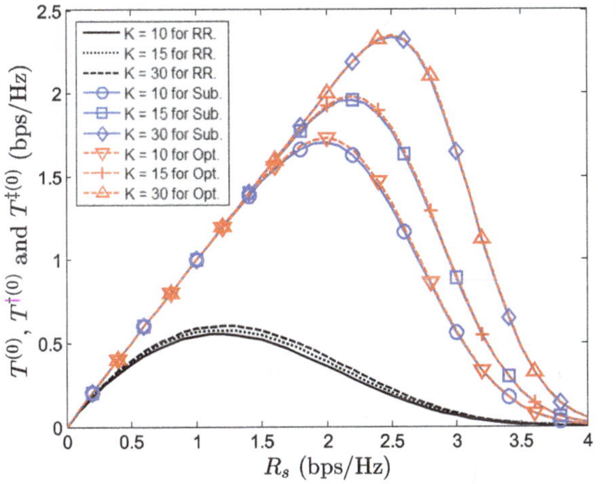

Fig. 5 Impacts of multi-user diversity on best-case-EST for round-robin scheme (RR.), suboptimal scheme (Sub.), and optimal scheme (Opt.)

Table 2 Maximum EST of round-robin scheme for different cases of n

SNR (dB)	$T^{(0)}_{\max}$	$T^{(1)}_{\max}$	$T^{(2)}_{\max}$	$T^{(3)}_{\max}$	$T^{(4)}_{\max}$
− 3	0.1716	0.1191	0.0881	0.0673	0.0540
0	0.3210	0.1873	0.1246	0.0893	0.0678
3	0.5540	0.2633	0.1575	0.1068	0.0781
10	1.4698	0.4033	0.2011	0.1260	0.0878
SNR (dB)	$T^{(5)}_{\max}$	$T^{(6)}_{\max}$	$T^{(7)}_{\max}$	$T^{(8)}_{\max}$	$T^{(9)}_{\max}$
− 3	0.0445	0.0368	0.0305	0.0259	0.0231
0	0.0533	0.0437	0.0360	0.0297	0.0248
3	0.0593	0.0477	0.0391	0.0322	0.0266
10	0.0663	0.0513	0.0419	0.0344	0.0283

which shows the multi-user diversity gain for $T^{\dagger(n)}$ is also very close to that for $T^{\ddagger(n)}$ for $n = 1, \ldots, K$.

5.3 Discussions

Although the performance of the suboptimal or optimal scheme precedes the round-robin scheme in achieving the maximum sum-EST for Bob, the values of EST for the users with bad channels are very low, which makes them be overheard easily. It seems unfair to these users. On the other hand, the absolutely fair round-robin is full of complexity in computing. Is there a scheduling scheme that can solve such a dilemma? It is easy to come up with a policy reversing the SIC order in the suboptimal scheme, i.e., sorting the SIC order from the highest channel gain to the lowest. We just call it *alternative scheme*. It is also difficult to derive the closed-form expressions of the EST for all cases of n for this scheme. Nevertheless, their simulation results can be obtained and demonstrated in Fig. 8, with the same simulation conditions as Fig. 6. We note, although the problem of fairness is improved, the performance is very bad such that not only the values of EST for the users with bad channel conditions are not improved too much but also the values of EST for the better-conditioned users are harmed significantly. The maximum EST for $n = 0$ is only about 0.06 bps/Hz, which indicates a good SIC order to a poor-conditioned user cannot improve its performance too much. On the other hand, the best channel condition in this scheme does not bring to the user high enough maximum EST (see curve with $n = 9$), which is less than that for $n = 4$, as its

Table 3 Comparison of $T^{\text{sum}}_{\max}$, $T^{\dagger(0)}_{\max}$, and $T^{\ddagger(0)}_{\max}$

SNR (dB)	$T^{\text{sum}}_{\max}$	$T^{\dagger(0)}_{\max}$	$T^{\ddagger(0)}_{\max}$
− 3	0.6609	0.6568	0.6771
0	0.9775	1.1046	1.1287
3	1.3646	1.7016	1.7305
10	2.5102	3.5042	3.5388

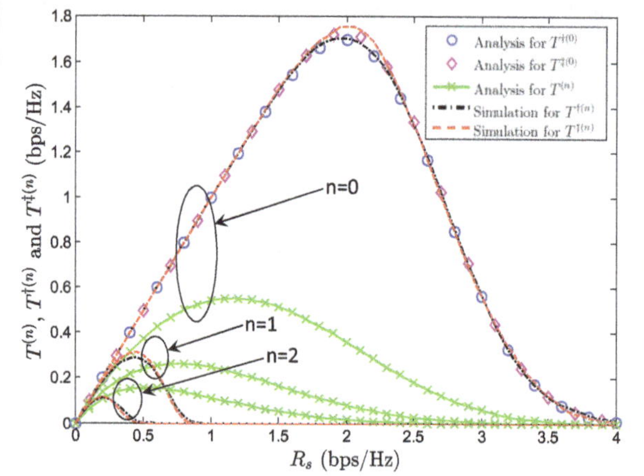

Fig. 6 Simulation results of EST for suboptimal scheme and optimal scheme on basis of SNR = 3 dB

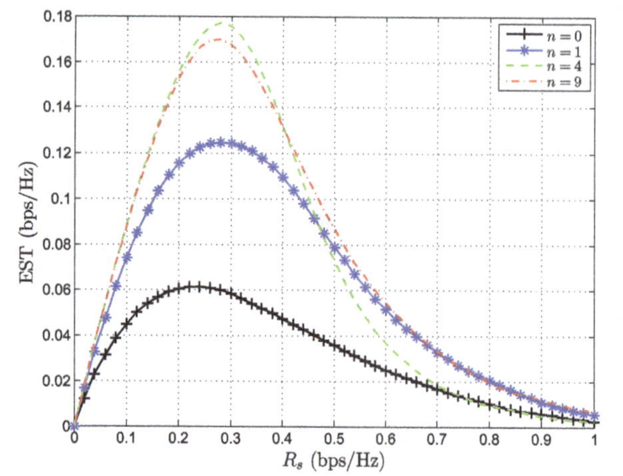

Fig. 8 Simulation results of EST for alternative scheme on basis of SNR = 3 dB

SIC order (decoded firstly) retards the performance. We also note the highest maximum EST occurs at $n = 4$ with a poor value less than 0.18 bps/Hz. All in all, the secrecy performance regarding the maximum sum-EST for this scheme is significantly worse than the other three schemes mentioned above.

Therefore, in this model of MAC-WT, to promote the security of the network, the users with bad conditions can play a role of jammers [21]. When their channel conditions improve, they can change their role of the jammers into the normal users, while other bad-conditioned users switch to the role of jammers.

Overall, the numerical results and observations in this section are consistent with our expectations.

6 Conclusions

In this work, we considered the quasi-static Rayleigh fading homogeneous MAC-WT with SIC decoding at the legitimate satellite and investigated the individual secrecy performance under an arbitrary SIC order by deriving the closed-form expressions of the secrecy outage probability and the effective secrecy throughput. We provided valuable insights into the impacts of the SIC order and the multi-user diversity on the secrecy performance. Three SIC order scheduling schemes were studied such that the suboptimal scheme and the optimal scheme have better performance in achieving the maximum sum-EST and the multi-user diversity gain while the round-robin scheme is a good option for fairness. In addition, the suboptimal scheme achieves the performance highly close to the ideal but usually impractical optimal scheme in terms of the EST and the multi-user diversity gain. We verified our analysis with the aid of numerical results.

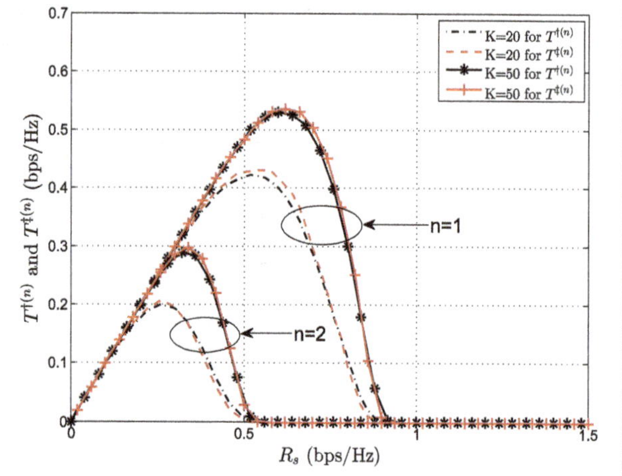

Fig. 7 Simulation results of multi-user diversity impacts on EST for suboptimal scheme and optimal scheme on basis of SNR = 3 dB

Endnotes

[1] The abbreviation of multiple access wiretap channel is a bit of confusion. It seems MAWC is more appropriate, yet, MAC-WT is more frequently used in the existing literature.

[2] The satellite channel model varies from different environments, which is beyond the scope of this paper. We just simplify it to be Rayleigh fading the same as some other literature, e.g., [52, 53] concerning low earth orbit (LEO) satellite communications.

[3] The superscript of $\dagger$ is used to specify corresponding notations for this scheme.

[4] The superscript of $\ddagger$ is used to differ corresponding notations from the other two schemes.

Appendix A

Assume a random variable $X \sim \text{Exp}(\lambda_0)$, and a random variable $Y_n = \sum_{i=1}^{n} \xi_i$, where $\xi_1, \xi_2, \ldots, \xi_n$ are independently and exponentially distributed, i.e., $\xi_i \sim \mathrm{Exp}\,(\lambda_i)$ $(i = 1, \ldots, n)$ and $\lambda_i \neq \lambda_j, \forall i \neq j$. Further assume X and Y_n are independent.

Thus, the CDF of a random variable $Z_n = \frac{X}{1+Y_n}$ can be expressed as

$$\begin{aligned} F_{Z_n}(z) &= \int_0^{\infty} \int_0^{zy+z} f_{Y_n}(y) f_X(x)\, dx dy \\ &= \int_0^{\infty} f_{Y_n}(y)\, F_X(zy+z)\, dy \\ &= 1 - \underbrace{\int_0^{\infty} f_{Y_n}(y) \exp\left(-\lambda_0 (zy+z)\right) dy}_{L_1}, \end{aligned} \tag{44}$$

where,

$$F_X(x) = \begin{cases} 1 - \exp(-\lambda_0 x), & x \geq 0 \\ 0, & x < 0. \end{cases} \tag{45}$$

The PDF of Y_n is derived in [54],

$$f_{Y_n}(y) = \begin{cases} f_1(y) \prod_{i=1}^{n} \lambda_i, & y > 0 \\ 0, & y \leq 0. \end{cases} \tag{46}$$

where

$$\begin{aligned} f_1(y) =& \frac{\exp(-\lambda_1 y)}{\prod_{i=2}^{n} (\lambda_i - \lambda_1)} + \cdots \\ &+ \frac{(-1)^{k-1} \exp(-\lambda_k y)}{\prod_{i=k+1}^{n} (\lambda_i - \lambda_k) \prod_{j=1}^{k-1} (\lambda_k - \lambda_j)} + \cdots \\ &+ (-1)^{n-1} \frac{\exp(-\lambda_n y)}{\prod_{j=1}^{n-1} (\lambda_n - \lambda_j)}. \end{aligned} \tag{47}$$

Substitute (47) into L_1, and obtain,

$$\begin{aligned} L_1 &= \left(\prod_{i=1}^{n} \lambda_i \right) \int_0^{\infty} f_1(y) \exp\left(-\lambda_0 (zy+z)\right) dy \\ &= \exp(-\lambda_0 z) \frac{\prod_{i=1}^{n} \lambda_i}{\prod_{i=1}^{n} \pi_i} \underbrace{\overbrace{\int_0^{\infty} f_2(y) \prod_{i=1}^{n} \pi_i\, dy}^{L_2}}_{L_3}, \end{aligned} \tag{48}$$

where $\pi_i = \lambda_i + \lambda_0 z$, $i = 1, \ldots, n$, and,

$$\begin{aligned} f_2(y) =& \frac{\exp(-\pi_1 y)}{\prod_{i=2}^{n} (\pi_i - \pi_1)} + \cdots \\ &+ \frac{(-1)^{k-1} \exp(-\pi_k y)}{\prod_{i=k+1}^{n} (\pi_i - \pi_k) \prod_{j=1}^{k-1} (\pi_k - \pi_j)} + \cdots \\ &+ (-1)^{n-1} \frac{\exp(-\pi_n y)}{\prod_{j=1}^{n-1} (\pi_n - \pi_j)}. \end{aligned} \tag{49}$$

We note that the function $f_2(y)$ has the same structure as $f_1(y)$. Hence, L_2 is similar to (46), which is a PDF. In particular, when applying a zero-to-infinity integral of this PDF, we get $L_3 = 1$, simplifying (48) into

$$\begin{aligned} L_1 &= \exp(-\lambda_0 z) \frac{\prod_{i=1}^{n} \lambda_i}{\prod_{i=1}^{n} \pi_i} \\ &= \exp(-\lambda_0 z) \frac{\prod_{i=1}^{n} \lambda_i}{\prod_{i=1}^{n} (\lambda_i + \lambda_0 z)}, \end{aligned} \tag{50}$$

Furthermore, substitute (50) into (44), achieving,

$$F_{Z_n}(z) = \begin{cases} 1 - \exp(-\lambda_0 z) \frac{\prod_{i=1}^{n} \lambda_i}{\prod_{i=1}^{n} (\lambda_i + \lambda_0 z)}, & z \geq 0 \\ 0, & z < 0. \end{cases} \tag{51}$$

Finally, substituting specific parameters into (51) obtains both (6) and (7).

By the way, during the derivation of the PDF of Y_n, we assume that the mean of ξ_i $(i = 1, \ldots, n)$ in Y_n is not equal to each other, that is, $\lambda_i \neq \lambda_j, \forall i \neq j$. Actually, we can loose the assumption to the general case that there are m of them with the same mean. We set $Y_n = Y'_{n-m} + Y''_m$, where Y'_{n-m} denotes the sum of all $n-m$ random variables with different means, while Y''_m denotes the sum of all m random variables with the same mean. The PDF of Y'_{n-m} can be obtained from (46), whereas $Y''_m \sim \chi^2_{2m}$. And the PDF of Y_n can be obtained by the convolution of the PDF of Y'_{n-m} and that of Y''_m,

$$f_{Y_n}(y) = \int_{-\infty}^{+\infty} f_{Y'_{n-m}}(x) f_{Y''_m}(y-x)\, dx. \tag{52}$$

As a result, substituting (52) into (44) yields the same result as (51).

Appendix B

According to the statistics of the variables ξ_i $(i \in \mathcal{K})$ and ρ, we first derive the secrecy outage probability of the user who has the best channel condition and is decoded at last (no cochannel interference at all),

$$\begin{aligned} &P_{\mathrm{so}}^{\dagger(0)}(R_s) \\ &= \Pr\left(\max_i \xi_i < 2^{R_s}\rho + 2^{R_s} - 1 \right) \\ &= \int_0^{\infty} \left(1 - e^{-\lambda(2^{R_s}\rho + 2^{R_s} - 1)}\right)^K f_\rho(\rho)\, d\rho \qquad (53) \\ &\overset{(a)}{=} \int_0^{\infty} \left(1 + \sum_{i=1}^{K} \binom{K}{i} \frac{(-1)^i}{e^{\lambda(2^{R_s}\rho + 2^{R_s} - 1)i}}\right) f_\rho(\rho)\, d\rho \\ &= 1 + \int_0^{\infty} \frac{(1+\rho)\mu + K - 1}{(1+\rho)^K e^{\mu\rho}} \sum_{i=1}^{K} \binom{K}{i} \frac{(-1)^i}{e^{\lambda(2^{R_s}\rho + 2^{R_s} - 1)i}}\, d\rho \\ &= 1 + \sum_{i=1}^{K} \binom{K}{i} (-1)^i e^{-\lambda(2^{R_s} - 1)i} \left\{ \Theta'_1(i) + \Theta'_2(i) \right\}, \end{aligned} \tag{54}$$

where $\Theta_1'(i)$ and $\Theta_2'(i)$ is given in (34) and (35), respectively. The step (a) is obtained by applying the Newton binomial theorem.

Therefore, the corresponding EST can be expressed as (33).

Acknowledgements
This work was supported by the National Natural Science Foundation of China (Grant Nos. 61572070 and 61371069) and the Specialized Research Fund for the Doctoral Program of Higher Education (Grant No. 20130009110015).

Authors' contributions
KJ conceived the idea of the system model and designed the proposed schemes. FZ performed simulations of the proposed schemes. YH and ZL provided substantial comments on the work. TJ supported and supervised the research. All of the authors participated in the project, and they read and approved the final manuscript.

Competing interests
The authors declare that they have no competing interests.

References

1. AD Wyner, The wire-tap channel. Bell Syst. Tech. J. **54**, 1355–1387 (1975)
2. I Csiszar, J Korner, Broadcast channels with confidential messages. IEEE Trans. Inf. Theory. **24**(3), 339–348 (1978)
3. SK Leung-Yan-Cheong, ME Hellman, The gaussian wire-tap channel. IEEE Trans. Inf. Theory. **IT-24**(4), 451–456 (1978)
4. J Barros, MRD Rodrigues, in *2006 IEEE International Symposium on Information Theory*. Secrecy capacity of wireless channels (IEEE, Seattle, 2006), pp. 356–360
5. PK Gopala, L Lai, HE Gamal, On the secrecy capacity of fading channels. IEEE Trans. Inf. Theory. **54**(10), 4687–4698 (2008)
6. N Yang, S Yan, J Yuan, R Malaney, I Land, Artificial noise: transmission optimization in multi-input single-output wiretap channels. IEEE Trans. Commun. **63**(5), 1771–1783 (2015)
7. M Jilani, T Ohtsuki, Joint svd-gsvd precoding technique and secrecy capacity lower bound for the mimo relay wire-tap channel. EURASIP Journal on Wireless Communications and Networking. **2012**(1), 361 (2012)
8. F Zhu, F Gao, M Yao, Zero-forcing beamforming for physical layer security of energy harvesting wireless communications. EURASIP J. Wireless Commun. Netw. **2015**(1), 58 (2015)
9. W Wu, B Wang, Robust secrecy beamforming for wireless information and power transfer in multiuser miso communication system. EURASIP J. Wireless Commun. Netw. **2015**(1), 161 (2015)
10. C Wang, HM Wang, DWK Ng, XG Xia, C Liu, Joint beamforming and power allocation for secrecy in peer-to-peer relay networks. IEEE Trans. Wirel. Commun. **14**(6), 3280–3293 (2015)
11. W Li, M Ghogho, B Chen, C Xiong, Secure communication via sending artificial noise by the receiver: Outage secrecy capacity/region analysis. IEEE Commun. Lett. **16**(10), 1628–1631 (2012)
12. Z Lin, Y Cai, W Yang, X Xu, Opportunistic relaying and jamming with robust design in hybrid full/half-duplex relay system. EURASIP J. Wireless Commun. Netw. **2016**(1), 129 (2016)
13. M Wiese, J Notzel, H Boche, A channel under simultaneous jamming and eavesdropping attack — correlated random coding capacities under strong secrecy criteria. IEEE Trans. Inf. Theory. **62**(7), 3844–3862 (2016)
14. Y Wang, Z Miao, R Sun, L Jiao, Distributed coalitional game for friendly jammer selection in ultra-dense networks. EURASIP J. Wireless Commun. Netw. **2016**(1), 211 (2016)
15. Z Li, T Jing, X Cheng, Y Huo, W Zhou, D Chen, in *2015 IEEE International Conference on Communications (ICC)*. Cooperative jamming for secure communications in mimo cooperative cognitive radio networks (IEEE, London, 2015), pp. 7609–7614
16. B Li, Z Fei, Robust beamforming and cooperative jamming for secure transmission in df relay systems. EURASIP J. Wireless Commun. Netw. **2016**(1), 68 (2016)
17. L Li, Y Xu, Z Chen, J Fang, Robust transmit design for secure af relay networks with imperfect csi. EURASIP J. Wireless Commun. Netw. **2016**, 142 (2016)
18. N Yang, M Elkashlan, TQ Duong, J Yuan, R Malaney, Optimal transmission with artificial noise in misome wiretap channels. IEEE Trans. Veh. Technol. **65**(4), 2170–2181 (2016)
19. M Bloch, J Barros, *Physical-layer security: from information theory to security engineering*. (Cambridge University Press, Cambridge, 2011)
20. D Tse, P Viswanath, *Fundamentals of wireless communication*. (Cambridge University Press, Cambridge, 2005)
21. E Tekin, A Yener, The general gaussian multiple-access and two-way wiretap channels: Achievable rates and cooperative jamming. IEEE Trans. Inf. Theory. **54**(6), 2735–2751 (2008)
22. E Tekin, A Yener, in *Proc. Annual Allerton Conf.* Secrecy sum-rates for the multiple-access wire-tap channel with ergodic block fading (IEEE, Illinois, 2007), pp. 856–863
23. H Zivari-Fard, B Akhbari, M Ahmadian-Attari, MR Aref, Multiple access channel with common message and secrecy constraint. IET Commun. **10**(1), 98–110 (2016)
24. H Zivari-Fard, B Akhbari, M Ahmadian-Attari, MR Aref, Imperfect and perfect secrecy in compound multiple access channel with confidential message. IEEE Trans. Inf. Forensics Secur. **11**(6), 1239–1251 (2016)
25. J Xie, S Ulukus, in *2013 IEEE International Symposium on Information Theory*. Secure degrees of freedom of the gaussian multiple access wiretap channel (IEEE, Istanbul, 2013), pp. 1337-1341
26. Y Fan, X Liao, Z Gao, L Sun, in *2016 IEEE International Conference on Communications Workshops (ICC)*. Physical layer security based on real interference alignment in k-user mimo y wiretap channels (IEEE, Kuala Lumpur, 2016), pp. 207–212
27. P Mukherjee, S Ulukus, in *2016 IEEE International Conference on Communications (ICC)*. Real interference alignment for the mimo multiple access wiretap channel (IEEE, Kuala Lumpur, 2016), pp. 1–6
28. RA Chou, MR Bloch, in *2014 IEEE Conference on Communications and Network Security*. Uniform distributed source coding for the multiple access wiretap channel (IEEE, San Francisco, 2014), pp. 127–132
29. M Hajimomeni, H Aghaeinia, IM Kim, K Kim, Cooperative jamming polar codes for multiple-access wiretap channels. IET Commun. **10**(4), 407–415 (2016)
30. YP Wei, S Ulukus, Polar coding for the general wiretap channel with extensions to multiuser scenarios. IEEE J. Selected Areas Commun. **34**(2), 278–291 (2016)
31. R Bassily, S Ulukus, Ergodic secret alignment. IEEE Trans. Inf. Theory. **58**(3), 1594–1611 (2012)
32. J Xie, S Ulukus, in *2013 Asilomar Conference on Signals, Systems and Computers*. Secure degrees of freedom region of the gaussian multiple access wiretap channel (IEEE, Pacific Grove, 2013), pp. 293–297
33. P Mukherjee, S Ulukus, in *2015 IEEE International Symposium on Information Theory (ISIT)*. Secure degrees of freedom of the multiple access wiretap channel with no eavesdropper csi (IEEE, Hong Kong, 2015), pp. 2311–2315
34. P Mukherjee, S Ulukus, in *2015 49th Asilomar Conference on Signals, Systems and Computers*. Secure degrees of freedom of the mimo multiple access wiretap channel (IEEE, Pacific Grove, 2015), pp. 554–558
35. H Jin, WY Shin, BC Jung, On the multi-user diversity with secrecy in uplink wiretap networks. IEEE Commun. Lett. **17**(9), 1778–1781 (2013)
36. Y Zou, J Zhu, G Wang, H Shao, in *2014 IEEE/CIC International Conference on Communications in China (ICCC)*. Secrecy outage probability analysis of multi-user multi-eavesdropper wireless systems (IEEE, Shanghai, 2014), pp. 309–313
37. Y Jiang, J Zhu, Y Zou, in *2015 IEEE 14th International Conference on Cognitive Informatics & Cognitive Computing (ICCI*CC)*. Secrecy outage analysis of multi-user cellular networks in the face of cochannel interference (IEEE, Beijing, 2015), pp. 441–446
38. Z Ni, J Fei, C Xing, D Zhao, N Wang, J Kuang, Secrecy balancing over two-user miso interference channels with rician fading. Int. J. Antennas Propag. **2013**(2013), 1–7

39. K Jiang, T Jing, Z Li, Y Huo, F Zhang, in *INFOCOM*. Analysis of secrecy performance in fading multiple access wiretap channel with sic receiver (IEEE, Atlanta, 2017), pp. 1602–1610
40. K Jiang, T Jing, F Zhang, Y Huo, Z Li, Zf-sic based individual secrecy in simo multiple access wiretap channel. IEEE Access. **5**, 7244–7253 (2017)
41. N Li, X Tao, Q Cui, J Xu, in *2015 IEEE Wireless Communications and Networking Conference (WCNC)*. Secure transmission with artificial noise in the multiuser downlink: Secrecy sum-rate and optimal power allocation (IEEE, New Orleans, 2015), pp. 1416–1421
42. AB Reid, AJ Grant, PD Alexander, List detection for the k-symmetric multiple-access channel. IEEE Trans. Inf. Theory. **51**(8), 2930–2936 (2005)
43. S Nitinawarat, in *2011 IEEE International Symposium on Information Theory Proceedings*. On maximal error capacity regions of symmetric gaussian multiple-access channels (IEEE, St. Petersburg, 2011), pp. 2258–2262
44. RA Chou, MR Bloch, J Kliewer, in *2014 IEEE Information Theory Workshop (ITW 2014)*. Low-complexity channel resolvability codes for the symmetric multiple-access channel (IEEE, Hobart, 2014), pp. 466–470
45. S Salehkalaibar, MR Aref, Lossy transmission of correlated sources over multiple-access wiretap channels. IET Commun. **9**(6), 754–770 (2015)
46. AS Bendary, YZ Mohasseb, H Dahshan, in *2016 IEEE Conference on Communications and Network Security (CNS)*. On the secure degrees of freedom for the k-user symmetric mimo wiretap mac channel (IEEE, Philadelphia, 2016), pp. 591–595
47. M Chraiti, A Ghrayeb, C Assi, in *2016 IEEE Global Conference on Signal and Information Processing (GlobalSIP)*. Achieving full secure degrees-of-freedom for the miso wiretap channel with an unknown eavesdropper, (2016), pp. 997–1001
48. A Chaaban, Z Rezki, B Alomair, MS Alouini, in *2016 IEEE Global Conference on Signal and Information Processing (GlobalSIP)*. The miso wiretap channel with channel uncertainty: Asymptotic perspectives (IEEE, Washington, 2016), pp. 959–963
49. Z Rezki, A Chaaban, B Alomair, MS Alouini, in *2016 IEEE Global Communications Conference (GLOBECOM)*. The miso wiretap channel with noisy main channel estimation in the high power regime (IEEE, Washington, 2016), pp. 1–5
50. E Tekin, A Yener, The gaussian multiple access wire-tap channel. IEEE Trans. Inf. Theory. **54**(12), 5747–5755 (2008)
51. IS Gradshteyn, IM Ryzhik, *Table of integrals, series and products*, 7th. (Academic Press, New York, 2007)
52. N Lebedev, JF Diouris, in *Conference record of the thirty-fourth Asilomar conference on signals, systems and computers*. Capacity study of a leo satellite link with multiple antennas user terminals (IEEE, Pacific Grove, 2000), pp. 511–515
53. LH Abderrahmane, DEB Hamed, M Benyettou, in *2008 IEEE Aerospace Conference*. Design of an adaptive communication system for implementation on board a future algerian leo satellite (IEEE, Big Sky, 2008), pp. 1–5
54. S Shulong, On distribution of sums of n independent random variables subject to exponential distribution. J. Liaoning Normal Univ. **13**(4), 51–58 (1990)

8

Performance analysis on joint channel decoding and state estimation in cyber-physical systems

Liang Li[1*], Shuping Gong[3], Ju Bin Song[2] and Husheng Li[1,2]

Abstract

We propose to use an mean square error (MSE) transfer chart to evaluate the performance of the proposed belief propagation (BP)-based channel decoding and state estimation scheme. We focus on two models to evaluate the performance of BP-based channel decoding and state estimation: the sequential model and the iterative model. The numerical results show that the MSE transfer chart can provide much insight about the performance of the proposed channel decoding and state estimation scheme.

Keywords: Channel coding, State estimation, EXIT chart, MSE transfer chart, CPSs

1 Introduction

Communication has been of great importance in cyber-physical systems (CPSs), which sends observations of the physical dynamics from the sensor to the controller as illustrated in Fig. 1. One promising way to improve the performance of physical dynamics (or system state) estimation is the BP-based joint channel decoding and system state estimation algorithm, which we have already developed in [1] to utilize time-domain redundancy of system state to assist channel decoding. For example, the quantized codeword before source/channel encoding at discrete time t, denoted as $\mathbf{b}(t)$, is generated by the observation of the physical dynamics, denoted as $\mathbf{y}(t)$, where t can be viewed as the beginning of tth time slot. Due to the time correlation of the system states, the observation $\mathbf{y}(t)$ is correlated with $\mathbf{y}(t-1)$, thus, $\mathbf{b}(t-1)$ can provide some information for decoding the quantized codeword, i.e., $\mathbf{b}(t)$ in discrete time t. Even though the effectiveness of the proposed joint channel decoding and system state estimation algorithm has been verified by numerical results in [1], the procedure of the given algorithm is still left unspecified. Contributing toward the previous work, this paper addresses the procedure of the message passing between the channel decoder, which processes the information of quantized bits, and the state estimator, which handles the information of continuous state values. We analyze the proposed algorithm from the following perspectives:

- Does the proposed algorithm converge and help to improve channel decoding and system state estimation?
- How much gain can be obtained by using redundancy of observations in time domain to assist channel decoding?

As pointed out before, the CPS is a hybrid system [2], which consists of system state $\mathbf{x}(t)$, observation $\mathbf{y}(t)$ with continuous values, and information bits $\mathbf{b}(t)$ transmitted in wireless communication with discrete values. The challenges in the channel decoding and system state estimation framework are that the priori information transmitted from a state estimator to a channel decoder is the prediction of $\mathbf{y}(t)$, while the channel decoder actually requires the priori information of each quantized bit of $\mathbf{y}(t)$, and that the output of the channel decoder is the extrinsic information of each quantized bit of $\mathbf{y}(t)$, while the state estimator actually requires the estimation of $\mathbf{y}(t)$ from the channel decoder. To handle these challenges, two models, the BP-based sequential model and the BP-based iterative model, are given to evaluate the performance of BP-based channel decoding and system state estimation

*Correspondence: liang.li.uestc@gmail.com
[1]Department of Electrical Engineering and Computer Science, the University of Tennessee, Knoxville, Knoxville, USA
Full list of author information is available at the end of the article

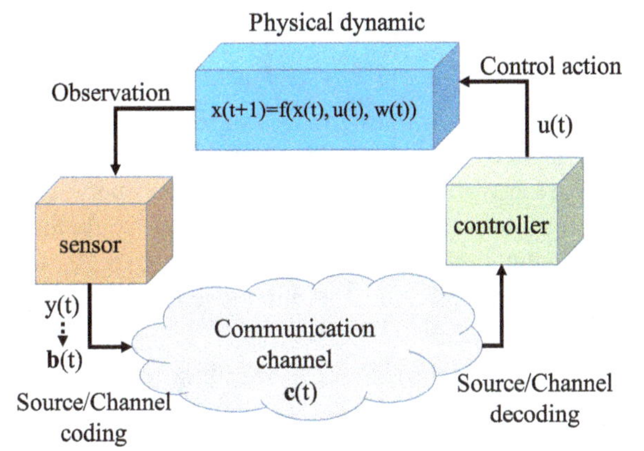

Fig. 1 An illustration of the components in CPSs

framework. The former can be used to evaluate the performance of system state estimation over multiple time slots, e.g., the gain by utilizing priori information from the previous time slot to assist channel decoding and state estimation at the current time slot. The latter can be used to check the following points:

1. Does the iterative channel decoding and estimation converge, and how many iterations are sufficient?
2. How much gain can be obtained by utilizing priori information from the previous time slots to assist state estimation at the current time slot?

In the area of wireless communication, the purpose of performance analysis for decoding scheme is to find out if, for any given encoder, decoder, and channel noise power, a message-passing iterative decoder can correct the errors or not.

To analyze the performance, in [3–5], Ten Brink proposed using an extrinsic information transfer (EXIT) chart to track the iterative decoding performance. Based on the assumption that the distribution of the extrinsic log-likelihood ratios (LLRs) is a Gaussian distribution, the EXIT chart tracks mutual information from the extrinsic LLRs through an iterative decoding process. Compared with the previously used method of density evolution, the EXIT chart is computationally simplified, and it also allows to visualize the evolution of mutual information through iterative decoding process in a graph. The details of the EXIT chart can be found in [6].

The EXIT chart has two useful properties as shown in [7]. One is the necessary condition for the convergence of iterative decoding that the flipped EXIT chart curve of the outer decoder for iterations lies below the EXIT chart curve of the inner coder. The other is that the area under the EXIT curve of outer code relates to the rate of inner coder. In [8], the authors demonstrated that if the priori channel is an erasure channel, for any outer code of rate R, the area under the EXIT curve is $1 - R$. To our best knowledge, the area property of the EXIT chart has been proved only for cases where the priori channels are erasure channels.

The mean square error (MSE) transfer chart improves the EXIT chart as shown in [7], because the area property of the MSE transfer chart corresponding to the area property of the EXIT charts has been proven in both erasure channels and AWGN channels. Instead of tracking mutual information, the MSE transfer chart, as an alternative to evaluate decoding performance, has been proposed in [9] to track the iterative decoding performance based on the relationship between mutual information and the minimum mean square error (MMSE) for the additive white Gaussian noise (AWGN) channel.

In this paper, we use the MSE transfer chart to analyze the message passing procedure of channel decoding and system state estimation by assuming that the priori information is an AWGN channel. Compared with [9], our hybrid model prioritizes practicality because the system states and observations considered are continuous values while the information transmitted in wireless system are quantized information bits. Unlike previous research, our algorithm addresses the message passing between continuous values from the state estimator and quantized information bits from the channel decoder with the condition that the system state is correlated over different time slots. In addition, in order to view the evolution of the estimation error, we analyze the performance of state estimation in two cases: within two time slots and more than two time slots.

Our work is also informed by other areas in wireless communication, which have faced similar issues in source coding (quantization) [10–13] and joint source and channel decoding [14–19]. The idea of source coding (quantization) in the context of this work is combining the side information available at the controller to assist system state estimation, and the route of joint source and channel decoding in [20–25] is to utilize redundancy in the source to assist channel decoding. Our work can also be considered as one special case of joint source and channel decoding. However, there are two major differences. One difference is that most works focus on the source with binary values and use the EXIT chart [26, 27] or the protograph EXIT (PEXIT) [16] for performance analysis, while [28, 29] considered the case with the source of non-binary values, but the performance analysis of decoding was not provided. The other difference is that most works, such as [29], considered the joint source and channel decoding within two time slots. For instance, only the estimation from the previous time slot is used to calculate the estimation of current time slot. In our work, the dynamic state changes over more than two time slots, and the performance of channel decoding and system estimation at the current time slots also impacts its performance in all future time slots. Therefore, we study the performance of iterative estimation and decoding across multiple time slots.

This paper is organized as following. Following the review of the literature on iterative decoding performance analysis method in Section 1, Section 2 briefly introduces on the EXIT chart and the MSE transfer chart for the performance evaluation of iterative channel decoding. Section 3 describes the system models for performance analysis. Section 4 presents the message passing framework between system observation and channel decoding. Section 5 describes the MSE transfer chart, and Section 6 presents how to use the MSE transfer chart to evaluate BP-based sequential and iterative channel decoding and state estimation. Finally, a brief conclusion is given in Section 7.

2 Preliminaries on the EXIT Chart

In this section, we review the concept of the EXIT chart and the MSE transfer chart by iterative decoding the output of a serially concatenated encoder. In Section 2.1, we use an example to illustrate the serially concatenated coding scheme and its iterative decoding process. Then, in Section 2.2, we review how to use the EXIT chart and the MSE transfer chart to analyze the performance of iterative decoding.

2.1 A serially concatenated encoding scheme and corresponding iterative decoding algorithm

Figure 2 shows a simple serially concatenated encoding scheme and its corresponding iterative decoding scheme.

At the transmitter, the source $\mathbf{S}$ with binary values is a vector with the length L_s, i.e., $\mathbf{S} = [S_1, \cdots, S_{L_s}]$. $\mathbf{S}$ is encoded by the outer channel encoder, which is a systematic convolutional encoder whose generator is g_{out} with an output of $\mathbf{B}_{\text{out}}$, a vector with the length L_{out}. Next, $\mathbf{B}_{\text{out}}$ is encoded by the inner channel encoder, also a systematic convolutional encoder whose the generator is g_{in} with an output of $\mathbf{B}_{\text{in}}$, a vector with the length L_{in}. Finally, $\mathbf{B}_{\text{in}}$ is modulated with an output of $\mathbf{B}_m$, i.e., $B_{m,i} = 2B_{\text{in},i} - 1, i = 1, \cdots, L_{\text{in}}$, and then sent over an AWGN channel with an output that can be calculated by

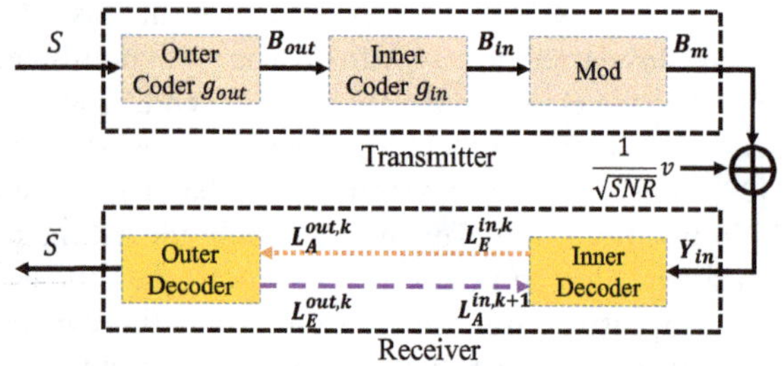

Fig. 2 A demo of concatenated encoding and iterative decoding

$$Y_{\text{in},i} = B_{m,i} + \frac{1}{\sqrt{\text{SNR}}} v_i \quad i = 1, \cdots, L_{\text{in}} \tag{1}$$

where $\text{SNR} = \frac{E_b}{N_0}$ is the signal power to noise power ratio and v_i is a zero mean and unit variance Gaussian noise.

At the receiver, decoding is done iteratively between the inner decoder and the outer decoder. The inputs for the inner channel decoder are the received signal $\mathbf{Y}_{\text{in}}$ and a priori information from the outer decoder, i.e., $\mathbf{L}_A^{\text{in},k} = \mathbf{L}_E^{\text{out},k-1}$, where $\mathbf{L}_E^{\text{out},k-1}$ is the extrinsic information of the outer decoder from $(k-1)$th decoding round, and the output for it is $\mathbf{L}_E^{\text{in},k}$, i.e.,

$$\mathbf{L}_{E,i}^{\text{in},k} = \text{LLR}\left(S_i | \mathbf{Y}_{\text{in}}, \mathbf{L}_{A,\bar{i}}^{\text{in},k}, g_{\text{in}}\right) \quad i = 1, \cdots, L_s \tag{2}$$

where $\mathbf{L}_{A,\bar{i}}^{\text{in},k}$ means the priori information of $\mathbf{L}_{A,i}^{\text{in},k}$ for all $\mathbf{S}$ except S_i. The input for the outer channel decoder is a priori information from the inner decoder, i.e., $\mathbf{L}_A^{\text{out},k} = \mathbf{L}_E^{\text{in},k}$, and the output for it is $\mathbf{L}_E^{\text{out},k}$, i.e.,

$$\mathbf{L}_{E,i}^{\text{out},k} = \text{LLR}\left(S_i | \mathbf{L}_{A,\bar{i}}^{\text{out},k}, g_{\text{out}}\right) \quad i = 1, \cdots, L_s \tag{3}$$

where $\mathbf{L}_{A,\bar{i}}^{\text{out},k}$ means the priori information of $\mathbf{L}_{A,i}^{\text{out},k}$ for all $\mathbf{S}$ except S_i.

2.2 The EXIT chart and the MSE transfer chart

The iterative decoding scheme in [30] can be analyzed by tracking the density evolution over iterations. However, the density evolution is complex as it requires to obtain probability density function (PDF) of extrinsic LLRs for each iteration; in addition, it does not provide much insight about the operations of iterative decoding. In order to overcome the drawbacks of density evolution, many transfer chart-based analysis frameworks have been proposed, such as the EXIT chart by [3–5] and the MSE transfer chart by [9]. The idea of these transfer chart-based analysis frameworks is to approximate the PDF of extrinsic LLRs exchanged between the inner decoder and outer decoder by a parameter, i.e.,

$$F(\mathbf{S}, \mathbf{L}) = \frac{1}{L_s} \sum_{i=1}^{L_s} F(S_i, \mathbf{L}_i), \quad i = 1, \cdots, L_s \tag{4}$$

where $\mathbf{L}_i$ is the extrinsic information, i.e. $\mathbf{L}_{E,i}^{\text{out},k}, \mathbf{L}_{E,i}^{\text{in},k}$, $\mathbf{L}_{A,i}^{\text{out},k}$, and $\mathbf{L}_{A,i}^{\text{in},k}$.

The measure used by the EXIT chart is mutual information, i.e., $F(\mathbf{S}, \mathbf{L}) = \mathbf{I}(\mathbf{S}, \mathbf{L})$, which is based on the observation that the PDF of extrinsic LLRs can be approximated by a Gaussian distribution [3–5].

The measure used by the MSE transfer chart is $F(\mathbf{S}, \mathbf{L}) = E\left[\tanh^2(L/2)\right]$, which is related to the MMSE estimation of $\mathbf{S}$ based on observation $\mathbf{Y}_{\text{in}}$ [9], i.e.,

$$\begin{cases} \text{MMSE}\left(\mathbf{S}|\mathbf{Y}_{\text{in}}\right) = E\left[\left(\mathbf{S}-\hat{\mathbf{S}}\right)^2 |\mathbf{Y}_{\text{in}}\right] = 1 - E\left[\tanh^2\left(\mathbf{L}/2\right)\right] \\ \hat{\mathbf{S}} = E\left[\mathbf{S}|\mathbf{Y}_{\text{in}}\right] = \tanh(\mathbf{L}/2) \\ \mathbf{L} = \text{LLR}\left(\mathbf{S}|\mathbf{Y}_{\text{in}}\right) = \log\left(\frac{P(\mathbf{S}=1,\mathbf{Y}_{\text{in}})}{P(\mathbf{S}=-1,\mathbf{Y}_{\text{in}})}\right) \end{cases} \tag{5}$$

The EXIT chart and the MSE transfer chart-based decoding frameworks for a serially concatenated encoding scheme are shown in Fig. 3 (a and b), respectively. The transfer chart includes two transfer curves. One curve is the measure for the priori information of inner decoder, i.e., $F_A^{\text{in}}(\mathbf{S},\mathbf{L})$, versus the measure for the extrinsic information of inner decoder, i.e., $F_E^{\text{in}}(\mathbf{S},\mathbf{L})$; the other curve is the measure for the priori information of outer decoder, i.e., $F_A^{\text{out}}(\mathbf{S},\mathbf{L})$, versus the measure for the extrinsic information of outer decoder, i.e., $F_E^{\text{out}}(\mathbf{S},\mathbf{L})$. The EXIT chart and the MSE transfer chart for the serially concatenated coding scheme are shown in Figs. 4 and 5, respectively. The predicted decoding path is also shown in these two figures. Since there is a decoding path found between the two curves, the iterative decoding converges.

3 System model

In this section, we describe the system model for the analysis.

3.1 Linear dynamic system and communication system

We consider a discrete time linear dynamic system, whose state evolution is given by

$$\begin{cases} \mathbf{x}(t+1) = \mathbf{A}\mathbf{x}(t) + \mathbf{B}\mathbf{u}(t) + \mathbf{n}(t) \\ \mathbf{y}(t) = \mathbf{C}\mathbf{x}(t) + \mathbf{w}(t) \end{cases} \tag{6}$$

where $\mathbf{x}(t)$ is the N-dimensional vector of system state at time slot t, $\mathbf{u}(t)$ is the M-dimensional control vector, $\mathbf{y}(t)$ is the K-dimensional observation vector, and $\mathbf{n}(t)$ and $\mathbf{w}(t)$ are noise vectors, which are assumed to be Gaussian distributions with zero mean and covariance matrix Σ_n and Σ_w, respectively. For simplicity, we do not consider $\mathbf{u}(t)$.

Additionally, we assume that the observation vector $\mathbf{y}(t)$ is obtained by a sensor[1], and the sensor quantizes each dimension of the observation vector $\mathbf{y}(t)$ using B bits, thus forming a KB-dimensional binary vector, which is given by

$$\mathbf{b}(t) = (b_1(t), b_2(t), \ldots, b_{KB}(t)) \tag{7}$$

The information bits $\mathbf{b}(t)$ are then put into an encoder to generate a codeword $\mathbf{c}(t)$. Suppose that the binary phase-shift keying (BPSK) is used for the transmission between the sensor and the controller, and $\mathbf{c}(t)$ is converted to alphabet $\{-1,+1\}$ with $\mathbf{s}(t) = 2\mathbf{c}(t)-1$. Next, the sequence $\mathbf{s}(t)$ is passed through a modulator and transmitted into an AWGN channel. Then, the received signal at the controller is given by

$$\mathbf{r}(t) = \mathbf{s}(t) + \mathbf{e}(t) \tag{8}$$

where the additive white Gaussian noise $\mathbf{e}(t)$ has a zero expectation and variance Σ_c. Note that we consider the AWGN channel, ignore the fading and normalize the transmit power to be 1. The algorithm and conclusion in this work can be easily extended to the cases with different channels and different types of fading.

3.2 Models for belief propagation based channel decoding and state estimation

In this section, we firstly introduce the Bayesian network structure and then use the following two models to evaluate BP-based channel decoding and state estimation: BP-based sequential processing and BP-based iterative processing.

3.2.1 *Bayesian network structure and the message passing*

The Bayesian network structure of the dynamic system and communication system is shown in Fig. 6, where the

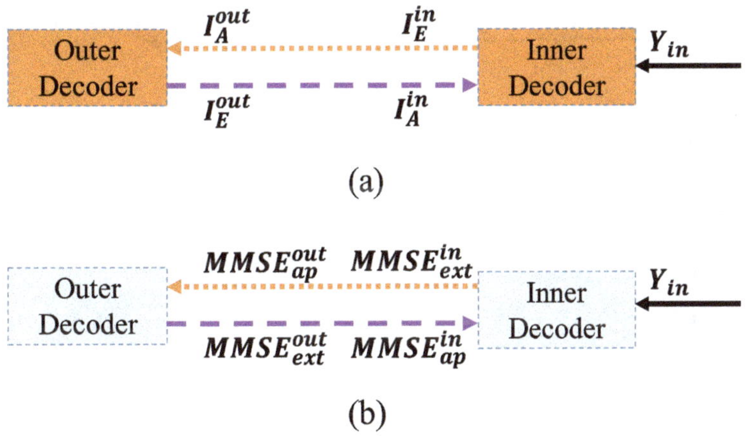

Fig. 3 The EXIT chart and the MSE transfer chart model for concatenated encoding and iterative decoding. **a** EXIT chart. **b** MSE based transfer chart

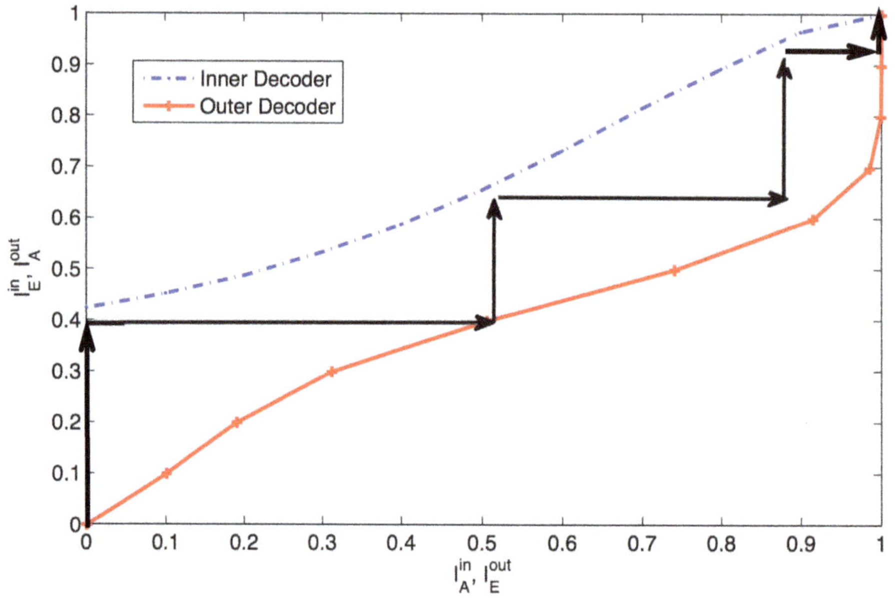

Fig. 4 An example of the EXIT chart for concatenated encoding and iterative decoding, $g_{out} = g_{in} = [1, 1, 0, 1; 1, 0, 0, 1]$, SNR $= -2$ dB

message passing in the system is illustrated by dotted arrows and dashed arrows in the figure with three time slots: $\mathbf{x}(t-2)$, $\mathbf{x}(t-1)$, and $\mathbf{x}(t)$. The dashed arrows transmit π-message from a parent to its children in Pearl's BP, and the details and formal description can be found in [31]. For instance, the message passed from $\mathbf{x}(t-2)$ to $\mathbf{x}(t-1)$ is $\pi_{\mathbf{x}(t-2),\mathbf{x}(t-1)}(\mathbf{x}(t-2))$, which is the priori information of $\mathbf{x}(t-2)$ given that all the information $\mathbf{x}(t-2)$ has been received. The dotted arrows transmit λ-message from a child to its parent. For instance, the message passed from $\mathbf{x}(t-1)$ to $\mathbf{x}(t-2)$ is $\lambda_{\mathbf{x}(t-1),\mathbf{x}(t-2)}(\mathbf{x}(t-2))$, which is the likelihood of $\mathbf{x}(t-2)$ given that the information $\mathbf{x}(t-1)$ has been received.

Note that the π-message and λ-message are passing in the form of probability distribution function (PDF), and based on the Bayesian network structure in Fig. 6 and Pearl's BP, the updating order and the message passing in one iteration is given as follows: step 1: $\mathbf{x}(t-1) \to \mathbf{y}(t-1)$; step 2: $\mathbf{y}(t-1) \to \mathbf{b}(t-1)$; step 3: $\mathbf{b}(t-1) \to \mathbf{y}(t-1)$; step 4: $\mathbf{y}(t-1) \to \mathbf{x}(t-1)$; step 5: $\mathbf{x}(t-1) \to \mathbf{x}(t)$; step 6: $\mathbf{x}(t) \to \mathbf{y}(t)$; step 7: $\mathbf{y}(t) \to \mathbf{b}(t)$; step 8: $\mathbf{b}(t) \to \mathbf{y}(t)$; step 9: $\mathbf{y}(t) \to \mathbf{x}(t)$; step 10: $\mathbf{x}(t) \to \mathbf{x}(t-1)$; and step 11:

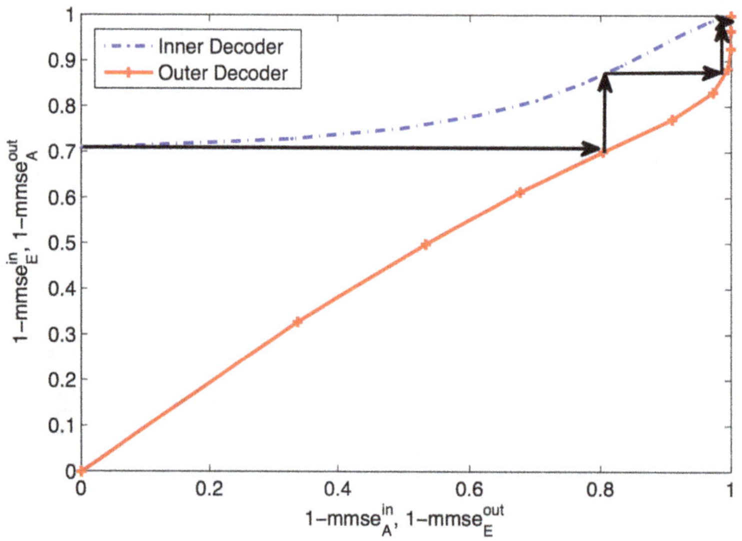

Fig. 5 An example of the MSE transfer chart for concatenated encoding and iterative decoding, $g_{out} = g_{in} = [1, 1, 0, 1; 1, 0, 0, 1]$, SNR $= -2$ dB

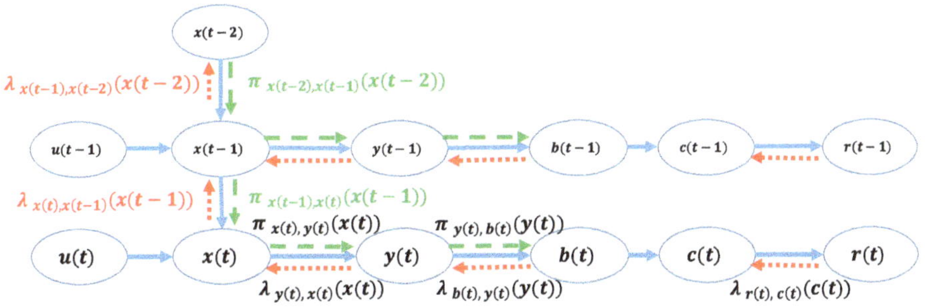

Fig. 6 Bayesian network structure and the message passing for CPSs

$\mathbf{x}(t-1)$ updates information. The key steps in the Pearl's BP have be derived as shown in Table 1, where the PDF of $\lambda_{\mathbf{b}(t),\mathbf{y}(t)}(\mathbf{y}(t))$ will be computed by iterative decoding provided in Section 4.

3.2.2 Model for BP-based sequential processing

Based on the Bayesian network structure in Fig. 6, we use the framework as shown in Fig. 7 (a) to evaluate sequential channel decoding and state estimation over more than two time slots. The priori information from the time slot $t-1$, i.e., $\pi_{\mathbf{x}(t-1),\mathbf{x}(t)}(\mathbf{x}(t-1))$, is used to assist channel decoding and state estimation at the time slot t, i.e., the estimated distribution of $\mathbf{x}(t-1)$, and we assume that it is also a Gaussian distribution with mean $\mathbf{x}_{\pi_x,t-1}$ and covariance matrix $\mathbf{P}_{\pi_x,t-1}$, i.e., $\mathcal{N}(\mathbf{x}_{t-1}, \mathbf{x}_{\pi_x,t-1}, \mathbf{P}_{\pi_x,t-1})$.

As noted in Section 1, two objectives here are to evaluate how much gain can be obtained by utilizing $\pi_{\mathbf{x}(t-1),\mathbf{x}(t)}(\mathbf{x}(t-1))$ to assist channel decoding and state estimation at time slot t and to evaluate the performance of state estimation over multiple time slots, i.e., the evolution of $\mathbf{P}_{\pi_x,t-1}$ as shown in Fig. 7(b).

3.2.3 Model for BP-based iterative processing between two time slots

Figure 8(a) illustrates the model used to evaluate BP-based iterative channel decoding and state estimation between two time slots. The inputs for this model include the received signals at the controller within the two time slots, i.e., $\mathbf{r}(t)$ and $\mathbf{r}(t+1)$, which can be found in Fig. 6, and the priori information from the previous time slot $t-1$, i.e., $\pi_{\mathbf{x}(t-1),\mathbf{x}(t)}(\mathbf{x}(t-1))$.

The goal is to evaluate the performance of iterative channel decoding and state estimation for different realizations of the two distributions for $\pi_{\mathbf{x}(t-1),\mathbf{x}(t)}(\mathbf{x}(t-1))$. For instance, when $\pi_{\mathbf{x}(t-1),\mathbf{x}(t)}(\mathbf{x}(t-1))$ (say, $\mathbf{P}_{\pi_x,t-1}$) equals to $0 \times \mathbf{I}$, $\mathbf{x}(t-1)$ is a determined state estimation. Therefore, this reference model can be converted to the model shown in Fig. 8 (b) by setting $\pi_{\mathbf{x}(t-1),\mathbf{x}(t)}(\mathbf{x}(t-1)$ (say, $\mathbf{P}_{\pi_x,t-1}$) as $0\times\mathbf{I}$. Or if $\pi_{\mathbf{x}(t-1),\mathbf{x}(t)}(\mathbf{x}(t-1))$ (say, $\mathbf{P}_{\pi_x,t-1}$) is set as $\infty \times \mathbf{I}$, $\mathbf{x}(t-1)$ is unknown. Then, the reference model is transformed to the model shown in Fig. 8 (c).

4 Message passing between state estimator and channel decoder

In this section, we describe the message passing between state estimator and channel decoder, which is the most challenging and vital part for the evaluation of BP-based channel decoding and state estimation. As stated in Table 1, $\pi_{\mathbf{y}(t),\mathbf{b}(t)}(\mathbf{y}(t))$ has been assumed to be a Gaussian distribution with the mean $\mathbf{y}_{\pi,t}$ and the covariance matrix $\mathbf{S}_{\pi,t}$ and $\lambda_{\mathbf{b}(t),\mathbf{y}(t)}(\mathbf{y}(t))$ with the mean $\mathbf{y}_{\lambda,t}$ and

Table 1 Message passing in BP-based channel decoding and state estimation system

Step	Distribution	Gaussian distribution	Details
$\mathbf{x}(t-1) \to \mathbf{x}(t)$	$\pi_{\mathbf{x}(t-1),\mathbf{x}(t)}(\mathbf{x}(t-1))$	$\mathcal{N}(\mathbf{x}_{t-1}, \mathbf{x}_{\pi_x,t-1}, \mathbf{P}_{\pi_x,t-1})$	*
$\mathbf{x}(t) \to \mathbf{y}(t)$	$\pi_{\mathbf{x}(t)}(\mathbf{x}(t))$	$\mathcal{N}(\mathbf{x}_t, \mathbf{x}_{I,t}, \mathbf{P}_{I,t})$	$\mathbf{x}_{I,t} = \mathbf{A}\mathbf{x}_{\pi_x,t-1} + \mathbf{B}\mathbf{u}_{t-1}$; $\mathbf{P}_{I,t} = \mathbf{A}\mathbf{P}_{\pi_x,t-1}\mathbf{A}^T + \Sigma_n$
–	$\pi_{\mathbf{x}(t),\mathbf{y}(t)}(\mathbf{x}(t))$	$\mathcal{N}(\mathbf{x}_t, \mathbf{x}_{\pi_y,t}, \mathbf{P}_{\pi_y,t})$	$\mathbf{x}_{\pi_y,t} = \mathbf{x}_{I,t}$; $\mathbf{P}_{\pi_y,t} = \mathbf{P}_{I,t}$
$\mathbf{y}(t) \to \mathbf{b}(t)$	$\pi_{\mathbf{y}(t)}(\mathbf{y}(t))$	$\mathcal{N}(\mathbf{y}_t, \mathbf{y}_{I,t}, \mathbf{S}_{I,t})$	$\mathbf{y}_{I,t} = \mathbf{C}\mathbf{x}_{\pi_y,t}$; $\mathbf{S}_{I,t} = \mathbf{C}\mathbf{P}_{\pi_y,t}\mathbf{C}^T + \Sigma_w$
–	$\pi_{\mathbf{y}(t),\mathbf{b}(t)}(\mathbf{y}(t))$	$\mathcal{N}(\mathbf{y}_t, \mathbf{y}_{\pi,t}, \mathbf{S}_{\pi,t})$	$\mathbf{y}_{\pi,t} = \mathbf{y}_{I,t}$; $\mathbf{S}_{\pi,t} = \mathbf{S}_{I,t}$
$\mathbf{b}(t) \to \mathbf{y}(t)$	$\lambda_{\mathbf{b}(t),\mathbf{y}(t)}(\mathbf{y}(t))$	$\mathcal{N}(\mathbf{y}_t, \mathbf{y}_{\lambda,t}, \mathbf{S}_{\lambda,t})$	The $\mathbf{y}_{\lambda,t}$ and $\mathbf{S}_{\lambda,t}$ is provided in Section 4.
$\mathbf{y}(t) \to \mathbf{x}(t)$	$\lambda_{\mathbf{y}(t),\mathbf{x}(t)}(\mathbf{x}(t))$	$\mathcal{N}(\mathbf{x}_t, \mathbf{x}_{\lambda_y,t}, \mathbf{P}_{\lambda_y,t})$	$\mathbf{x}_{\lambda_y,t} = \mathbf{C}^{-1} \times \mathbf{y}_{\lambda,t}$; $\mathbf{P}_{\lambda_y,t} = \mathbf{C}^{-1}(\mathbf{S}_{\lambda,t} + \Sigma_w)(\mathbf{C}^{-1})^T$
$\mathbf{x}(t) \to \mathbf{x}(t+1)$	$\pi_{\mathbf{x}(t),\mathbf{x}(t+1)}(\mathbf{x}(t))$	$\mathcal{N}(\mathbf{x}_t, \mathbf{x}_{\pi_x,t}, \mathbf{P}_{\pi_x,t})$	$\mathbf{P}_{\pi_x,t} = (\mathbf{P}_{I,t}^{-1} + \mathbf{P}_{\lambda_y,t}^{-1})^{-1}$; $\mathbf{x}_{\pi_x,t} = \mathbf{P}_{\pi_x,t}(\mathbf{P}_{I,t}^{-1}\mathbf{x}_{I,t} + \mathbf{P}_{\lambda_y,t}^{-1}\mathbf{x}_{\lambda_y,t})$

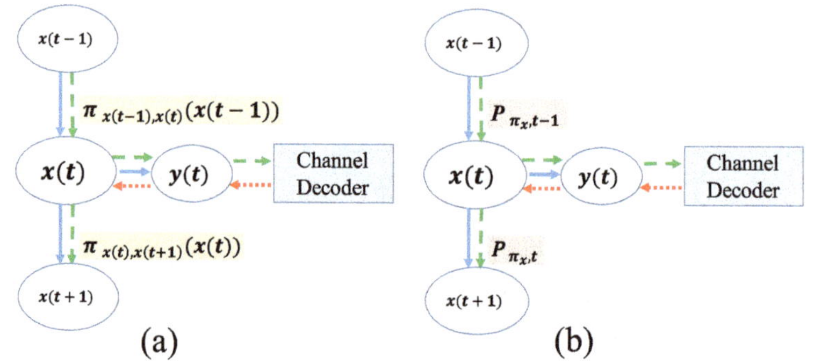

Fig. 7 a Sequential channel decoding and state estimation, **b** The measure of sequential channel decoding and state estimation

covariance matrix $\mathbf{S}_{\lambda,t}$. We can simplify the procedure of the message exchanging by substituting $\pi_{\mathbf{y}(t),\mathbf{b}(t)}(\mathbf{y}(t))$ and $\lambda_{\mathbf{b}(t),\mathbf{y}(t)}(\mathbf{y}(t))$ with $\mathbf{S}_{\pi,t}$ and $\mathbf{S}_{\lambda,t}$, respectively, as shown in Fig. 9. Below, we explain how the averaging $\mathbf{S}_{\lambda,t}$, i.e., $\bar{\mathbf{S}}_{\lambda,t}$, corresponding to $\mathbf{S}_{\pi,t}$ can be computed by using iterative decoding.

4.0.4 Quantizing the message between channel decoder and state estimator

The relationships among $\mathbf{y}_{\pi,t}$, $\mathbf{S}_{\pi,t}$, $\mathbf{y}'(t)$, and channel noise are shown in Fig. 10, where $\mathbf{y}'(t)$ is one realization of $\pi_{\mathbf{y}(t),\mathbf{b}(t)}(\mathbf{y}(t))$. Then, $\mathbf{y}'(t)$ is quantized, modulated, and transmitted over the wireless channel. We denote the corresponding quantized vector, modulated vector, channel noise vector, and received vector as $\mathbf{b}'(t)$, $\mathbf{c}'(t)$, $\mathbf{e}'(t)$ and $\mathbf{r}'(t)$, respectively.

The physical meaning of $\pi_{\mathbf{y}(t),\mathbf{b}(t)}(\mathbf{y}(t))$ is the PDF of $\mathbf{y}(t)$. Next, we can use $\pi_{\mathbf{y}(t),\mathbf{b}(t)}(\mathbf{y}(t))$ as the priori information to estimate each bit of $\mathbf{b}(t)$ based on its quantization scheme (the quantization scheme used for converting $\mathbf{y}(t)$ to $\mathbf{b}(t)$), i.e., $\mathbf{L}_A(\mathbf{y}_{\pi,t}, \mathbf{S}_{\pi,t})$. Note that $\mathbf{L}_A(\mathbf{y}_{\pi,t}, \mathbf{S}_{\pi,t})$ is determined by the PDF, i.e., $\pi_{\mathbf{y}(t),\mathbf{b}(t)}(\mathbf{y}(t))$, and also the quantization scheme. Thus, $\mathbf{L}_A(\mathbf{y}_{\pi,t}, \mathbf{S}_{\pi,t})$ is a *KB*-dimensional vector and its ith element can be calculated as

$$\mathbf{L}_{A,i}\left(\mathbf{y}_{\pi,t}, \mathbf{S}_{\pi,t}\right) = \log \frac{P\left(b_i(t) = 1 | \mathbf{y}(t) \in \mathcal{N}\left(\mathbf{y}_t, \mathbf{y}_{\pi,t}, \mathbf{S}_{\pi,t}\right)\right)}{P\left(b_i(t) = 0 | \mathbf{y}(t) \in \mathcal{N}\left(\mathbf{y}_t, \mathbf{y}_{\pi,t}, \mathbf{S}_{\pi,t}\right)\right)} \quad (9)$$

The inputs for the channel decoder are $\mathbf{L}_A(\mathbf{y}_{\pi,t}, \mathbf{S}_{\pi,t})$ and $\mathbf{r}'(t)$ shown in Fig. 10. The feedback from the channel decoder to the state estimator is the extrinsic LLR, which is denoted as $\mathbf{L}_E(\mathbf{L}_A(\mathbf{y}_{\pi,t}, \mathbf{S}_{\pi,t}), \mathbf{y}'(t), \mathbf{e}'(t))$. It is a *KB*-dimension vector and its ith element equals to

$$\mathbf{L}_{E,i}\left(\mathbf{L}_A\left(\mathbf{y}_{\pi,t}, \mathbf{S}_{\pi,t}\right), \mathbf{y}'(t), \mathbf{e}'(t)\right) = \log \frac{P\left(b_i(t) = 1 | \mathbf{L}_A\left(\mathbf{y}_{\pi,t}, \mathbf{S}_{\pi,t}\right), \mathbf{r}'(t)\right)}{P\left(b_i(t) = 0 | \mathbf{L}_A\left(\mathbf{y}_{\pi,t}, \mathbf{S}_{\pi,t}\right), \mathbf{r}'(t)\right)} \quad (10)$$

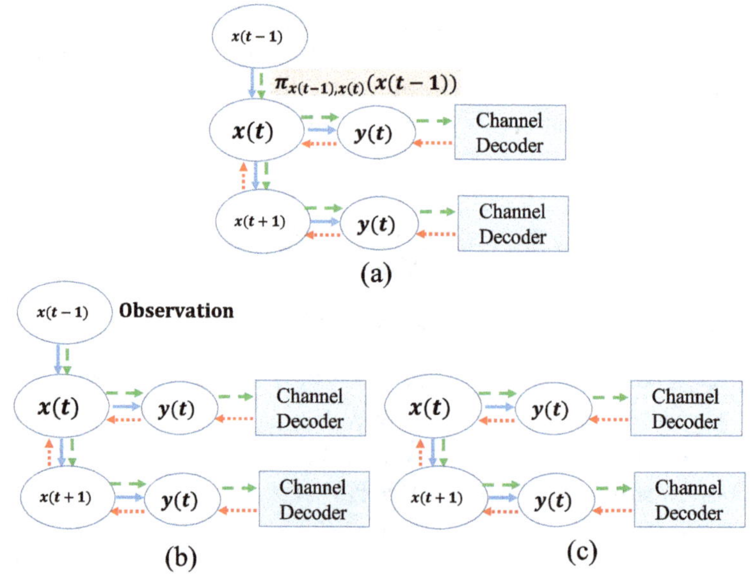

Fig. 8 Model for the iterative message passing with priori information from $\mathbf{x}(t-1)$. **a** Iterative channel decoding and state estimation, **b** Iterative channel decoding and state estimation with previous observation, **c** Iterative channel decoding and state estimation without previous observation

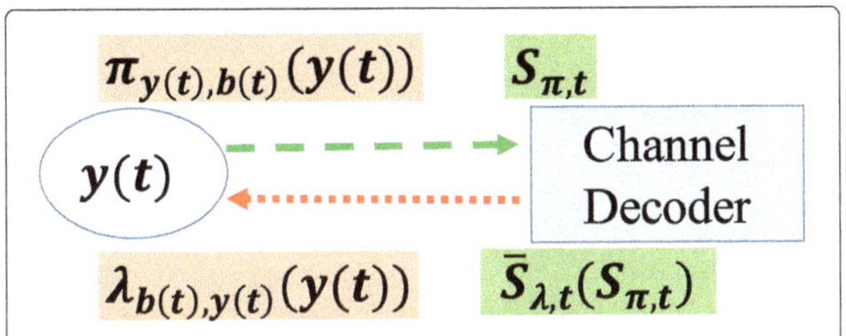

Fig. 9 Information exchange between channel decoder and state estimator

Note that $\mathbf{L}_E(\mathbf{L}_A(\mathbf{y}_{\pi,t}, \mathbf{S}_{\pi,t}), \mathbf{y}'(t), \mathbf{e}'(t))$ depends on $\mathbf{L}_A(\mathbf{y}_{\pi,t}, \mathbf{S}_{\pi,t})$, $\mathbf{y}'(t)$, and $\mathbf{e}'(t)$. Based on $\mathbf{L}_E(\mathbf{L}_A(\mathbf{y}_{\pi,t}, \mathbf{S}_{\pi,t}), \mathbf{y}'(t), \mathbf{e}'(t))$, if we assume $\mathbf{c}(t) = \mathbf{b}(t)$, based on [7], $\mathbf{b}_i(t)$ can be estimated as

$$\begin{aligned}\tilde{\mathbf{b}}_i(t) &= \frac{1}{2}\left(\tilde{\mathbf{s}}_i(t) + 1\right) \\ &= \frac{1}{2}\tanh\left\{\frac{L_{E,i}\left[\mathbf{L}_A\left(\mathbf{y}_{\pi,t}, \mathbf{S}_{\pi,t}\right), \mathbf{y}'(t), \mathbf{e}'(t)\right]}{2}\right\} + \frac{1}{2}\end{aligned} \tag{11}$$

where $\tilde{\mathbf{s}}_i(t)$ is the estimated-modulated bit and the MMSE in estimating $\mathbf{b}_i(t)$ from $\mathbf{L}_E(\mathbf{L}_A(\mathbf{y}_{\pi,t}, \mathbf{S}_{\pi,t}), \mathbf{y}'(t), \mathbf{e}'(t))$ is given as

$$\begin{aligned}&\text{MMSE}_{\mathbf{b}_i(t)} \\ &= \text{MMSE}\left(\mathbf{b}_i(t) | \mathbf{L}_E\left(\mathbf{L}_A\left(\mathbf{y}_{\pi,t}, \mathbf{S}_{\pi,t}\right), \mathbf{y}'(t), \mathbf{e}'(t)\right)\right) \\ &= \frac{1}{4} - \frac{1}{4}\tanh^2\left(\frac{L_{E,i}\left[\mathbf{L}_A\left(\mathbf{y}_{\pi,t}, \mathbf{S}_{\pi,t}\right), \mathbf{y}'(t), \mathbf{e}'(t)\right]}{2}\right)\end{aligned} \tag{12}$$

Then, $\mathbf{y}(t)$ can be estimated as

$$\tilde{\mathbf{y}}_k(t) = Q_I \sum_{i=1}^{B} \tilde{\mathbf{b}}_{(k-1)B+i}(t) 2^{i-1} + Q_{\min} \tag{13}$$

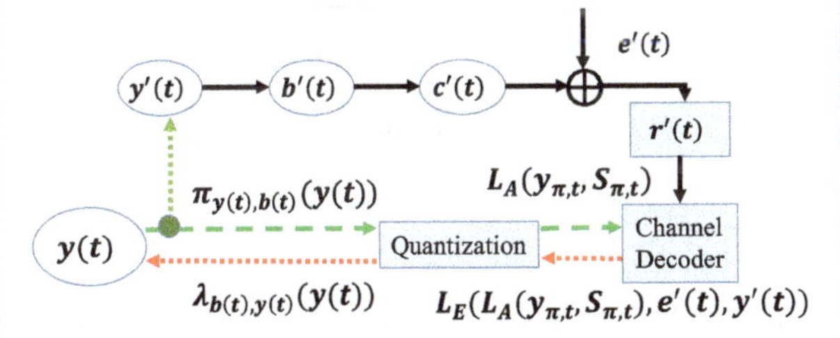

Fig. 10 Model for $\mathbf{S}_{\lambda,t}$ based on sample $\mathbf{y}_{\pi,t}$, $\mathbf{S}_{\pi,t}$, $\mathbf{y}'(t)$, and $\mathbf{e}'(t)$

and the MMSE in estimating $\mathbf{y}_k(t)$ is given by

$$\begin{aligned}&\text{MMSE}_{\mathbf{y}_k(t)} \\ &= \text{MMSE}\left(\mathbf{y}_k(t) | \mathbf{L}_E\left(\mathbf{L}_A\left(\mathbf{y}_{\pi,t}, \mathbf{S}_{\pi,t}\right), \mathbf{y}'(t), \mathbf{e}'(t)\right)\right) \\ &= (Q_I)^2 E\left\{\sum_{i=1}^{B}[\,\tilde{\mathbf{b}}_{(k-1)B+i}(t) - \mathbf{b}_{(k-1)B+i}(t)]\,2^{i-1}\right\}^2 \\ &= (Q_I)^2 \sum_{i=1}^{B} \text{MMSE}_{\mathbf{b}_i(t)} 2^{2(i-1)}\end{aligned} \tag{14}$$

where $[\,Q_{\min}, Q_{\max}]$ is the range for quantization and Q_I is the quantization interval, which is given as $Q_I = \frac{Q_{\max} - Q_{\min}}{2^B - 1}$. Note that when $i \neq j$, $E\{[\,\tilde{\mathbf{b}}_{(k-1)B+i}(t) - \mathbf{b}_{(k-1)B+i}(t)]\,[\,\tilde{\mathbf{b}}_{(k-1)B+j}(t) - \mathbf{b}_{(k-1)B+j}(t)]\,\} = 0$, which is obtained based on the independence of channel noise and required for the derivation of (14).

Then, for a given $\pi_{\mathbf{y}(t),\mathbf{b}(t)}(\mathbf{y}(t))$, which has PDF as $\mathcal{N}(\mathbf{y}_t, \mathbf{y}_{\pi,t}, \mathbf{S}_{\pi,t})$, a realization $\mathbf{y}'(t)$, and a channel noise $\mathbf{e}'(t)$, the feedback from the channel decoder to the node $\mathbf{y}(t)$ is $\lambda_{\mathbf{b}(t),\mathbf{y}(t)}(\mathbf{y}(t))$, corresponding to $\mathcal{N}(\mathbf{y}_t, \mathbf{y}_{\lambda,t}, \mathbf{S}_{\lambda,t})$, whose mean and covariance matrix are given as

$$\begin{aligned}\mathbf{y}_{\lambda,t} &= \left[\tilde{\mathbf{y}}_1(t), \cdots, \tilde{\mathbf{y}}_K(t)\right] \\ \mathbf{S}_{\lambda,t}(k,k) &= \text{MMSE}_{\mathbf{y}_k(t)}\end{aligned} \tag{15}$$

Thus, we finalize the computation of $\mathcal{N}(\mathbf{y}_t, \mathbf{y}_{\lambda,t}, \mathbf{S}_{\lambda,t})$ based on one realization, $\mathbf{y}'(t)$, which is obtained from the PDF of $\pi_{\mathbf{y}(t),\mathbf{b}(t)}(\mathbf{y}(t))$.

4.0.5 Approximation for the message passing between state estimator and channel decoder

As shown in Fig. 11 (a), for a given $\pi_{\mathbf{y}(t),\mathbf{b}(t)}(\mathbf{y}(t))$, the extrinsic information transferred from the channel decoder to the node $\mathbf{y}(t)$ can be computed as

$$\begin{aligned}&\mathbf{S}_{\lambda,t}\left(\mathbf{y}_{\pi,t}, \mathbf{S}_{\pi,t}\right) \\ &= E_{\mathbf{y}'(t) \in \pi_{\mathbf{y}(t),\mathbf{b}(t)}(\mathbf{y}(t)), \mathbf{e}'(t)}\left[\text{MMSE}_{\mathbf{y}_k(t)}\right]\end{aligned} \tag{16}$$

And, as shown in Fig. 11 (b), if we denote the PDF of $\mathbf{y}_{\pi,t}$ as $f_{\mathbf{y}_{\pi,t}}$, we can obtain the average $\mathbf{S}_{\lambda,t}$ corresponding to $\mathbf{S}_{\pi,t}$ by integrating $\mathbf{S}_{\lambda,t}(\mathbf{y}_{\pi,t}, \mathbf{S}_{\pi,t})$ over $\mathbf{y}_{\pi,t}$,

$$\begin{aligned}\mathbf{S}_{\lambda,t}(\mathbf{S}_{\pi,t}) &= E_{\mathbf{y}_{\pi,t}}\left[\mathbf{S}_{\lambda,t}\left(\mathbf{y}_{\pi,t}, \mathbf{S}_{\pi,t}\right)\right] \\ &= E_{\mathbf{y}_{\pi,t}}\left\{E_{\mathbf{y}'(t) \in \pi_{\mathbf{y}(t),\mathbf{b}(t)}(\mathbf{y}(t)), \mathbf{e}'(t)}\left[\text{MMSE}_{\mathbf{y}_k(t)}\right]\right\}\end{aligned} \tag{17}$$

Note that the computation of $\mathbf{S}_{\lambda,t}(\mathbf{S}_{\pi,t})$ in (17) would require the averaging over a sufficient number of realizations (i.e., large enough to show the probability distribution based on a fixed $\mathbf{y}_{\pi,t}$) of $\mathbf{y}'(t)$ and the averaging over all possible $\mathbf{y}_{\pi,t}$ based on its PDF $f_{\mathbf{y}_{\pi,t}}$ to be computed sequentially, i.e., *Forward Process* and *Backward Process* should be calculated sequentially. The Forward Process generally refers to the message passing from a node to

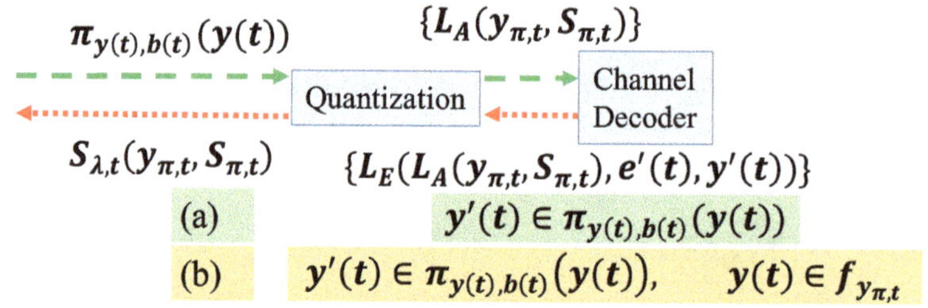

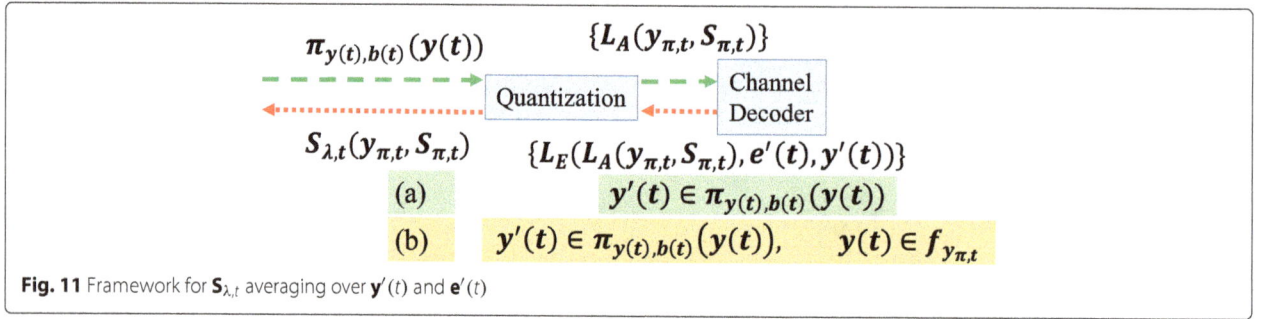

Fig. 11 Framework for $\mathbf{S}_{\lambda,t}$ averaging over $\mathbf{y}'(t)$ and $\mathbf{e}'(t)$

its children, and the Backward Process generally refers to the message passing from a node to its parents. This arises because $\mathbf{y}_{\pi,t}$ not only impacts priori information $\mathbf{L}_A(\mathbf{y}_{\pi,t}, \mathbf{S}_{\pi,t})$ but also defines the set of codewords which are generated by quantizing $\mathbf{y}'(t)$.

However, the computation as discussed above prohibits the independent evaluation of the message passing for Forward Process and Backward Process. Thus, to bypass this difficulty, we make the following approximations for the computation of $\mathbf{S}_{\lambda,t}(\mathbf{S}_{\pi,t})$, so that the Forward Process and Backward Process can be considered separately:

$$\begin{aligned}\mathbf{S}_{\lambda,t}(\mathbf{S}_{\pi,t}) &= E_{\mathbf{y}_{\pi,t}}\left[\mathbf{S}_{\lambda,t}\left(\mathbf{y}_{\pi,t}, \mathbf{S}_{\pi,t}\right)\right] \\ &= E_{\mathbf{y}_{\pi,t}}\left\{E_{\mathbf{y}'(t)\in \boldsymbol{U}(\mathbf{R}(\mathbf{y})),\mathbf{e}'(t)}\left[\mathrm{MMSE}_{\mathbf{y}_k(t)}\right]\right\} \\ &= E_{\mathbf{y}_{\pi,t},\mathbf{y}'(t)\in \boldsymbol{U}(\mathbf{R}(\mathbf{y})),\mathbf{e}'(t)}\left[\mathrm{MMSE}_{\mathbf{y}_k(t)}\right] \\ &= E_{\mathbf{L}_A(\mathbf{y}_{\pi,t},\mathbf{S}_{\pi,t}),\mathbf{y}'(t)\in \boldsymbol{U}(\mathbf{R}(\mathbf{y})),\mathbf{e}'(t)}\left[\mathrm{MMSE}_{\mathbf{y}_k(t)}\right]\end{aligned} \tag{18}$$

First, we approximate the PDF of the realizations $\mathbf{y}'(t)$ as a uniform distribution instead of $\mathcal{N}(\mathbf{y}_t, \mathbf{y}_{\pi,t}, \mathbf{S}_{\pi,t})$, and we denote the uniform distribution as $\boldsymbol{U}(\mathbf{R}(\mathbf{y}))$, where $\mathbf{R}(\mathbf{y})$ is the range of $\mathbf{y}(t)$. Given this approximation, the integration of $\mathbf{y}'(t)$ over $\boldsymbol{U}(\mathbf{R}(\mathbf{y}))$ is equivalent to the integration of all codewords with an equal probability, and hence, the realization $\mathbf{y}'(t)$ can be viewed as being independent of $\pi_{\mathbf{y}(t),\mathbf{b}(t)}(\mathbf{y}(t))$. Next, as shown in the final step of (18), the averaging over $\mathbf{y}_{\pi,t}$ can be changed to the averaging over $\mathbf{L}_A(\mathbf{y}_{\pi,t}, \mathbf{S}_{\pi,t})$. This exists because $\mathbf{y}_{\pi,t}$ impacts only priori information $\mathbf{L}_A(\mathbf{y}_{\pi,t}, \mathbf{S}_{\pi,t})$ and the set $\mathbf{L}_A(\mathbf{y}_{\pi,t}, \mathbf{S}_{\pi,t})$ includes all information from $\mathbf{y}_{\pi,t}$. The framework equivalent to (18) is shown in Fig. 12 (a), which is denoted as approximate framework 1, and the computation of $\mathbf{S}_{\lambda,t}(\mathbf{S}_{\pi,t})$ in (18) can be divided into the following three steps:

1. Compute the PDF of $\mathbf{L}_A(\mathbf{y}_{\pi,t}, \mathbf{S}_{\pi,t})$ and $f_{\mathbf{y}_{\pi,t}}$.
2. Compute the extrinsic information from the channel decoder, i.e., the PDF of $\mathbf{L}_E(\mathbf{L}_A(\mathbf{y}_{\pi,t}, \mathbf{S}_{\pi,t}), \mathbf{y}'(t), \mathbf{e}'(t))$.
3. Compute $\mathbf{S}_{\lambda,t}(\mathbf{S}_{\pi,t})$ from the extrinsic information $\mathbf{L}_E(\mathbf{L}_A(\mathbf{y}_{\pi,t}, \mathbf{S}_{\pi,t}), \mathbf{y}'(t), \mathbf{e}'(t))$.

Second, we can approximate the PDF of both $\mathbf{L}_A(\mathbf{y}_{\pi,t}, \mathbf{S}_{\pi,t})$ and $\mathbf{L}_E(\mathbf{L}_A(\mathbf{y}_{\pi,t}, \mathbf{S}_{\pi,t}), \mathbf{y}'(t), \mathbf{e}'(t))$ as the Gaussian distributions with zero means. Next, only the covariance matrixes need to be passed through the decoding process. Note that $\mathbf{L}_E(\mathbf{L}_A(\mathbf{y}_{\pi,t}, \mathbf{S}_{\pi,t}), \mathbf{y}'(t), \mathbf{e}'(t))$ for different bits might have a huge difference. This Gaussian approximation requires a good interleaver which can evenly distribute $\mathbf{L}_E(\mathbf{L}_A(\mathbf{y}_{\pi,t}, \mathbf{S}_{\pi,t}), \mathbf{y}'(t), \mathbf{e}'(t))$. Furthermore, if we represent the LLRs by mutual information or MMSE [7] as shown in Fig. 12(b), $\mathbf{S}_{\lambda,t}(\mathbf{S}_{\pi,t})$ can be computed in the following three steps:

1. Compute the mutual information I_A based on the PDF of $\mathbf{L}_A(\mathbf{y}_{\pi,t}, \mathbf{S}_{\pi,t})$ corresponding to $\mathbf{S}_{\pi,t}$ and $f_{\mathbf{y}_{\pi,t}}$.
2. Compute the extrinsic information from the channel decoder, i.e., mutual information I_E or $\mathrm{MMSE}^B_{\mathrm{ext}}$ based on the PDF of $\mathbf{L}_E(\mathbf{L}_A(\mathbf{y}_{\pi,t}, \mathbf{S}_{\pi,t}), \mathbf{y}'(t)), \mathbf{e}'(t))$.
3. Compute $\mathbf{S}_{\lambda,t}(\mathbf{S}_{\pi,t})$ from the extrinsic information of the channel decoder I_E or $\mathrm{MMSE}^B_{\mathrm{ext}}$.

5 The MSE transfer chart for channel decoding and state estimation

In this section, we show how to obtain the MSE transfer chart for evaluating BP-based sequential channel decoding and state estimation as shown in Fig. 7(b) and BP-based iterative channel decoding and state estimation as shown in Fig. 8(a).

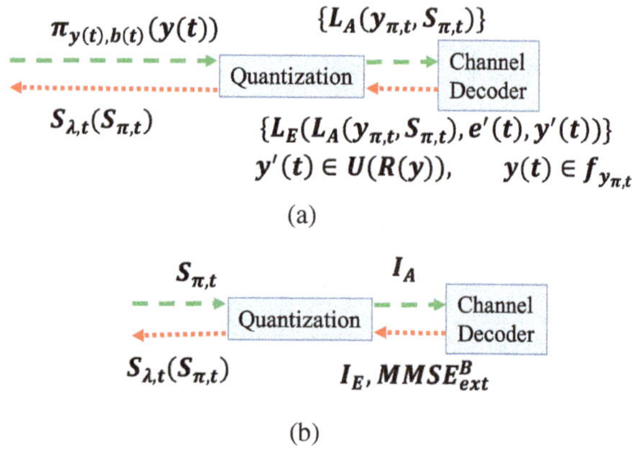

Fig. 12 Approximate framework for $\mathbf{S}_{\lambda,t}$ averaging over $\mathbf{y}'(t)$, $\mathbf{e}'(t)$ and $\mathbf{y}_{\pi,t}$. **a** Approximate Framework 1. **b** Approximate Framework 2

5.1 The MSE transfer chart for sequential channel decoding and state estimation

The corresponding model to Fig. 7(b) for the MSE transfer chart of the sequential message passing is illustrated in Fig. 13 (a). The MSE transfer chart for BP-based sequential channel decoding and state estimation shows the curves between $\mathrm{MMSE}^S_{\mathrm{ap}}$ versus $\mathrm{MMSE}^S_{\mathrm{ext}}$. As shown in Fig. 13 (a), $\mathrm{MMSE}^S_{\mathrm{ap}}$ is used to approximate $\mathbf{S}_{\pi,t}$(starting from node $\mathbf{y}(t)$), i.e.,

$$\mathbf{S}_{\pi,t} = \mathrm{MMSE}^S_{\mathrm{ap}}\mathbf{I}_K \tag{19}$$

and the PDF $\mathbf{y}_{\pi,t}$, i.e., $f_{\mathbf{y}_{\pi,t}}$, is assumed to keep the same in all time slots and all iterations. Similarly, $\mathrm{MMSE}^S_{\mathrm{ext}}$ represents the averaging $\mathbf{S}_{\pi,t+1}$ corresponding to $\mathbf{S}_{\pi,t}$ and $\mathbf{y}_{\pi,t}$ with the PDF of $f_{\mathbf{y}_{\pi,t}}$.

Compared with the model shown in Fig. 7 (b), the MSE transfer chart has two differences. First, the starting node for the message passing is changed from $\mathbf{x}(t-1)$ to $\mathbf{y}(t)$. The reason for this changing is to keep alignment with the structure of the MSE transfer chart for BP-based iterative channel decoding and state estimation. Note that since there is no extra information added from node $\mathbf{x}(t-1)$ to $\mathbf{y}(t)$, $\mathbf{x}(t-1)$ and $\mathbf{y}(t)$ provide the same amount of information for channel decoding in the context of information theory. From this point, these two models are equivalent. The second modification is that the scalar measures, to draw the MSE transfer chart $\mathrm{MMSE}^S_{\mathrm{ap}}$ and $\mathrm{MMSE}^S_{\mathrm{ext}}$, rather than the matrix measures are used to evaluate the performance of the sequential message passing.

The value of $\mathrm{MMSE}^S_{\mathrm{ext}}$ can be obtained by applying the results in Table 1, and the message passing flow is shown as

1. $\mathbf{y}(t) \rightarrow \mathbf{b}(t)$, we have $\mathbf{S}_{\pi,t} = \mathrm{MMSE}^S_{\mathrm{ap}}\mathbf{I}_K$ and the PDF of $\mathbf{y}_{\pi,t}$ is $f_{\mathbf{y}_{\pi,t}}$.
2. $\mathbf{b}(t) \rightarrow \mathbf{y}(t)$, $\mathbf{S}_{\lambda,t}$, the detailed derivation is provided in Section 4.
3. $\mathbf{y}(t) \rightarrow \mathbf{x}(t)$, we have $\mathbf{P}_{\lambda_y,t} = \mathbf{C}^{-1}(\mathbf{S}_{\lambda,t} + \Sigma_w)(\mathbf{C}^{-1})^T$.
4. $\mathbf{x}(t) \rightarrow \mathbf{x}(t+1)$, we have $\mathbf{P}_{\pi_x,t} = (\mathbf{P}^{-1}_{l,t} + \mathbf{P}^{-1}_{\lambda_y,t})^{-1}$, where $\mathbf{P}_{l,t} = \mathbf{A}\mathbf{P}_{\pi_x,t-1}\mathbf{A}^T + \Sigma_n$.
5. $\mathbf{x}(t+1) \rightarrow \mathbf{y}(t+1)$, we have $\mathbf{P}_{\pi_y,t+1} = \mathbf{A} \times \mathbf{P}_{\pi_x,t+1} \times \mathbf{A}^T + \Sigma_n$.
6. $\mathbf{y}(t+1) \rightarrow \mathbf{b}(t+1)$, we have $\mathbf{S}_{\pi,t+1} = \mathbf{C} \times \mathbf{P}_{\pi_y,t+1} \times \mathbf{C}^T + \Sigma_w$. Note that $\mathbf{S}_{\pi,t+1}$ is a matrix, $\mathrm{MMSE}^S_{\mathrm{ext}}$ is calculated by solving the following equation:

$$I_A(\mathrm{MMSE}^S_{\mathrm{ext}}\mathbf{I}_K) = I_A(\mathbf{S}_{\pi,t+1}) \tag{20}$$

where we first obtain the priori information I_A for each ith diagonal variance of $\mathbf{S}_{\pi,t+1}$, where $i \in \{1,..,K\}$. Next, to achieve (20), we compute the average value of all the variances, as $\mathrm{MMSE}^S_{\mathrm{ext}}$. An example of the calculation will be illustrated Section 6.1 in Fig. 15. The physical meaning of (20) is that $\mathrm{MMSE}^S_{\mathrm{ext}}$ is the value such that $\mathrm{MMSE}^S_{\mathrm{ext}}\mathbf{I}_K$ can provide the same amount of priori information for the channel decoder as $\mathbf{S}_{\pi,t+1}$.

5.2 The MSE transfer chart for BP-based iterative channel decoding and state estimation

In this section, we illustrate how the MSE transfer chart is modeled in order to evaluate the BP-based iterative channel decoding and state estimation as shown in Fig. 8 (a). The corresponding model for the MSE transfer chart is shown in Fig. 13(b), and it includes two flows of the message passing: one from $\mathbf{y}(t)$ to $\mathbf{y}(t+1)$ in Fig. 14(a) and the other from $\mathbf{y}(t+1)$ to $\mathbf{y}(t)$ in Fig. 14(b).

1. The flow from $\mathbf{y}(t)$ to $\mathbf{y}(t+1)$: the starting node is $\mathbf{y}(t)$, and $\mathrm{MMSE}^t_{\mathrm{ap}}$ represents the approximated $\mathbf{S}_{\pi,t}$, i.e.,

$$\mathbf{S}_{\pi,t} = \mathrm{MMSE}^t_{\mathrm{ap}}\mathbf{I}_K \tag{21}$$

and the PDF of $\mathbf{y}_{\pi,t}$, i.e., $f_{\mathbf{y}_{\pi,t}}$, is assumed to keep the same in all time slots and all iterations and is denoted by $f_{\mathbf{y}_\pi}$. $\mathrm{MMSE}^t_{\mathrm{ext}}$ represents the average $\mathbf{S}_{\pi,t+1}$ corresponding to $\mathbf{S}_{\pi,t} = \mathrm{MMSE}^t_{\mathrm{ap}}\mathbf{I}_k$ and $\mathbf{y}_{\pi,t}$ with the PDF, $f_{\mathbf{y}_\pi}$.
2. The flow from $\mathbf{y}(t+1)$ to $\mathbf{y}(t)$: the starting node is $\mathbf{y}(t+1)$, and $\mathrm{MMSE}^{t+1}_{\mathrm{ap}}$ represents the approximated $\mathbf{S}_{\pi,t+1}$, i.e.,

$$\mathbf{S}_{\pi,t+1} = \mathrm{MMSE}^{t+1}_{\mathrm{ap}}\mathbf{I}_K \tag{22}$$

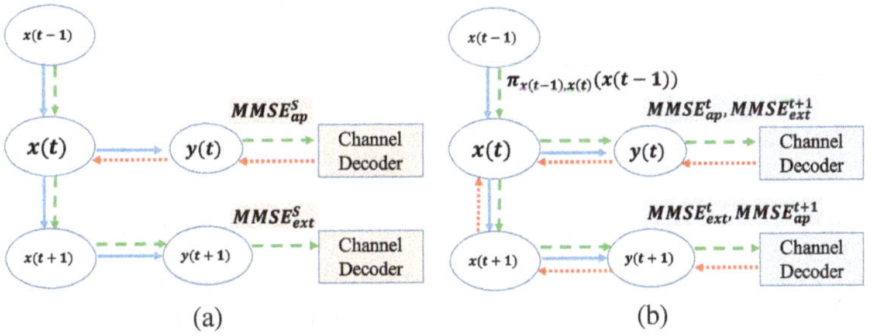

Fig. 13 The MSE transfer chart for the sequential message passing and iterative message passing. **a** Sequential message passing. **b** Iterative message passing

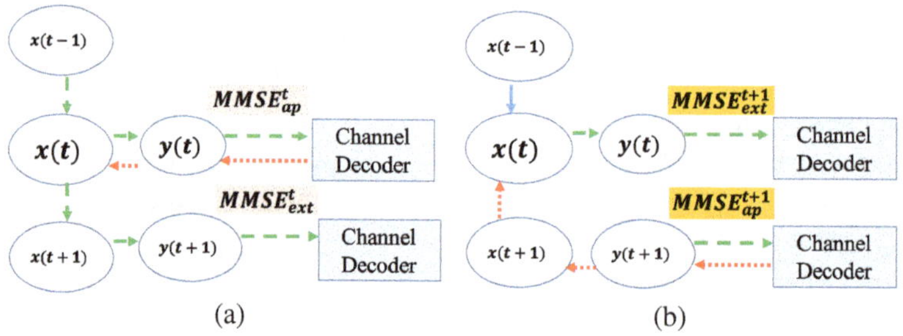

Fig. 14 Message passing over two time slots. **a** Time slot t to t+1. **b** Time slot t+1 to t

and the PDF of $\mathbf{y}_{\pi,t+1}$ is $f_{\mathbf{y}_\pi}$.
$\mathrm{MMSE}^{t+1}_{\mathrm{ext}}$ represents the averaging $\mathbf{S}_{\pi,t}$ corresponding to $\mathbf{S}_{\pi,t} = \mathrm{MMSE}^{t+1}_{\mathrm{ap}}\mathbf{I}_K$ and $\mathbf{y}_{\pi,t+1}$ with the PDF, $f_{\mathbf{y}_\pi}$.

Then, we obtain two curves: one curve with $\mathrm{MMSE}^t_{\mathrm{ap}}$ versus $\mathrm{MMSE}^t_{\mathrm{ext}}$ for the flow from $\mathbf{y}(t)$ to $\mathbf{y}(t+1)$; the other flipped curve with $\mathrm{MMSE}^{t+1}_{\mathrm{ext}}$ versus $\mathrm{MMSE}^{t+1}_{\mathrm{ap}}$ for the flow from $\mathbf{y}(t+1)$ to $\mathbf{y}(t)$. Finally, by following the steps in Table 1: (1) $\mathbf{y}(t) \to \mathbf{b}(t)$, (2) $\mathbf{b}(t) \to \mathbf{y}(t)$, (3) $\mathbf{y}(t) \to \mathbf{x}(t)$, (4) $\mathbf{x}(t) \to \mathbf{x}(t+1)$, (5) $\mathbf{x}(t+1) \to \mathbf{y}(t+1)$, (6) $\mathbf{y}(t+1) \to \mathbf{b}(t+1)$, (7) $\mathbf{y}(t+1) \to \mathbf{b}(t+1)$, (8) $\mathbf{b}(t+1) \to \mathbf{y}(t+1)$, (9) $\mathbf{y}(t+1) \to \mathbf{x}(t+1)$, (10) $\mathbf{x}(t+1) \to \mathbf{x}(t)$, (11) $\mathbf{x}(t) \to \mathbf{y}(t)$, and (12) $\mathbf{y}(t) \to \mathbf{b}(t)$, we can calculate the values of $\mathrm{MMSE}^t_{\mathrm{ext}}$ and $\mathrm{MMSE}^{t+1}_{\mathrm{ext}}$ by $I_A(\mathrm{MMSE}^t_{\mathrm{ext}}\mathbf{I}_K) = I_A(\mathbf{S}_{\pi,t+1})$ in step 5 and $I_A(\mathrm{MMSE}^{t+1}_{\mathrm{ext}}\mathbf{I}_K) = I_A(\mathbf{S}_{\pi,t})$ in step 11, respectively.

6 Numerical results

We consider an electric generator dynamic system for verification. Each dimension of the observation $\mathbf{y}(t)$ is quantized with 14 bits, and the dynamic range for quantization is $[-432, 432]$. A $\frac{1}{2}$-rate recursive systemic convolution (RSC) code is used as the channel encoding scheme, and the code generator is set as $g = [1,1,1;1,0,1]$.

The PDF of $\mathbf{y}_{\pi,t}$, i.e., $f_{\mathbf{y}_\pi}$, is Gaussian with zero mean, and the covariance matrix of $\mathbf{y}_{\pi,t}$ is obtained from the stationary distribution of $\mathbf{y}(t)$, namely

$$f_{\mathbf{y}_\pi} = \mathcal{N}(\mathbf{y}, \mathbf{0}, \Sigma_{\mathbf{y}_\pi}) \tag{23}$$

where $\Sigma_{\mathbf{y}_\pi}$ from the dynamic system is

$$\Sigma_{\mathbf{y}_\pi} = \begin{bmatrix} 2857 & 0 & 0 & 0 & 0 & 0 & 0 \\ 0 & 6056 & 0 & 0 & 0 & 0 & 0 \\ 0 & 0 & 747 & 0 & 0 & 0 & 0 \\ 0 & 0 & 0 & 1624 & 0 & 0 & 0 \\ 0 & 0 & 0 & 0 & 44 & 0 & 0 \\ 0 & 0 & 0 & 0 & 0 & 10 & 0 \\ 0 & 0 & 0 & 0 & 0 & 0 & 0.79 \end{bmatrix} \tag{24}$$

6.1 Message passing between state estimator and channel decoder

In this section, we show the performance results of the proposed channel decoding and state estimation algorithm, especially the message passing between the state estimator and channel decoder. In addition, we illustrate how much gain can be obtained by using the redundancy of system dynamics to assist channel decoding. To achieve above, the approximate framework 2 shown in Fig. 12 (b) is considered, but different from that we set $\mathbf{S}_{\pi,t}$ and $\mathbf{S}_{\lambda,t}$ as $\mathrm{MMSE}^Y_{\mathrm{ap}}\mathbf{I}_K$ and $\mathrm{MMSE}^Y_{\mathrm{ext}}\mathbf{I}_K$.

Figure 15 demonstrates the priori mutual information I_A provided by seven different $\mathrm{MMSE}^Y_{\mathrm{ap}}$. The curve labeled with D_i, corresponding to $f_{\mathbf{y}_\pi,i}$, complies to a zero-mean Gaussian distribution with variance as ith diagonal element of $\Sigma_{\mathbf{y}_\pi}$. The curve labeled with *mean*, corresponding to $f_{\mathbf{y}_\pi}$, is a zero-mean Gaussian distribution with the covariance matrix as $\Sigma_{\mathbf{y}_\pi}$. To calculate I_A for all eight curves, the curve I_A of curve mean is computed as the average of the other I_A labeled with $D1, \cdots, D7$. When $\mathrm{MMSE}^Y_{\mathrm{ap}}$ equals to 1.0e−1, 1.1, 1.0e+1, 1.0e+2, I_A equals to 0.55, 0, 4, 0.25, 0.15, respectively, and when $\mathrm{MMSE}^Y_{\mathrm{ap}}$ equals to 1.0e+5, the prediction of $\mathbf{y}(t)$ cannot provide any priori information for the channel decoder since I_A equals to 0. Note that although I_A can be as high as 0.8 when $\mathrm{MMSE}^Y_{\mathrm{ap}}$ equals to 1.0e−3, it is not achievable as the minimum value of $\mathrm{MMSE}^Y_{\mathrm{ap}}$ is limited by covariance matrix of state dynamics and system observation noise, i.e., Σ_n and Σ_w.

Figure 16 illustrates the relationship between $\mathrm{MMSE}^Y_{\mathrm{ap}}$ and $\mathrm{MMSE}^Y_{\mathrm{ext}}$. As shown in Fig. 15, the prediction of $\mathbf{y}(t)$ with $\mathrm{MMSE}^S_{\mathrm{ap}}$=1.0e+5 does not provide any priori information for $\mathbf{b}(t)$. Therefore, when $\mathrm{MMSE}^t_{\mathrm{ap}}$=1.0e+5, the corresponding $\mathrm{MMSE}^t_{\mathrm{ext}}$ is contributed by neither priori information $\mathbf{x}(t-1)$ nor the extrinsic information from the channel decoder at time slot t. The gain of $\mathrm{MMSE}^t_{\mathrm{ext}}$ from $\mathrm{MMSE}^S_{\mathrm{ap}}$ can be obtained by comparing it with the value of $\mathrm{MMSE}^t_{\mathrm{ext}}$ corresponding to $\mathrm{MMSE}^S_{\mathrm{ap}}$=1.0e+5. Following this flow, the gains of $\mathrm{MMSE}^Y_{\mathrm{ext}}$ with different $\mathrm{MMSE}^Y_{\mathrm{ap}}$ and $\frac{E_b}{N_0}$ in Fig. 15 are shown in Fig. 17.

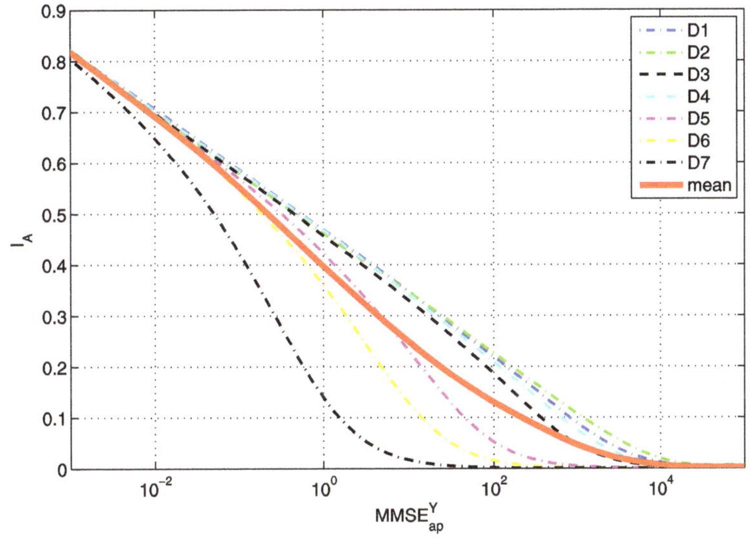

Fig. 15 Priori mutual information for $\mathbf{b}(t)$ from $\mathbf{y}(t)$ with different $\mathrm{MMSE}^Y_{\mathrm{ap}}$

6.2 Performance analysis for sequential channel decoding and state estimation

In this section, we show how to use the proposed MSE transfer chart to evaluate the performance of BP-based sequential channel decoding and state estimation over multiple time slots under different channel conditions, where the different channel conditions are modeled by different $\frac{E_b}{N_0}$. Figure 18 shows the relationship between $\mathrm{MMSE}^S_{\mathrm{ap}}$ and $\mathrm{MMSE}^S_{\mathrm{ext}}$ with different $\frac{E_b}{N_0}$. In this figure, we have the following observations:

1. When $\mathrm{MMSE}^S_{\mathrm{ap}}$ is less than 1, the corresponding $\mathrm{MMSE}^S_{\mathrm{ext}}$ for all $\frac{E_b}{N_0}$ are equal. Note that $\mathrm{MMSE}^S_{\mathrm{ap}}$ is used to model the amount of priori information from $\mathbf{x}(t-1)$. Therefore, the smaller $\mathrm{MMSE}^S_{\mathrm{ap}}$ is, the higher amount of priori information attained from $\mathbf{x}(t-1)$ is. Although the extrinsic information from the channel decoder at time slot t can also contribute to the prediction of $\mathbf{y}(t+1)$, with small $\mathrm{MMSE}^S_{\mathrm{ap}}$, the priori information from $\mathbf{x}(t-1)$ is dominant in the prediction of $\mathbf{y}(t+1)$. Therefore, the difference of gains from channel decoder with different $\frac{E_b}{N_0}$ is not seen in $\mathrm{MMSE}^S_{\mathrm{ext}}$.
2. With a large $\mathrm{MMSE}^S_{\mathrm{ap}}$, the higher $\frac{E_b}{N_0}$ is, the higher $\mathrm{MMSE}^S_{\mathrm{ext}}$ is. With the increasing of $\mathrm{MMSE}^S_{\mathrm{ap}}$, $\mathbf{x}(t-1)$ provides less amount of priori information

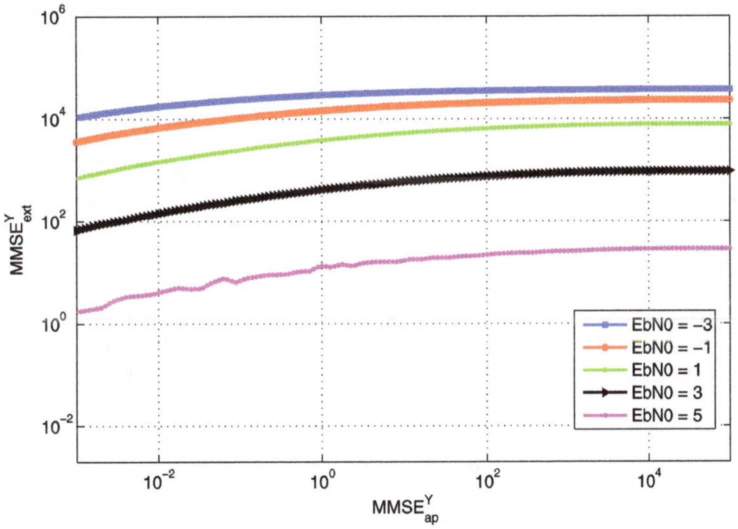

Fig. 16 $\mathrm{MMSE}^Y_{\mathrm{ext}}$ with different $\mathrm{MMSE}^Y_{\mathrm{ap}}$ and $\frac{E_b}{N_0}$

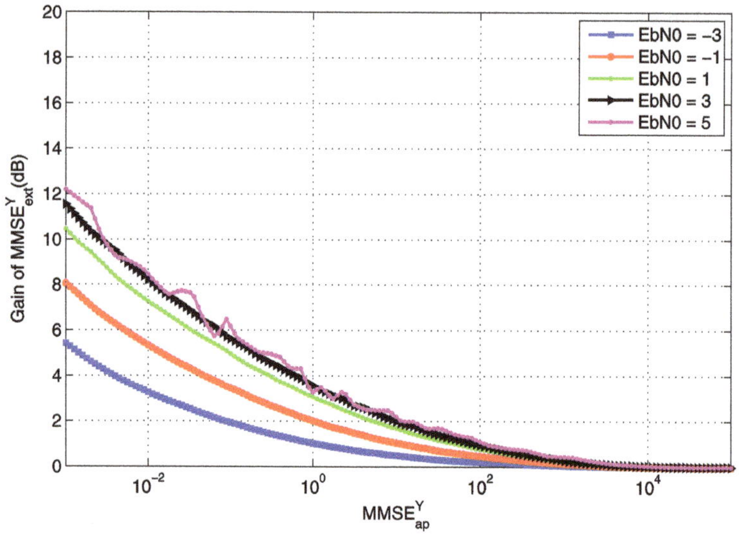

Fig. 17 Gain of MMSE$^Y_{ext}$ with different MMSE$^Y_{ap}$ with $\frac{E_b}{N_0}$

for the prediction $\mathbf{y}(t)$, and the extrinsic information from channel decoder becomes dominant in the prediction of $\mathbf{y}(t)$. Then, the channel gains with different $\frac{E_b}{N_0}$ are seen.

3. When MMSE$^S_{ap}$ is around 1.0e−2, the values MMSE$^S_{ext}$ are flat. This arises because, based on the information from the time slots priori to $t+1$, the prediction of $\mathbf{y}(t)$ is not reliable since state dynamics $\mathbf{n}(t)$ and the noise of system observation $\mathbf{w}(t)$ are not predictable. This leads that the minimum value of MMSE$^S_{ext}$ is limited by the covariance matrix of $\mathbf{n}(t)$ and $\mathbf{w}(t)$, i.e., Σ_n and Σ_w, respectively.

Following the same idea of the EXIT chart and the MSE transfer chart for iterative channel decoding, we use the MSE transfer chart to evaluate the performance of BP-based sequential channel decoding and state estimation over mutiple time slots. In the MSE transfer chart, we have two curves: one is MMSE$^{S,t}_{ap}$ versus MMSE$^{S,t}_{ext}$, which is equivalent to the curve for MMSE$^S_{ap}$ versus MMSE$^S_{ext}$(MMSE$^S_{ap}$ and MMSE$^S_{ext}$ at time slot t); the other is MMSE$^{S,t+1}_{ext}$ versus MMSE$^{S,t+1}_{ap}$, which is equivalent to the flipped curve for MMSE$^S_{ap}$ versus MMSE$^S_{ext}$(MMSE$^S_{ap}$ and MMSE$^S_{ext}$ at time slot $t+1$).

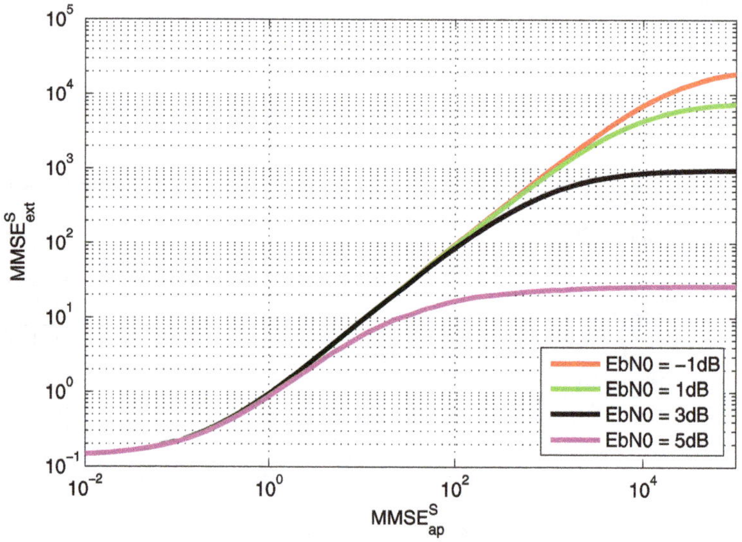

Fig. 18 Relationship between MMSE$^S_{ext}$ and MMSE$^S_{ap}$ for the sequential message passing with different $\frac{E_b}{N_0}$

The MSE transfer chart for $\frac{E_b}{N_0}$ =5 dB is shown in Fig. 19. There is one crossing/convergence point between these two curves, which means that it is not a successful decoding from $I_A = 0$ to $I_A = 1$. The arrows with black color show the trace of the sequential message passing(starting from $\text{MMSE}^S_{\text{ap}}$=1.0e−2 at time slot t). We can see that trace moves toward the convergence point at time slot $t+1$ and $t+2$. The trace shows that even if we have perfect knowledge of $\mathbf{y}(t)$ at time slot t, the performance of state estimation degrades toward the crossing point after running a few time slots. The arrows with red color show the other trace (starting from $\text{MMSE}^S_{\text{ap}}$=1.0e+5), at which there is no priori information from the prediction of $\mathbf{y}(t)$. Similarly, we can see the trace is moving toward the convergence point at time slot $t+1, t+2, \cdots$. Different from the above, this trace shows that even though there is no priori information of $\mathbf{y}(t)$ at time slot t, the performance of state estimation is improved after running a few time slots, and finally, the gain reaches the convergence point.

Figure 20 illustrates the MSE transfer chart as $\frac{E_b}{N_0}$ =3 dB with one crossing point at $\text{MMSE}^{S,t}_{\text{ap}}$=1.0e−0.15. In the range of [1.0e−0.15, 1.0e+2], the gap between the curve ($\text{MMSE}^{S,t}_{\text{ap}}$ versus $\text{MMSE}^{S,t}_{\text{ext}}$) and the flipped curve ($\text{MMSE}^{S,t+1}_{\text{ext}}$ versus $\text{MMSE}^{S,t+1}_{\text{ap}}$) is very small. This means that the convergence speed is slow and it takes many time slots to converge to the convergence point when the state estimator has no priori information of $\mathbf{y}(t)$ at time slot t.

Figure 21 shows the MSE transfer chart as $\frac{E_b}{N_0}$ =1 dB with one crossing point at $\text{MMSE}^{S,t}_{\text{ap}}$=1.0e−0.15. Note that after the crossing point the curve ($\text{MMSE}^{S,t}_{\text{ap}}$ versus $\text{MMSE}^{S,t}_{\text{ext}}$) almost overlaps with the flipped curve ($\text{MMSE}^{S,t+1}_{\text{ext}}$ versus $\text{MMSE}^{S,t+1}_{\text{ap}}$). This shows two conditions when the state estimator has no priori information of $\mathbf{y}(t)$ at time slot t: one is that it will gain a high MSE and cannot coverage to the crossing point; the other is that it will take many time slots to converge to the crossing point. Compared with previous two results, for $\frac{E_b}{N_0}$=1 dB, the MSEs of estimating $\mathbf{x}(t)$ and $\mathbf{y}(t)$ are much higher than that for $\frac{E_b}{N_0}$ equals to 3 and 5 dB. Similar results go with the case when $\frac{E_b}{N_0} = -1$ dB.

6.3 Performance analysis for iterative channel decoding and state estimation

In this section, we show how to use the proposed MSE transfer chart to evaluate the performance of BP-based iterative channel decoding and state estimation within two time slots. As we stated, the node $\mathbf{x}(t-1)$ and the node $\mathbf{y}(t)$ provide the same amount of information in the context of information theory. From this point, the priori information at the node $\mathbf{y}(t)$, as same as the node $\mathbf{x}(t-1)$, will be $\pi_{\mathbf{x}(t-1),\mathbf{x}(t)}(\mathbf{x}(t-1)) = \mathcal{N}(\mathbf{x}_{t-1}, \mathbf{x}_{\pi_x,t-1}, P_{\pi_x,t-1})$ with $\mathbf{x}_{\pi_x,t-1} = \mathbf{0}$ and $P_{\pi_x,t-1} = \text{MMSE}^{t-1,x}_{\text{ap}}\mathbf{I}_k$.

With the priori information as an initial input for iterative decoding, the gains of $\text{MMSE}^t_{\text{ext}}$(based on $\text{MMSE}^t_{\text{ap}}$) with different $\text{MMSE}^{t-1,x}_{\text{ap}}$ and $\frac{E_b}{N_0}$=5 dB are illustrated in Fig. 22, and from this figure, we have the following observations:

1. With the same $\text{MMSE}^{t-1,x}_{\text{ap}}$, the higher $\text{MMSE}^t_{\text{ap}}$ is, the lower gain of $\text{MMSE}^t_{\text{ext}}$ can be obtained from $\text{MMSE}^t_{\text{ap}}$.

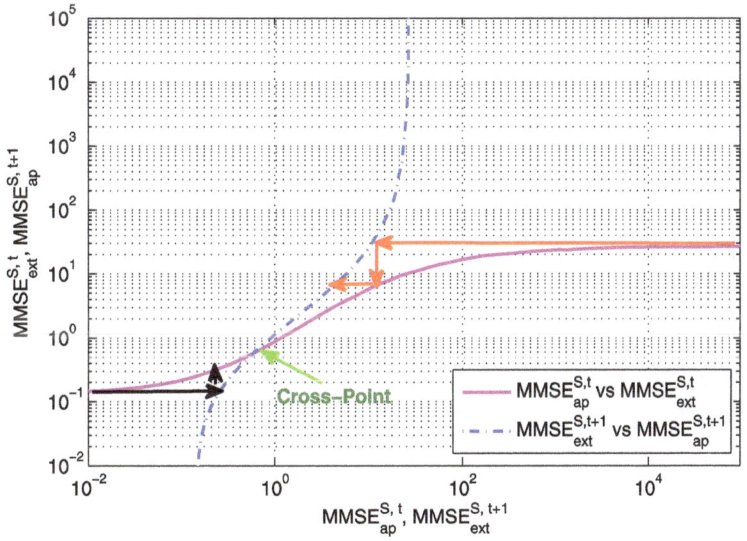

Fig. 19 The MSE transfer chart for sequential channel decoding and state estimation: $\frac{E_b}{N_0}$=5 dB

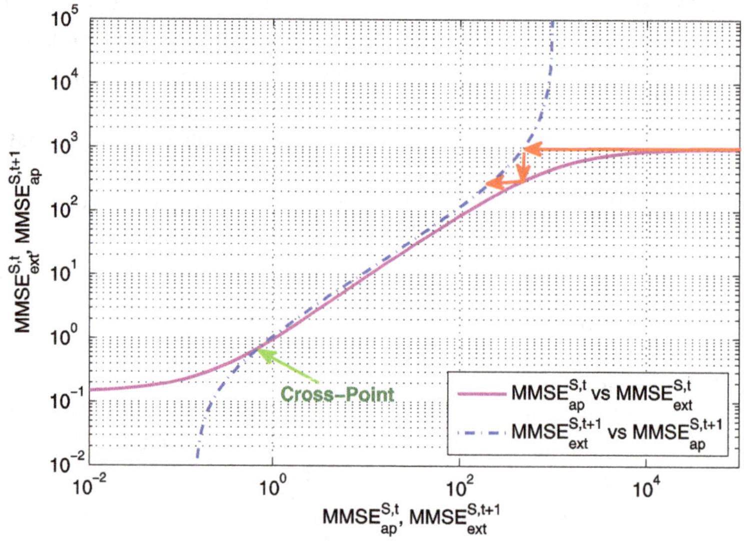

Fig. 20 The MSE transfer chart for sequential channel decoding and state estimation $\frac{E_b}{N_0}$=3 dB

2. The higher the $MMSE^{t-1,x}_{ap}$ is, the higher gain of $MMSE^{t}_{ext}$ can be obtained from $MMSE^{t}_{ap}$. This exists because the prediction of $\mathbf{y}(t+1)$ is contributed by both priori information from $\mathbf{x}(t-1)$ and the extrinsic information from the channel decoder at time slot t. When $MMSE^{t-1,x}_{ap}$ is small, the information from $\mathbf{x}(t-1)$ is dominant in predicting $\mathbf{y}(t+1)$. Although the prediction of $\mathbf{y}(t)$ can increase the amount of extrinsic information from channel decoder at time slot t, it cannot contribute much to the gain of $MMSE^{t}_{ext}$ as the priori information from $\mathbf{x}(t-1)$ is dominant.

Following the same idea, the gain of $MMSE^{t+1}_{ext}$ (based on $MMSE^{t+1}_{ap}$) is obtained and shown in Fig. 23. Similar observations as above are obtained:

1. With the same $MMSE^{t-1,x}_{ap}$, the higher $MMSE^{t+1}_{ap}$ is, the lower gain of $MMSE^{t+1}_{ext}$ can be obtained from $MMSE^{t}_{ap}$.
2. The higher the $MMSE^{t-1,x}_{ap}$ is, the higher gain of $MMSE^{t+1}_{ext}$ can be obtained from $MMSE^{t+1}_{ap}$.

Then, the MSE transfer chart for BP-based iterative channel decoding and state estimation is formed by the

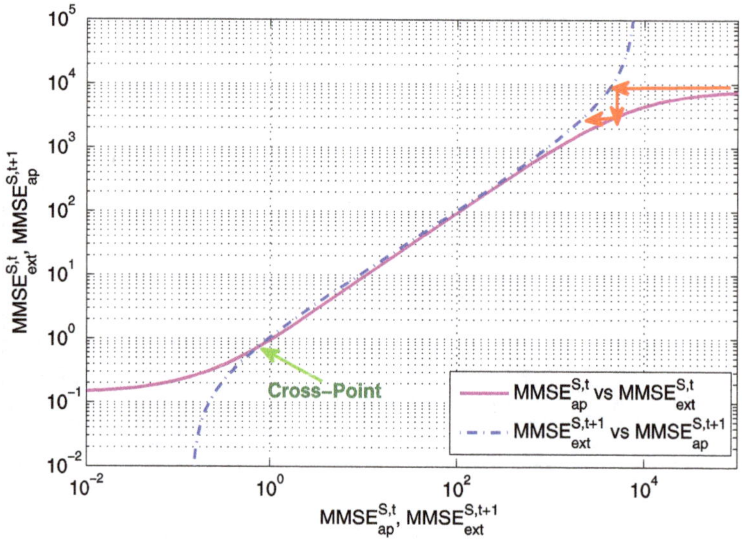

Fig. 21 The MSE transfer chart for sequential channel decoding and state estimation $\frac{E_b}{N_0}$=1 dB

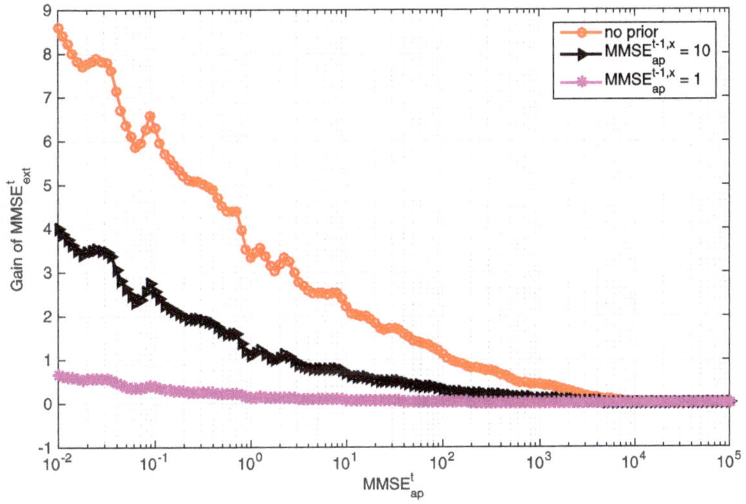

Fig. 22 Gain of MMSE$_{ext}^{t}$ from MMSE$_{ap}^{t}$ with different MMSE$_{ap}^{t-1,x}$ and $\frac{E_b}{N_0}$=5 dB

curve (MMSE$_{ap}^{t}$ versus MMSE$_{ext}^{t}$) and the flipped curve (MMSE$_{ap}^{t+1}$ versus MMSE$_{ext}^{t+1}$), and the results with different MMSE$_{ap}^{t-1,x}$ and $\frac{E_b}{N_0}$ =5 dB are shown in Fig. 24. In the following, we try to explain our observations based on the case where no priori information are considered.

1. BP-based iterative channel decoding can decrease MMSE$_{ext}^{t}$ and help improve the estimation of $\mathbf{x}(t+1)$. The values of the crossing point between the curve (MMSE$_{ap}^{t}$ versus MMSE$_{ext}^{t}$) and the flipped curve (MMSE$_{ap}^{t+1}$ versus MMSE$_{ext}^{t+1}$) for MMSE$_{ap}^{t}$ and MMSE$_{ext}^{t}$ are 1.0e+1.33 and 1.0e+1.25, respectively, while the values of MMSE$_{ap}^{t}$ and MMSE$_{ext}^{t}$ with no priori information are 1.0e+5 and 1.0e+1.43, respectively. Therefore, the total gain of MMSE$_{ext}^{t}$ for BP-based iterative channel decoding and state estimation is $10 * (1.43 - 1.25)$=1.8 dB.
2. The gain of MMSE$_{ext}^{t}$ with only three steps is close to the gain of MMSE$_{ext}^{t}$ at the convergence point. The trace of BP-based iterative channel decoding and state estimation with the mentioned three steps is shown by the arrows with blue color in the figure, and the details are listed as below:

   1) From $\mathbf{y}(t)$ to $\mathbf{y}(t+1)$: As there is no priori information, the starting point is MMSE$_{ap}^{t}$ with the value of 1.0e+5, and the corresponding value of MMSE$_{ext}^{t}$ is 1.0e+1.43.

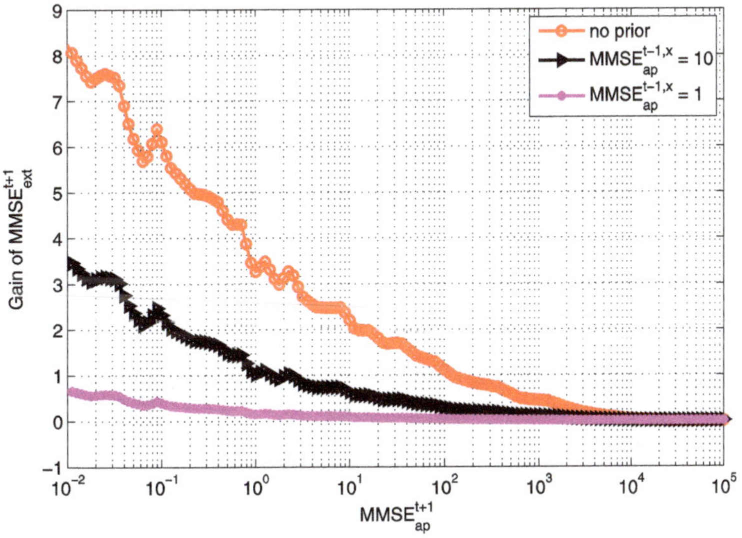

Fig. 23 Gain of MMSE$_{ext}^{t+1}$ from MMSE$_{ap}^{t+1}$ with different MMSE$_{ap}^{t-1,x}$ and $\frac{E_b}{N_0}$=5 dB

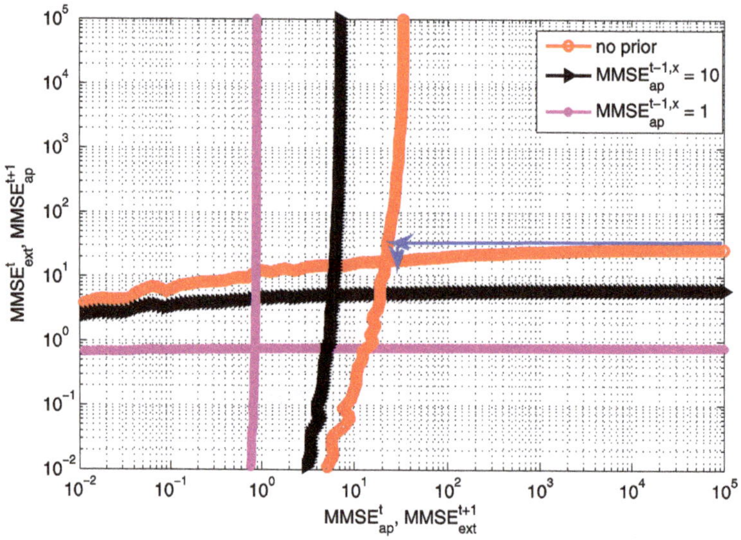

Fig. 24 MSE-based transfer chart for iterative channel decoding and system state estimation with different $\text{MMSE}_{ap}^{t-1,x}$ and $\frac{E_b}{N_0}$=5 dB

2) From $\mathbf{y}(t+1)$ to $\mathbf{y}(t)$: We have MMSE_{ap}^{t+1} with the value of 1.0e+1.43, and the corresponding value of MMSE_{ext}^{t+1} is 1.0e+1.34.
3) From $\mathbf{y}(t)$ to $\mathbf{y}(t+1)$: We have MMSE_{ap}^{t} with the value of 1.0e+1.34, and the corresponding value of MMSE_{ext}^{t} is 1.0e+1.255. Compared with step (1), $10 * (1.43 - 1.255)$=1.75 dB is obtained for MMSE_{ext}^{t} while it is 1.8 dB for MMSE_{ext}^{t} from convergence point. Thus, the gain loss for BP-based iterative channel decoding and state estimation with these three steps is just $1.8 - 1.75$=0.05 dB.

In summary, we can implement BP-based iterative channel decoding and state estimation with above three steps to obtain the gain of MMSE_{ext}^{t}, which is close to the gain at convergence point.

Note that when $\text{MMSE}_{ap}^{t-1,x}$ equals to 1 or 10, the BP-based iterative channel decoding and state estimation scheme cannot improve the performance of state estimation. This is because with a small $\text{MMSE}_{ap}^{t-1,x}$ the priori information from $\mathbf{x}(t)$ is dominant in estimating $\mathbf{y}(t+1)$, which leads that the prediction of $\mathbf{y}(t)$ in channel decoder at time slot t is negligible in predicting $\mathbf{y}(t+1)$.

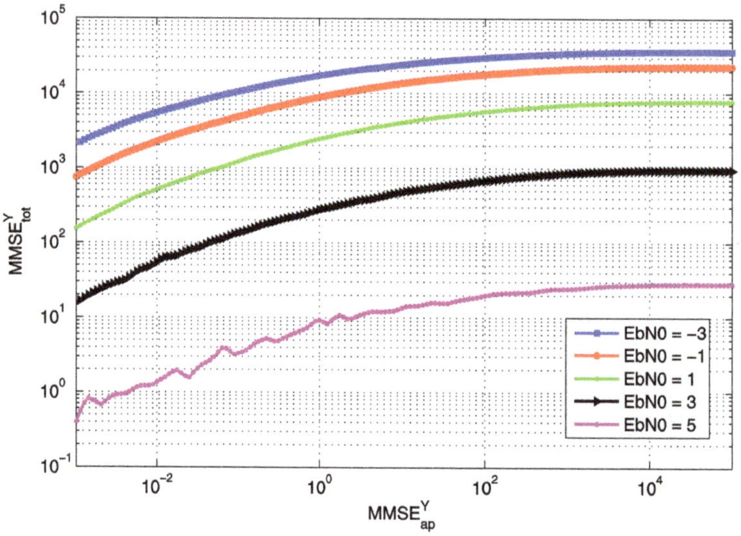

Fig. 25 MMSE_{tot}^{Y} with different MMSE_{ap}^{Y} and $\frac{E_b}{N_0}$

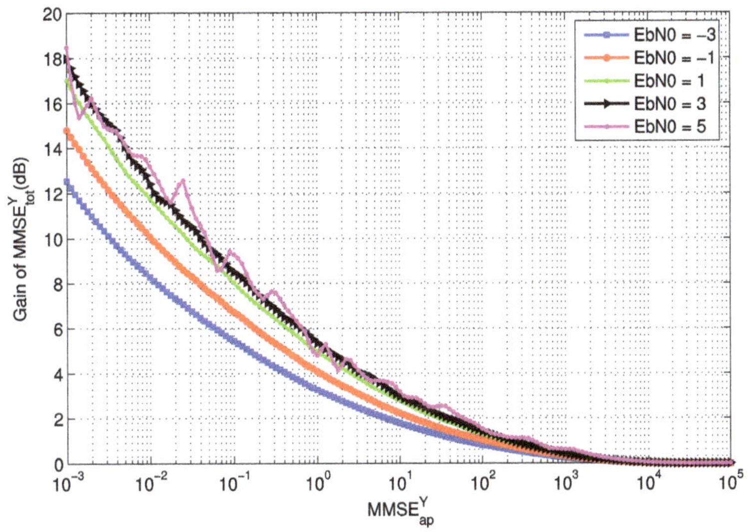

Fig. 26 Gain of $\mathrm{MMSE}^Y_{\mathrm{tot}}$ with different $\mathrm{MMSE}^Y_{\mathrm{ap}}$ and $\frac{E_b}{N_0}$

The MSE transfer charts for $\frac{E_b}{N_0} = 3$ and 1 dB have similar observations as that for $\frac{E_b}{N_0}$ =5 dB, so we will not list the results here.

6.4 Performance analysis for Kalman filtering-based heuristic approach

Similarly, by utilizing the redundancy of system dynamics, a Kalman filtering-based heuristic approach is evaluated in this section. In the Kalman filtering-based heuristic approach, the prediction of $\mathbf{y}(t)$ based on Kalman filtering is used as the priori information for $\mathbf{b}(t)$, and instead of using only the extrinsic information of the channel decoder to obtain a soft estimation of $\mathbf{y}(t)$, the total information including both the priori information and the extrinsic information generated by the channel decoder is used to obtain a hard estimation of $\mathbf{y}(t)$.

The corresponding framework used for the Kalman filtering-based heuristic approach is similar to the BP-based channel decoding and state estimation as shown in Fig. 12 (b). The priori information from $\mathbf{y}(t)$ is modeled with $\mathbf{S}_{\pi,t} = \mathrm{MMSE}^Y_{\mathrm{ap}}\mathbf{I}_K$, and the priori information for $\mathbf{b}(t)$ is represented by mutual information I_A. Finally, the total information from the channel decoder for estimating of $\mathbf{b}(t)$ is modeled by $\mathrm{MMSE}^B_{\mathrm{tot}}$, which means the MMSE of estimating $\mathbf{b}(t)$ based on the total information including both the priori information and the extrinsic information from the channel decoder.

Figure 25 illustrates the relationship between $\mathrm{MMSE}^Y_{\mathrm{ap}}$ and $\mathrm{MMSE}^Y_{\mathrm{tot}}$. The gains of $\mathrm{MMSE}^Y_{\mathrm{tot}}$ with different $\mathrm{MMSE}^Y_{\mathrm{ap}}$ and $\frac{E_b}{N_0}$ are shown in Fig. 26, which are further improved comparing to Fig. 17.

7 Conclusions

We propose to use the MSE transfer chart to evaluate the performance of BP-based channel decoding and state estimation. We focus on two models, the BP-based sequential processing model and the BP-based iterative processing model, for channel decoding and state estimation. The former can be used to evaluate the performance of sequential processing over multiple time slots, and the latter can be used to evaluate the performance of iterative processing within two time slots. The numerical results show by utilizing the MSE transfer chart the proposed channel decoding and state estimation algorithm can decrease the MSE and improve performance of channel decoding and state estimation. Specifically, a total 1.75 dB gain can be earned through three-step BP-based iterative channel decoding and state estimation process when no prior information is given.

Acknowledgements
The authors would like to thank the support of National Science Foundation under grants ECCS-1407679, CNS-1525226, CNS-1525418, and CNS-1543830.

Authors' contributions
HL conceived the Brainbow strategies for this work. JBS and HL supervised the project. SG built the initial constructs and LL validated them, analyzed the data, and wrote the paper. All authors read and approved the final manuscript.

Competing interests
The authors declare that they have no competing interests.

Author details
[1]Department of Electrical Engineering and Computer Science, the University of Tennessee, Knoxville, Knoxville, USA. [2]Sequans Communications, 1732, Deogyoung Road, 446701 Giheung, Yongin, South Korea. [3]Department of Electronic Engineering, Kyung Hee University, 90 Washington Valley Road, 07921 Bedminster, NJ, USA.

References

1. S Gong, H Li, L Lai, RC Qiu, in *2011 IEEE International Conference on Communications (ICC)*. Decoding the'nature encoded'messages for distributed energy generation control in microgrid (IEEE, Kyoto, 2011), pp. 1–5
2. H Li, *Communications for control in cyber physical systems: theory, design and applications in smart grids*. (Morgan Kaufmann, Cambridge, 2016)
3. S ten Brink, Convergence of iterative decoding. Electron. Lett. **35**(10), 806–808 (1999)
4. S Ten Brink, in *Proceedings of 3rd IEEE/ITG Conference on Source and Channel Coding, Munich, Germany*. Iterative decoding trajectories of parallel concatenated codes, (2000), pp. 75–80
5. S Ten Brink, Convergence behavior of iteratively decoded parallel concatenated codes. IEEE Trans. Commun. **49**(10), 1727–1737 (2001)
6. M El-Hajjar, L Hanzo, Exit charts for system design and analysis. IEEE Commun. Surveys Tutorials. **16**(1), 127–153 (2013)
7. K Bhattad, KR Narayanan, An MSE-based transfer chart for analyzing iterative decoding schemes using a Gaussian approximation. IEEE Trans. Inf. Theory. **53**(1), 22–38 (2007)
8. A Ashikhmin, G Kramer, S ten Brink, Extrinsic information transfer functions: model and erasure channel properties. IEEE Trans. Inf. Theory. **50**(11), 2657–2673 (2004)
9. D Guo, S Shamai, Verdú, Mutual information and minimum mean-square error in Gaussian channels. IEEE Trans. Inf. Theory. **51**(4), 1261–1282 (2005)
10. M Fu, CE de Souza, State estimation for linear discrete-time systems using quantized measurements. Automatica. **45**(12), 2937–2945 (2009)
11. S Yüksel, Baş,ar, Stochastic networked control systems. AMC. **10**, 12 (2013)
12. L Li, H Li, in *Global Communications Conference (GLOBECOM), 2016 IEEE*. Dynamic state aware source coding for networked control in cyber-physical systems (IEEE, Washington, DC, 2016), pp. 1–6
13. S Yüksel, On stochastic stability of a class of non-Markovian processes and applications in quantization. SIAM J. Control. Optim. **55**(2), 1241–1260 (2017)
14. N Ramzan, S Wan, E Izquierdo, Joint source-channel coding for wavelet-based scalable video transmission using an adaptive turbo code. EURASIP J. Image Video Process. **2007**(1), 1–12 (2007)
15. V Kostina, Verdú, Lossy joint source-channel coding in the finite blocklength regime. IEEE Trans. Inf. Theory. **59**(5), 2545–2575 (2013)
16. H Wu, L Wang, S Hong, J He, Performance of joint source-channel coding based on protograph LDPC codes over Rayleigh fading channels. IEEE Commun. Lett. **18**(4), 652–655 (2014)
17. X He, X Zhou, P Komulainen, M Juntti, T Matsumoto, A lower bound analysis of hamming distortion for a binary ceo problem with joint source-channel coding. IEEE Trans. Commun. **64**(1), 343–353 (2016)
18. Y Wang, M Qin, KR Narayanan, A Jiang, Z Bandic, in *Global Communications Conference (GLOBECOM), 2016 IEEE*. Joint Source-Channel Decoding of Polar Codes for Language-Based Sources (IEEE, Washington, DC, 2016), pp. 1–6
19. V Kostina, Y Polyanskiy, S Verd, Joint source-channel coding with feedback. IEEE Trans. Inf. Theory. **63**(6), 3502–3515 (2017)
20. L Yin, J Lu, Y Wu, in *Communication Systems, 2002. ICCS 2002. The 8th International Conference On*. LDPC-based joint source-channel coding scheme for multimedia communications, vol. 1 (IEEE, Singapore, 2002), pp. 337–341
21. J Garcia-Frias, W Zhong, LDPC codes for compression of multi-terminal sources with hidden Markov correlation. IEEE Commun. Lett. **7**(3), 115–117 (2003)
22. W Zhong, Y Zhao, J Garcia-Frias, in *Conference Record of the Thirty-Seventh Asilomar Conference on Signals, Systems and Computers*. Turbo-like codes for distributed joint source-channel coding of correlated senders in multiple access channels (IEEE, Pacific Grove, 2003), pp. 840–844
23. Z Mei, L Wu, Joint source-channel decoding of Huffman codes with LDPC codes. J Electron (China). **23**(6), 806–809 (2006)
24. X Pan, A Cuhadar, AH Banihashemi, Combined source and channel coding with JPEG2000 and rate-compatible low-density parity-check codes. IEEE Trans. Signal Process. **54**(3), 1160–1164 (2006)
25. L Pu, Z Wu, A Bilgin, MW Marcellin, B Vasic, LDPC-based iterative joint source-channel decoding for JPEG2000. IEEE Trans. Image Process. **16**(2), 577–581 (2007)
26. M Fresia, F Perez-Cruz, HV Poor, S Verdu, Joint source and channel coding. IEEE Signal Proc. Mag. **27**(6), 104–113 (2010)
27. E Koken, E Tuncel, Joint source-channel coding for broadcasting correlated sources. IEEE Trans. Commun. (2017)
28. B Girod, AM Aaron, S Rane, D Rebollo-Monedero, Distributed video coding. Proc. IEEE. **93**(1), 71–83 (2005)
29. Y Zhao, J Garcia-Frias, Joint estimation and compression of correlated nonbinary sources using punctured turbo codes. IEEE Trans. Commun. **53**(3), 385–390 (2005)
30. R Gallager, Low-density parity-check codes. IRE Trans. Informa. Theory. **8**(1), 21–28 (1962)
31. J Pearl, *Probabilistic Reasoning in Intelligent Systems: Networks of Plausible Inference*. (Morgan Kaufmann Publishers Inc., San Francisco, 1988)

9

Congestion-aware route selection in automatic evacuation guiding based on cooperation between evacuees and their mobile nodes

Yuki Kasai, Masahiro Sasabe* and Shoji Kasahara

Abstract

When a large-scale disaster occurs, evacuees have to evacuate to safe places quickly. For this purpose, an automatic evacuation guiding scheme based on cooperation between evacuees and their mobile nodes has been proposed. The previous work adopts shortest-distance based route selection and does not consider the impact of traffic congestion caused by evacuation guiding. In this paper, we propose congestion-aware route selection in the automatic evacuation guiding. We first adopt a traffic congestion model where each evacuee's moving speed on a road is determined by the population density of the road and his/her order among evacuees traveling in the same direction. Based on this congestion model, each evacuee's mobile node estimates the cost, i.e., traveling time, of each road in the area. Each mobile node collects information about blocked road segments and positions of other evacuees through communication infrastructures or other mobile nodes. Based on the obtained information, it calculates and selects the smallest-cost route. Through simulation experiments, we show that the congestion-aware route selection can reduce both average and maximum evacuation times compared to the shortest-distance-based route selection, especially under highly congested situations. Furthermore, we show that the congestion-aware route selection can work well even under highly damaged situations where only direct wireless communication among mobile nodes is available.

Keywords: Automatic evacuation guiding, Congestion-aware route selection, Mobile nodes

1 Introduction

In the 2011 Great East Japan Earthquake, tremendous damage to communication infrastructures made both fixed and mobile communication networks unavailable for a long time and in wide areas. As a result, it has been reported that evacuees and rescuers could not smoothly collect and distribute important information, e.g., safety information, evacuation information, and government information, even though they carried their own mobile nodes, e.g., cellular phones and smart phones [1]. In such situations, evacuees quickly have to grasp information about safe places and safe routes to those places. Although they can acquire static information, e.g., map and locations of safe places, in usual time, they cannot grasp dynamic information, e.g., damage situations in disaster areas, in advance. To tackle this problem, a scheme has been proposed that enables automatic evacuation guiding based on implicit interactions between evacuees and their mobile nodes [2]. In the automatic evacuation guiding scheme, the mobile node of an evacuee first calculates and presents an evacuation route, i.e., recommended route, to him/her. At the same time, it also traces his/her actual evacuation route as a trajectory by measuring his/her positions periodically. When it detects difference between the recommended route and the actual evacuation route, it can automatically estimate and record the corresponding blocked road segments. These discovered information can be shared among mobile nodes through direct wireless communication and/or communication via remaining communication infrastructures.

*Correspondence: sasabe@is.naist.jp
Graduate School of Information Science, Nara Institute of Science and Technology, 8916-5 Takayama-cho, Ikoma, 630-0192 Nara, Japan

In [2], the performance of the automatic evacuation guiding scheme is evaluated from the viewpoint of average/maximum evacuation time among evacuees, under the situations where the recommended route is given as a shortest path and traffic congestion does not occur. In actual situations, traffic congestion may be caused by evacuation behavior. For example, it has been reported that heavy traffic jam occurred due to the disruption of the transportation networks at the metropolitan area of Tokyo, in the case of the 2011 Great East Japan Earthquake [3]. Note that the automatic evacuation guiding tends to guide physically close evacuees to the same route, due to the sharing of information about blocked road segments among mobile nodes. Therefore, both the alleviation of traffic congestion and the safety of evacuation become important in the automatic evacuation guiding.

In this paper, we propose congestion-aware route selection in the automatic evacuation guiding. In general, the degree of congestion of a road can be modeled as a function of moving speed or traveling time with the number of people traveling on the road. Inspired by the traffic congestion model [4], we propose a traffic congestion model where each evacuee's moving speed on a road is determined by the population density of the road and his/her order among evacuees traveling in the same direction. Based on this congestion model, each evacuee's mobile node estimates the traveling time of each road in the area by collecting the number of evacuees on each road through direct wireless communication with other mobile nodes and/or a server via remaining communication infrastructures. The congestion-aware route selection selects the smallest-cost route based on the estimated traveling time of each road.

Trough simulation experiments, we examine the effectiveness of the proposed scheme in terms of average/maximum evacuation time. Since the performance of the propose scheme depends on the estimation accuracy of congestion state, which is affected by the communication environments, we evaluate the proposed scheme in two kinds of scenarios: global scenario and local scenario. The global scenario is an ideal case where each mobile node can always grasp the locations of all other nodes via communication infrastructures. On the contrary, in the local scenario, communication infrastructures are unavailable and each mobile node can collect the locations of other nodes only within its direct communication range.

The rest of this paper is organized as follows. Section 2 gives related work. Section 3 describes the automatic evacuation guiding. Section 4 describes the proposed congestion-aware route selection. The simulation results are shown in Section 5. Finally, Section 6 provides conclusions.

2 Related works

Various studies have been done for evacuation support in large-scale disasters and some of them focus on traffic congestion caused by evacuation behavior [2, 3, 5–9]. These are mainly divided into two types: analysis of evacuation behavior based on disaster simulations and field survey [3, 5], and evacuation guiding using information and communication technology [2, 6–9].

In [5], the impact of disruption of railway networks on road networks, e.g., geographical distribution of people, is estimated through railway-network simulation under the assumption that the Tokyo metropolitan earthquake occurs. In [3], evacuation behavior is analyzed by the trip data based on a questionnaire survey of the victims of the 2011 Great East Japan Earthquake. In these studies, it has been reported that traffic congestion caused by the disaster can be alleviated by controlling the timing of returning home among evacuees, e.g., making evacuees stay at their schools and offices.

In [6], the authors propose an evacuation guiding system that calculates evacuation routes with timings to suppress traffic congestion, where evacuees can almost always communicate with others through an ad-hoc network. In [7], the authors propose an evacuation guiding system using a delay tolerant network (DTN) [10]. When evacuees encounter blocked road segments or traffic congestion during their evacuations, they register these information to their mobile nodes and share it with other nodes through wireless communication. The shared information will improve the evacuation movement of evacuees. Since it may be difficult for evacuees to register the information to their mobile nodes in emergent situations, Komatsu et al. propose an automatic evacuation guiding based on implicit cooperation between evacuees and their mobile nodes [2]. Komatsu et al. also propose an information sharing scheme called On-Demand Direct Delivery for the automatic evacuation guiding, which can reduce the network load compared to the existing DTN routing [11]. In this paper, we try to combine congestion-aware route selection with the automatic evacuation guiding.

There have also been studied on congestion alleviation in general road networks [12–14]. Verroios and Kollias propose a congestion-aware route selection method for vehicles, with the help of a ad hoc network [12]. They assume that each vehicle is equipped with a Personal Digital Assistant (PDA). Each vehicle collects the position and speed information of other vehicles within its transmission range and estimates the required time to each segment in the target area. It also shares the estimated traveling time to each segment with other vehicles and dynamically determines a route to its destination based on the information. Lim and Rus propose a probabilistic route selection method to realize user equilibrium (Nash equilibrium) and social optimum in game theory [13].

Each agent in a transportation network has route candidates to its destination as a vector in which each element represents the probability to select. It may have some neighbors, each of which is the other agent that share some roads in their route candidates. Each agent tries to select an appropriate route from the candidates by taking account of the influence of route selection on neighbors.

Congestion-aware route selection schemes can be classified in terms of congestion models and sharing methods of congestion information. There are some kinds of congestion models: a model using cellular automata [6, 7], a model based on the function of traffic volume on each road [9, 13], and a model where moving speed depends o the density of moving objects on each road [4, 15]. In this paper, we propose a congestion model by extending the moving-speed based congestion model [4]. The sharing methods of congestion information can be classified in terms of network architectures. Lim and Rus assume that global information sharing among mobile agents can be achieved by communication infrastructures [13]. There are some studies that assume local information sharing can be achieved by ad hoc networks [6, 12] and DTNs [2, 7]. Since the proposed congestion-aware route selection is conducted over the automatic evacuation guiding [2], we apply the DTN-based information sharing.

In the evacuation, it is rational for each evacuee to select a evacuation route whose expected evacuation time is minimum among those of candidate routes. The phenomenon where each evacuee behaves according to such a rational decision is called selfish routing in game theory [16]. In general, such individually rational decisions of evacuees cannot necessarily result in the social optimum, where the average evacuation time among evacuees is minimized. Lim and Rus reveal which utility function to determine the route of each mobile agent results in the social optimum [13]. Note that they assume that the global information sharing among mobile agents can be achieved as mentioned above. In this paper, the proposed route selection is also a kind of selfish routing but the way of information sharing can vary depending on the communication environments, i.e., communication infrastructures or DTNs.

3 Automatic evacuation guiding system

Since the proposed scheme relies on the automatic evacuation guiding system [2], we give the overview of the system.

$G = (\mathcal{V}, \mathcal{E})$ denotes a graph representing the internal structure of the target region, where $\mathcal{V}$ is a set of vertices, i.e., intersections, and $\mathcal{E}$ is a set of edges, i.e., roads in the map. There are $K > 0$ evacuees in the region and each of them has a mobile node. $\mathcal{K} = \{1, 2, \ldots, K\}$ denotes the set of all the nodes. Each node $k \in \mathcal{K}$ measures its own locations by using GPS (Global Positioning System) at control intervals of $I_{\mathrm{M}} > 0$.

Figure 1 illustrates the flow of guiding one evacuee to a safe place. Note that the evacuee has to pre-install an application for the evacuation guiding into his/her mobile node before disasters occur. The application obtains static information, e.g., the surrounding map of the target region and the location information of the safe places, in usual time. When a disaster occurs, the application is invoked. The application first finds out the nearest safe place d from location s of node k. Then, it calculates evacuation route $\widehat{p}^{k,1}_{s,d}$ and presents him/her the route as a recommended route (Step 1 in Fig. 1).

The evacuee tries to move along the recommended route. When the evacuee encounters a blocked road segment during his/her evacuation along the recommended route $\widehat{p}^{k,1}_{s,d}$ (Step 2 in Fig. 1), he/she will take another route by his/her own judgment (Step 3 in Fig.1). As the same time, the application can trace his/her actual evacuation route as the trajectory by measuring his/her positions periodically with GPS. Thus, the application can detect the road segment $e \in \mathcal{E}$, which yields the difference

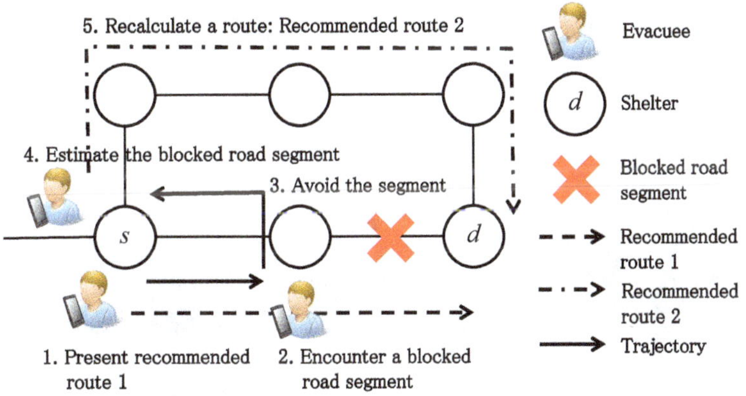

Fig. 1 The flow of the automatic evacuation guiding

between the recommended route and the actual evacuation route. The application adds the road segment e to the set $\mathcal{E}^k_{\mathrm{NG}}$ of blocked road segments (Step 4 in Fig. 1). After that, it also recalculates new evacuation route $\widehat{p}^{k,2}_{s,d}$, which does not include blocked road segments $\mathcal{E}^k_{\mathrm{NG}}$, and presents him/her the route (Step 5 in Fig. 1). Evacuation guiding finishes when the evacuee reaches the safe place or the application cannot find out any evacuation route to any safe place. In addition, the mobile node may obtain opportunities to obtain the information about blocked road segments from others, through direct wireless communication with other mobile nodes, e.g., Bluetooth and WiFi-Direct, and/or communication with communication infrastructures, e.g., 3G, 4G, and LTE. Sharing the information about blocked road segments among evacuees will improve their evacuation movements.

4 Congestion-aware route selection

In the previous work [2], the authors assume that the moving speeds of evacuees are constant, independently of traffic congestion. The recommended route is calculated by the shortest-distance based route selection with costs of static road length. When each evacuee selects the shortest path as the recommended route, many evacuees will concentrate on some roads and traffic congestion may occur. In this paper, we propose congestion-aware route selection that can alleviate the traffic congestion. Since the congestion-aware route selection is replaced with the shortest-distance based route selection in the automatic evacuation guiding, it is conducted at control intervals of I_{M}.

In what follows, we first describe a model of traffic congestion on roads. Next, we explain a method of acquiring information for route calculation and the definition of road costs. Finally, we describe the route calculation based on the obtained road costs.

4.1 Congestion model

The degree of congestion on a road is correlated with the number of evacuees passing through that road. In [4, 15], the relationship between the moving speed v [m/s] of pedestrians and population density κ [1/m^2] on a road is given by

$$v = 1.2 - 0.25\kappa. \tag{1}$$

In this paper, we extend this congestion model by taking account of the order of evacuee on a road. It is natural to assume that the congestion degree is different among evacuees on a road, depending on their positions on that road: the evacuee at the head of the road is not affected by other evacuees on that road while that at the tail of that road is affected by all other evacuees on that road.

Figure 2 illustrates our congestion model. In Fig. 2a, there is a group of four evacuees moving in the same direction on a road, and i-th evacuee is moving at speed v_i ($i = 1, \ldots, 4$). v_i is determined by (1) with the number of evacuees who are moving ahead of ith evacuee, i.e., $i - 1$, and the area of that road. Figure 2b illustrates the situation just after evacuee A leaves the road. At this moment, the moving speeds of the remaining evacuees B, C, and D change as shown in Fig. 2b. In what follows, we describe the congestion-aware route selection based on this congestion model.

4.2 Information retrieval of blocked road segments and evacuees' locations

In our congestion model, mobile node k needs to grasp the degree of congestion of each road segment, i.e., the number of evacuees. In addition, mobile node k also requires

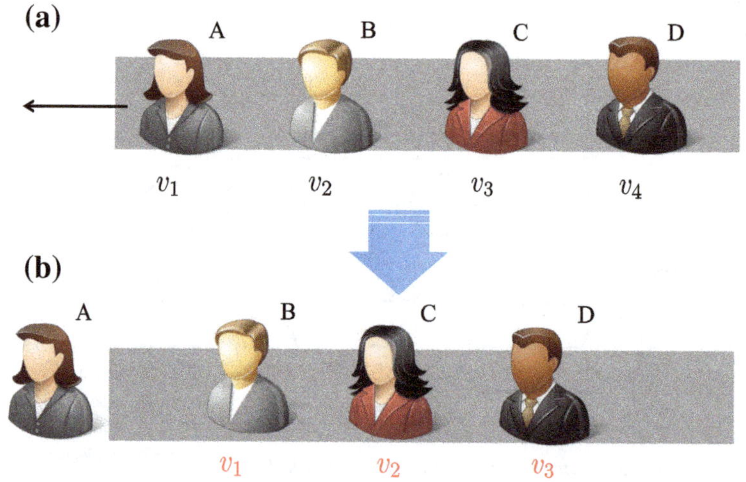

Fig. 2 Evacuees' moving speed model. **a** A situation where four evacuees, A, B, C, and D, move in the same direction on a road and **b** the situation just after evacuee A leaves the road

to grasp blocked road segments, which will be used in the automatic evacuation guiding scheme. Therefore, mobile node k tries to retrieve the information about block road segments and the current position from other mobile node at control intervals of I_M through communication infrastructure with transmission range R_I, e.g., cellular network and wireless LAN, or direct wireless communication with transmission range R_D, e.g., Wi-Fi Direct and Bluetooth.

In this paper, we assume that two extreme scenarios: global scenario and local scenario. In the global scenario, the communication infrastructure is sufficiently deployed and each mobile node can communicate with all other nodes at control intervals of I_M. In the local scenario, the communication infrastructure is fully damaged by the disaster and each mobile node can communicate with other node only through the direct wireless communication. Note that the range of flooding messages is limited to one hop in the local scenario, in order to alleviate the load on communication networks. We will show this limitation does not much affect the evacuation guiding with appropriate value of transmission range R_D (See Section 5.5).

Figure 3 illustrates the overview of information retrieval in global scenario and local scenario. In this figure, evacuee k moves on road segment e_1 from s to a and the number in each circle represents the moving order of the corresponding evacuee. As mentioned above, the movement speed of evacuee is determined by the moving order among evacuees traveling on the same road. Therefore, mobile node k requires the information about its moving order on each road segment when it travels the corresponding road segment. In the global scenario (the upside of Fig. 3), mobile node k can know the positions of all other nodes. On the contrary, mobile node k can know the positions of other nodes in transmission range of R_D in the local scenario (the downside of Fig. 3).

4.3 Road cost model

Let $o_{e,k}$ denote the moving order of evacuee k on road segment e. As for the road segment on which evacuee k travels, mobile node k can estimate $o_{e,k}$ directly from the obtained information, e.g., $o_{e_1,k} = 4$ in Fig. 3. As for the other road segment e, it is difficult for k to estimate the number of evacuees traveling on e when k arrives at e. Mobile node k estimates its moving order on e, $o_{e,k}$, as $n_e + 1$, where n_e is the number of evacuees that currently travels on e. For example, $o_{e_2,k} = 3 + 1 = 4$ and $o_{e_3,k} = 1 + 1 = 2$ in the global scenario of Fig. 3, $o_{e_2,k} = 0 + 1 = 1$ and $o_{e_3,k} = 1 + 1 = 2$ in the local scenario in Fig. 3.

We should note here that $o_{e,k}$ will change as time passes. As mentioned in Sec. 4.2, each mobile node tries to retrieve the positions of other mobile nodes at control intervals of I_M. To simplify the calculation, we define road e's cost for mobile node k, $c_{e,k}$, as follows:

$$c_{e,k} = \frac{l_e}{\frac{\sum_{i=1}^{o_{e,k}} v_i}{o_{e,k}}} = \frac{o_{e,k} l_e}{\sum_{i=1}^{o_{e,k}} v_i}, \quad (2)$$

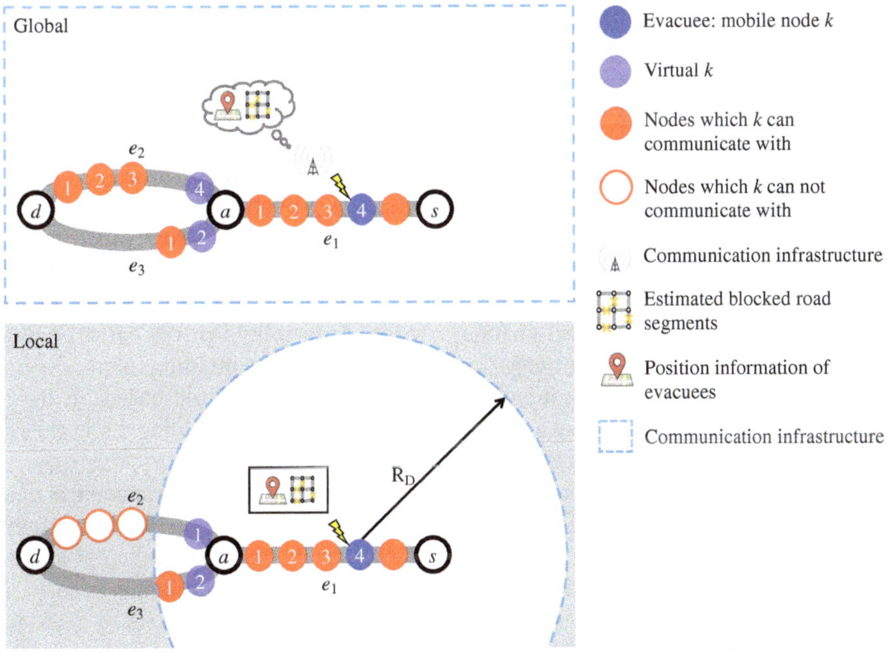

Fig. 3 Information retrieval of congestion-aware route selection

where l_e is the length of road e. (2) indicates the average traveling time of road e among evacuee k and his/her preceding evacuees.

4.4 Route calculation

Based on each road cost estimated by (1), mobile node $k \in \mathcal{K}$ selects route $\widehat{p}^k_{s,d}$ with the smallest total cost among set $\mathcal{P}^k_{s,d}$ of route candidates from current location $s \in \mathcal{V}$ to destination $d \in \mathcal{V}$, using existing graph search algorithms, e.g., Dijkstra's algorithm,

$$\widehat{p}^k_{s,d} = \arg\min_{p \in \mathcal{P}^k_{s,d}} \sum_{\forall e \in p} c_{e,k}. \tag{3}$$

For example, $\mathcal{P}^k_{s,d}$ is $\{\{e_1, e_2\}, \{e_1, e_3\}\}$ in Fig. 3. Suppose $l_{e_2} = l_{e_3}$. In the global (resp. local) scenario, mobile node k selects route $\{e_1, e_3\}$ (resp. $\{e_1, e_2\}$) because $o_{e_3,k} < o_{e_2,k}$ (resp. $o_{e_2,k} < o_{e_3,k}$).

When mobile node $k \in \mathcal{K}$ starts its evacuation, it calculates the first route. At control intervals of I_{M}, it also conducts the route calculation when at least one of the following conditions is satisfied.

- (C1) When mobile node k newly obtains information about a block road segment by the automatic evacuation guiding scheme.
- (C2) When the moving speed of mobile node k becomes slower than threshold $\theta_{\mathrm{V}} \geq 0$, due to the traffic congestion.

Condition (C1) ensures that mobile node k can avoid blocked road segment as in the existing automatic evacuation guiding. Note that not only the information about blocked road segments but also the latest information about each road cost will improve the evacuation route.

In addition to condition (C1), the congestion-aware route selection also aims to adapt to change of traffic condition if needed. The performance of congestion-aware route selection depends on the estimation accuracy of each road's cost given by (2). In the global scenario, each mobile node can obtain the number of evacuees on each road, regardless of their locations. However, it is not necessarily for each mobile node to grasp the congestion states of distant roads because they will change according to evacuees' movement. On the contrary, the estimated congestion states of near roads will not almost change when the evacuee arrive at them. Taking account of this characteristic, mobile node k evaluates the quality of current evacuation route based on that of current road segment, i.e., current moving speed, as in condition (C2).

This characteristic of estimation accuracy also indicates that the shortsighted estimation in the local scenario will be competitive with the farsighted estimation in the global scenario, depending on the transmission range R_{D} of direct wireless communication. We should also note here that the congestion-aware route selection tends to allocates evacuees to different routes, thus it will reduce the opportunities of information sharing among mobile nodes in the local scenario. Taking account of these points, we will examine how the congestion-aware route selection can work well even in the local scenario in Section 5.

5 Simulation experiments

In this section, we show the effectiveness of the congestion-aware route selection through simulation experiments.

5.1 Simulation model

We developed a simulator based on The ONE [17]. We used the map of southwest area of Nagoya station in Japan (Fig. 4). Table 1 shows the size and graph structure of the map. In case of Nagoya city, some useful information is available, i.e., population, area of road network, location of safe place, and road blockage probability.

The proposed congestion model given by (1) depends on population density on each road. The population of Nagoya city is about 2.3 million and the road area in the city is about 50 $[\mathrm{km}^2]$, according to the population and road statistics information [18]. Thus, the population density on the road network is 0.046 $[1/\mathrm{m}^2]$. Note that this population only includes the residents and is called the nighttime population. In the daytime, more people, e.g., business people, students, and shoppers, move into the urban areas. The proportion of the daytime population per 100 persons of nighttime population is called the ratio of daytime population to nighttime population, r. The average and maximum of r is 113.5 and 373.1 in Nagoya city, respectively [19]. In what follows, we evaluate the impact of degree of congestion by changing r in the range of $[100, 400]$.

As for the disaster scenario, we set part of roads to be blocked, i.e., red lines in Fig. 4, according to the road blockage probabilities. Nagoya city has been evaluating the regional risks, e.g., road blockage probabilities, caused by future large-scale disasters such as Nankai Trough Earthquake [20]. The road blockage probability is an estimated probability that the corresponding road is blocked due to collapse of buildings along the road under a certain disaster. It is calculated based on the degree of collapse and height of each building along the road, and the width of the road.

We assume that each evacuee has one mobile node and starts his/her evacuation from a random point on the map. We set a safe place to the actual location, i.e., green circle in Fig. 4. The simulation time is set to be 10,000 s. When the simulation starts, a disaster occurs and each evacuee starts evacuation. We set threshold of route recalculation θ_{V} to be 0 $[\mathrm{m/s}]$. As for the communication

Fig. 4 The map of Nagoya city

environments, we used the global scenario and the local scenario described in Section 4.2.

We evaluate the proposed scheme in terms of average evacuation time T_{avg}, maximum evacuation time T_{max}, and the evacuation ratio. Note that the evacuation time of each evacuee is defined as the time interval from his/her evacuation start to evacuation completion. The evacuation ratio is defined as the ratio of evacuees who arrive at the safe place to all evacuees. The following results are the average of 70 independent simulation results.

5.2 Sensitivity to control interval I_M

The performance of congestion-aware route selection depends on control interval I_M. Small I_M can achieve quick response to traffic congestion but will increase both computing and networking overhead. Figure 5 illustrates the relationship between I_M and average evacuation time in case of global scenario. Regardless of the ratio r of daytime population to nighttime population, we observe that the average evacuation time gradually increases with I_M, due to delays in detecting and avoiding congestion. In what follows, we set I_M to be 50 [s] by taking account of the balance between average evacuation time and overhead.

5.3 Effect of congestion-aware route selection

In this section, we show the performance comparison between congestion-aware route selection and shortest-distance based route selection in the global scenario.

Figure 6 represents the transition of evacuation ratio for congestion-aware route selection and shortest-distance based route selection when ratio of daytime population to nighttime population, r, is set to be 400. We observe that the congestion-aware route selection can increase the evacuation ratio faster than the shortest-distance based route selection. The average evacuation time becomes 2720 [s] for the congestion-aware route selection and 3599 [s] for the shortest-distance based route selection. Thus, the improvement ratio of average evacuation time is about 24%. On the contrary, the maximum evacuation time becomes 5846 [s] for the congestion-aware route selection and 9644 [s] for the shortest-distance based route selection. The improvement ratio of maximum evacuation time is about 39%.

We also find that both route selections show similar evacuation ratio at the early stage of evacuation but the congestion-aware route selection achieves much higher evacuation ratio than the shortest-distance based route selection during the remaining period. Since each evacuee starts from various points in the area, concentration of evacuees on specific roads is unlikely to occur at the early stage of evacuation. As a result, the congestion-aware route selection also tends to apply the shortest path. As time passes, the number of evacuees approaching the safe place gradually increases. As a result, heavy traffic congestion occurs in the case of shortest-distance based route selection. On the contrary, the congestion-aware route selection can alleviate the traffic congestion.

5.4 Impact of congestion degree

In this section, we evaluate the impact of congestion degree on both route selections in terms of average/maximum evacuation time in the global scenario.

Table 1 Information of the map of Nagoya city

Map information	
Map size	3,500 [m] ×2,300 [m]
Number of vertexes	3,972
Number of directed edges	7,918

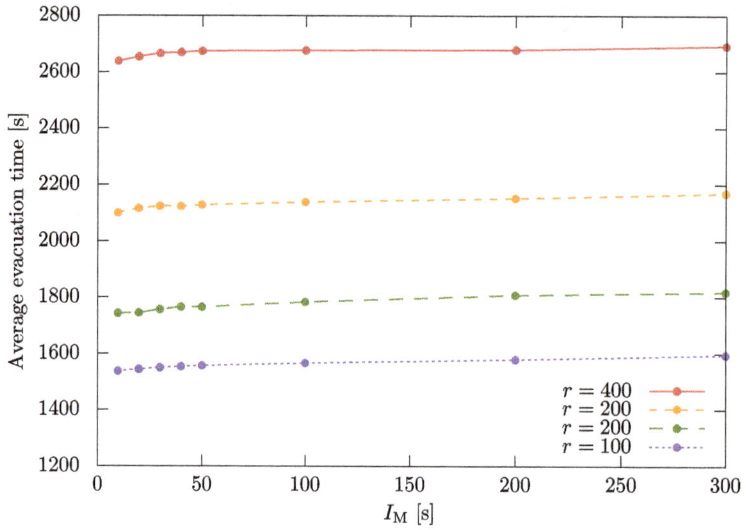

Fig. 5 Control interval I_M vs. average evacuation time (global scenario)

Figure 7 illustrates the relationship between ratio r of daytime to nighttime population and average evacuation time. We first observe that there is almost no difference between the results of two route selections when r is small, e.g., $r = 100$. When congestion degree r increases, the effectiveness of congestion-aware route selection increases.

Figure 8 shows the relationship between ratio r of daytime population to nighttime population and maximum evacuation time. As in case of average evacuation time, we observe that there is almost no difference between the results of two route selections when $r = 100$. Comparing Fig. 8 with 7, we find that the congestion-aware route selection can much improve the maximum evacuation time when r increases.

5.5 Impact of communication environments

In this section, we evaluate the impact of communication environments on average/maximum evacuation time for two route selections.

Figure 9 (resp. Fig. 10) illustrates the relationship between transmission range R_D of direct wireless communication and average (resp. maximum) evacuation time, when ratio r of daytime population to nighttime population is set to be 100 and 300. For comparison purpose, we also show the average (resp. maximum) evacuation time of congestion-aware route selection in case of global scenario in Fig. 9 (resp. Fig. 10).

First, we focus on the results of local scenario. We observe that there is almost no difference between two route selections when $r = 100$. In addition, transmission

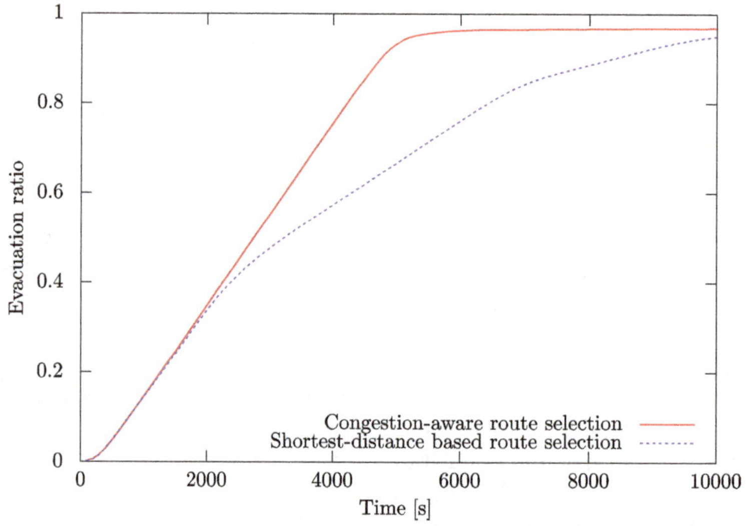

Fig. 6 Transition of evacuation ratio (global scenario)

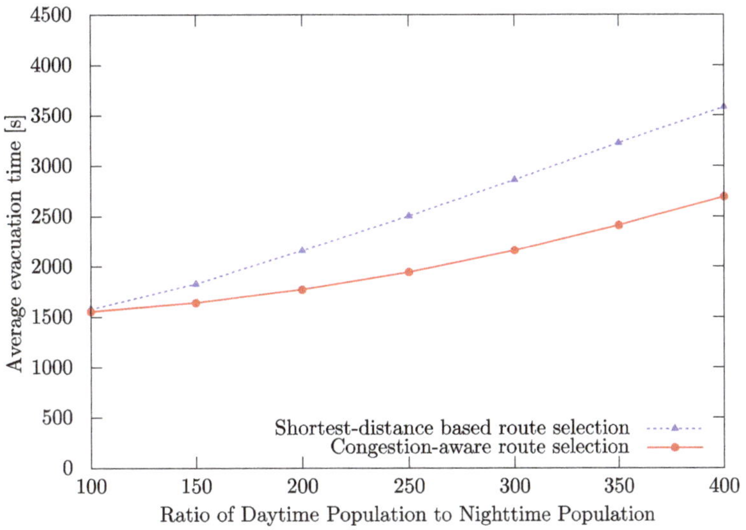

Fig. 7 The ratio of daytime population to nighttime population vs. average evacuation time (global scenario)

range R_D does not much affect the average/maximum evacuation time, regardless of route selection. This is because traffic congestion rarely occurs due to the low congestion degree, as mentioned in Section 5.3.

On the contrary, when the congestion degree increase, i.e., $r = 300$, we observer that the congestion-aware route selection (resp. the shortest-distance based route selection) can (resp. not) improve the average/maximum evacuation time with increase of R_D. In the automatic evacuation guiding, regardless of the route selection, the wide transmission range will results in increase of information retrieval of blocked road segments. As for the shortest-distance based route selection, each evacuee selects the shortest-distance route, thus route concentration among evacuees frequently occurs. When the congestion degree increases, the impact of traffic congestion dominates the average/maximum evacuation time in case of shortest-distance based route selection.

Focusing on the congestion-aware route selection in case of $r = 300$, we observe that the average/maximum evacuation time is improved with increase of transmission range. In particular, we find that large improvement occurs when $R_D = 100$. This indicates that Bluetooth whose transmission range is dozens of meters is not sufficient but Wi-Fi Direct whose transmission range can be one hundred meter is sufficient. In case of the road

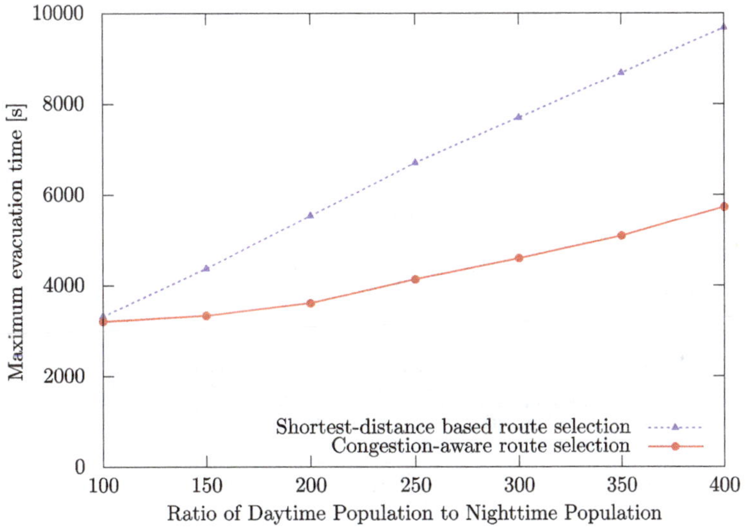

Fig. 8 The ratio of daytime population to nighttime population vs. maximum evacuation time (global scenario)

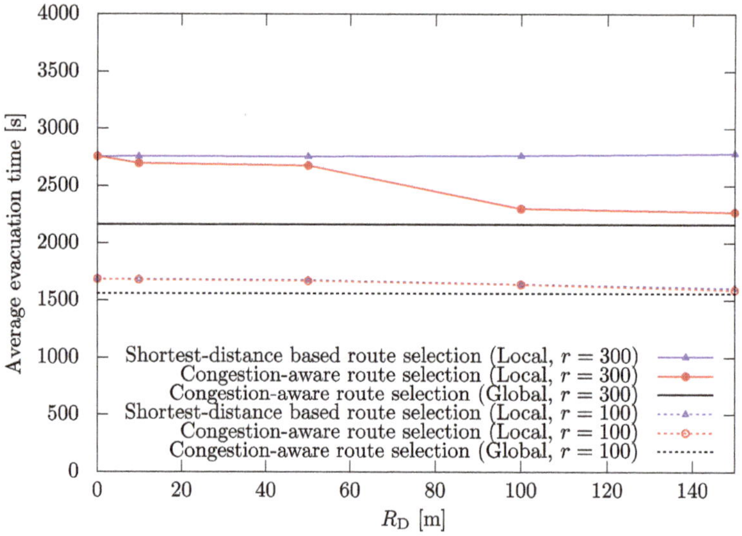

Fig. 9 Transmission range vs. average evacuation time

network in Fig. 4, the average road length is about 60 [m]. Thus, the congestion-aware route selection can work well even if each mobile node only grasps the number of evacuees on the current road and next road candidates. We also find that the average evacuation time in the local scenario with $R_D = 100$ is competitive with that in the global scenario.

These results indicate that autonomous and decentralized congestion alleviation can be realized even in the poor communication environments where communication infrastructures are unavailable and each mobile node can communicate with other nodes only within its direct communication range.

6 Conclusions

In this paper, we consider the problem of traffic congestion caused by the automatic evacuation guiding system. We first developed a traffic congestion model where an evacuee's moving speed on a road is determined based on the population density and his/her moving order among evacuees on that road. Next, we proposed the congestion-aware route selection where each evacuees selects an evacuation route with the minimum cost based on the cost model.

Through several simulation experiments, we showed that (1) the congestion-aware route selection works better than the shortest-distance based route selection at

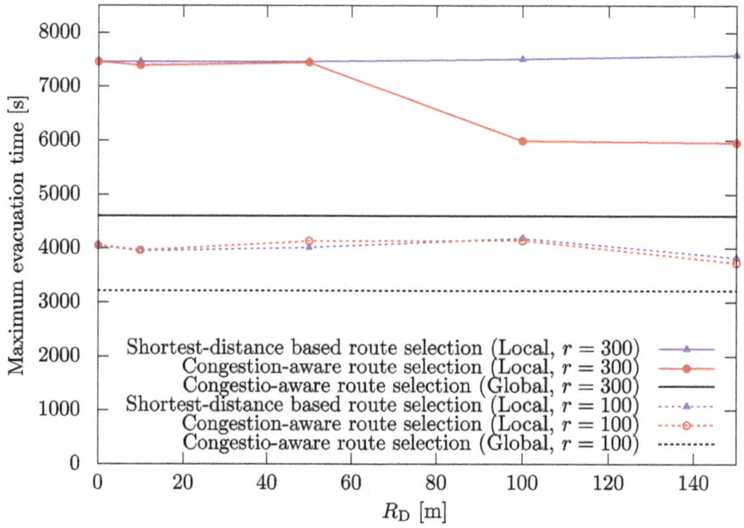

Fig. 10 Transmission range vs. maximum evacuation time

the middle and late stage of evacuation by alleviating the traffic congestion around the safe place, (2) the effect of congestion-aware route selection becomes high with increase of congestion degree, and (3) the congestion-aware route selection with local information sharing through direct wireless communications among mobile nodes can be competitive with that with global information sharing through communication infrastructures under the realistic transmission range of direct wireless communication, e.g., $R_D = 100$.

Authors' contributions

The authors have contributed to this paper and all authors read and approved the final manuscript.

Competing interests

The authors declare that they have no competing interests.

References

1. Ministry of Internal Affairs and Communications. 2011 WHITE PAPER Information and Communications in Japan. http://www.soumu.go.jp/johotsusintokei/whitepaper/eng/WP2011/2011-index.html. Accessed 2 Oct 2017
2. N Komatsu, M Sasabe, J Kawahara, S Kasahara, Automatic Evacuation Guiding Scheme Based on Implicit Interactions between Evacuees and Their Mobile Nodes. GeoInformatica, 1–15 (2016). https://link.springer.com/article/10.1007/s10707-016-0270-1#citeas
3. U Hiroi, N Sekiya, R Nakajima, S Waragai, H Hanahara, Questionnaire Survey Concerning Stranded Commuters in Metropolitan Area in the East Japan Great Earthquake. Ann. Inst. Soc. Saf. Sci. **15**, 343–353 (2011)
4. M Okada, *Ergonomics of Architecture and City: Mechanism of Space and Human Behavior (in Japanese)*. (Kajima Institute Publishing, Tokyo, 1977)
5. S Toriumi, M Kawaguchi, A Taguchi, Damage Assumption of Railway Commuters under the Tokyo Metropolitan Earthquake (in Japanese). Commun. Oper. Res. Soc. Jpn. **53**(2), 111–118 (2008)
6. Y Iizuka, K Yoshida, K Iizuka, in *Proc. of International Conference on Human-Computer Interaction*. An Effective Disaster Evacuation Assist System Utilized by an Ad-Hoc Network, (2011), pp. 31–35. https://link.springer.com/chapter/10.1007/978-3-642-22095-1_7#citeas
7. A Fujihara, H Miwa, in *Big Data and Internet of Things: A Roadmap for Smart Environments*, ed. by N Bessis, C Dobre. Disaster Evacuation Guidance Using Opportunistic Communication: The Potential for Opportunity-Based Service (Springer, Cham, 2014), pp. 425–446
8. CH Lin, PY Chen, WT Chen, in *Proc. of 2013 IEEE 77th Vehicular Technology Conference (VTC Spring)*. An Adaptive Guiding Protocol for Crowd Evacuation Based on Wireless Sensor Networks, (2013), pp. 1–5. doi:10.1109/VTCSpring.2013.6691867
9. S Li, A Zhan, X Wu, P Yang, G Chen, Efficient Emergency Rescue Navigation with Wireless Sensor Networks. J. Inf. Sci. Eng. **27**(1), 51–64 (2011)
10. K Fall, in *Proc. of SIGCOMM'03*. A Delay-Tolerant Network Architecture for Challenged Internets, (2003), pp. 27–34. https://dl.acm.org/citation.cfm?id=863960
11. N Komatsu, M Sasabe, S Kasahara, in *Mobile Web and Intelligent Information Systems (MobiWis 2016), Lecture Notes in Computer Science*, ed. by M Younas, I Awan, N Kryvinska, C Strauss, and D Thanh. On Information Sharing Scheme for Automatic Evacuation Guiding System Using Evacuees' Mobile Nodes, vol. 9847 (Springer, Cham, 2016), pp. 213–221
12. V Verroios, K Kollias, PK Chrysanthis, A Delis, in *Proc. of the 5th International Conference on Pervasive Services*. Adaptive Navigation of Vehicles in Congested Road Networks, (2008), pp. 47–56. https://dl.acm.org/citation.cfm?id=1387277
13. S Lim, D Rus, Congestion-Aware Multi-Agent Path Planning: Distributed Algorithm and Applications. Comput. J. **57**(6), 825–839 (2014)
14. W Chen, S Zhu, D Li, in *Proc of The 7th IEEE International Conference on Mobile Ad-hoc and Sensor Systems (IEEE MASS 2010)*. VAN: Vehicle-Assisted Shortest-Time Path Navigation, (2010), pp. 442–451. doi:10.1109/MASS.2010.5663936
15. C-H Liu, Y Oeda, T Sumi, A Model for Pedesterian Movement with Obstacle Evasion using Personal Space Concept. J. Jpn. Soc. Civil Eng. **64**(4), 513–524 (2008)
16. T Roughgarden, É Tardos, How Bad Is Selfish Routing?. J. ACM. **49**(2), 236–259 (2002)
17. A Keränen, J Ott, T Kärkkäinen, in *Proc. of the 2nd International Conference on Simulation Tools and Techniques*. The ONE Simulator for DTN Protocol Evaluation, (2009), pp. 55–15510. https://dl.acm.org/citation.cfm?id=1537683
18. Greenification and Public Works Bureau, Nagoya City. Road Statistics of Nagoya City (in Japanese). http://www.douroninteizu.city.nagoya.jp/dourotoukei.html. Accessed 2 Oct 2017
19. Statistics Bureau, Ministry of Internal Affairs and Communications. Final Report of The 2010 Population Census: Population and Households of Japan (in Japanese). http://www.stat.go.jp/data/kokusei/2010/final/pdf/01-11_4.pdf. Accessed 2 Oct 2017
20. Housing and City Planning Bureau, Nagoya City. City Development Policies for Earthquakes (in Japanese). http://www.city.nagoya.jp/jutakutoshi/cmsfiles/contents/0000002/2717/honpen.pdf. Accessed 2 Oct 2017

10

Modeling and performance analysis for mobile cognitive radio cellular networks

Anum L. Enlil Corral-Ruiz[1], Felipe A. Cruz-Perez[1], S. Lirio Castellanos-Lopez[2*], Genaro Hernandez-Valdez[2] and Mario E. Rivero-Angeles[3]

Abstract

In this paper, teletraffic performance and channel holding time characterization in mobile cognitive radio cellular networks (CRCNs) under fixed-rate traffic with hard-delay constraints are investigated. To this end, a mathematical model to capture the effect of interruption of ongoing calls of secondary users (SUs) due to the arrival of primary users (PUs) is proposed. The proposed model relies on the use of an independent potential interruption time associated with the instant of possible interruption for each ongoing call in every visited cell. Then, a Poisson process is used to approximate the secondary users' call interruption process due to the arrival of PUs. Based on this model and considering that unencumbered service time (UST) and cell dwell time (CDT) of SUs are independent generally distributed random variables, analytical formulae for both the probability distributions of channel holding times and inter-cell handoff attempts rate are derived. Also, a novel approximated closed-form mathematical expression for call forced termination probability of SUs is derived under the restriction that the UST is exponentially distributed. Additionally, by considering all the involved time variables exponentially distributed and employing fractional channel reservation to prioritize intra- and inter-cell handoff call attempts over new call requests, a queuing analysis to evaluate the call-level performance of CRCNs in terms of the maximum Erlang capacity is developed. The accuracy of our proposed mathematical models is extensively investigated under a variety of different evaluation scenarios for all the considered call-level performance metrics. Numerical results demonstrate that channel holding time statistics are highly sensitive to both interruption probability of ongoing secondary calls and type of probability distribution functions used to model CDT and UST. From the teletraffic perspective, numerical results reveal that the system Erlang capacity largely depends on the relative value of the mean secondary service time to the mean primary service time and the primary channels' utilization factor. Also, the obtained results show that there exists a critical utilization factor of the primary resources from which it is no longer possible to guarantee the required quality of service of SUs and, therefore, services with hard-delay constraints cannot be even supported in CRCNs.

Keywords: Cognitive radio cellular networks, Erlang capacity, Call admission control, Cell dwell time, Channel holding time, Call forced termination probability, New call blocking probability, Handoff failure probability, Intra- and inter-cell handoff

1 Introduction

Cognitive radio (CR), which enables secondary users (SUs) to temporarily utilize in an opportunistic fashion the non-in-use spectrum bands allocated to the primary users (PUs) of the spectrum, has been proposed as a promising technology to improve spectrum usage and solving the problem of heterogeneity of radio devices [1]. In Table 1, a list of the acronyms used in the manuscript is provided. In CR networks (CRNs), spectrum handoff allows interrupted secondary calls to be switched to an idle channel, if one is available, to continue its service. Due to the great interest in providing multimedia service with stringent quality of service (QoS) over CR networks [2–5], it is highly desirable for CRNs to support real-time traffic with QoS provisioning. In this research direction, there is a lot of research focused on the QoS provisioning for the secondary sessions; see [2–6] and the references therein.

Recently, implementation of cognitive radio cellular networks (CRCNs) has been investigated to solve both

* Correspondence: salicalo@correo.azc.uam.mx
[2]Electronics Department, UAM, Mexico City, Mexico
Full list of author information is available at the end of the article

Table 1 List of acronyms

Acronyms	Description
CCDF	Complementary cumulative distribution function
CDT	Cell dwell time: the time that a mobile station spends in the jth (for $j = 0, 1, \ldots$) handed off cell irrespective of whether it is engaged in a call or not
CDTr	Residual cell dwell time: the time between the instant that a new call of a user is initiated and the instant that the user is handed off to another cell
CHT	Channel holding time: the amount of time a call holds a channel in a cell
CHTh	Handoff call channel holding time
CHTn	New call channel holding time
CoV	Coefficient of variation
CR	Cognitive radio
CRCN	Cognitive radio cellular networks
FR-HDC	Fixed-rate with hard-delay constraints
IT	Interruption time
LT	Laplace transform
MS	Mobile station
NSH	Without spectrum handoff
PU	Primary user
QoS	Quality of service
RV	Random variable
SH	Spectrum handoff
Sk	Skewness
SU	Secondary user
UCIT	Unencumbered call interruption time: the period of time from the instant a secondary call establish a radio link with a cell until the time it would be interrupted due to the arrival of a PU assuming that the network has unlimited resources and the service time is of infinite duration
UST	Unencumbered service time: the amount of time that the call would remain in progress if it experiences no forced termination
USTr	Residual unencumbered service time: the remaining service time after an ongoing SU has been handed off to another cell

spectrum inefficiency and spectrum scarcity [4, 5]. The main differences between traditional cellular networks and CRCNs lie in the so-called spectrum handoff, interruption due to the arrival of PUs (including the possibility of interruption of multiple sessions of SUs due to non-homogeneous bandwidth of PU and SU channels), and the fluctuating nature of spectrum resource because of the arrival and departure of PUs. In CRCNs, SUs are allowed to perform spectrum handoff within a given CR cell (a.k.a. intra-cell handoff) as well as inter-cell handoff among different CR cells [7]. Developing analytical models for the performance analysis of CRCNs that effectively capture relevant aspects of these complex networks represents a very challenging and important topic of research. This is the topic of research of the present paper. Specifically, teletraffic performance and channel holding time characterization of CRCNs under fixed-rate traffic with stringent-delay constraints are investigated.

Channel holding time (CHT), unencumbered service time (UST), and cell dwell time (CDT) are fundamental time variables to model and analyze the performance of mobile cellular networks. To realistically capture the overall effects of users' mobility and traffic flow characteristics of a real system, the use of general distributions for modeling UST and CDT has been highlighted in previous works [8–12]. These time variables are of primary importance as they can be used to derive other key performance metrics such as new call blocking, handoff failure, and call forced termination probabilities, as well as inter-cell handoff arrival rate and system Erlang capacity. However, none of the previous studies that investigate the functional relationship between CHT, UST, CDT, and their impact on system performance have considered the fluctuating nature of spectrum availability in CRCNs. Moreover, the effect of mobile users across multiple cells has not received enough attention in CRCNs in part because it is a recent topic of research [7, 13, 14]. Thus, service interruption of ongoing secondary sessions due to the arrival of a PU, spectrum handoff, and user mobility must be jointly captured in the teletraffic model for the performance analysis of CRCNs, which represents an important but unexplored topic of research in CR networks to date.

In this paper, the call-level performance analysis of mobile CRCNs under fixed-rate with hard-delay constraints (FR-HDC) traffic[1] is investigated. This issue is addressed in the following manner: (a) as a first step, a novel mathematical model to approximately capture the effect of interruption of SU calls due to the arrival of primary sessions is proposed. In particular, it is proposed to use an independent potential time (referred hereafter as *unencumbered call interruption time*) associated with the instant of possible interruption for each call in every visited cell. (b) Then, the SUs' call interruption process due the arrival of primary users is approximated by a Poisson process.[2] (c) Building on this model and under the assumption that both CDT and UST have general probability distribution functions, analytical formulae for the probability distribution functions and the first three standardized moments of new call channel holding time (CHTn) and handoff call channel holding time (CHTh) as well as mathematical expressions for inter-cell and intra-cell handoff arrival rates are derived. Additionally, for exponentially distributed UST and generally distributed CDT, a novel closed-form mathematical expression for call forced termination probability of SUs (due to either inter-cell or intra-cell handoff failures) is derived. (d) On the other hand, considering all the involved time variables (i.e., unencumbered call interruption time, unencumbered service time, inter-arrival time, and cell dwell time) exponentially distributed and using fractional channel reservation to prioritize intra (due to spectrum handoff)- and inter (due to users' mobility)-cell handoff call attempts over new call requests, a queuing analysis to evaluate the call-level performance of CRCNs with FR-HDC traffic in terms of Erlang capacity is developed. (Erlang capacity is obtained by optimizing the number of reserved channels.) (e) Finally, channel holding time statistics and the effects of the relative value of the mean secondary service time to the mean primary channel holding time, users' mobility, the use of spectrum handoff (SH), and the primary channel utilization factor on the maximum Erlang capacity of the system are investigated.

Traffic modeling and call-level performance evaluation of wireless cellular communication systems have been widely investigated in the literature [9, 10, 15, 16] (and the references therein). Different performance metrics have been investigated (channel holding time probability distribution, completion probability, forced termination probability, new call blocking probability, among others); however, most published works on traffic modeling in wireless cellular communication systems have considered exponentially distributed UST. Exceptions of this are references [9, 10, 15, 16]. In [9, 15, 16], channel holding times for new and handoff calls are analyzed considering general distributed CDT. In [10], it is additionally considered that the UST is generally distributed.

On the other hand, teletraffic analysis of wireless communication systems taking into account both resource insufficiency and link unreliability is addressed in [9, 17–19]. In [17], Zhang et al. derived mathematical expressions for the probability that a call complete successfully considering the concurrent impacts of the bad quality in the channel and the lack of radio resources. In [18], Zhang studied the impact of Rayleigh fast fading on various teletraffic QoS metrics in wireless networks taking into account carrier frequency. In [19], a closed-form formula for the call completion probability is developed under the generalized wireless channel model and the general CHT distribution using theory of complex variable and transform techniques (Laplace-Stieltjes transform and z-transform). The mathematical models considered in [17–19] are based on link-level statistics.[3]

Contrary to [17–19], in our early work [9], a teletraffic analysis is performed taking into account both resource insufficiency and channel unreliability through a system-level model by introducing a simple call interruption process to model the effect of wireless channel unreliability. Mathematical expressions for the call forced termination probability and inter-cell handoff arrival rate are derived for general distributed CDT and unencumbered call interruption time. Nonetheless, in [9], the UST is assumed to be exponentially distributed.

It is important to notice that none of the described papers so far have considered the particular features of CRNs. In particular, neither call interruption due to the arrival of PUs, intra-cell spectrum handoff, nor the fluctuating nature of spectrum resource was considered. As stated in [7], there has been little investigation so far on the opportunistic spectrum access performance in the presence of inter-cell handoff calls. Most studies have focused considering a single CRN, and therefore, users' mobility and inter-cell handoffs have been neither modeled nor evaluated. In [7], by considering a homogeneous cognitive radio multi-cellular system, both inter-cell handoff rate and forced termination probability of SUs are determined. Reference [20] extends the reported work in [7] by considering that instead of channel reservation, the queueing buffer is used to give both intra-handoff and inter-handoff SU calls priority over the initial SU calls. In [21], the performance of secondary users at the packet level in a cognitive radio cellular network was analyzed, where the transmitter and receiver are not able to communicate directly. In [21], it is considered that packets are delay sensitive and can tolerate a limited amount of waiting time. Additionally, the effect of errors in spectrum sensing on the performance of SUs is addressed in [21]. On the other hand, in [22], multirate cognitive radio cellular networks where the users of each class have a different bandwidth requirement are considered. Two call admission control (CAC) policies are studied, and the system is modeled by considering as a multi-dimensional Markov process. An iterative algorithm to find the steady-state probability distribution and compute the performance measures is proposed. Finally, in our previous related work [23], the impact of mobility of both primary and secondary users on the performance of cognitive radio mobile cellular networks with real-time traffic is investigated. In [23], it is considered that all the involved time variables are modeled by exponentially negative random variables for both secondary and primary calls. However, no closed-form mathematical expressions are obtained and only exponentially distributed CDT and UST are considered in [7, 20–23]. Furthermore, channel holding times are not analyzed, and Erlang capacity is not addressed in [7, 20–23].

In [13, 24], a non-homogeneous cognitive radio network with multiple cells (i.e., cognitive radio base stations) is proposed and analytically evaluated. Probabilities of dropping and blocking for PUs and SUs and forced termination for SUs are investigated. Additionally, a resource planning method to compensate for the uneven traffic load distribution is proposed in [24]. Due to the non-homogeneous characteristic of the considered network, a pair of variables is needed to represent the state of each cell. To avoid the state space explosion, a small topology of cell is considered. All the involved times are considered exponentially distributed, and no mathematical expression for the considered performance metrics are obtained in [13, 24]. Finally in [14], a spectrum-aware mobility management scheme for CRCNs is proposed. Firstly, a novel CRCN architecture based on the spectrum pooling to mitigate heterogeneous spectrum availability is introduced. Considering spectrum mobility management, user mobility, and inter-cell resource allocation, a unified mobility management framework is developed. Cell capacity, handoff rate, drop rate, and link efficiency are evaluated by simulation. However, neither mathematical modeling nor analysis is presented in [14].

It is important to notice that the mathematical analysis developed in the present paper is based on the analytical framework proposed by Fang in [25, 26] and by Wang and Fan in [10] for the performance evaluation for wireless networks and mobile computing systems. Nevertheless, in [10, 25, 26], particular features of CRNs are not considered and forced call termination is considered to occur due only to resource insufficiency.

The remainder of this paper is structured as follows. In Section 2, the cognitive radio cellular system model is described. Also, in Section 2, the proposed modeling for the call interruption process (due to the arrival of primary users) is introduced. General mathematical expressions for CHT are derived in Section 3, while analytical formulae for inter-handoff rate and forced termination probability are derived in Section 4. In Section 5, a queuing analysis to evaluate the call-level performance of CRCNs in terms of the maximum Erlang capacity is developed. In Section 6, numerical evaluations are carried out for the performance analysis of CRCNs. Finally, Section 7 shows the conclusions.

2 System and proposed call interruption modeling

In this section, the general guidelines for the mathematical analysis of cognitive radio cellular networks under fixed-rate secondary traffic with hard-delay constraints are presented. As stated before, for this type of traffic, it is considered that both blocked and interrupted (due to either inter-cell or intra-cell handoff failures) calls are clear from the system. That is, a secondary type of traffic that has the most stringent QoS requirements (such as the unsolicited grant service class in mobile WiMAX) is considered. Hence, the principal performance metrics are new call blocking, call interruption, and forced call termination probabilities. A homogeneous infrastructure-based mobile CRCN with omni-directional antennas located at the center of each cell is assumed. That is, the structures of all cells within the CRCN are statistically identical. It is assumed that two types of radio users coexist: the licensed (primary) holders of the spectrum and the unlicensed (secondary) users that opportunistically share the spectrum resources with the PUs within the service area of the CRCN, while causing negligible interference to the PUs. PUs are not aware of the existence of SUs' activities, and SUs can detect activities of PUs through spectrum sensing. In this environment, PUs have absolute priority over SUs. The spectrum consists of M primary bands, and each primary band is divided into N sub-bands. Thus, there exist NM sub-bands that are shared by the primary and secondary users. SUs are only allowed to transmit when PUs are not using the spectrum. This resource allocation strategy is shown in Fig. 1. Thus, when an idle primary channel is detected, SUs may temporally occupy this unused channel. If a PU decides to access a primary channel, all SUs using this channel must relinquish their transmission immediately. These unfinished secondary calls are directed to other available channel (this process is called spectrum handoff).[4] If no vacant channels are available, interrupted secondary sessions are dropped. Some other relevant assumptions and definitions are given below. In Table 2, a list of the variables used in the manuscript is provided.

As in [7, 20–22], it is considered that secondary users only use empty channels (resources not being used by primary users) for their data transmission. Since primary users are using a cellular system where co-channel interference is tightly controlled and surveyed in order to guarantee an adequate QoS of cellular users, then co-channel interference is also controlled for the secondary users. In other words, the co-channel interference generated by the secondary users is the same interference generated by the primary users.

Modeling the spectrum sensing error has been widely addressed in the open literature [4, 27–32]. For instance, some previous works advice the use of sensor nodes placed in the serving area to perform this task or even secondary nodes can perform a collaborative sensing by

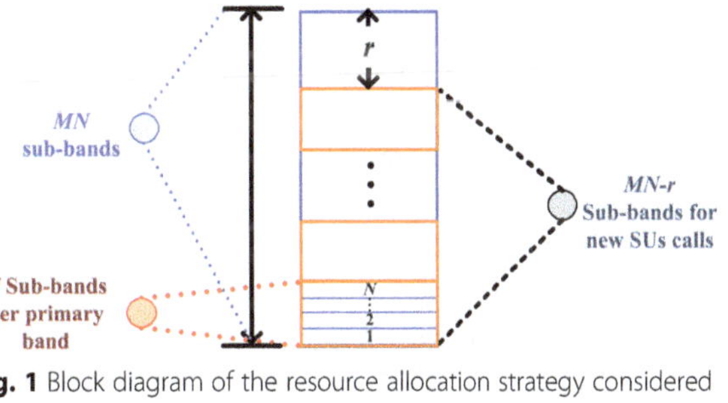

Fig. 1 Block diagram of the resource allocation strategy considered

Table 2 List of variables

Variables	Description
$\lambda^{(P)}$	Arrival rate per cell of PUs' calls
λ_h	Inter-cellular handoff rate
$\lambda_h{}^{(\text{in})}$	Incoming handoff rate
$\lambda_h{}^{(\text{out})}$	Outgoing handoff rate
λ_n	Arrival rate per cell of SUs' calls
$f_{X_a^{(P)}}(t)$	Probability density function of the inter-arrival time of PUs' calls
$f_{X_c^{(n)}}(t)$	Probability density function of the CHTn
$f_{X_c^{(h)}}(t)$	Probability density function of the CHTh
$f_{X_{\text{Int\_PU}}}(t)$	Probability density function of the unencumbered call interruption time
$f_{X_r}(t)$	Probability density function of the CDTr
$f_X^*(s)$	Laplace transform of the probability density function (pdf) of the random variable *X*
$F_X^*(s)$	Laplace transform of the cumulative distribution function (CDF) of the random variable *X*
$f_{X_s}(t)$	Probability density function of the UST
$f_{X_{sr}}(t)$	Probability density function of the USTr
k	The number of active cognitive users that have to relinquish their respective sub-channel due to the arrival of a PU service request
M	Number of primary bands in the system
N	Number of sub-bands which a primary band is divided; *N* sub-bands are equivalent to one primary band
$N^{(P)}(t)$	It is a RV used to represent the number of PU's arrivals in the interval [0, *t*].
p	Probability to reserve $\lfloor r \rfloor + 1$ sub-bands
$P(\mathbf{x})$	The state **x** stationary probability
$P_b{}^{(S)}$	The new call blocking probability for SUs
P_{b_tr}	Maximum acceptable values of the new call blocking
P_{ft}	Forced call termination probability
P_{ft_tr}	Maximum acceptable values of the forced call termination
P_{ft}^{H}	Probability that a SU attempts a handoff procedure as it leaves its current cell and there are no available resources in the target cell
P_{ft}^{Int}	Probability that an interruption occurs due to an arrival of a PU that requires the resources used by the SU
P_{hi}	Probability of intra-cell handoff failure (due to either that an interrupted SU cannot find available resources to perform spectrum handoff or spectrum handoff is not employed)
P_{hl}	Inter-cellular handoff failure probability: the probability that the handoff fails due to the lack of resources as a secondary user moves from one cell to another
P_{Int}	Probability that the analyzed call is interrupted by the first arrival of a PU after the beginning of the call
r	Number of reserved sub-channels to prioritize both intra (due to spectrum handoff)- and inter (due to users' mobility)-cell handoff call attempts over new call requests
x_i	State variable to represent the number of PUs when $i = 0$ and the number of SUs when $i = 1$
$\mathbf{x} = (x_0, x_1)$	Vector of state variables
$x_d{}^{(j)}$	It is a variable to represent a given/particular value of the CDT in the *j*th handed off cell
x_r	It is a variable to represent a given/particular value of the CDTr
x_s	It is a variable to represent a given/particular value of the UST
x_{sr}	It is a variable to represent a given/particular value of the USTr
$X_a{}^{(P)}$	It is a RV used to represent the inter-arrival time of PUs calls
$X_c^{(n)}$	It is a RV used to represent the CHTn of secondary users
$X_c^{(h)}$	It is a RV used to represent the CHTh of secondary users
$X_d{}^{(j)}$	It is a RV used to represent the CDT of secondary users in the *j*th handed off cell
X_{int_PU}	It is a RV used to represent the unencumbered call interruption time
X_r	It is a RV used to represent the CDTr of secondary users
X_s	It is a RV used to represent the UST of secondary users
X_{sr}	It is a RV used to represent the USTr of secondary users
$1/\mu$	$E\{X_s\} = 1/\mu$ is the mean value of X_s (UST of secondary users)
$1/\mu^{(P)}$	Mean value of UST of primary users

sharing their own information regarding the status of the channel in order to make a more accurate decision. Note that these alternatives are open issues and we believe that lay outside the scope of this work. In the submitted manuscript, it is considered that ongoing secondary users can sense the activity of primary users instantaneously and reliable (i.e., ideal spectrum sensing is assumed). In this sense, the obtained performance metrics can be considered as an upper bound [33]. It has been shown in the literature [4, 27–30] that the effect of false alarm and misdetection can be plugged with no major problem into most of the developed mathematical analysis of cognitive radio networks (CRNs). However, in this paper, to keep our mathematical analysis simple and to concentrate our study into the pure performance of the cognitive cellular radio network, we assume ideal spectrum sensing (spectrum sensing is error free). This represents a reasonable assumption in CRNs where a sensor network having enough sensors performing collaborative sensing of the radio environment in space and time is employed, as it is considered in [34, 35]. Modeling spectrum sensing error in CRNs has been addressed in [4, 27–30]. Additionally, the effect of unreliable spectrum sensing in CRNs under voice over Internet Protocol traffic is investigated in our early work [4]. Also as in [4], it is assumed that service events are unlikely to happen during the sensing period since the sensing periods are relatively small. Several works has evaluated (quantitatively and qualitatively) the extent by which spectrum sensing errors affect the performance of CRNs [4, 27–30].

First, the UST per call of SUs x_s is the amount of time that the call would remain in progress if it experiences no forced termination. Unless otherwise specified, this quantity is modeled by a generally distributed random variable (RV). The RV used to represent this time is X_s, and its mean value is $E\{X_s\} = 1/\mu$. The pdf of the UST is represented by $f_{X_s}(t)$. On the other hand, the mean UST of PUs is $1/\mu^{(P)}$.

The residual unencumbered service time (USTr) x_{sr} is defined as the remaining service time after an ongoing SU has been handed off to another cell. The RV used to represent residual UST is X_{sr}. The pdf of the USTr is represented by $f_{X_{sr}}(t)$, and it is derived in Section 3.2.

Now, CDT $x_d^{(j)}$ is defined as the time that a mobile station (MS) spends in the jth (for $j = 0, 1,...$) handed off cell irrespective of whether it is engaged in a call or not. The RVs used to represent these times are $X_d^{(j)}$ (for $j = 0, 1,...$) and are assumed to be independent and generally distributed. For homogeneous cellular systems, this assumption has been widely accepted in the literature [10].

The residual CDT (CDTr) x_r is defined as the time between the instant that a new call of a user is initiated and the instant that the user is handed off to another cell. Notice that always $x_d^{(0)} > x_r$ because the beginning of residual cell dwell time is randomly chosen within a cell dwell time interval. The RV used to represent CDTr is X_r. Thus, the pdf of the CDTr, $f_{X_r}(t)$, can be calculated in terms of the cell dwell time using the well-known residual life theorem [9] as follows:

$$f_{X_r}(t) = \frac{1-F_{X_d}(t)}{E\{X_d\}} \tag{1}$$

and its corresponding Laplace transform is given by

$$f^*_{X_r}(s) = \frac{\left[1-f^*_{X_d}(s)\right]}{E\{X_d\}s} \tag{2}$$

CHT is defined as the amount of time a call holds a channel in a cell. Let $X_{c^{(n)}}$ and $X_{c^{(h)}}$ denote, respectively, the CHTn and the CHTh with their corresponding pdfs $f_{X_c^{(n)}}(t)$ and $f_{X_c^{(h)}}(t)$. The relationship between the above defined times is illustrated in Fig. 2.

Call arrivals of PUs and SUs are assumed to follow independent Poisson processes with mean arrival rate $\lambda^{(P)}$ and λ_n arrivals per second per cell, respectively. Since the arrival process for the PUs follows a Poisson process, then the pdf of the inter-arrival time of PUs $X_a^{(P)}$ is given by $f_{X_a^{(P)}}(t) = \lambda^{(P)}e^{-\lambda^{(P)}t}$.

2.1 Proposed model for the call interruption time

In this sub-section, a simple but fundamental mathematical model to capture the effect of interruption of SU's calls due to the arrival of primary users is proposed. The proposed model relies on the use of an independent potential interruption time (referred as unencumbered call interruption time and it is denoted by $X_{\text{int\_PU}}$, with pdf $f_{X_{\text{int\_PU}}}(t)$) associated with the instant of possible interruption for each ongoing call in every visited cell. The relationship between this time and the rest of the times is illustrated in Fig. 3.

The unencumbered call interruption time is defined as the period of time from the instant a secondary call establish a radio link with a cell until the time it would be interrupted due to the arrival of a PU assuming that the network has unlimited resources and the service time is of infinite duration. Physically, this time represents the period of time that a secondary call last before it would be abruptly terminated due to a primary call arrival under the assumption that both the cell dwell time and the unencumbered service time are of infinite duration. The call interruption time is said to be "unencumbered" because the interruption of a secondary call in progress by a primary user arrival can or cannot occur, depending on the values of the CDT and the UST. Thus, a particular secondary call is interrupted in the current cell if and only if its potential call interruption time is smaller than both the unencumbered service time and residual cell dwell time (or both the unencumbered residual service time and cell dwell time). Therefore, notice that this call interruption time is not a measurable physical quantity. Indeed, in case that the user leaves the cell or completes its call before the call is interrupted, it is not possible to know the actual value of this interruption time since, in this case, call interruption due to the arrival of primary users actually never happens.

The probability distribution probability of the unencumbered call interruption time (IT) is related to the arrival process of PUs and the stochastic processes of the numbers of PUs and SUs with ongoing calls. In this work, in order to obtain a tractable mathematical analysis, it is proposed to approximate the secondary users' call interruption process due the arrival of primary users by a Poisson one. This proposed model is detailed next.

As stated above, the IT of an analyzed SU call due to an arrival of a PU is the elapsed time from the beginning

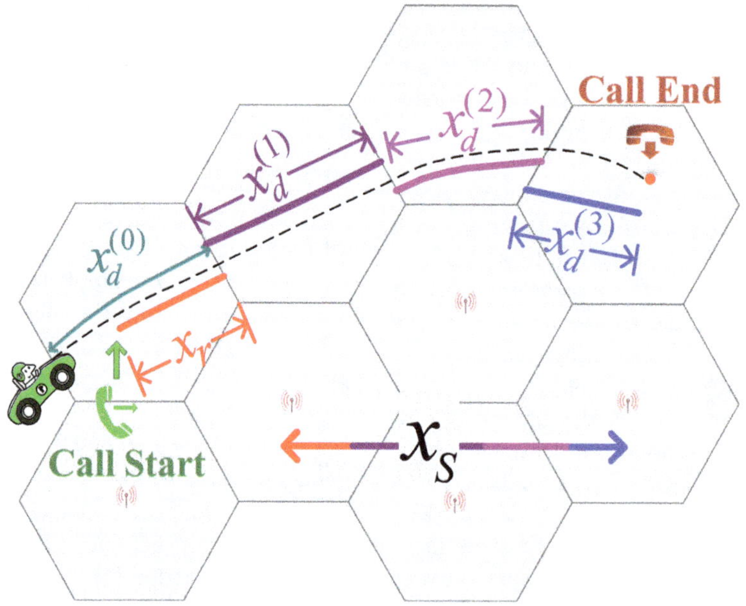

Fig. 2 Relationship between the cell residence and service times for successfully completed calls

of the call until the arrival of a PU that causes the interruption of the call. Building on this, let us define P_{Int} as the probability that an arrival of a PU interrupts the analyzed call.[5] As an approximation and for the sake of mathematical tractability, it is considered that the interruption probability is independent for each individual user.[6] As shown below, with this approximation, the call interruption process of SUs due the arrival of PUs is a Poisson process. However, as it is shown in Appendix 1, the call interruption process of SUs due the arrival of PUs does not actually follow a Poisson one. In Section 6, the inaccuracy introduced by this approximation in the numerical results is extensively investigated under a variety of different evaluation scenarios for all the considered call-level performance metrics.

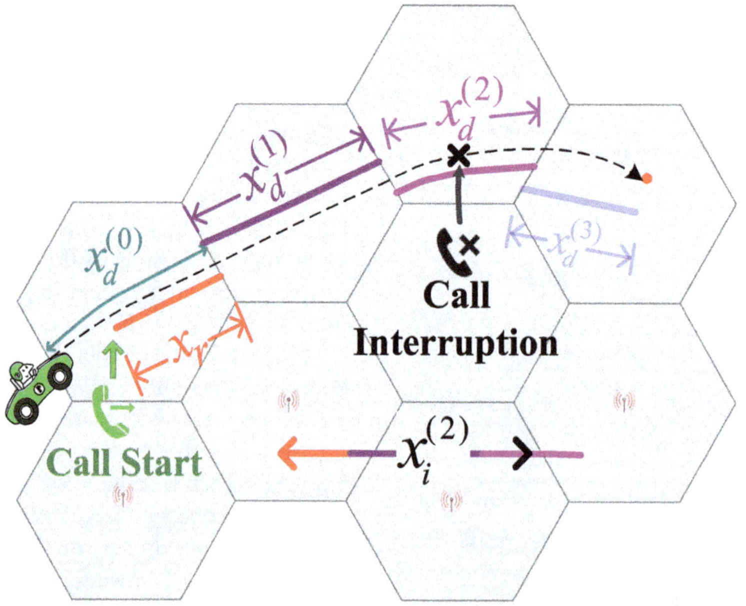

Fig. 3 Relationship between the cell residence and interruption times for interrupted calls

Due to the memory-less property of the exponential distribution of the primary call inter-arrival time, for the case when the analyzed call is interrupted by the first arrival of a PU after the beginning of the call (which occurs with probability P_{Int}), the elapsed time until the interruption is $X_{a(1)}{}^{(P)}$. In the case when the analyzed SU call is interrupted by the arrival of the second PU after the beginning of the SU call, the elapsed time is $X_{a(1)}{}^{(P)} + X_{a(2)}{}^{(P)}$ and occurs with probability $(1 - P_{\text{Int}})P_{\text{Int}}$, and so on and so forth. Then, the pdf of $X_{\text{int\_PU}}$ is given by

$$f_{X_{\text{int\_PU}}}(t) = P_{\text{Int}}\sum_{i=0}^{\infty}(1-P_{\text{Int}})^{i} f_{\sum_{j=1}^{i+1} X_{a(j)}^{(P)}}(t) = P_{\text{Int}}\sum_{i=0}^{\infty}(1-P_{\text{Int}})^{i}\frac{\left(\lambda^{(P)}\right)^{i+1} t^{i}}{i!}e^{-\lambda^{(P)}t}$$
$$= P_{\text{Int}}\lambda^{(P)}e^{-P_{\text{Int}}\lambda^{(P)}t} \tag{3}$$

From (3), it is clear that the elapsed time from the beginning of the secondary call until it is interrupted due to the arrival of a primary call is exponentially distributed with mean $E\{X_{\text{int\_PU}}\} = 1/(P_{\text{Int}}\lambda^{(P)})$.

In this paper, the pdf, the cumulative distribution, and the Laplace transform (LT) of a given RV, say X, are denoted by $f_X(t)$, $F_X(t)$, and $f_X^*(s)$, respectively. It is assumed that $f_X^*(s)$ is a rational function in the analysis developed in Sections 3 and 4. In the following analysis, we focus on the performance of secondary users.

3 Channel holding time statistics

In this section, the probability distributions of the CHTn and CHTh in CRCNs are derived for general distributions of both unencumbered service time and cell dwell times.

3.1 New call channel holding time

In this section, the pdf of the CHTn is derived. To this end, notice that SUs that access the system release their assigned resources in the same cell that they were assigned for the following reasons: (a) the SU completes its service (in this case, the new call channel holding time equals the service time, X_s); (b) the SU leaves the cell where it arrived (in this case, the new call channel holding time equals the residual cell dwell time, X_r); or (c) the SU's call is interrupted due to a primary arrival (in this case, the CHTn equals $X_{\text{int\_PU}}$). As such, the CHTn can be expressed as follows:

$$X_c^{(n)} = \min\left(X_s,\ X_r, X_{\text{int\_PU}}\right) \tag{4}$$

Considering (3), the Laplace transform of the CDF is given by

$$\begin{aligned} F^*_{X_c^{(n)}}(s) &= \frac{f^*_{X_s}(s)}{s} + \frac{f^*_{X_r}(s)}{s} + \frac{f^*_{X_{\text{int\_PU}}}(s)}{s} \\ &- \int_0^\infty F_{X_{\text{int\_PU}}}(t)\ F_{X_s}(t)\ e^{-st}dt \\ &- \int_0^\infty F_{X_{\text{int\_PU}}}(t)\ F_{X_r}(t)\ e^{-st}dt \\ &- \int_0^\infty F_{X_s}\ (t)\ F_{X_r}\ (t)\ e^{-P_{\text{Int}}\lambda^{(P)}t}e^{-st}dt \end{aligned} \tag{5}$$

In [10], it is shown that

$$\int_0^\infty F_X(t)\ F_Y(t)\ e^{-st}dt = \frac{f^*_Y(s)}{s} - \sum_{p\in\Omega_X}\underset{\xi=s+p}{\mathbf{Res}}\frac{f^*_Y(\xi)}{\xi}\ \frac{f^*_X(s-\xi)}{s-\xi} \tag{6}$$

where Ω_X is the set of poles of $f^{*}(-s)$ and $\underset{\xi=s+p}{\mathbf{Res}}$ denotes the residue at pole $\xi = s + p$. Hence, substituting (6) in (5) and simplifying,

$$\begin{aligned} F^*_{X_c^{(n)}}(s) &= \frac{f^*_{X_{\text{int\_PU}}}(s)}{s} \\ &+ \sum_{p\in\Omega_{X_{\text{int\_PU}}}}\underset{\xi=s+p}{\mathbf{Res}}\frac{f^*_{X_s}(\xi)f^*_{X_{\text{int\_PU}}}(s-\xi)}{\xi\quad s-\xi} \\ &+ \sum_{p\in\Omega_{X_{\text{int\_PU}}}}\underset{\xi=s+p}{\mathbf{Res}}\frac{f^*_{X_r}(\xi)f^*_{X_{\text{int\_PU}}}(s-\xi)}{\xi\quad s-\xi} \\ &- \int_0^\infty F_{X_s}(t)F_{X_r}(t)e^{-P_{\text{Int}}\lambda^{(P)}}te^{-st}dt \end{aligned} \tag{7}$$

Since $f^*_{X_s}(0) = 1$, the last part of (7) can be expressed as

$$\begin{aligned} -\int_0^\infty F_{X_s}(t)F_{X_r}(t)e^{-P_{\text{Int}}\lambda^{(P)}}te^{-st}dt &= -\frac{f^*_{X_r}\left(s + P_{\text{Int}}\lambda^{(P)}\right)}{s + P_{\text{Int}}\lambda^{(P)}} \\ &+ \sum_{p\in\{\Omega_{X_s}\}}\underset{\xi=s+P_{\text{Int}}\lambda^{(P)}+p}{\mathbf{Res}}\frac{f^*_{X_r}(\xi)f^*_{X_s}\left(s + P_{\text{Int}}\lambda^{(P)}-\xi\right)}{\xi\quad s + P_{\text{Int}}\lambda^{(P)}-\xi} \end{aligned} \tag{8}$$

Substituting (8) in (7) we obtain

$$F^*_{X_{c(n)}}(s) = \frac{f^*_{X_{\text{int\_PU}}}(s)}{s} + \sum_{p \in \Omega_{X_{\text{int\_PU}}}} \underset{\xi=s+p}{\text{Res}} \frac{f^*_{X_s}(\xi) f^*_{X_{\text{int\_PU}}}(s-\xi)}{\xi \quad s-\xi}$$
$$+ \sum_{p \in \Omega_{X_{\text{int\_PU}}}} \underset{\xi=s+p}{\text{Res}} \frac{f^*_{X_r}(\xi) f^*_{X_{\text{int\_PU}}}(s-\xi)}{\xi \quad s-\xi} - \frac{f^*_{X_r}\left(s+P_{\text{Int}}\lambda^{(P)}\right)}{s+P_{\text{Int}}\lambda^{(P)}}$$
$$+ \sum_{p \in \{\Omega_{X_s}\}} \underset{\xi=s+P_{\text{Int}}\lambda^{(P)}+p}{\text{Res}} \frac{f^*_{X_r}(\xi) f^*_{X_s}\left(s+P_{\text{Int}}\lambda^{(P)}-\xi\right)}{\xi \quad s+P_{\text{Int}}\lambda^{(P)}-\xi} \tag{9}$$

Considering the Laplace Transform of the pdf of $X_{\text{Int\_PU}}$, shown in Appendix 2, (9) can be written as

$$F^*_{X_c^{(n)}}(s) = \frac{f^*_{X_{\text{Int\_PU}}}(s)}{s} - \frac{f^*_{X_r}\left(s+P_{\text{Int}}\lambda^{(P)}\right)}{s+P_{\text{Int}}\lambda^{(P)}}$$
$$+ \lim_{\xi \to s+\lambda^{(P)} P_{\text{Int}}} \left(\xi - s - \lambda^{(P)} P_{\text{Int}}\right) \frac{f^*_{X_s}(\xi)}{\xi} \frac{\lambda^{(P)} P_{\text{Int}}}{(s-\xi)\left(\lambda^{(P)} P_{\text{Int}} + s - \xi\right)}$$
$$+ \lim_{\xi \to s+\lambda^{(P)} P_{\text{Int}}} \left(\xi - s - \lambda^{(P)} P_{\text{Int}}\right) \frac{f^*_{X_r}(\xi)}{\xi} \frac{\lambda^{(P)} P_{\text{Int}}}{(s-\xi)\left(\lambda^{(P)} P_{\text{Int}} + s - \xi\right)}$$
$$+ \sum_{p \in \{\Omega_{X_s}\}} \underset{\xi=s+P_{\text{Int}}\lambda^{(P)}+P}{\text{Res}} \frac{f^*_{X_r}(\xi)}{\xi} \frac{f^*_{X_s}\left(s+P_{\text{Int}}\lambda^{(P)}-\xi\right)}{s+P_{\text{Int}}\lambda^{(P)}-\xi} \tag{10}$$

After some algebraic manipulation and simplifying,

$$F^*_{X_c^{(n)}}(s) = \frac{\lambda^{(P)} P_{\text{Int}}}{s\left(\lambda^{(P)} P_{\text{Int}} + s\right)} + \frac{f^*_{X_s}\left(s+\lambda^{(P)} P_{\text{Int}}\right)}{s+\lambda^{(P)} P_{\text{Int}}}$$
$$+ \sum_{p \in \{\Omega_{X_s}\}} \underset{\xi=s+P_{\text{Int}}\lambda^{(P)}+p}{\text{Res}} \frac{f^*_{X_r}(\xi) f^*_{X_s}\left(s+P_{\text{Int}}\lambda^{(P)}-\xi\right)}{\xi \quad s+P_{\text{Int}}\lambda^{(P)}-\xi} \tag{11}$$

The Laplace transform of the pdf of CHTn is given by

$$f_{X_c^{(n)}}{}^*(s) = \frac{\lambda^{(P)} P_{\text{Int}}}{\left(\lambda^{(P)} P_{\text{Int}} + s\right)} + \frac{s f^*_{X_s}\left(s+\lambda^{(P)} P_{\text{Int}}\right)}{s+\lambda^{(P)} P_{\text{Int}}}$$
$$+ s \sum_{p \in \{\Omega_{X_s}\}} \underset{\xi=s+P_{\text{Int}}\lambda^{(P)}+p}{\text{Res}} \frac{f^*_{X_r}(\xi) f^*_{X_s}\left(s+P_{\text{Int}}\lambda^{(P)}-\xi\right)}{\xi \quad s+P_{\text{Int}}\lambda^{(P)}-\xi} \tag{12}$$

3.2 Handoff call channel holding time

In this section, the pdf of the CHTh is derived. In the following analysis, we consider that the inter-cellular handoff failure probability P_{hI} (defined as the probability that the handoff fails due to the lack of resources as a secondary user moves from one cell to another) is known. However, this probability is studied in detail in Section 5. To derive the CHTh, notice that an ongoing SUs is successfully handed off to a new cell and the system releases the radio resources assigned in the new cell for the following reasons: (a) the SU completes its service (in this case, the handoff call channel holding time equals the residual service time, X_{sr}); (b) the SU leaves the current cell (in this case, the CHTh equals the cell dwell time, X_d); or (c) the SU's call is interrupted due to a PU arrival (in this case, the CHTh equals $X_{\text{int\_PU}}$). As such, the CHTh can be expressed as follows:

$$X_C^{(h)} = \min\left(X_{sr}, X_d, X_{\text{int\_PU}}\right) \tag{13}$$

Thus, we need to derive the pdf of the USTr. To this end, let us define Y_m as the elapsed time from the beginning of the call to the mth handoff performed for the mobile of interest, then $Y_m = X_r + X_{d(1)} + X_{d(2)} + \ldots + X_{d(m-1)}$ with its Laplace transform given by

$$f^*_{Y_m}(s) = f^*_{X_r}(s)\left[f^*_{X_d}(s)\right]^{m-1} \tag{14}$$

where we have used the fact that $X_{d(i)}$ for $i = 1, 2, \ldots, m - 1$ are considered independent and identically distributed (*i.i.d.*) RVs. Now, let us denote by $N^{(P)}(t)$ the RV that represents the number of PU's arrivals in the interval $[0, t]$. Also, let us define the following events: $\{B_m\}$ is the event that the secondary call has experienced m successful handoffs before entering the current cell and {No Int by PU until Y_m} is the event that the call of interest has not been interrupted by PUs until time Y_m. Then,

$$P\{B_m\} - (1-P_{hI})^m P\{X_s > Y_m\} P\{\text{No Int by PU until } Y_m\} \tag{15}$$

Considering a given value for Y_m, say $Y_m = y_m$, the probability of k arrivals in the interval $[0, y_m]$ is

$$P\left\{N^{(P)}(y_m) = k \middle| Y_m = y_m\right\} = \frac{\left(\lambda^{(P)} y_m\right)^k}{k!} e^{-\lambda^{(P)} y_m} \tag{16}$$

and the probability that none of these PU's arrivals interrupts the ongoing SU call is given by $(1 - P_{\text{Int}})^k$. Thus, the probability that the ongoing SU call of interest is not interrupted by the arrivals of PUs until time y_m is given by

$$P\{\text{No Int by PU until } Y_m | Y_m = y_m\}$$
$$= \sum_{k=0}^{\infty} (1-P_{\text{Int}})^k \frac{\left(\lambda^{(P)} y_m\right)^k}{k!} e^{-\lambda^{(P)} y_m} = e^{-\lambda^{(P)} P_{\text{Int}} y_m} \tag{17}$$

According to the total probability formula, we have

$$P\{\text{No Int by PU until } Y_m\} = \int_0^{\infty} e^{-P_{\text{Int}}\lambda^{(P)} y_m} f_{Y_m}(y_m) dy_m \tag{18}$$

Using (18) in (15), we have

$$P\{B_m\} = (1-P_{hI})^m P\{X_s > Y_m\} \int_0^{\infty} e^{-P_{\text{Int}}\lambda^{(P)} t} f_{Y_m}(t) dt \tag{19}$$

Applying the Laplace transform and then the inverse Laplace transform to the last part of (19)

$$P\{B_m\} = (1-P_{hI})^m P\{X_s > Y_m\} \times \int_{t=0}^{\infty} \frac{1}{2\pi j} \int_{\phi=-\infty}^{\infty} e^{\phi t} f^*_{X_r} \times \left(\phi + P_{\text{Int}}\lambda^{(P)}\right) \left[f^*_{X_d}\left(\phi + P_{\text{Int}}\lambda^{(P)}\right)\right]^{m-1} d\phi dt \tag{20}$$

Since

$$\int_{t=0}^{\infty} e^{-P_{\text{Int}}\lambda^{(P)} t} f_{Y_m}(t) dt = \int_{t=0}^{\infty} L^{-1}\left\{f^*_{X_r}\left(\phi + P_{\text{Int}}\lambda^{(P)}\right) \times \left[f^*_{X_d}\left(\phi + P_{\text{Int}}\lambda^{(P)}\right)\right]^{m-1}\right\} dt \tag{21}$$

where $L^{-1}\{\cdot\}$ represents the inverse Laplace transform whose independent variable is ϕ. From [10],

$$P(X > Y) = -\sum_{p \in \Omega_X} \underset{s=p}{\mathbf{Res}} \frac{f^*_Y(s)}{s} f^*_X(-s) \tag{22}$$

Using the proposition (22) in (20), we have

$$P\{B_m\} = -(1-P_{hI})^m \sum_{p \in \Omega_{X_s}} \underset{\zeta=p}{\mathbf{Res}} \frac{f^*_{Y_m}(\zeta)}{\zeta} f^*_{X_s}(-\zeta) \int_{t=0}^{\infty} \frac{1}{2\pi j} \int_{\phi=-\infty}^{\infty} e^{\phi t} f^*_{X_r}\left(\phi + P_{\text{Int}}\lambda^{(P)}\right) \times \left[f^*_{X_d}\left(\phi + P_{\text{Int}}\lambda^{(P)}\right)\right]^{m-1} d\phi dt \tag{23}$$

Defining $\Delta = \sum_{m=1}^{\infty} P\{B_m\}$, it is calculated as

$$\Delta = -\sum_{m=1}^{\infty} \left\{ (1-P_{hI})^m \sum_{p \in \Omega_{X_s}} \underset{\zeta=p}{\mathbf{Res}} \frac{f^*_{X_r}(\zeta)\left[f^*_{X_d}(\zeta)\right]^{m-1}}{\zeta} f^*_{X_s}(\zeta) \int_{t=0}^{\infty} \frac{1}{2\pi j} \int_{\phi=-\infty}^{\infty} e^{\phi t} f^*_{X_r}\left(\phi + P_{\text{Int}}\lambda^{(P)}\right) \left[f^*_{X_d}\left(\phi + P_{\text{Int}}\lambda^{(P)}\right)\right]^{m-1} d\phi dt \right\} \tag{24}$$

Interchanging the order of the summations and integrals, and simplifying, we have

$$\Delta = 1-(1-P_{hI}) \sum_{p \in \Omega_{X_s}} \underset{s=p}{\mathbf{Res}} \frac{f^*_{X_r}(\zeta) f^*_{X_s}(-\zeta)}{\zeta} \int_{t=0}^{\infty} L^{-1}\left\{ \frac{f^*_{X_r}\left(\phi + P_{\text{Int}}\lambda^{(P)}\right)}{1-(1-P_{hI}) f^*_{X_d}(\zeta) f^*_{X_d}\left(\phi + P_{\text{Int}}\lambda^{(P)}\right)} \right\} dt \tag{25}$$

Note that the last term in (25) is a constant. Now, notice that the complementary cumulative distribution function (CCDF) of the USTr of the accepted calls can be expressed as

$$1-F_{X_{sr}}(t) = \frac{1}{\Delta} \sum_{m=1}^{\infty} P\{X_{sr} > t,\ B_m\} \tag{26}$$

Using (19) in (26), we have

$$1-F_{X_{sr}}(t) = \frac{1}{\Delta} \sum_{m=1}^{\infty} (1-P_{hI})^m \int_{y=0}^{\infty} e^{-P_{Int}\lambda^{(P)} y} f_{Y_m}(y) dy \int_{x=0}^{\infty} [1-F_{X_s}(t+x)] f_{Y_m}(x) dx \tag{27}$$

Then, the pdf of the residual unencumbered service time is given by

$$f_{X_{sr}}(t) = \frac{1}{\Delta} \sum_{m=1}^{\infty} (1-P_{hI})^m \int_{y=0}^{\infty} e^{-P_{Int}\lambda^{(P)} y} f_{Y_m}(y) dy \int_{x=0}^{\infty} f_{X_s}(t+x) f_{Y_m}(x) dx \tag{28}$$

and its corresponding Laplace transform is given by

$$f^*_{X_{sr}}(s) = \frac{1}{\Delta} \sum_{m=1}^{\infty} (1-P_{hI})^m \int_{y=0}^{\infty} e^{-P_{\text{Int}}\lambda^{(P)} y} f_{Y_m}(y) dy \int_{t=0}^{\infty} e^{-st} \int_{x=0}^{\infty} f_{X_s}(t+x) f_{Y_m}(x) dx dt \tag{29}$$

In [10], it was demonstrated that

$$\int_{t=0}^{\infty} \int_{x=0}^{\infty} f_{X_s}(t+x) f_{Y_m}(x) e^{-st} dx dt = -\sum_{p \in \Omega_{X_s}} \underset{z=p}{\mathbf{Res}} \frac{f^*_{Y_m}(z)}{s+z} f^*_{X_s}(-z) \tag{30}$$

Using (30) in (29), interchanging the orders of the summations and integrals, and simplifying,

$$f^*_{X_{sr}}(s) = -\frac{1}{\Delta}\sum_{p\in\Omega_{X_s}} \underset{z=p}{\mathbf{Res}} \frac{f^*_{X_s}(-z)}{s+z} \sum_{m=1}^{\infty} f^*_{Y_m}(z)(1-P_{hI})^m \int_{y=0}^{\infty} e^{-P_{\text{Int}}\lambda^{(P)}y} f_{Y_m}(y)dy \quad (31)$$

Using (21) in (31), we have

$$\begin{aligned} f^*_{X_{sr}}(s) &= -\frac{1}{\Delta}\sum_{p\in\Omega_{X_s}} \underset{z=p}{\mathbf{Res}} \frac{f^*_{X_s}(-z)}{s+z} \\ &\times \sum_{m=1}^{\infty} f^*_{Y_m}(z)(1-P_{hI})^m \int_{t=0}^{\infty} L^{-1}\Big\{f^*_{X_r}\left(\xi + P_{\text{Int}}\lambda^{(P)}\right) \\ &\times \left[f^*_{X_d}\left(\xi + P_{\text{Int}}\lambda^{(P)}\right)\right]^{m-1}\Big\}dt \\ &= -\frac{(1-P_{hI})}{\Delta}\sum_{p\in\Omega_{X_s}} \underset{z=p}{\mathbf{Res}} \frac{f^*_{X_s}(-z)}{s+z} f^*_{X_r}(z) \\ &\times \int_{t=0}^{\infty} L^{-1}\left\{\frac{f^*_{X_r}\left(\xi + P_{\text{Int}}\lambda^{(P)}\right)}{1-(1-P_{hI})f^*_{X_d}(z)f^*_{X_d}\left(\xi + P_{\text{Int}}\lambda^{(P)}\right)}\right\}dt \end{aligned} \quad (32)$$

Using (25) in (32), we have

$$f^*_{X_{sr}}(s) = -\frac{(1-P_{hI})\sum_{p\in\Omega_{X_s}} \underset{z=p}{\mathbf{Res}} \frac{f^*_{X_s}(-z)}{s+z} f^*_{X_r}(z)\delta}{-(1-P_{hI})\sum_{p\in\Omega_{X_s}} \underset{z=p}{\mathbf{Res}} \frac{f^*_{X_r}(\zeta)f^*_{X_s}(-\zeta)}{\zeta}\gamma} \quad (33)$$

where $\delta = \int_{t=0}^{\infty} L^{-1}\left\{\frac{f^*_{X_r}\left(\xi+P_{\text{Int}}\lambda^{(P)}\right)}{\left(s+P_{\text{Int}}\lambda^{(P)}\right)\left[1-(1-P_{hI})f^*_{X_d}(z)f^*_{X_d}\left(\xi+P_{\text{Int}}\lambda^{(P)}\right)\right]}\right\}dt;\gamma$

$= \int_{t=0}^{\infty} L^{-1}\left\{\frac{f^*_{X_r}\left(\phi+P_{\text{Int}}\lambda^{(P)}\right)}{\left(s+P_{\text{Int}}\lambda^{(P)}\right)\left[1-(1-P_{hI})f^*_{X_d}(\varsigma)f^*_{X_d}\left(\phi+P_{\text{Int}}\lambda^{(P)}\right)\right]}\right\}dt.$

Now, we have all the elements to derive the pdf of the handoff call channel holding time. Thus, using (5), (13), and (33) and following a similar procedure as the one used in Section 3.1 for deriving the LT of the CHTn, the LT of the pdf of the CHTh is found as

$$\begin{aligned} f^*_{X_c^{(h)}}(s) &= \frac{\lambda^{(P)}P_{\text{Int}}}{\left(\lambda^{(P)}P_{\text{Int}}+s\right)} + \frac{sf^*_{X_{sr}}\left(s+\lambda^{(P)}P_{\text{Int}}\right)}{s+\lambda^{(P)}P_{\text{Int}}} \\ &+s\sum_{p\in\{\Omega_{X_{sr}}\}} \underset{\xi=s+P_{\text{Int}}\lambda^{(P)}+p}{\mathbf{Res}} \frac{f^*_{X_d}(\xi)f^*_{X_{sr}}\left(s+P_{\text{Int}}\lambda^{(P)}-\xi\right)}{\xi\left(s+P_{\text{Int}}\lambda^{(P)}-\xi\right)} \end{aligned} \quad (34)$$

An equivalent expression is found using the poles of the CDT. This expression is given by

$$\begin{aligned} f^*_{X_c^{(h)}}(s) &= \frac{\lambda^{(P)}P_{\text{Int}}}{\left(\lambda^{(P)}P_{\text{Int}}+s\right)} + \frac{sf^*_{X_d}\left(s+\lambda^{(P)}P_{\text{Int}}\right)}{s+\lambda^{(P)}P_{\text{Int}}} \\ &+s\sum_{p\in\Omega_{X_d}} \underset{\xi=s+P_{\text{Int}}\lambda^{(P)}+p}{\mathbf{Res}} \frac{f^*_{X_{sr}}(\xi)f^*_{X_d}\left(s+P_{\text{Int}}\lambda^{(P)}-\xi\right)}{\xi\left(s+P_{\text{Int}}\lambda^{(P)}-\xi\right)} \end{aligned} \quad (35)$$

From (12) and (35), it is straightforward to obtain the lth moment of both CHTn and CHTh by obtaining the lth derivative of $f^*_{X_c^{(n)}}(-s)$ and $f^*_{X_c^{(h)}}(-s)$ at $s = 0$, respectively.

4 Forced call termination and inter-cellular handoff arrival rate

In this section, the expressions of both the inter-cell handoff arrival rate and forced termination probability of secondary users are derived.

4.1 Inter-cellular handoff rate for secondary users

For the analysis in this sub-section, both the UST and CDT are considered to be generally distributed. First, the outgoing handoff rate $\lambda_h^{(\text{out})}$ of a given cell can be expressed as follows:

$$\begin{aligned} \lambda_h^{(\text{out})} &= \lambda_n\left(1-P_b^{(s)}\right)P\{X_r < \min(X_{\text{int\_PU}}, X_s)\} \\ &+\lambda_h^{(\text{in})}(1-P_{hI})P\{X_d < \min(X_{\text{int\_PU}}, X_{sr})\} \end{aligned} \quad (36)$$

where $P_b^{(S)}$ is the new call blocking probability for SUs and it is considered to be known. This probability is studied in detail in Section 5.

The rationale behind (36) is that in order to have an outgoing inter-cellular handoff for SUs, first, a new call of a SU must arrive and not to be blocked, then it has to leave the cell before its service be either interrupted by the arrival of a PU or successfully completed. Also, incoming calls from other cells (with rate $\lambda_{h^{(\text{in})}}$) must be accepted (a successful handoff) and the users with those calls have to leave the cell before their respective (residual) services be either interrupted by the arrival of a PU or completed. In a homogenous system in steady state, the following condition must be hold $\lambda_h = \lambda_h^{(\text{out})} = \lambda_h^{(\text{in})}$. Then, from (36), we get

$$\lambda_h = \frac{\lambda_n\left(1-P_b^{(s)}\right)p\{X_r < \min(X_{\text{int\_PU}}, X_s)\}}{1-(1-P_{hI})P\{X_d < \min(X_{\text{int\_PU}}, X_{sr})\}} \quad (37)$$

(a) For the case when X_s and X_d have a general pdf, in Appendix 3, it is shown that

$$P\{X_r < \min(X_{\text{int\_PU}}, X_s)\} = -\sum_{p \in \Omega_{X_r}} \underset{s=p}{\textbf{Res}} \frac{\left[1-f^*_{X_s}\left(\lambda^{(P)} P_{\text{Int}} + s\right)\right] f^*_{X_r}(-s)}{\lambda^{(P)} P_{\text{Int}} + s} \tag{38}$$

and

$$P\{X_d < \min\ (X_{\text{int\_PU}}, X_{sr})\} = -\sum_{p \in \Omega X_d} \underset{s=p}{\textbf{Res}} \frac{\left[1-f^*_{X_{sr}}\left(\lambda^{(P)} P_{\text{Int}} + s\right)\right] f^*_{X_d}(-s)}{\lambda^{(P)} P_{\text{Int}} + s} \tag{39}$$

Using (38) and (39) in (37), we have

$$\lambda_n = \frac{\lambda_n \left(1-P_b^{(s)}\right)\left(-\sum_{p \in \Omega_{X_r}} \underset{s=p}{\textbf{Res}} \frac{\left[1-f^*_{X_s}\left(\lambda^{(P)} P_{\text{Int}}+s\right)\right] f^*_{X_r}(-s)}{\lambda^{(P)} P_{\text{Int}}+s}\right)}{1 + (1-P_{hI}) \sum_{p \in \Omega_{X_d}} \underset{s=p}{\textbf{Res}} \frac{\left[1-f^*_{X_{sr}}\left(\lambda^{(P)} P_{\text{Int}}+s\right)\right] f^*_{X_d}(-s)}{\lambda^{(P)} P_{\text{Int}}+s}} \tag{40}$$

(b) For the particular case when X_s is exponentially distributed, in Appendix 3, it is shown that

$$P\{X_r < \min\ (X_{\text{int\_PU}}, X_s)\} = f^*_{X_r}\left(\mu + \lambda^{(P)} P_{\text{Int}}\right) \tag{41}$$

Using (41) in (37), we have

$$\lambda_h = \frac{\lambda n \left(1-P_b^{(s)}\right) f^*_{X_r}\left(\mu + \lambda^{(P)} P_{\text{Int}}\right)}{1 + (1-P_{hI}) \sum_{p \in \Omega_{X_d}} \underset{s=p}{\textbf{Res}} \frac{\left[1-f^*_{X_{sr}}\left(\lambda^{(P)} P_{\text{Int}}+s\right)\right] f^*_{X_d}(-s)}{\lambda^{(P)} P_{\text{Int}}+s}} \tag{42}$$

For the sake of clarity and completeness, an alternative mathematical expression for calculation of the inter-cellular handoff arrival rate based on the cell departure rate is shown in Section 5, considering both UST and CDT exponentially distributed.

4.2 Forced call termination probability for secondary users

In this section, considering that UST is exponentially distributed and CDT is generally distributed, a closed analytical expression for the forced call termination probability in CRCNs is obtained. In mobile communication networks, forced call termination is due to two fundamental features: resource insufficiency and link unreliability [9]. Without loss of generality, in the following analysis, forced call termination probability due to link unreliability because of the propagation impairments is ignored. Instead, due to the nature of CRNs, a new type of forced call termination is introduced: forced termination due to intra-cellular handoff failure triggered by the arrival of a PU.

Note that $P_{\text{Int}} = P_O P_{hi}$ gives the interruption probability of a particular SU's call due to the arrival of a PU, where P_O is the probability that the SU's call of interest occupies the claimed PU's resources and P_{hi} represents the probability of intra-cell handoff failure (due to either that an interrupted SU cannot find available resources to perform spectrum handoff or spectrum handoff is not employed). It is important to notice that when spectrum handoff is not considered $P_{hi} = 1$. In order to find the forced call termination probability for SUs, the main reasons for this are identified as follows: (a) when a SU attempts a handoff procedure as it leaves its current cell and there are no available resources in the target cell (which occurs with probability P_{ft}^H) and (b) when an interruption occurs due to an arrival of a PU that requires the resources used by the SU (which occurs with probability P_{ft}^{Int}) and no available channels there exist in the system. First, P_{ft}^H is studied in detail. A new call (handoff call) requires a handoff when its residual cell dwell (cell dwell) time in the current cell is smaller than both its unencumbered service (residual unencumbered service) time and the unencumbered interruption time. Let us consider that a particular handoff call fails with probability P_{hI}. As such, the forced call termination probability due to an inter-cellular handoff attempt failure is expressed as

$$\begin{aligned} P_{ft}^H &= P\left\{X_r < \min\left(X^{(0)}_{\text{int\_PU}}, X_s\right)\right\} P_{hI} \\ &+ P\left\{X_r < \min\left(X^{(0)}_{\text{int\_PU}}, X_s\right), X_d^{(1)} < \min\left(X^{(1)}_{\text{int\_PU}}, X_s - X_r\right)\right\}(1-P_{hI}) P_{hI} \\ &+ P\left\{X_r < \min\left(X^{(0)}_{\text{int\_PU}}, X_s\right), X_d^{(1)} < \min\left(X^{(1)}_{\text{int\_PU}}, X_s - X_r\right), X_d^{(2)} < \min \right. \\ &\left. \times \left(X^{(2)}_{\text{int\_PU}}, X_s - X_r - X_d^{(1)}\right)\right\}(1-P_{hI})^2 P_{hI} + \ldots \end{aligned} \tag{43}$$

Since here X_s is assumed to be exponentially distributed, using the memory-less property and considering that random variables $X_d^{(i)} \left\{X^{(j)}_{\text{int\_PU}}\right\}$ (for i = 1, 2,...) {for j = 0, 1, ...} are *i.i.d.*, we have

$$P_{ft}^H = P\{X_r < \min(X_{\text{int\_PU}}, X_s)\} P_{hI} \frac{1}{1-P\{X_d < \min(X_{\text{int\_PU}}, X_s)\}(1-P_{hI})} \tag{44}$$

In Appendix 3, it is shown that

$$P\{X_r < \min(X_{\text{int\_PU}}, X_s)\} = f^*_{X_r}\left(\mu + \lambda^{(P)}P_{\text{Int}}\right) \tag{45}$$

and

$$P\{X_d < \min(X_{\text{int\_PU}}, X_s)\} = f^*_{X_d}\left(\mu + \lambda^{(P)}P_{\text{Int}}\right) \tag{46}$$

Using (45) and (46) in (44), we have

$$P^H_{ft} = \frac{P_{hI}f^*_{X_r}\left(\mu + \lambda^{(P)}P_{\text{Int}}\right)}{1-(1-P_{hI})f^*_{X_d}\left(\mu + \lambda^{(P)}P_{\text{Int}}\right)} \tag{47}$$

Now, P^{Int}_{ft} is studied in detail. A new call (handoff call) is terminated due to a PU arrival when the unencumbered interruption time of such user is smaller than both its residual cell dwell (cell dwell) time in the current cell and to the residual service time. Hence, it can be expressed as

$$\begin{aligned} P^{\text{Int}}_{ft} = {} & P\left\{X^{(0)}_{\text{int\_PU}} < \min\ (X_r,\ X_s)\right\} \\ & + P\{X_r < \min\left(X^{(0)}_{\text{int\_PU}}, X_s\right), X^{(1)}_{\text{int\_PU}} \\ & < \min\ \left(X^{(1)}_d, X_s - X_r\right)\}(1-P_{hI}) \\ & + P\{X_r < \min\ \left(X^{(0)}_{\text{int\_PU}},\ X_s\right),\ X^{(1)}_d \\ & < \min\ \left(X^{(1)}_{\text{int\_PU}},\ X_s - X_r\right),\ X^{(2)}_{\text{int\_PU}} \\ & < \min\ \left(X^{(2)}_d,\ X_s - X_r - X^{(1)}_d\right)\}(1-P_{hI})^2 + \ldots \end{aligned} \tag{48}$$

When X_s is assumed to be exponentially distributed, using the memory-less property and considering that the variables $X^{(i)}_d \left\{X^{(j)}_{\text{int\_PU}}\right\}$ (for i = 1, 2, ...) {for j = 0, 1, ...} are *i.i.d.*, we have

$$P^{\text{Int}}_{ft} = P\{X_{\text{int\_PU}} < \min(X_r, X_s)\} + \frac{(1-P_{hI})P\{X_r < \min(X_{\text{int\_PU}}, X_s)\}P\{X_{\text{int\_PU}} < \min(X_d, X_s)\}}{1-(1-P_{hI})P\{X_d < \min(X_{\text{int\_PU}}, X_s)\}} \tag{49}$$

In Appendix 3, it is shown that

$$P\{X_{\text{int\_PU}} < \min(X_r,\ X_s)\} = \frac{\lambda^{(P)}P_{\text{Int}}\left[1 - f^*_{X_r}\left(\mu + \lambda^{(P)}P_{\text{Int}}\right)\right]}{\mu + \lambda^{(P)}P_{\text{Int}}} \tag{50}$$

$$P\{X_r < \min(X_{\text{int\_PU}}, X_s)\} = f^*_{X_r}\left(\mu + \lambda^{(P)}P_{\text{Int}}\right) \tag{51}$$

$$P\{X_{\text{int\_PU}} < \min\ (X_d,\ X_s)\} = \frac{\lambda^{(P)}P_{\text{Int}}\left[1 - f^*_{X_d}\left(\mu + \lambda^{(P)}P_{\text{Int}}\right)\right]}{\mu + \lambda^{(P)}P_{\text{Int}}} \tag{52}$$

and

$$P\{X_d < \min(X_{\text{int\_PU}}, X_s)\} = f^*_{X_d}\left(\mu + \lambda^{(P)}P_{\text{Int}}\right) \tag{53}$$

Using (50)–(53) in (49) and simplifying, we have

$$P^{\text{Int}}_{ft} = \frac{\lambda^{(P)}P_{\text{Int}}\left(1 - f^*_{X_r}\left(\mu + \lambda^{(P)}P_{\text{Int}}\right)\right)}{1-(1-P_{hI})f^*_{X_d}\left(\mu + \lambda^{(P)}P_{\text{Int}}\right)} + \frac{(1-P_{hI})f^*_{X_r}\left(\mu + \lambda^{(P)}P_{\text{Int}}\right)\lambda^{(P)}P_{\text{Int}}\left[1 - f^*_{X_d}\left(\mu + \lambda^{(P)}P_{\text{Int}}\right)\right]}{\left(\mu + \lambda^{(P)}P_{\text{Int}}\right)\left[1-(1-P_{hI})f^*_{X_d}\left(\mu + \lambda^{(P)}P_{\text{Int}}\right)\right]} \tag{54}$$

Using the fact that $f^*_{X_r}\left(\mu + \lambda^{(P)}P_{\text{Int}}\right) = \eta\left[1 - f^*_{X_d}\left(\mu + \lambda^{(P)}P_{\text{Int}}\right)\right] / \left(\mu + \lambda^{(P)}P_{\text{Int}}\right)$, (54) can be rewritten as

$$P^{\text{Int}}_{ft} = \lambda^{(P)}P_{\text{Int}}\left(\frac{\left(\lambda^{(P)}P_{\text{Int}} + \mu - \eta P_{hI}\right)}{\left(\mu + \lambda^{(P)}P_{\text{Int}}\right)^2\left[1-(1-P_{hI})f^*_{X_d}\left(\mu + \lambda^{(P)}P_{\text{Int}}\right)\right]} - \frac{\left[\left(\lambda^{(P)}P_{\text{Int}} + \mu\right)(1-P_{hI}) - \eta P_{hI}\right]f^*_{X_d}\left(\mu + \lambda^{(P)}P_{\text{Int}}\right)}{\left(\mu + \lambda^{(P)}P_{\text{Int}}\right)^2\left[1-(1-P_{hI})f^*_{X_d}\left(\mu + \lambda^{(P)}P_{\text{Int}}\right)\right]}\right) \tag{55}$$

Thus, the global forced call termination probability is given by

$$P_{ft} = \frac{P_{hI}f^*_{X_r}\left(\mu + \lambda^{(P)}P_{\text{Int}}\right)}{1-(1-P_{hI})f^*_{X_d}\left(\mu + \lambda^{(P)}P_{\text{Int}}\right)} + \frac{\lambda^{(P)}P_{\text{Int}}\left[1 - f^*_{X_r}\left(\mu + \lambda^{(P)}P_{\text{Int}}\right)\right]}{\mu + \lambda^{(P)}P_{\text{Int}}} + \frac{(1-P_{hI})f^*_{X_r}\left(\mu + \lambda^{(P)}P_{\text{Int}}\right)\lambda^{(P)}P_{\text{Int}}\left[1 - f^*_{X_d}\left(\mu + \lambda^{(P)}P_{\text{Int}}\right)\right]}{\left(\mu + \lambda^{(P)}P_{\text{Int}}\right)\left[1-(1-P_{hI})f^*_{X_d}\left(\mu + \lambda^{(P)}P_{\text{Int}}\right)\right]} \tag{56}$$

Finally, it is important to mention that the methodology proposed in this section to derive the forced call termination probability turns mathematically intractable when non-exponential distributions for the UST are considered. Nonetheless, when phase-type probability distributions are considered to model the UST, the call forced termination probability can be calculated as the ratio of the rate of forced termination calls and the rate of accepted calls as it is done in [9]. However, in this case, it is necessary to calculate a quantity which is closely related to the steady-state probabilities (as in [9, 11]) and no closed analytical expressions can be obtained. For the sake of clarity and completeness, an alternative mathematical expression for the forced call termination probability (computed as the ratio of the rate of forced termination calls and the rate of accepted calls) is presented in the next section, considering both UST and CDT exponentially distributed.

5 Teletraffic analysis

In this section, the queuing analysis for the performance evaluation of mobile CRCNs with FR-HDC traffic is developed. For the sake of simplicity, it is considered that both UST and CDT of SUs are exponentially distributed.[7] Also, it is assumed that the primary channel holding time is exponentially distributed with mean $1/\mu^{(P)}$. To maximize system Erlang capacity and for the adequate and fair performance comparison of the different considered scenarios, fractional channel reservation [1, 38, 39] to prioritize both intra (due to spectrum handoff)- and inter (due to users' mobility)-cell handoff call attempts over new call requests is considered in the call admission control strategy. To this end, a number of $\lfloor r \rfloor + 1$ sub-bands is reserved with probability $p = r - \lfloor r \rfloor$ and a number of $\lfloor r \rfloor$ sub-bands is reserved with probability $(1 - p)$. Thus, a bi-dimensional birth and death process is required for modeling this system. Each state variable is denoted by x_i (for $i = 0, 1$). x_0 represents the number of primary users, and x_1 represents the number of SUs. To simplify mathematical notation, the following vector of state variables is defined $\mathbf{x} = (x_0, x_1)$. k represents the number of active cognitive users that have to relinquish their respective sub-channel due to the arrival of a PU service request. As explained in [36], when no spectrum handoff is used, k is a random variable ($\mathbf{k}$), and on the other hand, when spectrum handoff is used, k is a deterministic value. Thus, the probability mass function (pmf) of the random variable $\mathbf{k}$, when spectrum handoff is not used, is given by [36]

$$p_{\mathrm{NSH}}(\mathbf{x}, \mathbf{k} = k) = \frac{\binom{N}{k}\binom{N\left(M - (x_0+1)\right)}{x_1 - k}}{\binom{N\left(M - x_0\right)}{x_1}} \tag{57}$$

for $k = 0, 1, \ldots, \min(x_1, N)$ and, when spectrum handoff is used, it is given by

$$p_{\mathrm{SH}}(\mathbf{x}, \mathbf{k} = k) = \begin{cases} 1\ ; (M-1)N < x_1 + Nx_0 \le MN \\ 0; \text{otherwise} \end{cases} \tag{58}$$

for $k = x_1 + N(x_0 + 1) - MN$. The valid state space Ω is

$$\Omega = \{\mathbf{x} | 0 \le x_0 \le M, 0 \le x_1 \le MN - Nx_0\}$$

5.1 System without spectrum handoff

When no spectrum handoff is allowed, if a PU decides to access a primary channel, all SUs using that channel must relinquish their transmission immediately. Let us represent by $r_{\mathbf{xy}}$ and $\mathbf{e}_i$ the transition rate from state $\mathbf{x}$ to state $\mathbf{y}$ ($\mathbf{x} \in \Omega$) and a two-dimensional vector with position i set to 1 and the other position set to 0, respectively. Then, the steady-state probabilities balance equation can be written as

$$P(\mathbf{x}) \sum_{\forall \mathbf{y} \in \Omega} r_{\mathbf{xy}} = \sum_{\forall \mathbf{y} \in \Omega} P(\mathbf{y}) r_{\mathbf{xy}} \quad \forall \mathbf{x} \in \Omega \tag{59}$$

where $P(\mathbf{x})$ is the state $\mathbf{x}$ stationary probability and the transition rate $r_{\mathbf{xy}}$ from states $\mathbf{x}$ to $\mathbf{y}$ is given by

$$r_{\mathbf{xy}} = \begin{cases} a_0(\mathbf{x})\lambda^{(P)} & ; \mathbf{y} = \mathbf{x} + \mathbf{e}_0 - k\mathbf{e}_1; \quad \mathbf{y} \in \Omega \\ \lambda_n + \lambda_h & ; \mathbf{y} = \mathbf{x} + \mathbf{e}_1; \ x_1 \ge 0; \ Nx_0 + x_1 < MN - \lfloor r \rfloor - 1 \\ (1-p)\lambda_n + \lambda_h & ; \mathbf{y} = \mathbf{x} + \mathbf{e}_1; \ x_1 \ge 0; \ Nx_0 + x_1 = MN - \lfloor r \rfloor - 1 \\ \lambda_h & ; \mathbf{y} = \mathbf{x} + \mathbf{e}_1; x_1 \ge 0; \ MN > Nx_0 + x_1 > MN - \lfloor r \rfloor - 1 \\ x_0 \mu^{(P)} & ; \mathbf{y} = \mathbf{x} - \mathbf{e}_0; \quad \mathbf{y} \in \Omega \\ x_1(\mu + \eta) & ; \mathbf{y} = \mathbf{x} - \mathbf{e}_1; \quad \mathbf{y} \in \Omega \\ 0 & \text{otherwise} \end{cases} \tag{60}$$

where $a_0(\mathbf{x}) = p_{\mathrm{NSH}}(\mathbf{x}, \mathbf{k})$. The state transition diagram of this system is shown in Fig. 4. The values of $P(\mathbf{x})$ are obtained from (59) and the normalization equation.

It is important to note that the previous teletraffic analysis represents an approximation approach for the performance evaluation of the considered system. Specifically, the steady-state probabilities are obtained considering that the interruption time of SUs' calls due to the arrival of PUs follows a Poisson process. As

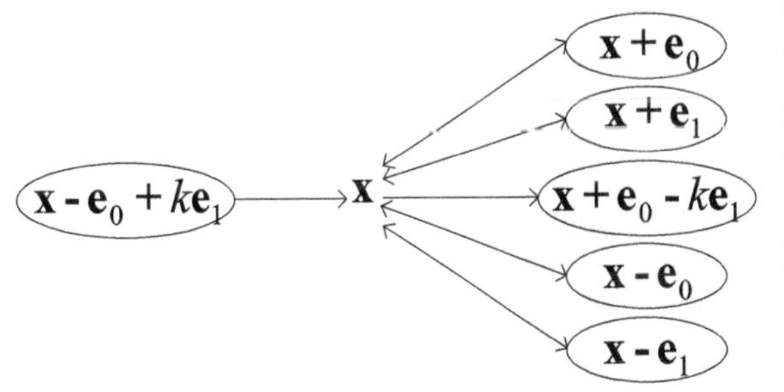

Fig. 4 State transition diagram

proven in Appendix 1, this is not true. However, this approximation renders accurate numerical results for most of the scenarios and performance metrics of interest, as shown in Section 6.

Then, the inter-cellular handoff failure probability P_{hI} and new call blocking probability $P_b^{(S)}$ can be computed as follows

$$P_{hI} = P_b^{(S)} = \sum_{x_0=0}^{M} P(x_0, MN\text{-}Nx_0) \tag{61}$$

When no spectrum handoff is considered, a random number of **k** cognitive users will be forced to terminate their service due to the arrival of a PU in a specific primary channel which is occupied by **k** SUs. The interruption probability due to the arrival of PU, P_{Int}, is given by

$$P_{\text{Int}} = \frac{\sum_{x_0=0}^{M}\sum_{x_1=1}^{MN-Nx_0}\Gamma(\mathbf{x})P(\mathbf{x})}{\sum_{x_0=0}^{M}\sum_{x_1=1}^{MN-Nx_0}P(\mathbf{x})} \tag{62}$$

where

$$\Gamma(\mathbf{x}) = \sum_{\substack{k=0 \\ x_1+Nx_0 \leq N(M-1)}}^{min(x_1,N)} k p_{\text{NSH}}(\mathbf{x},\mathbf{k}) + \sum_{\substack{k=x_1+N(x_0+1)-MN \\ N(M-1)<x_1+Nx_0\leq MN}}^{\min(x_1,N)} k p_{\text{NSH}}(\mathbf{x},\mathbf{k}) \tag{63}$$

Considering (42) and the fact that CDT is considered exponentially distributed in this section, the inter-cellular handoff rate λ_h can be calculated as

$$\lambda_h = \frac{\eta\lambda_n\left(1\text{-}P_b^{(s)}\right)}{\eta P_{hI} + \mu + \lambda^{(P)}P_{\text{Int}}} \tag{64}$$

Alternatively, inter-cellular handoff rate λ_h can be calculated using the cell departure rate as follows

$$\lambda_h = \sum_{x_0=0}^{M}\sum_{x_1=0}^{MN-Nx_0} x_1\eta P(\mathbf{x})$$

The fixed point iteration method is employed to iteratively calculate the inter-cellular handoff rate as explained in [40].

Considering exponentially distributed cell dwell time, the forced call termination probability given by (56) can be written as

$$P_{ft} = \frac{\eta P_{hI} + \lambda^{(P)}P_{\text{Int}}}{\eta P_{hI} + \mu + \lambda^{(P)}P_{\text{Int}}} \tag{65}$$

Notice that (65) is a closed-form approximated expression to compute forced call termination probability. Equation (65) is an approximated expression in the sense that it was derived in Section 4 considering that call interruption times follow a Poisson process. Moreover, expression (65) depends on P_{hI} and P_{Int}, which are derived in this section considering, also, that call interruption process is a Poisson one. Thus, we can say that (65) is based on a "double approximation." Nonetheless, the importance of (65) lies on the fact that it depends on parameters that can be easily obtained from statistics that can be collected in real networks.

An alternative expression for calculating forced call termination probability can be obtained as the ratio of the rate of forced terminated calls and the rate of accepted calls. Mathematically,

$$P_{ft} = \frac{\lambda^{(P)}\sum_{x_0=0}^{M-1}\sum_{x_1=N(M-1)-Nx_0}^{MN}[N(x_0+1)+x_1\text{-}MN]P(\mathbf{x})}{\lambda_n\left(1\text{-}P_b^{(S)}\right)} + \frac{\lambda_h P_{hI}}{\lambda_n\left(1\text{-}P_b^{(S)}\right)} \tag{66}$$

Notice that for the computation of (66), it is needed to know the steady-state probabilities. However, compared to (65), the unique source of imprecision of (66) is the fact that steady-state probabilities are derived considering that call interruption process is a Poisson one.

5.2 System with spectrum handoff

When spectrum handoff is considered, interrupted SUs are allowed to move to other vacant channels (if available). It is not difficult to observe that when spectrum handoff is used, $a_0(\mathbf{x}) = p_{\text{SH}}(\mathbf{x},\mathbf{k})$ in Eq. (58).

In this case (i.e., for cognitive radio networks with spectrum handoff), the mathematical expressions for the new call blocking and forced call termination probabilities and the inter-cellular handoff rate remain unchanged relative to the case when no spectrum handoff is considered; they are given by (61), (65), and (64), respectively. However, in this case, the interruption probability due to the arrival of PU, P_{Int}, is given by

$$P_{\text{Int}} = \frac{\sum_{x_0=0}^{M} \sum_{x_1=1}^{MN-Nx_0} \frac{(x_1+N(x_1+1)-MN)}{x_1} P(\mathbf{x})}{\sum_{x_0=0}^{M} \sum_{x_1=1}^{MN-Nx_0} P(\mathbf{x})} \quad (67)$$

5.3 Erlang capacity maximization

The maximum Erlang capacity is computed as the maximum value of the offered traffic for which all the QoS requirements are still met. Maximum Erlang capacity is obtained by optimizing the number of reserved channels to prioritize intra- and inter-cell handoff call attempts over new call requests. The optimal value of the number of reserved channels is systematically searched by using the fact that new call blocking (handoff failure) probability is a monotonically increasing (decreasing) function of the number of reserved channels. Once a value of the number of reserved channels for which the new call blocking probability and/or handoff failure probability achieves its maximum acceptable value is found, another offered traffic load is tested. The capacity maximization procedure ends when both the new call blocking probability and forced call termination probability achieve their maximum acceptable values.

6 Performance evaluation

The goal of the numerical evaluations presented in this section is to verify the applicability as well as the accuracy and robustness of our developed mathematical models. In particular, in this section, numerical results for performance evaluation of mobile CRCN in the presence of both intra-cellular e inter-cellular handoff mechanisms are shown. The analytical numerical results presented in this section were verified by a wide set of discrete-event computer simulation results for a variety of evaluation scenarios. Part of this numerical validation is shown in our previous work [23], as commented in Section 6.2. In Section 6.1, channel holding time statistics are analyzed, while in Section 6.2, the effect of secondary session interruption due to the arrival of PUs on the performance of CRCN is investigated. In this section, we define the mobility parameter as the ratio between the mean value of the UST and the mean value of the CDT, that is, the mobility parameter is given by η/μ.

6.1 Influence of the interruption probability and the probability distributions used to model the cell dwell and unencumbered service times on the channel holding time statistics

In this section, numerical results are presented to investigate the extent by which the probability distributions of both CDT and UST as well as the interruption probability (P_{Int}) affect the statistics of both new call channel holding time (CHTn) and handoff call channel holding time (CHTh). Different phase-type distributions to model CDT and UST are used. Remember that the random variable used to model the elapsed time between call interruptions is approximated by an exponential distribution with parameter $P_{\text{Int}}\lambda^{(P)}$. As it is explained in the last paragraph of Section 3, the statistics of the CHTn and CHTh are obtained using (12) and (35), respectively.

Figures 5, 6, and 7 present the mean, coefficient of variation (CoV), and skewness (Sk) of the CHTn (represented by solid lines) and CHTh (represented by dashed lines)

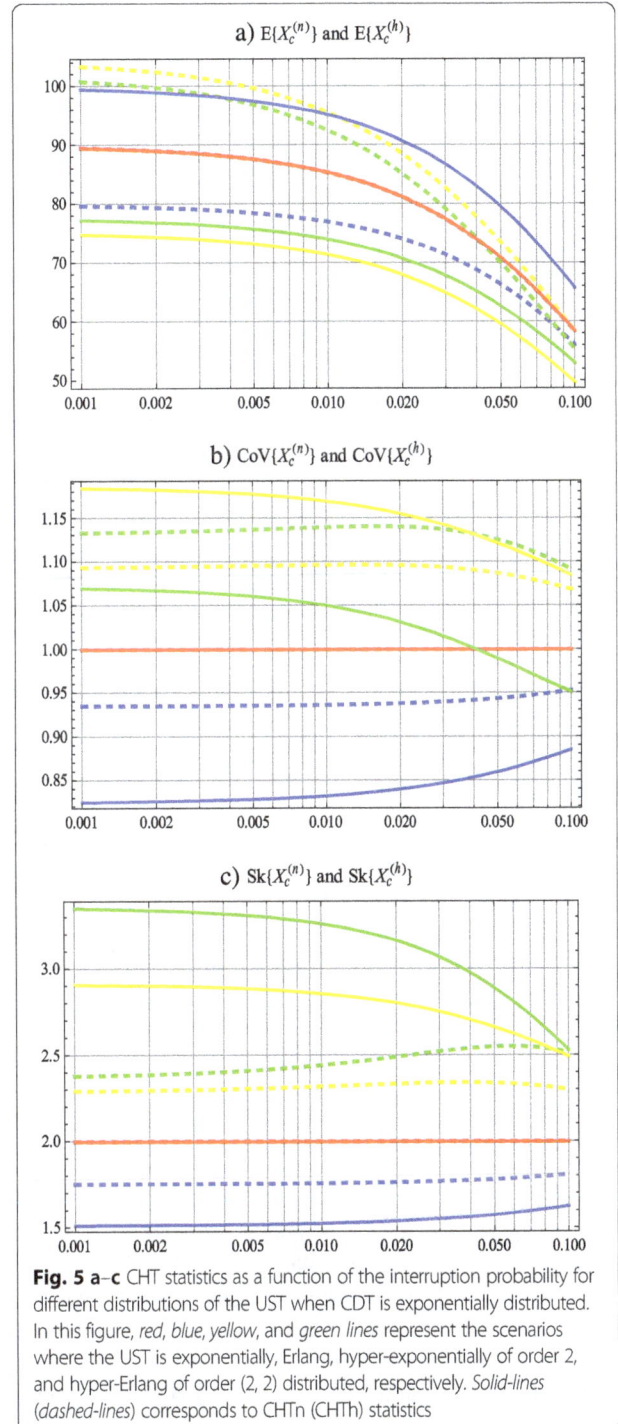

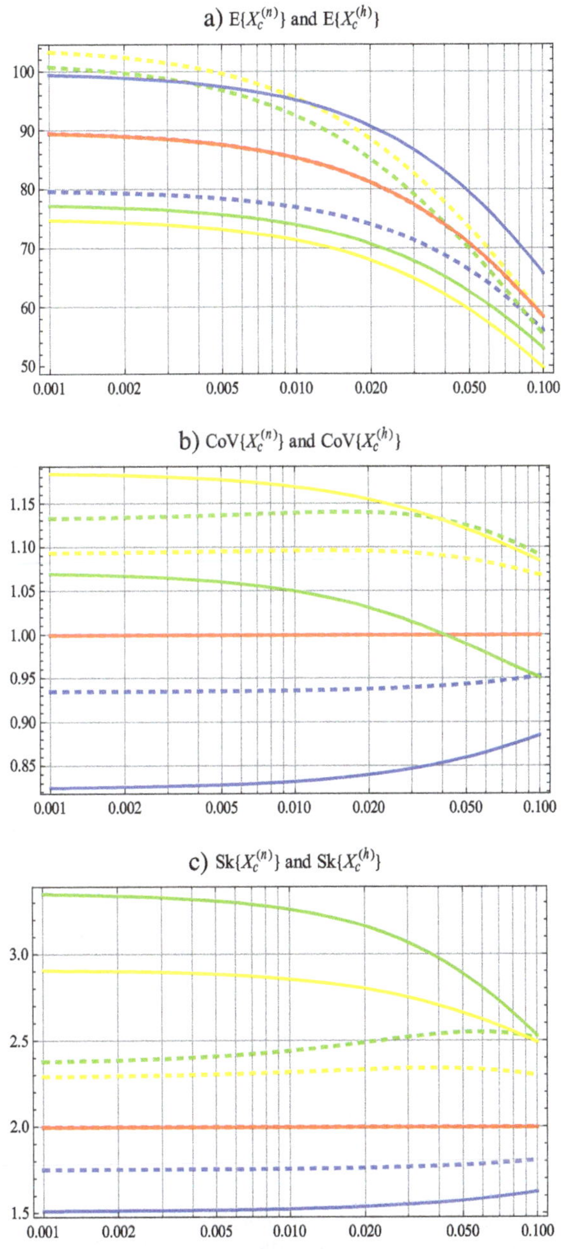

Fig. 5 a–c CHT statistics as a function of the interruption probability for different distributions of the UST when CDT is exponentially distributed. In this figure, *red, blue, yellow*, and *green lines* represent the scenarios where the UST is exponentially, Erlang, hyper-exponentially of order 2, and hyper-Erlang of order (2, 2) distributed, respectively. *Solid-lines (dashed-lines)* corresponds to CHTn (CHTh) statistics

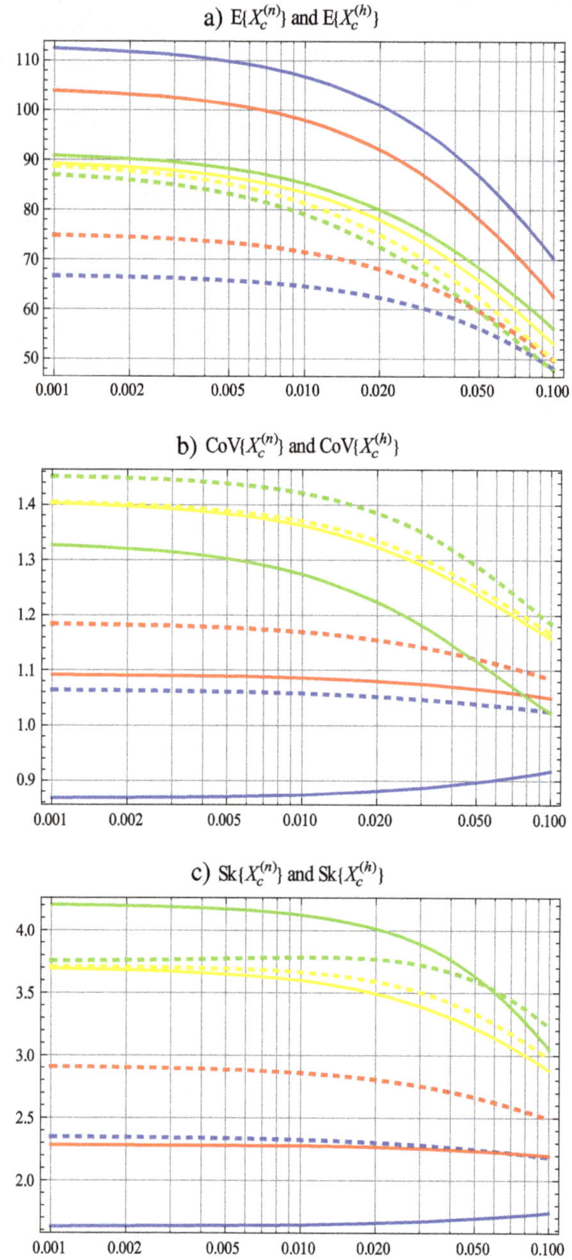

Fig. 6 a–c CHT statistics as a function of the interruption probability for different distributions of the UST when CDT is hyper-exponentially distributed. In this figure, *red*, *blue*, *yellow*, and *green lines* represent the scenarios where the UST is exponentially, Erlang, hyper-exponentially of order 2, and hyper-Erlang of order (2, 2) distributed, respectively. *Solid-lines* (*dashed-lines*) corresponds to CHTn (CHTh) statistics

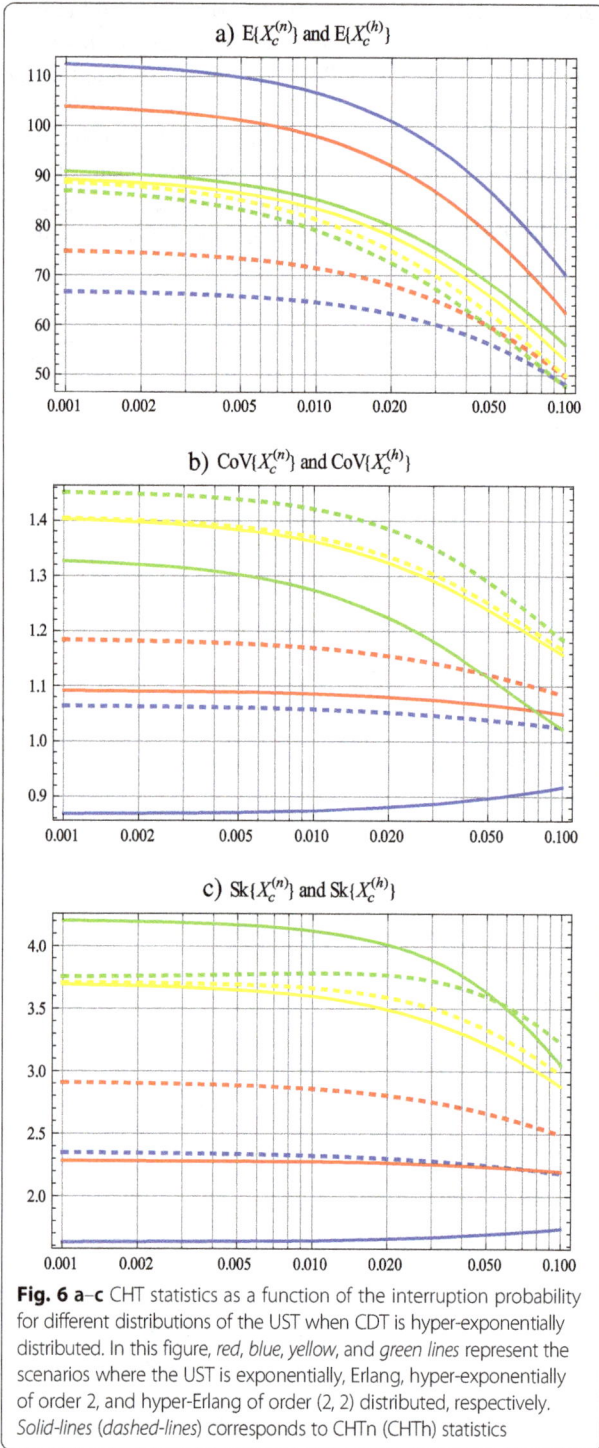

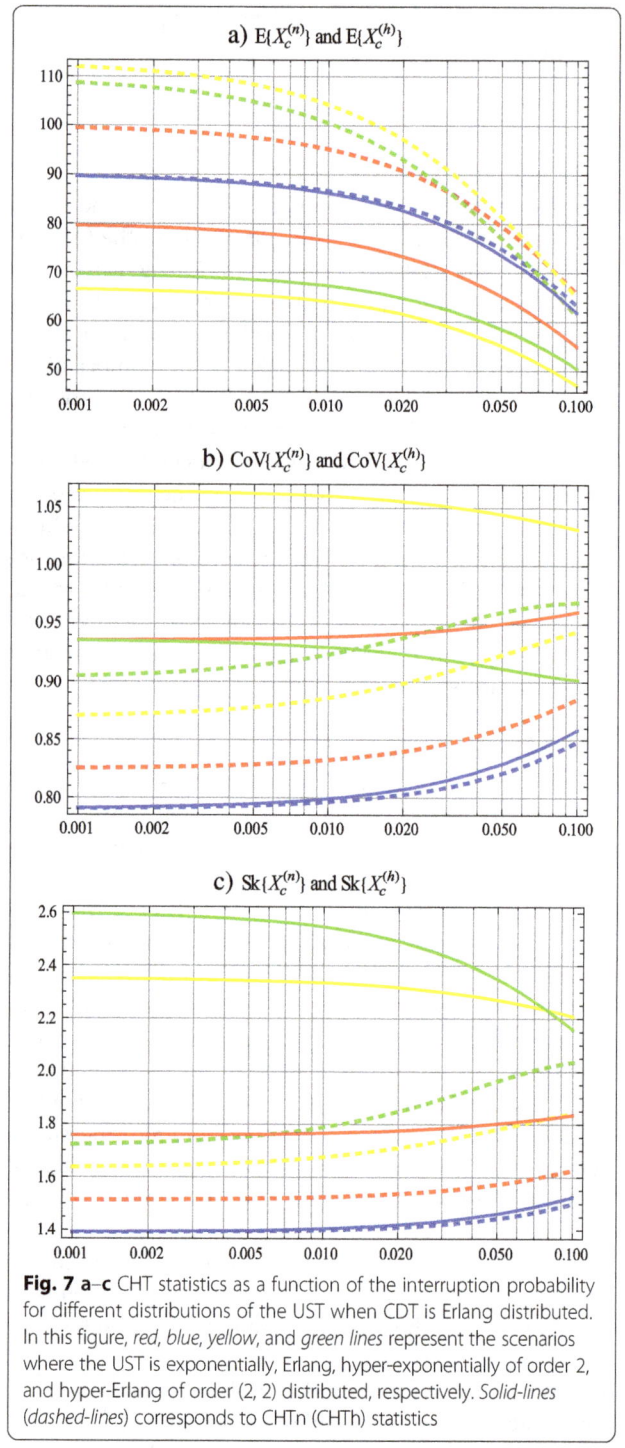

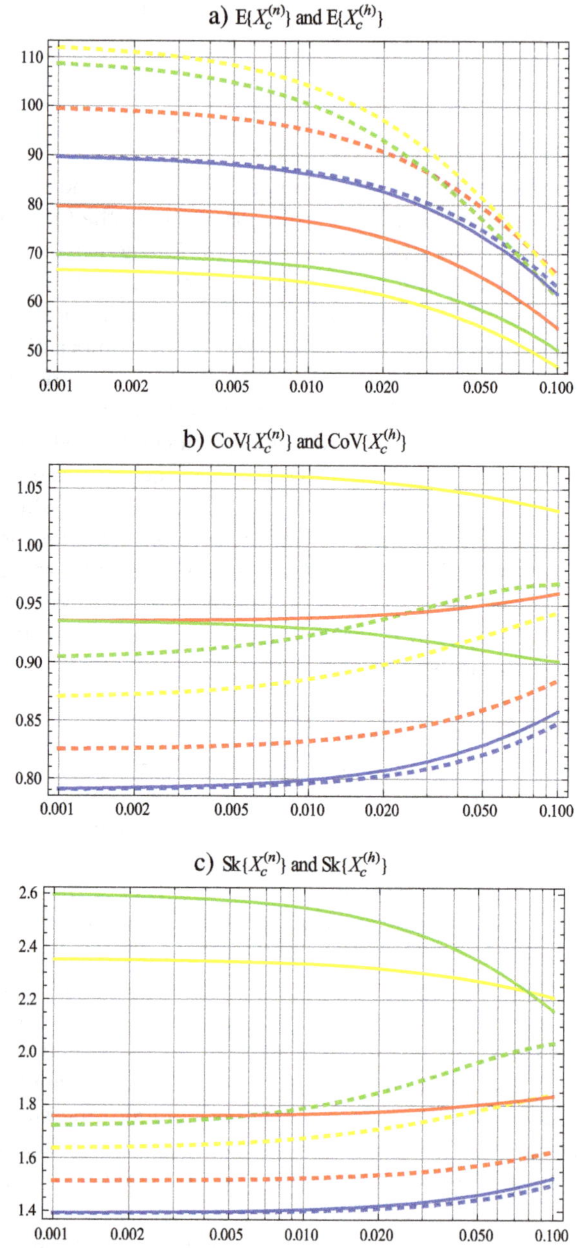

Fig. 7 a–c CHT statistics as a function of the interruption probability for different distributions of the UST when CDT is Erlang distributed. In this figure, *red*, *blue*, *yellow*, and *green lines* represent the scenarios where the UST is exponentially, Erlang, hyper-exponentially of order 2, and hyper-Erlang of order (2, 2) distributed, respectively. *Solid-lines* (*dashed-lines*) corresponds to CHTn (CHTh) statistics

versus the interruption probability P_{Int} for different distributions of the UST; specifically, in Figs. 5, 6, and 7, red, blue, yellow, and green lines represent, respectively, the scenarios where the UST is exponentially, Erlang, hyper-exponentially of order 2, and hyper-Erlang of order (2, 2) distributed. Referring to the CDT, Figs. 5, 6, and 7 assume that the CDT is exponentially, hyper-exponentially of order 2, and Erlang distributed, respectively. Referring to the different probability distributions considered in this section, it is important to point out that the exponential distribution is completely characterized by its mean value, the Erlang distribution is completely characterized by its first two moments (i.e., mean value and coefficient of variation), and the hyper-exponential and hyper-Erlang distributions are both completely characterized by their first

Table 3 Mean of the CHTn for different distributions of the UST when CDT is exponentially distributed

P_{Int}	UST distribution							
	Exp-Neg		Erlang		Hyper-exponential order 2		Hyper-Erlang (2,2)	
	Simulation	Analytical	Simulation	Analytical	Simulation	Analytical	Simulation	Analytical
0.001	89.17	89.52	99.15	99.5	75.55	74.89	77.22	77.35
0.02	81.04	81.23	90.81	90.78	68.36	68.02	70.91	70.79
0.05	70.68	70.87	79.42	79.57	59.34	59.65	62.84	62.73
0.1	58.21	58.44	65.67	65.74	49.98	49.79	53.25	53.06

three moments (i.e., mean value, coefficient of variation, and skewness). As long as possible, the same values of the first three moments of the different probability distributions are considered. In all the plots of this section, it is considered that $P_{hI} = 0.01$ and the following representative values are considered: $E\{X_s\} = 1/\mu = 180$ s, $\text{CoV}\{X_s\} = 1.58$, $\text{Sk}\{X_s\} = 3.54$, $E\{X_d\} = 1/\eta = 180$ s, $\text{CoV}\{X_d\} = 1.58$, $\text{Sk}\{X_d\} = 3.54$, and $\lambda^{(P)} = 0.06$ arrivals per second. Notice that, in this scenario, the mobility parameter (i.e., η/μ) equals 1.0, which corresponds to a relatively high users' mobility scenario. It is important to note that the analytical results have been compared to Monte Carlo simulation results in order to validate our numerical results. In Tables 3, 4, and 5, we can see a perfect match between these two models (analytical and simulation) for each scenario considered in this work.

Several interesting observations can be extracted from Figs. 5, 6, and 7. For instance, from Fig. 5, it is observed that, when CDT and UST are modeled by the (unrealistic) exponential distribution and for any value of the interruption probability, the corresponding mean, CoV, and Sk of both CHTn and CHTh have the same value (this scenario is represented by the solid-red and dashed-red lines in Fig. 5). This is an expected behavior that validates our mathematical formulation and can be explained as follows. Due to the memory-less property of the exponential distribution, when CDT (UST) is exponentially distributed, the residual CDT (residual UST) is also exponentially distributed with the same parameter [12]. Thus, the random variables that represent CHTn and CHTh (whose probability distributions are given by Eqs. (12) and (35), respectively) are also exponentially distributed with the same expected value, and therefore, the CoV and Sk are, respectively, 1 and 2 for both CHTn and CHTh. Also, from Figs. 5a, 6a, and 7a, it is observed that the mean value of both CHTn and CHTh is a monotonically decreasing function of the interruption probability. It is an intuitively understandable behavior due to the fact that as the probability of interruption increases, more ongoing secondary calls are prematurely terminated in detrimental of the mean value of both CHTn and CHTh.

However, the most relevant result that can be extracted from the plots presented in Figs. 5, 6, and 7 is the fact that the CHT statistics are highly sensitive to the distribution type of both CDT and UST. To exemplify this, let us quantify the impact of the distribution type used to model the UST on the statistics of the CHTn (the behavior of the CHTn is represented by the solid lines in Figs. 5, 6, and 7). To this end, let us consider the case when both the UST and CDT are exponentially distributed as the reference case (this scenario is represented by the red solid-line in Fig. 5). Thus, considering the scenario where $P_{\text{Int}} = 0.01$, Fig. 5a shows that (with respect to the reference case) the mean value of the CHTn increases 11.5% when the UST is Erlang distributed and decreases 16.3 and 13.3% when the UST is hyper-exponentially and hyper-Erlang distributed, respectively. Under the same scenario, Fig. 5b shows that (with respect to the reference case) the CoV of CHTn decreases 16.7% when the UST is Erlang distributed and increases 17 and 5% when the UST is hyper-exponentially and hyper-Erlang distributed, respectively. Similarly, Fig. 5c shows that (with respect to the reference case) the skewness of the CHTn decreases 23.7% when the UST is Erlang distributed and increases 42.9 and 63% when the UST is hyper-exponentially and hyper-Erlang distributed, respectively. Similar results are obtained when the CHTh is considered.

Table 4 Coefficient of variation of the CHTn for different distributions of the UST when CDT is exponentially distributed

P_{Int}	UST distribution							
	Exp-Neg		Erlang		Hyper-exponential order 2		Hyper-Erlang (2,2)	
	Simulation	Analytical	Simulation	Analytical	Simulation	Analytical	Simulation	Analytical
0.001	1	1	0.826	0.825	1.18	1.18	1.07	1.07
0.02	1	1	0.84	0.84	1.16	1.155	1.04	1.03
0.05	1	1	0.86	0.86	1.12	1.12	0.99	0.99
0.1	1	1	0.88	0.885	1.09	1.09	0.95	0.95

Table 5 Skewness of the CHTn for different distributions of the UST when CDT is exponentially distributed

P_{int}	UST distribution							
	Exp-Neg		Erlang		Hyper-exponential order 2		Hyper-Erlang (2,2)	
	Simulation	Analytical	Simulation	Analytical	Simulation	Analytical	Simulation	Analytical
0.001	1.99	2	1.49	1.51	2.98	2.90	3.22	3.35
0.02	2.02	2	1.52	1.53	2.91	2.80	3.25	3.16
0.05	1.99	2	1.56	1.57	2.71	2.66	2.74	2.89
0.1	2.00	2	1.63	1.62	2.47	2.49	2.55	2.53

On the other hand, Figs. 5a, 6a, and 7a show that, for a given CDT probability distribution and for a given value of the interruption probability, the quantitative difference between the mean value of the CHTn when the UST is hyper-exponentially distributed and the mean value of the CHTn when the UST is hyper-Erlang distributed is small but not negligible (differences not greater than 6.6% are found). Referring to the CoV {skewness} of the CHTn, Figs. 5b, c, 6b, c, and 7b, c show important differences between the case when the UST is hyper-exponential and the case when the UST is hyper-Erlang distributed (differences up to 12.75 and 15.20% are observed for the CoV and Sk, respectively). Similar results are obtained when the CHTh is considered. Due to the fact that both distributions (i.e., hyper-exponential and hyper-Erlang) for modeling the UST have the same first three standardized moments, we conclude that the (not negligible) quantitative difference among the mean value, CoV, and Sk of both CHTn and CHTh when the UST is hyper-exponentially distributed compared to the case when the UST is hyper-Erlang distributed is due to moments higher than the third one. Analyzing the impact of moments of UST (and CDT) higher than the third one on channel holding time statistics represents a topic of our current research.

Now let us study the impact of the variability of both UST and CDT on CHT statistics. To this end, it is important to notice that the CoV of the Erlang, exponential, and hyper-exponential/hyper-Erlang distributions are, respectively, smaller than 1, equal to 1, and larger than 1. Thus, the variability of UST increases as its probability distribution moves from the Erlang to the exponential and from the exponential to the hyper-exponential/hyper-Erlang. The same observation applies for the CDT. Considering this fact, the following observation can be extracted from Figs. 5, 6, and 7. Figures 5, 6, and 7 show that, in general terms, as the variability of the UST (CDT) increases (decreases), the mean value of the CHTn decreases and at the same time the mean value of the CHTh increases. This behavior can be explained as follows. First, note that as the CoV of UST (CDT) increases, the variability of UST (CDT) increases; that is, the values of UST (CDT) spread out over a larger range with respect to its mean value. Consequently, the service time (cell residence time) of calls that end their service in the cell where they were originated is, in general, considerable smaller (greater) than the mean UST (CDT), resulting in a diminution (increase) on the mean value of the CHTn. On the other hand, the service time (cell residence time) of calls that are handed off to another cell is, in general, considerably greater (lower) than the mean UST (CDT), resulting in an augment (diminution) on the mean value of the CHTh. The combined effects of these facts lead us to the behavior explained above and observed in Figs. 5, 6, and 7. Many others interesting observations can be extracted from Figs. 5, 6, and 7; however, the most important ones are summarized as follows. The mean values of both CHTn and CHTh are strongly sensitive to the interruption probability. For instance, when the UST is hyper-Erlang distributed and the CDT is exponential distributed (notice that this scenario corresponds to the green lines of Fig. 5a), the mean value of CHTn (CHTh) decreases from 77.35 to 53.057 s (101 to 55.4 s) as the interruption probability goes from 0.001 to 0.1. Similar behaviors are observed when the CDT is modeled by either an Erlang or a hyper-exponential distribution. Also, from Figs. 5, 6, and 7, it is evident that the first three moments of both CHTn and CHTh are highly sensitive to the probability distribution of both CDT and UST. Thus, selecting a suitable model that effectively captures the realistic statistics of both CDT and UST is of paramount importance when analyzing the performance of mobile cognitive radio cellular networks.

6.2 Effects of the secondary service time, users' mobility, and the primary channel utilization factor on system performance

In this section, the effects of the value of the mean secondary service time relative to the mean primary service time (hereafter called *relative service time*), mobility parameter (defined as the ratio between the mean service time and the mean cell residence time), the use or not of spectrum handoff (SH), and the primary channel utilization factor on the system Erlang capacity are evaluated. Additionally, the accuracy of the proposed approximate teletraffic model is evaluated for different evaluation scenarios. Specifically, four different scenarios are considered: (S1) low mobility with small relative service time, (S2) low mobility with large relative service time, (S3) high mobility with small relative service time, and (S4) high mobility with large relative service time. Table 4 summarizes the values of the parameters used in these scenarios. Unless otherwise specified, the following values of system parameters were used for the numerical evaluations shown in this section: mean service time is $1/\mu^{(S)} = 1/0.82$ s; as indicated in Table 6, two different values have been used for both the relative service time and mobility parameter, say, $\mu^{(P)}/\mu^{(S)} = \{0.1, 1\}$ and $\eta/\mu^{(S)} = \{0.2, 1\}$, respectively; total number of identical primary channels per cell $M = 3$; the number of identical sub-channels per primary channel $N = 6$ (that is, non-homogeneous bandwidth of PU and SU channels are considered), including the possibility of interruption of multiple SUs.

The analytical numerical results presented in this subsection were verified by a wide set of discrete-event computer simulation results for a variety of evaluation scenarios. Part of this numerical validation is shown in our previous work [23]. Nevertheless, for clarity proposes and easy visualization and comparison of the different scenarios, simulation results for the metrics presented in Figs. 8 and 9 are omitted. Please note that the teletraffic analysis (and the correspondent considered system model) of Section 5.2 is identical to that developed in [23]. In this paper, new call blocking and forced call termination probabilities of secondary users are employed to obtain both the maximum Erlang capacity and the optimum number of reserved channels shown in Figs. 8 and 9, respectively. Figures 2 and 3 of [23] show simulation and analytical numerical results of new call blocking and forced call termination probabilities of secondary users as function of the mobility factor of both primary and secondary users with the secondary traffic load as parameter. From the curves presented in Figures 2 and 3 of [23], it is observed perfect agreement between analytical and simulation results. As perfect agreement between analytical and simulation results is observed for these probabilities in [23], no additional validation is needed for the analytical results presented in Figs. 8 and 9 of this paper. On the other hand, note that perfect agreement is observed between analytical and simulation results for all the evaluation scenarios shown in Fig. 10.

Table 6 Parameters for the considered scenarios

Scenario	Description	Mobility parameter ($\eta/\mu^{(S)}$)	Relative service time ($\mu^{(P)}/\mu^{(S)}$)
S1	Low mobility, small relative service time	0.2	0.1
S2	Low mobility, large relative service time	0.2	1
S3	High mobility, small relative service time	1	0.1
S4	High mobility, large relative service time	1	1

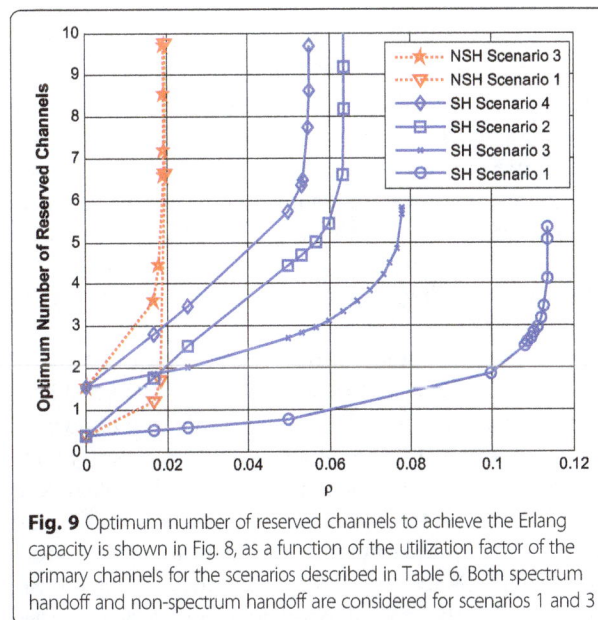

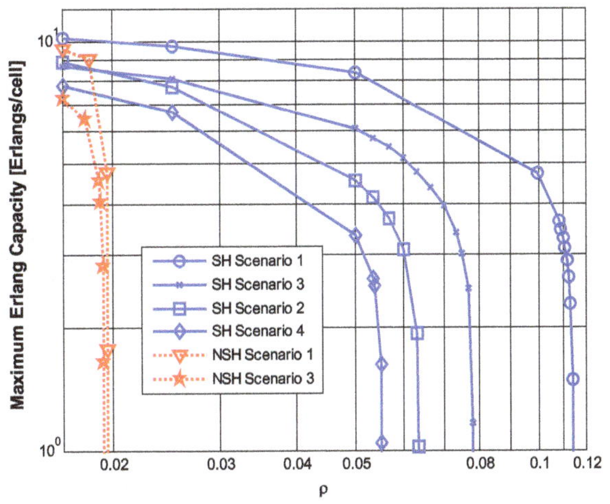

Fig. 8 Erlang capacity as a function of the utilization factor of the primary channels for the scenarios described in Table 6. Both spectrum handoff and non-spectrum handoff cases are considered for scenarios 1 and 3

Figure 8 depicts Erlang capacity as function of the primary channel utilization factor (defined as the ratio between the primary carried load and the total number of primary channels, i.e., $\rho = a_{c^{(P)}}/M$) for the different scenarios described in Table 6. Both spectrum handoff (SH) and non-spectrum handoff (NSH) cases are considered. Erlang capacity shown in Fig. 8 is obtained using the optimization procedure described in the last paragraph of Section 5.3. Specifically, the optimal number of reserved channels is computed in such a way that the system capacity is maximized, while the QoS requirements,

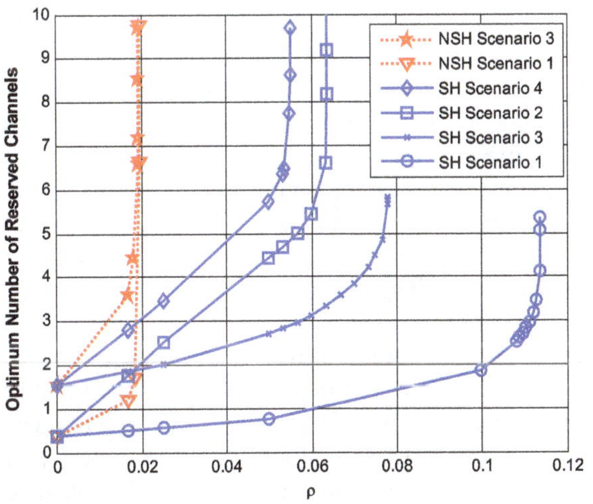

Fig. 9 Optimum number of reserved channels to achieve the Erlang capacity is shown in Fig. 8, as a function of the utilization factor of the primary channels for the scenarios described in Table 6. Both spectrum handoff and non-spectrum handoff are considered for scenarios 1 and 3

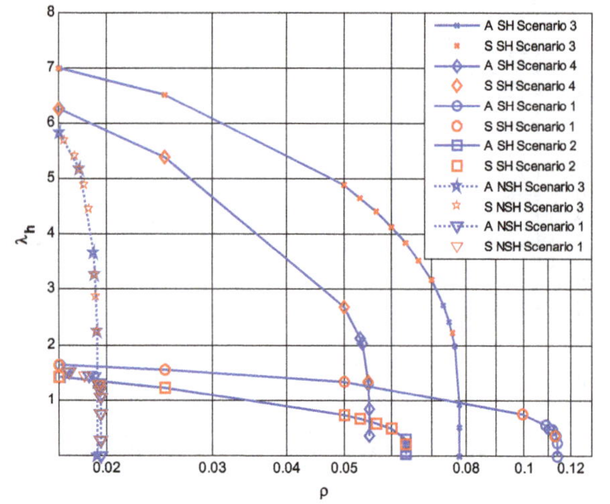

Fig. 10 Inter-cell handoff arrival rate for the Erlang capacity is shown in Fig. 8, as a function of the utilization factor of the primary channels for the scenarios described in Table 6. Both spectrum handoff and non-spectrum handoff are considered for scenarios 1 and 3

in terms of new call blocking and forced call termination probabilities, are achieved. New call blocking and forced call termination probabilities are evaluated using (61) and (66), respectively. Both of these probabilities are function of the interruption probability, which is evaluated using (62) and (67) for the NSH and SH cases, respectively. Maximum acceptable values of the new call blocking (P_{b_tr}) and forced call termination (P_{ft_tr}) probabilities are fixed to 2 and 0.2%, respectively.

From Fig. 8, it is observed that, for all the scenarios, as the utilization factor of primary channels (denoted by ρ) increases, system Erlang capacity decreases. This is an intuitively understandable result because as ρ increases, the average number of available channels for both new and (inter- and intra-) handed off secondary sessions decreases in detrimental of system capacity. Moreover, as ρ increases, more secondary calls are interrupted due to the arrival of primary sessions in detrimental of system performance. Let us consider, for example, the scenario S1 for the SH case. Figure 8 shows that as ρ moves from 0.0 (notice that this value of ρ corresponds to the conventional cellular system) to 0.1, Erlang capacity diminishes 52%.

Figure 8 also shows that there exists a critical value of the utilization factor of the primary channels (critical operational point) at which it is no longer possible to guarantee the QoS of the admitted secondary users (Erlang capacity abruptly decreases toward zero). This behavior is more evident for the NSH case. For instance, for the scenario S1, Fig. 8 indicates that this critical value is about 0.018 for the NSH case and it is about 0.106 for the SH case; that is, the critical operational point is increased about 488% by using spectrum handoff. From these results, we can see that the utilization factor of primary channels for which the Erlang capacity is zero when SH is enabled is $\rho = 0.106$ and when NSH is enabled is $\rho = 0.018$. As such, the utilization factor is increased as by $\Delta\rho = (0.106 - 0.018) = 0.088$. However, relative percentage difference to the case when NSH is being used is given by $\frac{100(0.106-0.018)}{0.018}\% = 488\%$. In other words, these results reveal that the use of spectrum handoff is essential for maximizing the arrival rate of PUs at which it is still possible to provide the QoS demanded by SUs. Similarly, Fig. 8 shows that the critical operational point decreases as either relative service time or mobility parameter increases. For instance, for the high mobility case (scenarios S3 and S4), the critical operational point decreases 30% as the relative service time moves from 0.1 (scenario S3) to 1.0 (scenario S4). On the other hand, for the large relative service time case (scenarios S2 and S4), the critical operational point decreases 13.12% as the mobility parameter goes from 0.2 (scenario S2) to 1.0 (scenario S4). This is because the utilization factor of primary channels for which the Erlang capacity is zero in scenarios S2 and S4 with SH is $\rho = 0.6346$ and $\rho = 0.5513$ for scenarios S2 and S4, respectively. Hence, there is an absolute decrease of $\Delta\rho = (0.6346 - 0.5513) = 0.0833$, but the relative percentage difference to the utilization factor of scenario S2 is $\frac{100(0.6346-0.5513)}{0.6346}\% = 13.12\%$. In general, Fig. 8 shows that Erlang capacity is more sensitive to the relative service time than to the mobility parameter. For instance, let us consider the scenario S1 for the SH case as the scenario of reference. Figure 8 shows that, for $\rho = 0.05$, Erlang capacity decreases 44% as the relative service time changes from 0.1 (scenario S1) to 1.0 (scenario S2), while Erlang capacity decreases 28% as the mobility parameter goes from 0.2 (scenario S1) to 1.0 (scenario S3). Figure 8 presents that for scenario S2, when the utilization factor of primary channels is $\rho = 0.05$, the Erlang capacity is 4.7 Erlangs. For scenario S1 for the same value of ρ, it has 8.4 Erlangs which corresponds to a reduction of $\frac{100(8.4-4.7)}{8.4}\% = 44.04\%$. In a similar way, for scenario S3, when $\rho = 0.05$, the Erlang capacity is 6.05 Erlangs. Hence, the Erlang capacity reduction relative to scenario S1 is $\frac{100(8.4-6.05)}{8.4}\% = 27.98\%$.

This behavior is due to the following two facts. First, the handoff arrival rate increases as the user mobility increases in detrimental of call forced termination probability. Thus, in order to achieve the required forced call termination probability (i.e., 0.2%), as the user mobility increases, more channels to prioritize intra- and intercell handoff call attempts over new call requests need to be reserved in detrimental of new call blocking probability. Consequently, in order to achieve the required new call blocking probability (i.e., 2%), system Erlang capacity is sacrificed (mobility/capacity conversion [41]). Second, increasing the relative service time implies that the mean value of the secondary service time relative to the mean value of the primary service time increases. In this sense, as the relative service time increases (while the utilization factor of the primary channels remains unchanged), it is more likely that an ongoing secondary call is interrupted due to the arrival of primary sessions in detrimental of forced call termination probability. Because of its larger duration, each secondary call is exposed to a larger number of interruptions. Thus, as explained above, in order to guarantee the required QoS, in terms of the forced call termination, more channels need to be reserved in detrimental of the maximum achievable system Erlang capacity. Also, for a given value of the utilization factor of the primary channels, as the relative service time increases, the departure rate of successfully terminated calls decreases relative to the arrival rate of primary sessions; thus, the average number of idle channels decreases in detrimental of the new call blocking probability and, consequently, in detrimental of system capacity. The joint effect of these facts leads us to the behavior explained above and illustrated in Fig. 8.

Figure 9 plots the optimal number of reserved channels needed to maximize the system Erlang capacity. The maximum achieved Erlang capacity as function of the primary channel utilization factor for the scenarios described in Table 6 (both SH and NSH cases are considered) is shown in Fig. 8. Figure 9 confirms that as either mobility parameter or relative service time increases, more channels to prioritize intra- and inter-cell handoff call attempts over new call requests are reserved to guarantee the required QoS of secondary users.

On the other hand, for offered traffic loads given by the Erlang capacity values presented in Fig. 8, Fig. 10 plots, the inter-cell handoff arrival rate as a function of the utilization factor of the primary channels for the scenarios described in Table 6 (both SH and NSH cases are considered). Figure 10 presents both simulation (denoted by label "S") and analytical (denoted by label "A") results. Excellent agreement between analytical and simulation results is observed in Fig. 10. As expected, Fig. 10 shows that, for a given value of ρ and while system capacity is maximized, handoff arrival rate increases as the mobility parameter increases. For instance, let us consider the spectrum handoff scenarios with small relative service time (i.e., scenarios S1 and S3). Figure 10 shows that, for $\rho = 0.05$, the inter-cell handoff arrival rate increases 262% as the mobility parameter goes from 0.2 (scenario S1) to 1.0 (scenario S3). Figure 10 also shows that, for a given value of ρ and while system capacity is maximized, handoff arrival rate decreases as the relative service time increases. This behavior is due to the fact that as the relative service time increases, it is more likely that an ongoing secondary call to be interrupted due to the arrival of a primary session, consequently, a lower average number of ongoing calls are handed-off to adjacent cells (that is, the handoff rate decreases as the relative service time increases). Finally, Table 7 presents the maximum percentage difference of analytical results relative to simulation results for both new call blocking and forced call termination probabilities in order to achieve the Erlang capacity presented in Fig. 8. Two approaches for evaluating forced termination probability are considered in Table 7: the closed-form approximated expression (65) and the rate ratio approach represented by (66). Table 7 indicates that the maximum percentage difference between analytical and simulation results found for new call blocking probability is 1.4%. Similar results were found for the handoff rate, interruption probability, and inter-cellular handoff failure probability. On the other hand, Table 7 reveals that, for low mobility scenarios and irrespective of the relative service time (i.e., scenarios S1 and S2), the maximum percentage difference between analytical and simulation results found for the forced call termination probability is 2.4% {25%} if expression (66) {(65)} is used for evaluating this probability. In general, from Table 7, it can be concluded that expression (66) is the best option to evaluate forced call termination probability, except for the case when SH is used and scenario S3 is considered.

7 Conclusions

In this paper, teletraffic performance and channel holding time characterization in mobile cognitive radio cellular networks (CRCNs) under fixed-rate traffic with hard-delay constraints was investigated. To this end, a fundamental mathematical model to capture the effect of interruption of secondary users' calls due to the arrival of primary users was developed. Based on this model, closed-form mathematical expressions for the probability distribution function of channel holding time for new calls (CHTn) and handed-off calls (CHTh), call forced termination probability, and inter-cell handoff attempt rate were derived. Additionally, a teletraffic analysis for the performance evaluation of CRCNs in terms of the Erlang capacity was developed. The accuracy, applicability, and robustness of our proposed mathematical models were extensively investigated under a variety of different evaluation scenarios for all the considered call-level performance metrics. Although numerical results are extracted from particular scenarios with certain set of parameter values, our contribution clearly shows that there exist relevant sensitive issues

Table 7 Maximum relative difference between analytical and simulation numerical results for both forced call termination and new call blocking probabilities

Network type	Scenario	Forced call termination probability		New call blocking probability
		Using (67) (%)	Using (68) (%)	Using (63) (%)
With spectrum handoff (SH)	S1	15	2.4	0.7
	S2	25	0.8	0.4
	S3	7	12	1.4
	S4	15	4	1.0
Without spectrum handoff (NSH)	S1	1	0.5	0.8
	S3	1	0.5	1.3

concerning interruption time (IT), cell dwell time (CDT), unencumbered service time (UST), and certain parameters of the primary network (i.e., mean service time, channels usage) that significantly influence the performance of the secondary network. Numerical results reveal that the first three standardized moments of both CHTn and CHTh are highly sensitive to both interruption probability and type of probability distribution functions used to model CDT and UST. From the teletraffic point of view, numerical results demonstrate that system Erlang capacity strongly depends on the relative value of the mean secondary service time to the mean primary service time, users' mobility, whether or not spectrum handoff (SH) is used, and the primary channel utilization factor. Furthermore, numerical results reveal that there exists a critical utilization factor of the primary resources from which it is no longer possible to guarantee the required QoS of SUs and, therefore, delay sensitive services cannot be even supported in CRCNs. Thus, it is of paramount importance to develop mechanisms that effectively mitigate the adverse effects of service interruption of SUs in cognitive radio networks. Also, improvement of the accuracy of the forced call termination probability metric is a research topic of interest.

8 Endnotes

[1]For this type of traffic, it is considered that both blocked and interrupted (due to either inter-cell or intra-cell handoff failures) sessions are clear from the system. That is, a secondary type of traffic that has the most stringent QoS requirements (such as the unsolicited grant service class in mobile WiMAX) is considered.

[2]The accuracy of this proposed mathematical approach is extensively investigated under a variety of different evaluation scenarios for all the considered call-level performance metrics.

[3]System-level analysis involves statistics such as channel holding times for successful terminated calls and for forced terminated calls which are easily obtained at real networks while link-level statistics involve channel characterization (i.e., in terms of channel states characteristics and the probability distributions of the channel states duration), which are not easily obtained at real cellular networks.

[4]It is important to point out that in order to guarantee quality of service of secondary real-time traffic, in the literature, it has been proposed to reserve spectrum resources for exclusive use of the secondary users [13] but this is not considered in this manuscript.

[5]Considering Poisson arrivals for PUs, mathematical expressions for P_{Int} are derived in [36] and [37]. It is assumed that the UST for SUs is exponentially [36] and two-phase Coxian [37] distributed. For completeness, in Section 5, the mathematical expression of P_{Int} for exponentially distributed UST of SUs is shown.

[6]In general, the interruption probability is not independent. Indeed, when the traffic load is high, whenever a SU is interrupted due to the arrival of a PU, it is highly probable that the next arrival of a PU also causes the interruption of another secondary session because resource occupancy conditions at the different time instants (primary arrival epochs) are not independent. Conversely, in a low traffic load scenario, if the arrival of a PU does not cause an interruption of a SU, it is unlikely that the arrival of another PU in the same conditions causes a SU interruption. Hence, a certain degree of correlation between the interruption probabilities is present, and therefore, the secondary users' call interruption process actually is not a Poissonian one.

[7]The mathematical analysis for the phase-type probability distributions of both UST and CDT can be developed in a similar way, considering more state variables.

9 Appendix 1

In this Appendix, we show that the call interruption process is not a Poisson process. For this, the Bayes theorem is employed and we assume that the call interruption time due to arrival of PUs $X_{\text{int\_PU}}$ follows an exponential distribution with parameter $\lambda^{(P)}P_{\text{Int}}$. Additionally, it is assumed that both the unencumbered service time and cell dwell time for SUs are exponentially distributed with parameters μ and η, respectively.

Notice that the call interruption time $X_{\text{int\_PU}}$ is not a measurable physical quantity because it is a potential time (when the interruption of a call does not occur, it is not possible to know its value), as explained in Section 2.1. Then, to show whether or not the call interruption process is Poissonian, we are interested on finding the probability distribution of a measurable physical quantity that can be obtained from our discrete-event computer simulator. In particular, we are interested on the conditional call interruption time given that calls are interrupted by the arrival of PUs which pdf is represented by $f_{X_{\text{int\_PU}}}|X_{\text{int\_PU}} < \min(X_d, X_s)(t)$.

Let $X_c = \min(X_d, X_s)$ represent the channel holding time, then

$$f_{X_{\text{int\_PU}}}|X_{\text{int\_PU}<X_c}(t) = \frac{P(X_{\text{int\_PU}} < X_c | X_{\text{int\_PU}} = t) f_{X_{\text{int\_PU}}}(t)}{P(X_{\text{int\_PU}} < X_c)} \tag{68}$$

From this, it is easy to say that

$$P(X_{\text{int\_PU}} < X_c | X_{\text{int\_PU}} = t) = \int_t^{\infty} f_{X_c}(x)dx = e^{-(\eta+\mu)t} \tag{69}$$

On the other hand,

$$P(X_{\text{int\_PU}} < X_c) = \frac{\lambda^{(P)}P_{\text{Int}}}{\mu + \eta + \lambda^{(P)}P_{\text{Int}}} \tag{70}$$

Substituting (69) and (70) in (68), we get

$$f_{X_{\text{int}_{\text{PU}}}}|X_{\text{int}_{\text{PU}}} < X_c(t) = \left(\mu+\eta+\lambda^{(P)}P_{\text{Int}}\right)e^{-\left(\mu+\eta+\lambda^{(P)}P_{\text{Int}}\right)t} \tag{71}$$

Therefore, it has been shown that if the call interruption time due to arrival of PUs $X_{\text{int\_PU}}$ follows an exponential distribution, then the conditional call interruption time given that calls are interrupted by the arrival of PUs must be also exponentially distributed with mean

$$E\{X_{\text{int\_PU}} | X_{\text{int\_PU}} < X_c\} = \frac{1}{\mu + \eta + \lambda^{(P)}P_{\text{Int}}} \tag{72}$$

and with coefficient of variation (CoV) equal to 1.

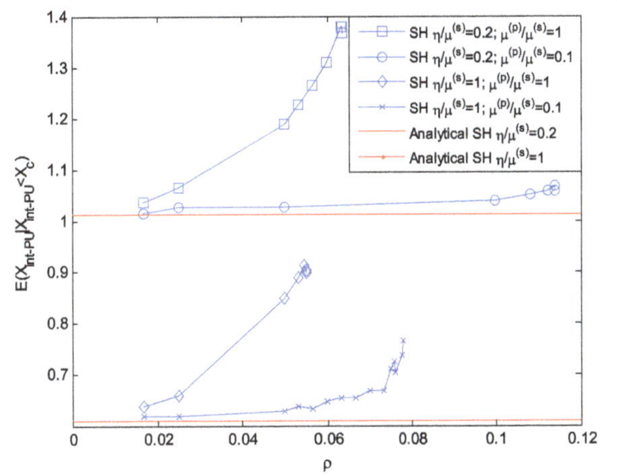

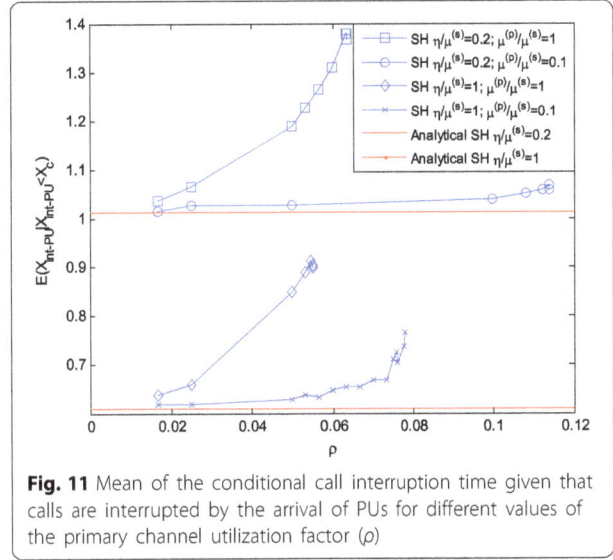

Fig. 11 Mean of the conditional call interruption time given that calls are interrupted by the arrival of PUs for different values of the primary channel utilization factor (ρ)

Again, this result is obtained assuming that the call interruption process is a Poisson process. Now, in Figs. 11 and 12, we compare, respectively, the mean and coefficient of variation of the conditional call interruption time given that calls are interrupted by the arrival of PUs obtained analytically (red symbols) and by simulation (blue lines) for the evaluation scenarios and conditions used in Section 5. From these numerical results, it is clear that the coefficient of variation differs from 1. Hence, the call interruption process is not actually a Poisson process.

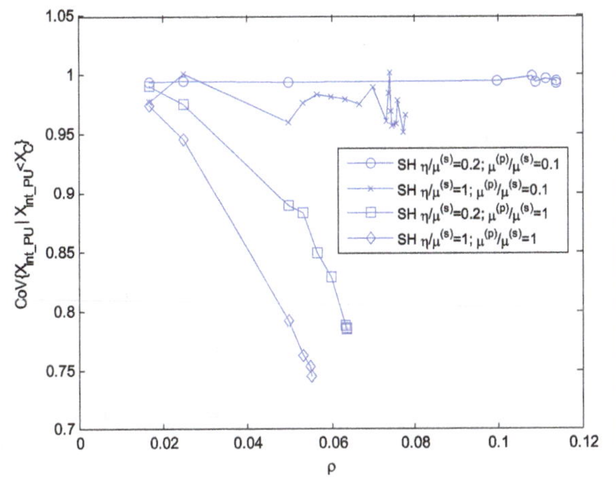

Fig. 12 Coefficient of variation of the conditional call interruption time given that calls are interrupted by the arrival of PUs for different values of the primary channel utilization factor (ρ)

10 Appendix 2

The Laplace Transform of the pdf of $X_{\text{Int\_PU}}$, given by (3), is calculated as

$$f^*_{X_{\text{int\_PU}}}(s) = \frac{\lambda^{(P)}P_{\text{Int}}}{\lambda^{(P)}P_{\text{Int}} + s} \tag{73}$$

It is clear that the unique pole of $f_{X\text{int\_PU}}{}^*(-s)$ is $\Omega_{X_{\text{int\_PU}}} = \{\lambda^{(P)}P_{\text{Int}}\}$.

11 Appendix 3

In this Appendix, different statements used in the mathematical analysis are proven.

1. The statement $P\{X_{r,d} < \min(X_{\text{int\_PU}}, X_s)\} = f^*_{X_{r,d}}(\mu + \lambda^{(P)}P_{\text{Int}})$ (where $X_{r,d}$ is either X_r or X_d) is proven as follows. Since X_{int_PU} and X_s are exponentially distributed, then the minimum between these variables is also exponentially distributed with parameter $\mu + \lambda^{(P)}P_{\text{Int}}$. As such,

$$f^*_{\min(X_{\text{int\_PU}},X_s)}(s) = \frac{\mu + \lambda^{(P)}P_{\text{Int}}}{\mu + \lambda^{(P)}P_{\text{Int}} + s} \tag{74}$$

Using proposition 2 of [9],

$$P\{X_{r,d} < \min(X_{\text{int\_PU}}, X_s)\} = -\lim_{s \to \mu + \lambda^{(P)}P_{\text{Int}}} \frac{(s - \mu - \lambda^{(P)}P_{\text{Int}}) f^*_{X_{r,d}}(s)}{s} \cdot \frac{\mu + \lambda^{(P)}P_{\text{Int}}}{\mu + \lambda^{(P)}P_{\text{Int}} - s} \tag{75}$$

From this, it follows that

$$P\{X_{r,d} < \min(X_{\text{int\_PU}}, X_s)\} = f_{X_{r,d}}{}^*(\mu + \lambda^{(P)}P_{\text{Int}}) \tag{76}$$

2. The statement $P\{X_{\text{int\_PU}} < \min(X_{d,r}, X_s)\} = \frac{\lambda^{(P)}P_{\text{Int}}\left[1 - f^*_{X_{d,r}}(\mu + \lambda^{(P)}P_{\text{Int}})\right]}{\mu + \lambda^{(P)}P_{\text{Int}}}$ (where $X_{d,r}$ is either X_d or X_r) is proven as follows. Let us define $Z_{d,r} = \min(X_{d,r}, X_s)$. Calculating the Laplace transform using proposition 2 of [9] we get

$$f^*_{Z_{d,r}}(s) = \frac{\mu}{\mu + s} + s\frac{f^*_{X_{d,r}}(\mu + s)}{\mu + s} = \frac{\mu + s f^*_{X_{d,r}}(\mu + s)}{\mu + s} \tag{77}$$

Since $P\{X_{\text{int\_PU}} < \min(X_{d,r}, X_s)\} = P\{X_{\text{int\_PU}} < Z_{d,r}\}$, using proposition 1 of [9], we obtain

$$P\{X_{\text{int\_PU}} < Z_{d,r}\} = \frac{\lambda^{(P)}P_{\text{Int}}\left[1 - f^*_{X_{d,r}}(\mu + \lambda^{(P)}P_{\text{Int}})\right]}{\mu + \lambda^{(P)}P_{\text{Int}}} \tag{78}$$

3. Now, let us prove the statement $P\{X_d < \min(X_{\text{int\_PU}}, X_{sr})\} = -\sum_{p \in \Omega_{X_d}} \operatorname{Res}_{s=p} \frac{[1 - f^*_{X_{sr}}(\lambda^{(P)}P_{\text{Int}} + s)] f^*_{X_d}(-s)}{\lambda^{(P)}P_{\text{Int}} + s}$.

To this end, the variable $\boldsymbol{Z}_{sr}$ is defined as follows $Z_{sr} = \min(X_{\text{int\_PU}}, X_{sr})$. The Laplace transform of the pdf of $\boldsymbol{Z}_{sr}$ is calculated using proposition 2 of [9] as

$$f^*_{Z_{sr}}(s) = \frac{\lambda^{(P)}P_{\text{Int}} + s f^*_{X_{sr}}(\lambda^{(P)}P_{\text{Int}} + s)}{\lambda^{(P)}P_{\text{Int}} + s} \tag{79}$$

Since $P\{X_d < \min(X_{\text{int\_PU}}, X_{sr})\} = P\{X_d < Z_{sr}\}$, using proposition 1 of [9], we obtain

$$P\{X_d < Z_{sr}\} = -\sum_{p \in \Omega_{X_d}} \operatorname{Res}_{s=p} \frac{\left[1 - f^*_{X_{sr}}(\lambda^{(P)}P_{\text{Int}} + s)\right] f^*_{X_d}(-s)}{\lambda^{(P)}P_{\text{Int}} + s} \tag{80}$$

4. Finally, in this section, the statement $P\{X_r < \min(X_{\text{int\_PU}}, X_s)\} = -\sum_{p \in \Omega_{X_r}} \operatorname{Res}_{s=p} \frac{[1 - f^*_{X_s}(\lambda^{(P)}P_{\text{Int}} + s)] f^*_{X_r}(-s)}{\lambda^{(P)}P_{\text{Int}} + s}$ is proven. To this end, the variable $\boldsymbol{Z}_s$ is defined as follows: $Z_s = \min(X_{\text{int\_PU}}, X_s)$. The Laplace transform of the pdf of $\boldsymbol{Z}_s$ is calculated using proposition 2 of [9] as

$$f^*_{Z_s}(s) = \frac{\lambda^{(P)}P_{\text{Int}} + s f^*_{X_s}(\lambda^{(P)}P_{\text{Int}} + s)}{\lambda^{(P)}P_{\text{Int}} + s} \tag{81}$$

Since $P\{X_r < \min(X_d, X_s)\} = P\{X_r < Z_s\}$, using proposition 1 of [9], we obtain

$$P\{X_r < Z_s\} = -\sum_{p \in \Omega_{X_r}} \operatorname{Res}_{s=p} \frac{\left[1 - f^*_{X_s}(\lambda^{(P)}P_{\text{Int}} + s)\right] f^*_{X_r}(-s)}{\lambda^{(P)}P_{\text{Int}} + s} \tag{82}$$

Acknowledgements

No one.

Funding

PRODEP contract 916040 and project no. 22411508 as well as UAM-A project no. EL001-13 provide us resources to acquire computing equipment for the numerical evaluation of the studied systems.

Authors' contributions

ALECR developed mathematical analysis for the channel holding times. He also obtained the numerical results obtained to evaluate the performance of the system in terms of the channel holding time. SLCL developed the teletraffic analysis and the optimization algorithm to maximize system Erlang capacity. She also participated in the mathematical analysis derived in this work and the numerical results obtained to evaluate the performance of the system. FACP was involved in the development of the different mathematical models as well as identifying, analyzing, and studying the different factors that affect system performance. GHV was involved in the development of the different mathematical models as well as identifying, analyzing, and studying the different factors that affect system performance. MERA was involved in the mathematical analysis as well as in the numerical evaluation of the considered system. All authors read and approved the final manuscript.

Competing interests
The authors declare that they have no competing interests.

Author details
[1]Electrical Engineering Department, CINVESTAV-IPN, Mexico City, Mexico. [2]Electronics Department, UAM, Mexico City, Mexico. [3]Communication Networks Laboratory, CIC-Instituto Politécnico Nacional, Mexico City, Mexico.

References
1. X. Zhu, L. Shen, T.-S.P. Yum, Analysis of cognitive radio spectrum access with optimal channel reservation. IEEE Commun. Lett. **11**(4), 304–306 (2007)
2. Y.Y. Mihov, B.P. Tsankov, in *Proc. IEEE International Conference on Microwaves, Communications, Antennas and Electronics Systems (COMCAS 2011), Tel Aviv, Israel*. QoS provisioning via channel reservation in cognitive radio networks (2011)
3. S.L. Castellanos López, F.A. Cruz Pérez, M.E. Rivero-Ángeles, G. Hernández Valdez, in *Proc. 7th International Conference on Cognitive Radio Oriented Wireless Networks (CROWNCOM'2012), Stockholm, Sweden*. Impact of the primary resource occupancy information on the performance of cognitive radio networks with VoIP traffic (2012)
4. S.L. Castellanos-López, F.A. Cruz-Pérez, M.E. Rivero-Ángeles, G. Hernández-Valdez, Joint connection level and packet level analysis of cognitive radio networks with VoIP traffic. IEEE J. Select. Areas Commun. **32**(3), 601–614 (2014)
5. S.L. Castellanos-López, F.A. Cruz-Pérez, M.E. Rivero-Ángeles, G. Hernández-Valdez, Performance analysis of coordinated cognitive radio networks under fixed-rate traffic with hard delay constraints. J. Commun Netw Spec Issue Cogn Netw **16**(2), 130–139 (2014)
6. X. Mao, H. Ji, V.C.M. Leung, M. Li, in *Proc. IEEE GLOBECOM'2010, Miami, Florida, USA*. Performance enhancement for unlicensed users in coordinated cognitive radio networks via channel reservation (2010)
7. Y. Yao, S.R. Ngoga, D. Erman, A. Popescu, in *Proc. IEEE ICC'2012, Ottawa, Canada*. Performance of cognitive radio spectrum access with intra- and inter-handoff (2012), pp. 1549–1554
8. A.L.E. Corral-Ruiz, F.A. Cruz-Pérez, G. Hernández-Valdez, in *Proc. IEEE WCNC'2011, Cancun, QR, Mexico*. Channel holding time in mobile cellular networks with heavy-tailed distributed cell dwell time (2011), pp. 2065–2070
9. C.B. Rodríguez-Estrello, G. Hernández-Valdez, F.A. Cruz-Pérez, System-level analysis of mobile cellular networks considering link unreliability. IEEE Trans. Veh. Technol. **58**(2), 926–940 (2009)
10. X. Wang, P. Fan, Channel holding time in wireless cellular communications with general distributed session time and dwell time. IEEE Commun. Lett. **11**(2), 158–160 (2007)
11. P.V. Orlik, S.S. Rappaport, A model for teletraffic performance and channel holding time characterization in wireless cellular communication with general session and dwell time distributions. IEEE J. Sel. Areas Commun. **16**(5), 788–803 (1998)
12. A.L.E. Corral-Ruiz, F.A. Cruz-Pérez, G. Hernández-Valdez, in *Proc. IEEE GLOBECOM'2010, Miami, Florida*. On the functional relationship between channel holding time and cell dwell time in mobile cellular networks (2010)
13. D. Cavdar, H.B. Yilmaz, T. Tugcu, F. Alagöz, in *Proc. 6th Advanced International Conference on Telecommunications (AICT'2010), Barcelona, Spain*. Analytical modeling and performance evaluation of cognitive radio networks (2010), pp. 35–40
14. W.-Y. Lee, I.F. Akyildiz, Spectrum-aware mobility management in cognitive radio cellular networks. IEEE Trans. Mob. Comput. **11**(4), 529–542 (2012)
15. Y. Fang, I. Chlamtac, Y.B. Lin, Channel occupancy times and handoff rate for mobile computing and PCS networks. IEEE Trans. Comput. **47**(6), 679–692 (1998)
16. Y. Fang, I. Chlamtac, Teletraffic analysis and mobility modeling of PCS networks. IEEE Trans. Commun. **47**(7), 1062–1072 (1999)
17. Y. Zhang, B. Soong, Performance of mobile network with wireless channel unreliability and resource insufficiency. IEEE Trans. Wirel. Commun. **5**(5), 990–995 (2006)
18. Y. Zhang, M. Fujise, Performance analysis of wireless networks over Rayleigh fading channel. IEEE Trans. Veh. Technol. **55**(5), 1621–1632 (2006)
19. Y. Zhang, M. Ma, M. Fujise, Call completion in wireless networks over lossy link. IEEE Trans. Veh. Technol. **56**(2), 929–942 (2007)
20. Y. Yao, A. Popescu, A. Popescu, On prioritised opportunistic spectrum access in cognitive radio cellular networks. Trans. Emerg. Telecommun. Technol. **27**(2), 294–310 (2016)
21. A. Homayounzadeh, M. Mahdavi, *Quality of Service Analysis for the Real-Time Secondary Users in Cognitive Radio Cellular Networks, Accepted for Its Publication in Wireless Personal Communications, First Online* (2017). https://doi.org/10.1007/s11277-017-4340-y
22. Hau-luen Chiou, "Performance Evaluation of Multirate Cognitive Radio Cellular Networks with Channel Aggregation", Master's Thesis, Department of Electrical Engineering, Date of Defense: 2014
23. J. Serrano-Chavez, G. Hernandez-Valdez, F.A. Cruz-Pérez, S.L. Castellanos-Lopez, E.A. Andrade-Gonzalez, M. Reyes-Ayala, J.R. Miranda-Tello, in *Proc. WSEAS Recent Advances on Systems, Signals, Control, Communications and Computers (DNCOCO'15), Budapest, Hungary*. Impact of mobility on the performance of cognitive radio mobile cellular networks with real-time traffic (2015), pp. 236–242
24. D. Cavdar, H.B. Yilmaz, T. Tugcu, F. Alagöz, Analytical modeling and resource planning for cognitive radio systems. Wirel. Commun. Mob. Comput. **12**(3), 277–292 (2012)
25. Y. Fang, Performance evaluation of wireless cellular networks under more realistic assumptions. Wirel. Commun. Mob. Comput. **5**(8), 867–885 (2005)
26. Y. Fang, Modeling and performance analysis for wireless mobile networks: A new analytical approach. IEEE/ACM Trans. Networking **13**(5), 989–902 (2005)
27. S. Tang, B.L. Mark, Modeling and analysis of opportunistic spectrum sharing with unreliable spectrum sensing. IEEE Trans. Wirel. Commun. **8**(4), 1934–1943 (2009)
28. I. Suliman, J. LehtomLki, T. BrLysy, K. Umebayashi, in *Proc. IEEE PIMRC'2009, Tokyo, Japan*. Analysis of cognitive radio networks with imperfect sensing (2009), pp. 1616–1620
29. X. Gelabert, O. Salient, J. Prez-Romero, R. Agust, Spectrum sharing in cognitive radio networks with imperfect sensing: A discrete-time Markov model. Comput. Netw. **54**(14), 2519–2536 (2010)
30. S. Tang, A general model of opportunistic spectrum sharing with unreliable sensing. Int. J. Commun. Syst. **27**(1), 31–34 (2014)
31. J. Martinez-Bauset, V. Pla, J.R. Vidal, L. Guijarro, Approximated analysis of cognitive radio systems using time-scale separation and its accuracy. IEEE Commun. Lett. **17**(1), 35–38 (2013)
32. O. Salameh, K. de Turck, D. Fiems, H. Bruneel, and S. Wittevrongel, "Performance Analysis of a Cognitive Radio Network with Imperfect Spectrum Sensing," IEICE Trans. on Communications, Advance Publication: Released June 22, 2017. https://doi.org/10.1587/transcom.2017EBP3037
33. J.-W. Wang, R. Adriman, in *Proc. (IMIS'2013), Taichung, Taiwan*. Analysis of cognitive radio networks with imperfect sensing and backup channels (2013), pp. 626–631
34. F. Wang, J. Huang, Y. Zhao, Delay sensitive communications over cognitive radio networks. IEEE Trans. Wirel. Commun. **11**(4), 1402–1411 (2012)
35. L. Duan, J. Huang, B. Shou, Investment and pricing with spectrum uncertainty: A cognitive operator's perspective. IEEE Trans. Mob. Comput. **10**(11), 1590–1604 (2011)
36. Y. Zhang, in *Proc. IEEE ICC'2008, Beijing, China*. Dynamic spectrum access in cognitive radio wireless networks (2009), pp. 4927–4932
37. S.L. Castellanos-Lopez, F.A. Cruz-Perez, G. Hernandez-Valdez, in *Proc. 7th IEEE International Conference on Wireless and Mobile Computing, Networking and Communications (WiMob'2011), Shangai, China*. Performance of cognitive radio networks under resume and restart retransmission strategies (2011), pp. 51–59
38. F.A. Cruz-Pérez, D. Lara-Rodríguez, M. Lara, Fractional channel reservation in mobile communication systems. IEE Electron. Lett. **35**(23), 2000–2002 (1999)
39. R. Ramjee, R. Nagarajan, D. Towsley, On optimal call admission control in cellular networks. ACM/Baltzer Wirel. Netw. J. **3**(1), 29–41 (1997)
40. D. Sarkar, T. Jewell, S. Ramakrishnan, Convergence in the calculation of the handoff arrival rate: A log-time iterative algorithm. EURASIP J. Wirel. Commun. Netw. **2006**(1), 1–11 (2006)
41. M. Zhang, C.T. Lea, On the mobility/capacity conversion in wireless networks. IEEE Trans. Veh. Technol. **53**(3), 734–746 (2004)

11

Performance limit of AOA-based localization using MIMO-OFDM channel state information

Tanee Demeechai* and Pratana Kukieattikool

Abstract

Wireless communication networks are increasingly based on the ubiquitous multiple-input multiple-output orthogonal frequency division multiplexing (MIMO-OFDM) modulation scheme. Their channel state information is generally obtained each time by a base station receiver as soon as a data packet is successfully received from a mobile device. As it has been shown recently that the MIMO-OFDM channel state information can be used for angle of arrival-based localization, this paper presents a theoretical investigation of the localization performance. The method of computing the Cramer-Rao lower bound, which represents the performance of a minimum variance unbiased estimator, is presented and then used for insightful investigation purposes by means of inspecting the viability of the system requirements and the design properties.

Keywords: Localization, Angle of arrival, MIMO-OFDM, Channel state information, Multipath propagation, Cramer-Rao lower bound

1 Introduction

Wireless localization is currently a research topic of primary interest in the field of wireless communications due to providing a large variety of location-based services for increasingly ubiquitous smart mobile devices. For wireless localization, a method requiring hardware of just an existing wireless network is of particular interest, because no extra hardware is required for both the infrastructure and the mobile devices (MDs). Unfortunately, although many studies have investigated wireless localization [1–5], it is not straightforward to apply their results without requiring dedicated hardware or affecting the network protocol structure.

A few localization methods requiring hardware of just an existing wireless network have been proposed in the literature [6–9] and implemented commercially. Most of them [6–8] adopt fingerprint matching as the basic scheme of location determination. In this scheme, prior to the localization operation, algorithm parameters are determined by a training process. The parameters are determined in principle as a representation of the characteristics of the observation data associated with a number of MD reference locations. In the localization operation, the location is determined as a linear combination of the reference locations, based on the observation data and the trained parameters. One major drawback of this scheme is that the training process is usually a costly measurement process. In addition, because the radio wave propagation characteristics that characterize the observation data are sensitive to the environment, the training process will be required every time the environment of the localization region has changed.

We realize that the angle of arrival (AOA) estimation [10, 11] is an interesting study that can be applied for localization. Recently, it has been shown that multiple-input multiple-output (MIMO) orthogonal frequency division multiplexing (OFDM) channel state information (CSI) may be used as the observation data for localization based on AOA [9]. For localization based on AOA, it generally requires that direct radio paths between the MD and the base stations (BSs) are available sufficiently for triangulation. However, using MIMO-OFDM CSI as the observation data for localization based on AOA is of particular interest here, due to the following reasons.

*Correspondence: tanee.demeechai@nectec.or.th
National Electronics and Computer Technology Center, Phahonyothin Road, 12120 Pathum Thani, Thailand

AOA-based localization does not require a costly training process. In addition, the MIMO-OFDM technique is currently the top air interface technology prevalently employed for very high-speed wireless networks.

The purpose of this paper is to investigate the performance characteristics of AOA-based localization methods that employ the MIMO-OFDM CSI as the observation data. This may be done by using the Cramer-Rao lower bound (CRLB), which is the variance of the estimation error of a minimum variance unbiased estimator (MVUE), as it has been successfully used to investigate theoretical performance characteristics for a number of related estimation problems [12–14]. In this paper, a method of computing the CRLB in general for the AOA-based localization that uses MIMO-OFDM CSI as the observation data is presented. Numerical results are then obtained for a specific case detailing about the geometry and placement of the BS antenna arrays and the localization region. The results are discussed, respecting the impact of radio propagation characteristics and system infrastructures on the average system performance and providing fundamental insights on the system requirements and design principles.

2 System and observation models

Channel frequency response (CFR) for a used OFDM subcarrier is a necessary information for demodulation of data carried by that subcarrier. Conventionally, the transmitter needs to facilitate CFR estimation at the receiver by provision of pilot bits [15]. It is then reasonable to assume that once a receiver can successfully detect the data carried by the used OFDM subcarriers, it can also provide the CFR for the set of used OFDM subcarriers, i.e., the CSI, at no overhead. For localization, the CSI may be obtained, based on either uplink or downlink transmission, for sending to a localization processor. In this paper, we are only interested in localization with processor residing in the network infrastructure to avoid burdening the generally resource-limited MDs. In this case, uplink CSI estimated by a BS receiver may be sent to the processor at low cost via an existing non-radio connection. On the other hand, sending the downlink CSI estimated by a MD receiver to the processor inevitably requires some radio resource that can be costly. Reducing the cost of sending the downlink CSI to the processor may be done with a chunk-based scheme [16–18], where an average magnitude of CFR for a block of contiguous subcarriers is sent instead of the complex values of the CFR of all subcarriers within the block. However, AOA-based localization requires the CFR in complex form [9]. Therefore, strictly following [16–18] may not be appropriate for our problem. In this regard, we consider that a similar chunk-based scheme that sends an average complex CFR for the block instead of the average magnitude may be adopted for AOA-based localization. This is because the average CFR obtained by this scheme is essentially same as that obtained by a conventional subcarrier-based scheme with the same total bandwidth but larger subcarrier spacing. However, this scheme therefore provides CFR with less details along the frequency axis. Accordingly, the processor will perform worse on separating the direct path from reflection paths, leading eventually to lower performance.

As using the uplink CSI for the localization processor residing in the infrastructure is less cumbersome compared with using the downlink CSI, we assume that the location of a MD is determined by processing the CSI obtained by the MIMO-OFDM receivers of all BSs upon successful reception of a data packet from that MD. Assume that each BS employs an array of antennas, with the spacing between them on the order of half wavelength. Then, the CFR for the q-th BS respecting the k-th subcarrier and m-th antenna can be expressed by [9]

$$H_{q,k,m} = \sum_{n=0}^{N_{\mathrm{P}}-1} g_{q,n} e^{-j(2\pi t_{q,n} k\Delta + \phi_{q,n,m})} + w_{q,k,m}, \tag{1}$$

where N_{P} is the number of paths in the radio wave propagation model, Δ is the OFDM subcarrier spacing, $g_{q,n}$ is the complex gain with magnitude representing the strength of the n-th path for the q-th BS, $t_{q,n}$ is the propagation delay associated with the n-th path between the MD antenna and the center of the antenna array of the q-th BS and subtracted by a constant delay specific to the q-th BS, $\phi_{q,n,m}$, $\forall m$ are the phase terms obeying $\sum_{\forall m} \phi_{q,n,m} = 0$ and sharing a relationship that depends on the q-th BS antenna array geometry and the AOA to the array from the n-th path, and $w_{q,k,m}$ is the estimation error that is assumed to behave as additive white Gaussian noise (AWGN). Assume that the terms in (1) have been sorted in ascending order of $t_{q,n}$, i.e., $t_{q,n_1} \leq t_{q,n_2}$ if $n_1 \leq n_2$. In addition, it is required that direct path from the MD antenna to the BS antenna array exists as the shortest path. Thus, it is the relationship between $\phi_{q,0,m}$, $\forall m$ that contains the AOA-based location information of the MD.

We have also assumed that the OFDM time synchronizer and provision of the guard interval that houses the cyclic prefix of the OFDM symbol are perfect on their roles. Hence, the following can be deduced in subsequent. First, by perfect provision of the guard interval, the maximum delay spread is not greater than the guard interval length, i.e., $t_{q,N_{\mathrm{P}}-1} - t_{q,0} \leq t_{\mathrm{g}}$, where t_{g} is the guard interval length. As a result, letting the l-th OFDM symbol excluding cyclic prefix begin arriving at the receiver at time instant t_{d}, we can say that the contribution of the $(l-1)$-th OFDM symbol arriving at the receiver ends accordingly at time instant $t_{\mathrm{d}} - t_{\mathrm{m}}$, where $t_{\mathrm{m}} \geq 0$ and $t_{\mathrm{m}} = t_{\mathrm{g}} - (t_{q,N_{\mathrm{P}}-1} - t_{q,0})$. Then, by perfect time synchronization, the receiver should be able to avoid inter-symbol interference (ISI) by beginning the demodulation of the l-th OFDM symbol at t_0 with $t_d - t_m \le t_0 \le t_d$. Then, by virtue of the cyclic prefix, the orthogonality between the subcarriers is still preserved, and the direct path manifests itself to the synchronized demodulator as having delay of $t_{q,0} = t_d - t_0$. Note that a good practice of OFDM time synchronization is to have t_0 close to t_d. Accordingly, $t_{q,0}$ is usually a small non-negative number compared with the guard interval length.

3 CRLB computation

The computation of CRLB can be based on the following theoretical assumptions. The number of antennas which are included in the array corresponds to M_q for the q-th BS. Then, the observation data from the q-th BS may be expressed in a matrix form as

$$\mathbf{H}_{q,k} = \mathbf{S}_{q,k} + \mathbf{W}_{q,k}, \tag{2}$$

where $\mathbf{W}_{q,k} = \left[w_{q,k,1}\ w_{q,k,2} \cdots w_{q,k,M_q} \right]^{\mathrm{T}}$, with $(\cdot)^{\mathrm{T}}$ denoting the matrix transpose, $\mathbf{S}_{q,k} = \mathbf{A}_q \mathbf{G}_{q,0} \mathbf{R}^k_{q,0}$, and we have used the following notations.

$$\mathbf{A}_q = \left[\mathbf{a}_q(\theta_{q,0})\ \mathbf{a}_q(\theta_{q,1}) \cdots \mathbf{a}_q(\theta_{q,N_P-1}) \right], \tag{3}$$

where

$$\mathbf{a}_q(\theta_{q,n}) = \left[e^{-j\phi_{q,n,1}}\ e^{-j\phi_{q,n,2}} \ldots e^{-j\phi_{q,n,M_q}} \right]^{\mathrm{T}}. \tag{4}$$

Note that $\mathbf{a}_q(\cdot)$ is a vector-valued function that depends on the q-th BS antenna array geometry, and $\theta_{q,n}$ denotes the AOA to the array from the n-th radio path. In addition, we have assumed that an AOA can be represented by a single number, and hence, two-dimensional localization is implied.

$$\mathbf{G}_{q,i} = \mathrm{diag}\left\{ g_{q,n} \right\}_{n=i}^{N_P-1}, \tag{5}$$

$$\mathbf{R}^k_{q,i} = \left[e^{-j\omega_{q,i}k}\ e^{-j\omega_{q,i+1}k} \ldots e^{-j\omega_{q,N_P-1}k} \right]^{\mathrm{T}}, \tag{6}$$

where $\omega_{q,n} = 2\pi t_{q,n}\Delta$. Denote the number of involved BSs as Q. Then, since $\theta_{q,0}$, $1 \le q \le Q$ are uniquely determined by the MD location given the geometry and placement of all the BS antenna arrays, we may define the row vector of independent parameters for the observation data as

$$\mathbf{u} = \left[x_0\ y_0\ \mathbf{v}_1\ \mathbf{v}_2 \cdots \mathbf{v}_Q \right], \tag{7}$$

where x_0 and y_0 represent the MD location in a Cartesian coordinate system, and

$$\mathbf{v}_q = \left[\Omega_q\ \Re(\mathbf{g}_q)\ \Im(\mathbf{g}_q)\ \Theta_q \right], \tag{8}$$

with $\Omega_q = \left[\omega_{q,0}\,\omega_{q,1} \cdots \omega_{q,N_P-1}\right]$, $\mathbf{g}_q = \left[g_{q,0}\, g_{q,1} \cdots g_{q,N_P-1}\right]$, and $\Theta_q = \left[\theta_{q,1}\theta_{q,2} \cdots \theta_{q,N_P-1}\right]$. For the CRLB computation, we necessarily evaluate the derivative of $\mathbf{S}_{q,k}$ with respect to each parameter, and we then obtain the following results.

$$\mathbf{D}^{(x)}_{q,k} = \frac{\partial \mathbf{S}_{q,k}}{\partial x_0} = \frac{\partial \mathbf{a}_q(\theta_{q,0})}{\partial x_0} g_{q,0} e^{-j\omega_{q,0}k}, \tag{9}$$

$$\mathbf{D}^{(y)}_{q,k} = \frac{\partial \mathbf{S}_{q,k}}{\partial y_0} = \frac{\partial \mathbf{a}_q(\theta_{q,0})}{\partial y_0} g_{q,0} e^{-j\omega_{q,0}k}, \tag{10}$$

$$\mathbf{D}^{(\Omega)}_{q,k} = \frac{\partial \mathbf{S}_{q,k}}{\partial \Omega_q} = -jk\mathbf{A}_q\mathbf{G}_{q,0}\mathrm{diag}\left\{\mathbf{R}^k_{q,0}\right\}, \tag{11}$$

$$\mathbf{D}^{(\mathbf{g})}_{q,k} = \left[\frac{\partial \mathbf{S}_{q,k}}{\partial \Re(\mathbf{g}_q)}\ \frac{\partial \mathbf{S}_{q,k}}{\partial \Im(\mathbf{g}_q)} \right] = \left[\mathbf{A}_q\mathrm{diag}\left\{\mathbf{R}^k_{q,0}\right\}\ j\mathbf{A}_q\mathrm{diag}\left\{\mathbf{R}^k_{q,0}\right\} \right], \tag{12}$$

and

$$\mathbf{D}^{(\Theta)}_{q,k} = \frac{\partial \mathbf{S}_{q,k}}{\partial \Theta_q} = \left[\frac{\partial \mathbf{a}_q(\theta_{q,1})}{\partial \theta_{q,1}}\ \frac{\partial \mathbf{a}_q(\theta_{q,2})}{\partial \theta_{q,2}} \cdots \frac{\partial \mathbf{a}_q(\theta_{q,N_P-1})}{\partial \theta_{q,N_P-1}} \right] \mathbf{G}_{q,1}\mathrm{diag}\left\{\mathbf{R}^k_{q,1}\right\}. \tag{13}$$

Let $\mathbf{S}_k$ denote the column vector of complete observation data for the k-th subcarrier:

$$\mathbf{S}_k = \left[\mathbf{S}^{\mathrm{T}}_{1,k}\ \mathbf{S}^{\mathrm{T}}_{2,k} \cdots \mathbf{S}^{\mathrm{T}}_{Q,k} \right]^{\mathrm{T}}. \tag{14}$$

Also, let $\mathbf{D}^{(*)}_{q,k} = \left[\mathbf{D}^{(\Omega)}_{q,k}\ \mathbf{D}^{(\mathbf{g})}_{q,k}\ \mathbf{D}^{(\Theta)}_{q,k} \right]$, and $\mathbf{E}_k = \partial\mathbf{S}_k/\partial\mathbf{u}$. Thus,

$$\mathbf{E}_k = \begin{bmatrix} \mathbf{D}^{(x)}_{1,k} & \mathbf{D}^{(y)}_{1,k} & \mathbf{D}^{(*)}_{1,k} & 0 & \cdots & 0 \\ \mathbf{D}^{(x)}_{2,k} & \mathbf{D}^{(y)}_{2,k} & 0 & \mathbf{D}^{(*)}_{2,k} & 0 & \vdots \\ \vdots & \vdots & \vdots & 0 & \ddots & 0 \\ \mathbf{D}^{(x)}_{Q,k} & \mathbf{D}^{(y)}_{Q,k} & 0 & \cdots & 0 & \mathbf{D}^{(*)}_{Q,k} \end{bmatrix}. \tag{15}$$

Then, based on [19], the Fisher information matrix for the parameter vector $\mathbf{u}$ can be computed by

$$\mathbf{I} = \sum_{\forall k} \frac{2}{\sigma^2}\Re(\mathbf{E}^{\mathrm{H}}_k\mathbf{E}_k), \tag{16}$$

where $(\cdot)^{\mathrm{H}}$ denotes the matrix conjugate and transpose, and σ^2 is the variance of $w_{q,k,m}$. Then, the CRLB for the location error can be computed by

$$\mathrm{CRLB} = \left[\mathbf{I}^{-1}\right]_{1,1} + \left[\mathbf{I}^{-1}\right]_{2,2}, \tag{17}$$

where $[\mathbf{K}]_{i,j}$ denotes the (i,j) element of matrix $\mathbf{K}$.

4 Numerical results

4.1 Assumptions

To gain fundamental insights on the system requirements and design principles, we obtain numerical results specific to the following assumptions.

4.1.1 Signal model

The following are assumed for the signal model. An antenna gain of 0 dB is assumed for the MD antenna as well as for each array element of a BS. The variance of $w_{q,k,m}$ depends only on the system noise, i.e., $E[\,|H_{q,k,m} - w_{q,k,m}|^2]\,/E[\,|w_{q,k,m}|^2]$ equals the average signal-to-noise ratio per OFDM subcarrier. The noise figure of the OFDM receiver is 4 dB. The number of antennas in an antenna array is a constant, i.e., $M_q = M, \forall q$. For the OFDM signal, the center frequency is 5.25 GHz, the subcarrier spacing Δ follows that employed in [20], i.e., $\Delta = 312.5$ kHz, and the set of used subcarrier indexes is $\{k| - K \leq k \leq K\}$, where K is an integer controlling the bandwidth. We consider the bandwidth without accounting a guard band, i.e., $B = (2K+1)\Delta = (0.625K + 0.3125)$ MHz.

It should be noted that the total bandwidth B_{T} practically allocated for the OFDM system is bigger than B. This is because guard bands are necessarily allocated to avoid interference from system of nearby spectrum. The guard bands may be implemented by having blocks of subcarriers with zero amplitudes at both ends of the active spectrum. Hence, $B_{\mathrm{T}} = (N_{\mathrm{u}} + N_{\mathrm{z}})\Delta$, where N_{u} is the number of used subcarriers $2K+1$, and N_{z} is the number of zero amplitude subcarriers. Then, we may note that while both the guard band and guard interval are necessarily allocated to deal with interference, they necessarily reduce the spectrum efficiency. Assume that the guard interval corresponds to cyclic prefix of N_{c} sample, where the sampling rate is the nominal one for OFDM, i.e., $1/B_{\mathrm{T}}$. Then, by the proper anticipation of guard band and interval, the average rate of effectively carried information symbols is down-scaled by $N_{\mathrm{u}}/(N_{\mathrm{u}} + N_{\mathrm{z}} + N_{\mathrm{c}})$ as noted in [21].

4.1.2 Geometrical models

A linear array with half-wavelength spacing between elements is assumed for each BS antenna array. Accordingly, the following geometrical reference model is applied. Cartesian coordinates (x_0, y_0) and (x_q, y_q) are used for referring to the location of respectively the MD and the array center of the q-th BS. The relation between the two locations is represented as shown in Fig. 1 by a polar coordinate $(r_q, \theta_{q,0})$. The orientation of an antenna array is represented by the normal looking direction of the array defined as follows. The normal looking direction is determined by first setting the phase shifts of all the antenna elements to be zero and then determining the main radiation direction that is closest to the direction of arrival of the MD. We use γ_q to represent the normal looking direction for the q-th BS, as a counterclockwise angle relative to the x-axis. For example, if the placement line of the q-th BS antenna array is on the x-axis in Fig. 1, γ_q will be $\pi/2$. Assuming r_q is much greater than the spacing between

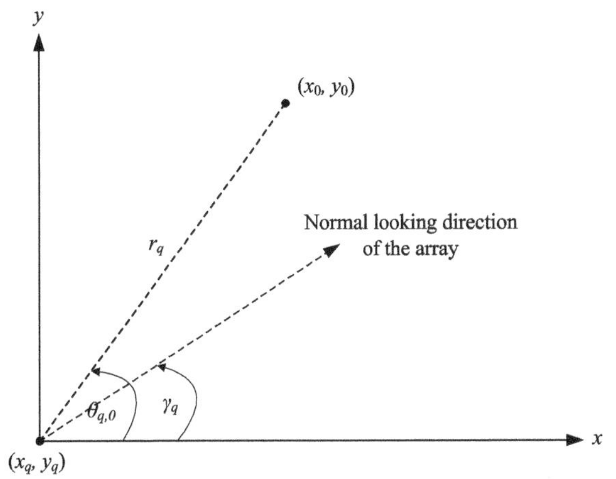

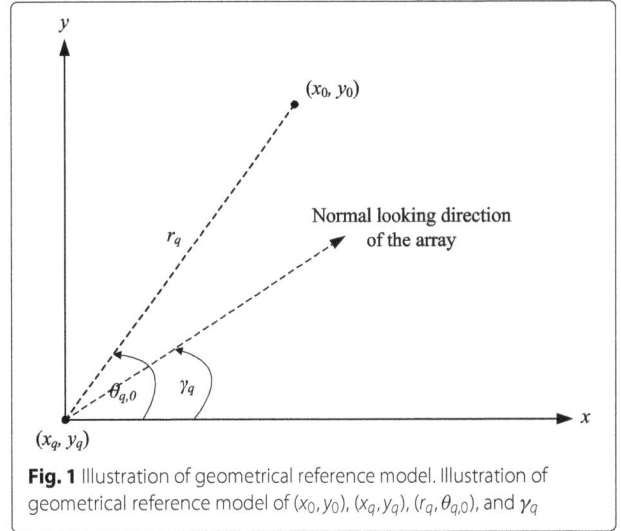

Fig. 1 Illustration of geometrical reference model. Illustration of geometrical reference model of (x_0, y_0), (x_q, y_q), $(r_q, \theta_{q,0})$, and γ_q

the antenna array elements, then $\phi_{q,n,m}, \forall(q,n,m)$ can be approximated by

$$\phi_{q,n,m} = \pi(m - (M+1)/2)\sin(\theta_{q,n} - \gamma_q). \tag{18}$$

Thus,

$$\frac{\partial \mathbf{a}_q(\theta_{q,n})}{\partial \theta_{q,n}} = -j\pi\cos(\theta_{q,n} - \gamma_q)\mathrm{diag}\,\{m - (M+1)/2\}_{m=1}^{M}\,\mathbf{a}_q(\theta_{q,n}), \tag{19}$$

$$\frac{\partial \mathbf{a}_q(\theta_{q,0})}{\partial x_0} = j\pi r_q^{-1}\sin(\theta_{q,0})\cos(\theta_{q,0} - \gamma_q)\;\mathrm{diag}\,\{m - (M+1)/2\}_{m=1}^{M}\,\mathbf{a}_q(\theta_{q,0}), \tag{20}$$

and

$$\frac{\partial \mathbf{a}_q(\theta_{q,0})}{\partial y_0} = -j\pi r_q^{-1}\cos(\theta_{q,0})\cos(\theta_{q,0} - \gamma_q)\;\mathrm{diag}\,\{m - (M+1)/2\}_{m=1}^{M}\,\mathbf{a}_q(\theta_{q,0}). \tag{21}$$

The performance results will be obtained by averaging the computed CRLBs over various values of the parameter vector $\mathbf{u}$. Figure 2 shows the geometrical assumptions, on which the variation of $\mathbf{u}$ is based. It shows that the region of interest for localization is a square with area of 400 m^2, there is a BS located at each corner, and the normal looking directions of the four arrays are pointed towards the region center.

4.1.3 Multipath propagation model

Radio wave propagation is subjected to multipath characteristics, which depend on the environment. In this paper, the multipath characteristics are varied, respecting density of the multipaths and relationship between the powers of direct and reflection paths, for investigation of their impacts on the CRLB performance. In this regard, the following statistical model is employed.

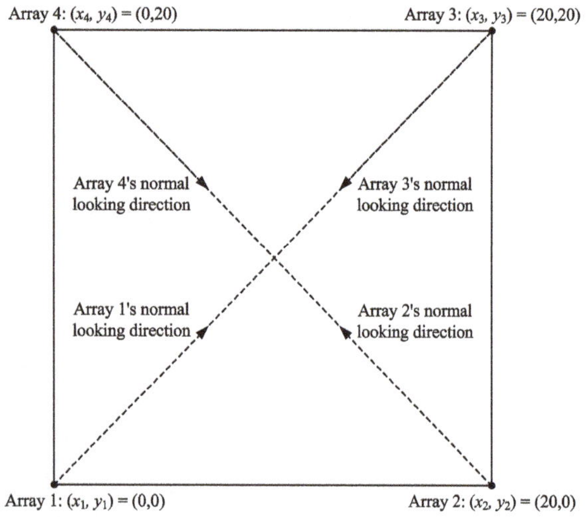

Fig. 2 Illustration of geometrical simulation assumptions. The 400-m^2 region of interest for localization and the normal looking directions of the antenna arrays of four BSs located at the corners

The direct path The direct path parameters for the q-th BS consist of only $g_{q,0}$ and $\omega_{q,0}$, because $\theta_{q,0}$ can be determined form x_0 and y_0. In this paper, $\angle g_{q,0}$ is drawn from a uniform distribution over $[0, 2\pi)$, and $|g_{q,0}|$ is obtained, based on the free-space path loss model. Assume the variance of $w_{q,k,m}$ is $\sigma^2 = 1$. Also, recall that the gain of each antenna is 0 dB. Thus,

$$|g_{q,0}|^2 = P_t \left(\frac{c}{4\pi f_c r_q}\right)^2 / (\kappa T_e B), \tag{22}$$

where P_t is the MD transmitted power, c is the light speed, $f_c = 5.25E+9$ Hz as assumed, κ is the Boltzmann constant, and T_e is the effective noise temperature of the BS receiver. For the value of $\omega_{q,0}$, we set $t_{q,0} = 0$, based on the discussion in Section 2.

The reflection paths In this paper, $\angle g_{q,n}$ and $\theta_{q,n}$, $\forall n > 0$ are independently drawn from a uniform distribution over $[0, 2\pi)$. In addition, $|g_{q,n}|$ and $\omega_{q,n}$, $n > 0$ are set to obey

$$\frac{|g_{q,n-1}|^2}{|g_{q,n}|^2} = \rho, \tag{23}$$

and

$$t_{q,n} - t_{q,n-1} = \delta, \tag{24}$$

where ρ reflects a constant power-decay condition, and δ^{-1} reflects the multipath density. The magnitude of a reflection path is not modeled to have a random characteristic, as it is treated as a controlled variable of the study. However, as the reflection paths are delayed relative to the direct path and their complex amplitudes have identical and independent random phases, the frequency response over the frequency band of the signal has a random frequency-selective fading characteristic. In this regard, we may note from (1) that the CFR for a specific subcarrier is a summation of independent complex zero mean circularly symmetric random variables. Hence, for large N_P, the frequency response over a subcarrier could exhibit frequency flat-fading characteristics with Rayleigh or Rician distribution, the common characteristics in practical radio propagation environments [22]. The Rayleigh distribution could occur when $|g_{q,0}|^2 \ll \sum_{n=1}^{N_P-1} |g_{q,n}|^2$, which is possible with $\rho \approx 1$. This is because when $\rho = 1$, we have $|g_{q,n}| = |g_{q,0}|$ for all $n > 0$ and accordingly the Rician K factor [22] is $1/(N_P-1) \approx 0$. On the other hand, the Rician distribution could occur when $|g_{q,0}|^2 \gg \sum_{n=1}^{N_P-1} |g_{q,n}|^2$, which is possible with $\rho \gg 1$.

4.2 Discussion

The mean square error (MSE) performance results presented here are obtained by averaging the computed CRLBs over various values of the parameter vector $\mathbf{u}$. In this regard, the MD is uniformly and randomly located in the 400-m^2 region, except the locations that are close to the center of an antenna array by less than 1 m. The performance will be presented for $P_t = -20$ dBm, respecting the infrastructure conditions (B and M) and the radio propagation characteristics (ρ, δ, and N_P). The performance respecting P_t is not presented, because it is obvious that the MSE-versus-P_t curve in this case is a line with a slope of -10 dB/decade. Note also that a study of the performance respecting geometry may be useful in determining the positions of BS antenna arrays [23]. Such study is beyond the scope of our work, and we believe it merits further investigation.

Figures 3 and 4 show the MSE performance in relation to B and M, for $\delta = 10$ ns, $\rho = 0$ dB, and $N_P = 5$. Note that changing M affects the quantity of observation

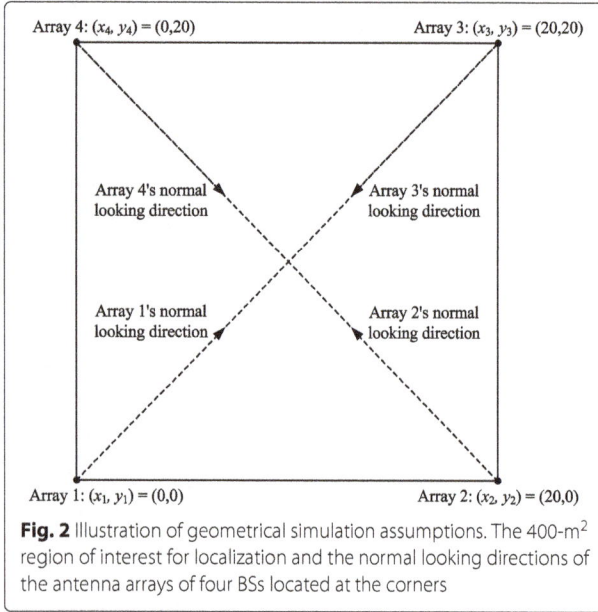

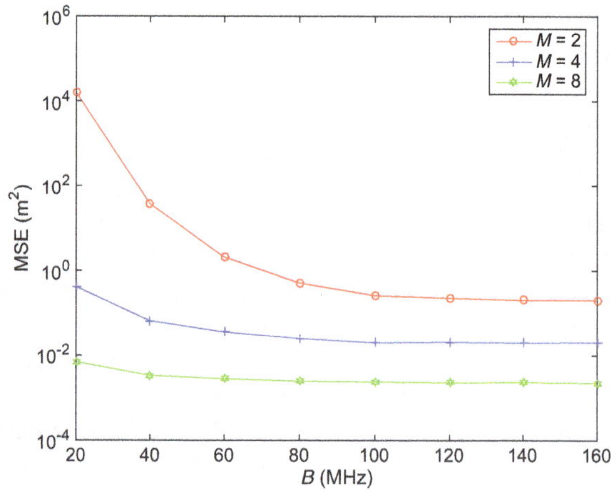

Fig. 3 Numerical result 1. MSE versus B with M as a parameter, for $\delta = 10$ns, $\rho = 0$ dB, and $N_P = 5$

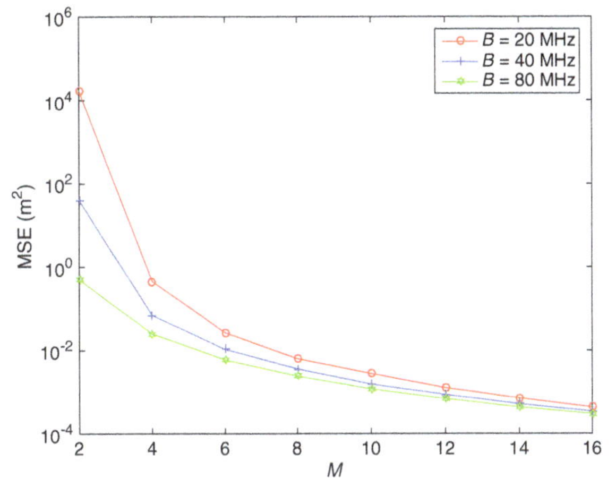

Fig. 4 Numerical result 2. MSE versus M with B as a parameter, for $\delta = 10$ ns, $\rho = 0$ dB, and $N_P = 5$

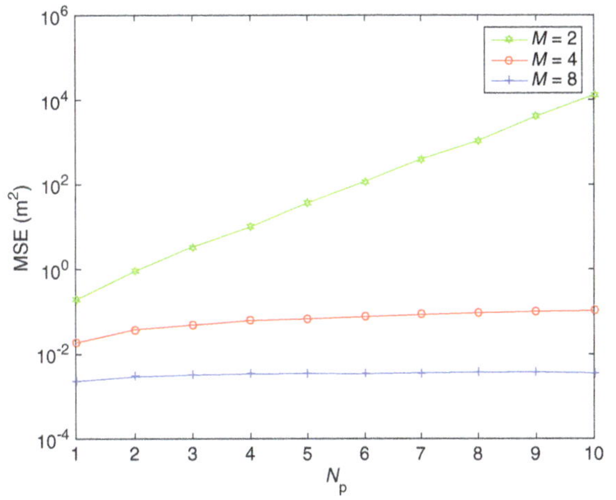

Fig. 5 Numerical result 3. MSE versus N_P with M as a parameter, for $B = 40$ MHz, $\delta = 10$ ns, and $\rho = 0$ dB

data and accordingly the number of rows of E_k that equals QM. Also, changing B affects the quantity of observation data and accordingly the number of summands in (16) that equals $2K+1$. In addition, by changing B, we must recompute $g_{q,n}, \forall(q,n)$, according to (22) and (23), for use in the construction of new E_k. Figure 3 shows that the MSE may be reduced by increasing B, but there is a limit where the MSE cannot be reduced further. According to the figure, that limit is found to be reachable at smaller B for larger M. That limit is also found to depend on M, as the limits for $M = 2, 4$, and 8 are respectively about 0.2094, 0.0201, and 0.0023 m^2. Figure 4 shows that the MSE may be reduced by increasing M. However, as M becomes larger, the MSE always decreases while the importance of B becomes lesser, as the MSE limit discussed above is being reached. Hence, the above observations may be summarized as follows. The MSE performance can be generally improved by increasing either B or M, but increasing B can only reduce the MSE to a limit depending on M. Increasing M alone can push the MSE to the limit, which is also smaller as M increases. In addition, with a larger M, a smaller B would be required for the MSE to approach the limit.

Figure 5 shows the MSE performance in relation to N_P and M, for $B = 40$ MHz, $\delta = 10$ ns, and $\rho = 0$ dB, which includes the free-space results ($N_P = 1$). Note that changing N_P affects the number of parameters for estimation and accordingly the number of columns of E_k that equals $(4N_P - 1)Q + 2$. Interestingly, it should be first noted that the free-space results in Fig. 5 and the MSE limits in Fig. 3 are same. This suggests that the MSE limit previously discussed corresponds to the performance virtually obtained from free-space channel. Hence, in this paper, we will refer to the free-space channel result as the free-space limit. This limiting characteristic of the MSE is consistent with the following principles. First, note from (1) that the contribution of a particular path is a complex sinusoidal function of k with a particular frequency depending on the path delay. In this regard, we should note that direct-path AOA estimation basically requires separating the direct-path contribution from reflection-path contributions, where large B is of great help. This is because of that the key difference between the paths is the sinusoidal frequency and that a less ambiguous result on such frequency determination basically requires observing the CFR over a larger range of k. Then, once B is large enough for perfect path separation, further increasing B while retaining the total transmitted power as assumed here is not seen to further improve the localization performance. This indicates that AOA estimation performance for a perfectly separated path depends on the total transmitted signal power, regardless of the signal bandwidth. Hence, increasing B by more spreading of the fixed transmitted power over the spectrum can only help the MVUE separate the multipaths, which eventually pushes the performance to approach the free-space limit. Figure 5 also shows that the logarithm of the MSE rises almost linearly with N_P for $M = 2$, while for other M, the MSE remains almost constant even if N_P is close to 10. The results for $N_P = 5$ in Fig. 5 are consistent with the results of the same case, i.e., $B = 40$ MHz, in Fig. 3. The results of the mentioned case from both figures consistently indicate that the MSE is far from the free-space limit for $M = 2$, while for other M, the MSE is already close to the free-space limit. Hence, if the MSE is already close to the free-space limit due to an appropriate provision of B and M to deal with the multipath environment, the MVUE will be robust against variation of N_P. Otherwise, the corresponding MSE will rise exponentially with N_P.

Figure 6 shows the MSE performance in relation to N_P and B, for $M = 2$, $\delta = 10$ ns, and $\rho = 0$ dB.

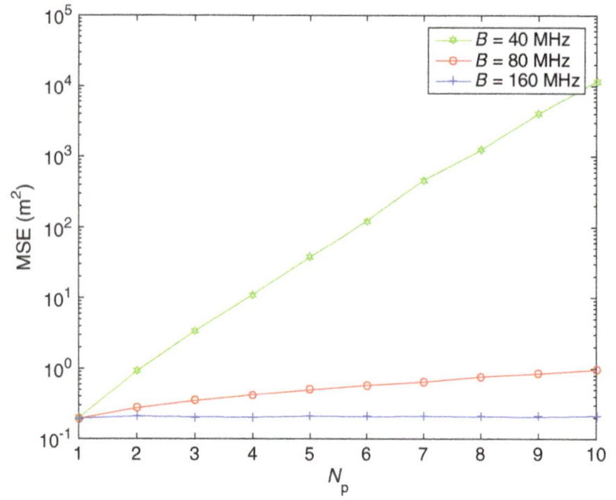

Fig. 6 Numerical result 4. MSE versus N_P with B as a parameter, for $M = 2$, $\delta = 10$ ns, and $\rho = 0$ dB

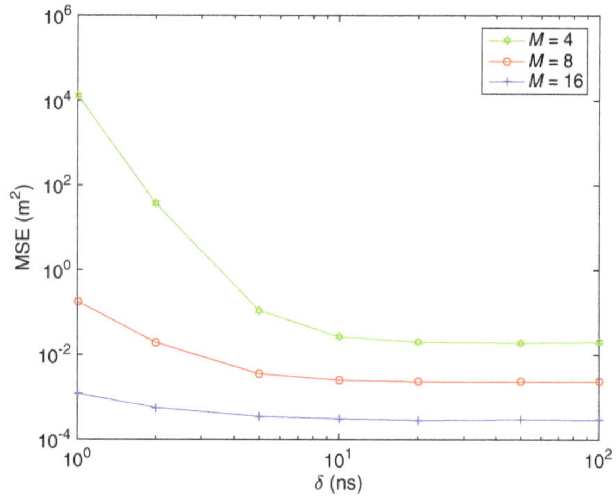

Fig. 7 Numerical result 5. MSE versus δ with M as a parameter, for $B = 80$ MHz, $\rho = 0$ dB, and $N_P = 10$

The figure provides same observations as previously discussed, regarding the characteristics of MSE against N_P and impact of increasing B on the MSE performance for multipath environments. Interestingly, it also shows that the free-space results do not depend on B. Theoretically, this may be justified by the following considerations. For $N_P = 1$, (1) becomes $H_{q,k,m} = g_{q,0}e^{-j(\omega_{q,0}k+\phi_{q,0,m})} + w_{q,k,m}$. The AOA-based localization basically relies on the estimation of $\phi_{q,0,m}$, $\forall(q,m)$, in which the AOA information is contained. As a result of [24], we note that the CRLB of $\phi_{q,0,m}$ based on the $(2K+1)$ sample observation of $H_{q,k,m}$ is inversely proportional to $|g_{q,0}|^2(2K+1)$ for large K. Recall that $B = (2K+1)\Delta$, and $|g_{q,0}|^2B$ is proportional to P_t according to (22). Therefore, $|g_{q,0}|^2(2K+1)$ is also proportional to P_t. In addition, the CRLB of $\phi_{q,0,m}$ depends on P_t, instead of B. Note also that by first-order approximation, the association between a small error of $\phi_{q,0,m}$ and the resulting localization error is approximately linear. Therefore, the free-space localization CRLB results also depend on P_t, instead of B.

Finally, regarding comparison between the impacts of B and M on the MSE performance, we note from Figs. 3, 4, 5 and 6 that doubling M gives greater performance benefit than doubling B for all cases observable from the figures. This should come from the following principles of AOA-based localization using (1). Increasing B while maintaining P_t only helps the MVUE separate the multipaths. On the other hand, increasing M not only provides more observation data that help multipath separation but also improves AOA sensitivity and therefore the free-space limit.

Figure 7 shows the MSE performance in relation to δ and M, for $B = 80$ MHz, $\rho = 0$ dB, and $N_P = 10$, while Fig. 8 shows the MSE performance in relation to δ and B, for $M = 4$, $\rho = 0$ dB, and $N_P = 10$. Note that changing δ does not affect either the size of E_k or the number of summands in (16), because it does not affect either quantity of observation data or the number of parameters for estimation. By changing δ , we only recompute $\omega_{q,n}$, $n > 0, \forall q$ to obey (24) for use in the construction of new E_k. Below are the reported various observations from the figures. The multipath density has a profound impact on the MSE performance, as the performance tends to get severely worse than the free-space limit when δ is lower than a certain boundary value that can be reduced by increasing B or M. For example, it can be seen from Fig. 7 that for $M = 4$, the MSE starts to get worse than the free-space limit when the δ is decreasing below a value between 5 and 10 ns, while for a larger M, such characteristic occurs at a smaller δ. It can be also seen from Fig. 8

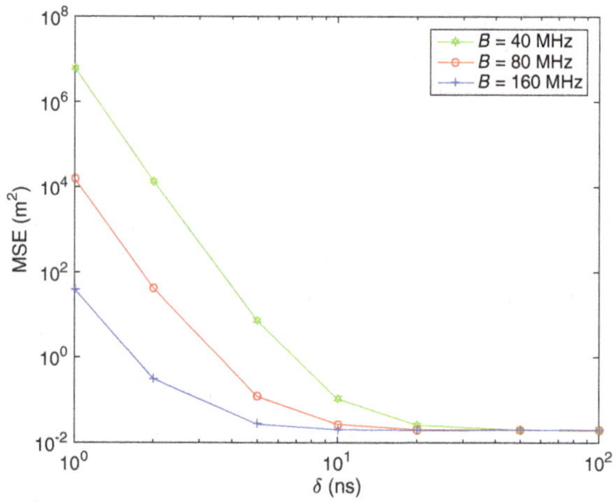

Fig. 8 Numerical result 6. MSE versus δ with B as a parameter, for $M = 4$, $\rho = 0$ dB, and $N_P = 10$

that for $B = 40$ MHz, the MSE starts to get worse than the free-space limit when the δ is decreasing below a value between 10 and 20 ns, while for a larger B, such characteristic occurs at a smaller δ. It may be also noted from Fig. 7 that in addition to reducing the boundary value, provision for a larger M also reduces the rate of deterioration when the δ decreases below beyond the boundary value. Again, regarding comparison between the impacts of B and M on the MSE performance, we see that increasing M provides greater performance benefit.

Figure 9 shows the MSE performance in relation to ρ and M, for $B = 80$ MHz, $\delta = 2$ ns, and $N_P = 10$. Note that changing ρ does not affect either the size of E_k or the number of summands in (16). By changing ρ, we only recompute $g_{q,n}$, $n > 0, \forall q$ to obey (23) for use in the construction of new E_k. The figure shows that all curves are essentially flat, and therefore, the MSE seems to be independent of ρ. Comparing the MSE results in Fig. 9 and the free-space limits observable from Fig. 7, we see that the MSE at $\delta = 2$ ns is farthest to the free-space limit when $M = 4$, while it is nearest when $M = 16$. The flat characteristic of the curves in Fig. 9 exists, no matter how far from the free-space limit the MSE is. This may sound surprising, because the channel at $\rho = 20$ dB is very much alike a free-space channel, and for this condition, one can easily design an algorithm that performs very close to the free-space limit. An algorithm that simply treats the channel as a free-space channel should be able to achieve such performance, because that insignificant model mismatch should only cause insignificant bias to the estimation result. This also means that when the modeled reflection paths are weak, the MVUE could perform surprisingly as poor as the case where the paths are strong. Such behavior of the MVUE, which embraces all modeled paths into its idealized unbiased estimation task, will be mathematically justified soon in this section. Hence, we suggest that reflection paths with insignificant amplitude should not be included explicitly in the channel model. This is to maintain that the CRLB may be also meaningful in a study of biased estimation algorithms.

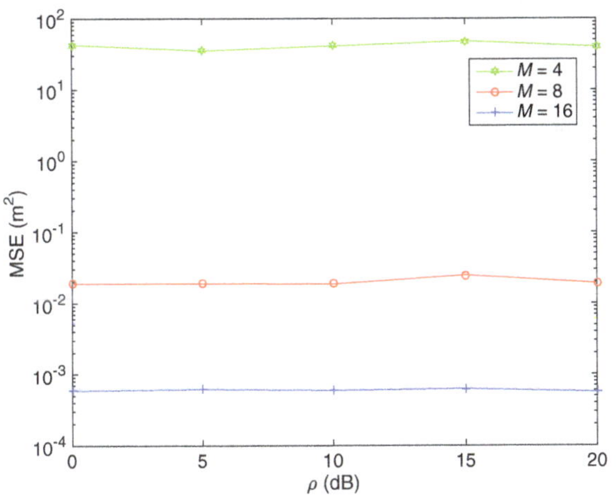

Fig. 9 Numerical result 7. MSE versus ρ with M as a parameter, for $B = 80$ MHz, $\delta = 2$ ns, and $N_P = 10$

We may claim more strictly from the results in Fig. 9 that the localization performance of the MVUE is independent of the strength of a reflection path. This can be mathematically justified by the following considerations. Note that the gain of a particular reflection path appears as a constant factor consistently for all non-zero elements along only two columns of E_k. For a specific $g_{q,n}$, $n > 0$, such two columns correspond to respectively the $(n+1)$-th column of $\mathbf{D}_{q,k}^{(\Omega)}$ and the n-th column of $\mathbf{D}_{q,k}^{(\Theta)}$. The indexes of them within E_k are always greater than two, because they all belong to $\mathbf{D}_{q,k}^{(*)}$ in (15). Therefore, by scaling the value of the $g_{q,n}$ with a real factor, E_k can be recomputed accordingly by post-multiplying itself with a real constant diagonal matrix, where all diagonal elements are equal to one except the elements associated with the two indexes. The inverse of the Fisher information matrix can then be recomputed by pre- and post-multiplying itself with the inverse of that diagonal matrix. Such operation will not affect the localization performance of the MVUE computed by (17), because the two indexes are always greater than two.

We have been realizing from the results of Figs. 7 and 8 that the multipath density reflected by δ is crucial to the effectiveness of the infrastructure reflected by B and M. However, we consider that care may be only required for the second path that has delay closest to the direct path, because it contributes most to the difficulty of separating the direct path from other paths. Thus, it may be more interesting to investigate the case of channels where the path delay rule in (24) is replaced by

$$t_{q,n} - t_{q,n-1} = \begin{cases} \delta_0, & \text{if } n = 1 \\ \delta_1, & \text{otherwise.} \end{cases} \qquad (25)$$

Figure 10 shows the MSE performance in relation to δ_1 and B, for $M = 4$, $\rho = 0$ dB, $N_P = 10$, $\delta_0 = 10$ ns, and $\delta_1 \leq 10$ ns, which can then be compared with the cases $\delta_0 = \delta_1 \leq 10$ ns presented in Fig. 8. We can then see that the performance degradation due to moving from $\delta_0 = \delta_1 = 10$ ns to $\delta_0 = 10$ ns with $\delta_1 = 1$ ns is much less than the degradation due to moving from the same original condition to $\delta_0 = \delta_1 = 1$ ns. Hence, allowing the reflection paths to be close in delay between themselves could not degrade the performance so much, compared to allowing significant reflection paths to be close in delay to the direct path. It is then very important to give much care for each significant reflection path with delay close to that of the direct path. Note that a position of scatterer that could produce such a reflection path can only be around the straight line connecting the

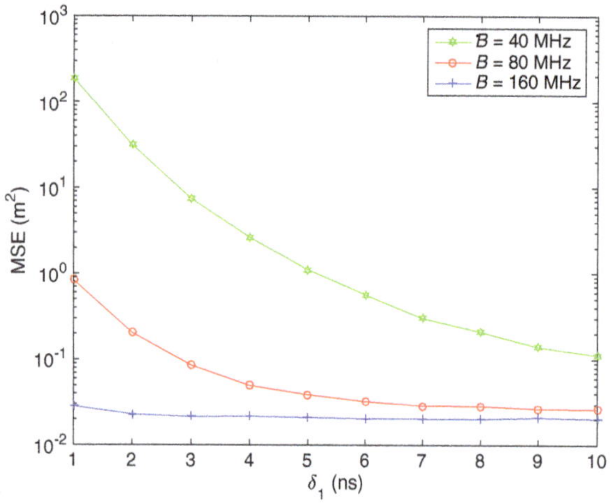

Fig. 10 Numerical result 8. MSE versus δ_1 with B as a parameter, for $M = 4$, $\rho = 0$ dB, $N_P = 10$, and $\delta_0 = 10$ ns

MD and the BS. For such a position, we may further note that everyone around the BS is most cumbersome. This is because the corresponding reflection path AOA can be considerably different from the direct-path AOA, and then, the reflection path needs to be modeled explicitly as having such different AOA but similar delay to that of the direct path. On the other hand, for such a scatterer position rather than those around the BS, the corresponding reflection path AOA cannot be considerably different from the direct-path AOA. Accordingly, by neglecting the seen insignificant difference on both the delay and AOA, the reflection path can be approximately modeled by merging itself with the direct path based on only complex path-amplitude addition. Therefore, it is important to avoid the most cumbersome scatterers by ensuring that the BS antenna arrays are carefully placed not close to any significant reflective materials, and in some cases, using radio-absorptive materials may be helpful.

5 Conclusions

We studied the problem of AOA-based localization using MIMO-OFDM channel state information as observation data. In particular, we derived a method to compute the CRLB which is a fundamental limit on the localization MSE and applied it to obtain fundamental insights into the problem. We found that the CRLB is independent of the strength of a reflection path. Therefore, we suggested that reflection paths with insignificant amplitude should not be included explicitly in the channel model. Provided that the mobile device-transmitted signal power is fixed, the following may be concluded from our investigation. The MSE localization performance can be improved by increasing the transmitted signal bandwidth B alone. However, there is a limit where further increasing B cannot further improve the performance. The limit is found to equal the performance virtually obtained from the free-space radio propagation assumption. The performance can be considerably improved also by increasing the base station antenna array size M alone, as, when M increases, not only does the free-space limit decrease but also the MSE gets closer to the decreasing limit. In particular, scaling M always has greater impact on the MSE than scaling B. Provided that the MSE is close to the free-space limit, the optimum unbiased estimator will be robust to variation on the number of multipaths N_P. Otherwise, the MSE will rise exponentially with N_P. The multipath density that is inversely proportional to δ has a profound impact on the MSE performance, as the performance tends to get severely worse than the free-space limit when the density is higher than a value proportional to B and M. Finally, when significant reflection paths with AOAs very different from that of the direct path have similar delays to that of the direct path, the localization performance could be severely degraded.

Acknowledgements

The authors would like to thank the anonymous reviewers for their constructive comments, which helped a lot to improve the presentation of this paper.

Funding

Not applicable.

Authors' contributions

Both authors contributed to the searching of the literature. TD contributed to theoretical analysis. Both authors contributed to the design and implementation of computer simulation experiments and interpretation of the results. Both authors have read and approved the final manuscript.

Competing interests

The authors declare that they have no competing interests.

References

1. H Liu, H Darabi, P Banerjee, J Liu, Survey of wireless indoor positioning techniques and systems. IEEE Trans. Syst. Man Cybern. Part C Appl. Rev. **37**(6), 1067–1080 (2007)
2. Z Farid, R Nordin, M Ismail, Recent advances in wireless indoor localization techniques and system. J. Comput. Netw. Commun. **2013**(185138), 1–12 (2013)
3. K Yu, I Sharp, YJ Guo, *Ground-based wireless positioning*, 1st edn. (Wiley, West Sussex, 2009)
4. Y Zhao, K Liu, Y Ma, Z Li, An improved k-NN algorithm for localization in multipath environments. EURASIP J. Wirel. Commun. Netw. **2014**(208), 1–10 (2014)
5. MB Zeytinci, V Sari, FK Harmanci, E Anarim, M Akar, Location estimation using RSS measurements with unknown path loss exponents. EURASIP J. Wirel. Commun. Netw. **2013**(178), 1–14 (2013)
6. P Bahl, VN Padmanabhan, in *IEEE INFOCOM*. RADAR: an in-building RF-based user location and tracking system (The Institute of Electrical and Electronics Engineers (IEEE), Piscataway, 2000), pp. 775–784

7. M Youssef, A Agrawala, in *Proceedings of the 3rd International Conference on Mobile Systems, Applications, and Services*. The Horus WLAN Location Determination System (Association for Computing Machinery (ACM), New York, 2005), pp. 205–218
8. K Wu, J Xiao, Y Yi, D Chen, X Luo, LM Ni, CSI-based indoor localization. IEEE Trans. Parallel Distrib. Syst. **24**(7), 1300–1309 (2013)
9. T Demeechai, P Kukieattikool, T Ngo, T-G Chang, Localization based on standard wireless lan infrastructure using mimo-ofdm channel state information. EURASIP J. Wirel. Commun. Netw. **2016**(146), 1–16 (2016)
10. NB Rejeb, I Bousnina, MBB Salah, A Samet, Joint mean angle of arrival, angular and Doppler spreads estimation in macrocell environments. EURASIP J. Adv. Signal Process. **2014**(133), 1–10 (2014)
11. SO Al-Jazzar, A Muchkaev, A Al-Nimrat, M Smadi, Low complexity and high accuracy angle of arrival estimation using eigenvalue decomposition with extension to 2D AOA and power estimation. EURASIP J. Wirel. Commun. Netw. **2011**(123), 1–13 (2011)
12. Y Shen, MZ Win, Fundamental limits of wideband localization: Part I: a general framework. IEEE Trans. Inf. Theory. **56**(10), 4956–4980 (2010)
13. H Godrich, AM Haimovich, RS Blum, Target localization accuracy gain in mimo radar-based systems. IEEE Trans. Inf. Theory. **56**(6), 2783–2803 (2010)
14. K Witrisal, E Leitinger, S Hinteregger, P Meissner, Bandwidth scaling and diversity gain for ranging and positioning in dense multipath channels. IEEE Wirel. Commun. Lett. **5**(4), 396–399 (2016)
15. H Minn, N Al-Dhahir, Optimal training signals for MIMO OFDM channel estimation. IEEE Trans. Wirel. Commun. **5**(5), 1158–1168 (2006)
16. H Zhu, J Wang, Chunk-based resource allocation in OFDMA systems—part I: chunk allocation. IEEE Trans. Commun. **57**(9), 2734–2744 (2009)
17. H Zhu, J Wang, Chunk-based resource allocation in OFDMA systems—part II: Joint chunk, power and bit allocation. IEEE Trans. Commun. **60**(2), 499–509 (2012)
18. H Zhu, Radio resource allocation for OFDMA systems in high speed environments. IEEE J. Sel. Areas Commun. **30**(4), 748–759 (2012)
19. SM Kay, *Fundamentals of statistical signal processing: estimation theory*. (Prentice Hall, New Jersey, 1993), p. 47
20. IEEE Std 802.11ac - 2013, IEEE Standard for Information Technology-Telecommunications and information exchange between systemsŮlocal and metropolitan area networks-specific requirements, Part 11: wireless LAN medium access control (MAC) and physical layer (PHY) specifications, Amendment 4: enhancements for very high throughput for operation in bands below 6 GHz. (The Institute of Electrical and Electronics Engineers, Inc., New York, NY, USA, 2013)
21. SK Chronopoulos, C Votis, V Raptis, G Tatsis, P Kostarakis, in *AIP Conference Proceedings 1203*. in depth analysis of noise effects in orthogonal frequency division multiplexing systems, utilising a large number of subcarriers (American Institute of Physics (AIP), New York, 2010), pp. 967–972
22. SK Chronopoulos, V Christofilakis, G Tatsis, P Kostarakis, Performance of turbo coded OFDM under the presence of various noise types. Wirel. Pers. Commun. **87**(4), 1319–1336 (2016)
23. W Shi, X Qi, J Li, S Yan, L Chen, Y Yu, X Feng, Simple solution to the optimal deployment of cooperative nodes in two-dimensional TOA-based and AOA-based localization system. EURASIP J. Wirel. Commun. Netw. **2017**(79), 1–16 (2017)
24. DC Rife, RR Boorstyn, Single-tone parameter estimation from discrete-time observations. IEEE Trans. Inf. Theory. **20**(5), 591–598 (1974)

12

A feature selection method based on synonym merging in text classification system

Haipeng Yao[1*], Chong Liu[1], Peiying Zhang[1,3] and Luyao Wang[2]

Abstract

As an important step in natural language processing (NLP), text classification system has been widely used in many fields, like spam filtering, news classification, and web page detection. Vector space model (VSM) is generally used to extract feature vectors for representing texts which is very important for text classification. In this paper, a feature selection algorithm based on synonym merging named SM-CHI is proposed. Besides, the improved CHI formula and synonym merging are used to select feature words so that the accuracy of classification can be improved and the feature dimension can be reduced. In addition, for feature words selected by SM-CHI, this paper presented three weight calculation algorithms to explore the best feature weight update method. Finally, we designed three comparative experiments and proved the classification accuracy is the highest when choosing the improved CHI formula 2, set the threshold α to 0.8 and use the largest weight among the synonyms to update the feature weight, respectively.

Keywords: Text classification, Feature selection, Synonym merging, Feature weights calculation

1 Introduction

With the development of the Internet, the amount of Chinese text information shows an exponential growth trend. How to effectively manage the massive Chinese documents and mine the information contained in the documents has become a critical research problem. Automatic text classification can complete the work of text processing effectively. It also plays an important role in natural language processing (NLP) and data mining.

The most common method used in text classification is the vector space model (VSM). It represents text as a feature vector. The specific process is shown in Fig. 1.

From Fig. 1, we know that the first step in Chinese text classification is to preprocess the text, including word segmentation, part of speech tagging, and removal of stop words. The purpose is to remove the useless words and only leave the nouns, adjectives, and verbs that contain category information. After this, the text can be represented as a vector to form VSM. Then, we use the feature selection method to select the feature words that can symbolize the text categories, and merge the synonym to reduce dimensions. Next, TF-IDF [1] method is used to calculate the weight of each feature of each text to transform the text into a feature vector. Last but not least, by using the Bayesian classifier to train the sample data, we can get the final text classifier.

Feature selection is the most important step because the selected feature words directly affect the accuracy of the classifier. In VSM, the best feature selection method is χ^2 statistics (CHI) [2, 3]. But the defect is high-dimensional feature vectors selected by CHI may cause dimension disaster. The writer consider to merge the synonyms among the feature words selected by CHI so that the dimension of feature space can be reduced. Then, in the next step, an improved TF-IDF method is used to calculate the feature weights for each word to generate the feature vector of each text. This paper mainly studies the influence of feature selection and synonym merging on the accuracy of classification in automatic text classification. Synonym merging can reduce the dimension of the feature space and improve the classification performance. The study of feature selection algorithm and synonym merging has a strong practical significance. The main contributions of this paper are as follows:

*Correspondence: yaohaipeng@bupt.edu.cn
[1] State Key Laboratory of Networking and Switching Technology, Beijing University of Posts and Telecommunications, Haidian District Xitucheng Road 10, Beijing 100876, People's Republic of China
Full list of author information is available at the end of the article

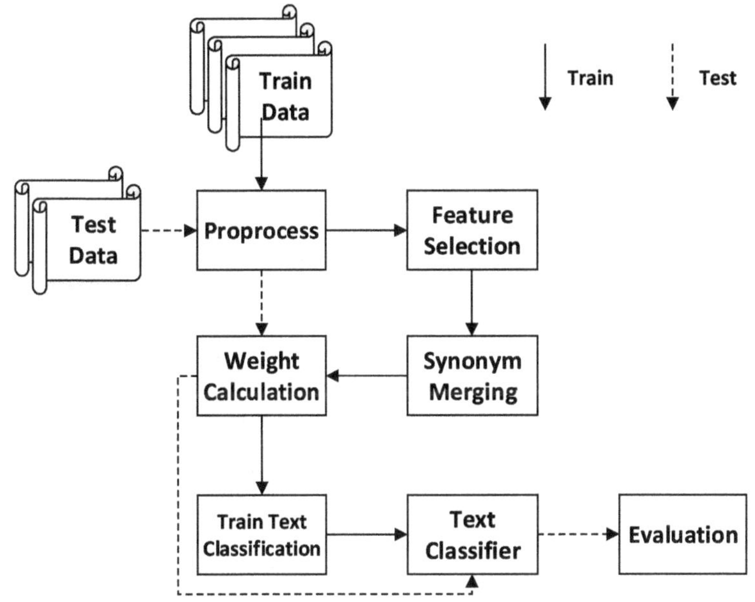

Fig. 1 Text classification model framework

1. We presented a new feature selection algorithm named SM-CHI based on an improved CHI [4] formula and synonym merging to achieve efficient feature selection and dimension reduction.
2. We found that the original CHI formula multiplied by a log term has the best classification performance comparing with the original CHI and two improved CHI algorithms [4, 5].
3. The choice of thresholds α ($0 \leqslant \alpha \leqslant 1$) is critical. Only the most similar feature words will be merged when α is close to 1, so we use grid search method to find the optimal α. The result show that the classification accuracy is highest when α is equal to 0.8.
4. We proposed three improved weight calculation methods based on TF-IDF. The experimental results show that using the maximum value of the synonym group as the feature weight is the best way.

The organization of this paper is as follows. Next section discusses some related works. The text classification system based on semantic similarity is introduced in Section 3. We present the details of SM-CHI and the corresponding weight calculation methods in Section 4. In Section 5, we show and discuss the experimental results. We conclude our work and describe future work in Section 6.

2 Related works

Text classification has been widely used in the classification and labeling of news and web pages. There are many works on text classification. The classification methods based on synonym merging have appealled much attention in the past 2 years. Next, we briefly introduce some of the existing research results.

2.1 Classification model

Nowadays, most of the text classification methods are based on VSM where the texts are represented in the form of (feature vector, label). In this way, we can transform each text into feature vectors to construct the training and testing dataset. It is a long existing problem of machine learning to train a classifier for text classification. Many machine learning algorithms can be used as the classifier, such as Naive Bayes Method [6], k-Nearest Neighbor [7], Decision Trees [8], Support Vector Machines (SVM) [9], Markov Model [10], and Neural Networks [11]. KNN as one of the best classifiers in VSM is simple, effective, nonparameter. So we choose it as the classification method in this paper.

2.2 Feature selection algorithm

In order to construct a valid classifier, feature words selection is another main problem. Common feature selection methods include document frequency (DF), mutual information (MI) [12], a χ^2 statistics (CHI), information gain (IG) [6, 13], and term strength (TS) [14, 15]. The works in [2, 3] show that CHI is the best feature selection method through contrast experiments. However, CHI feature selection method also has shortcomings [16]. For example, the CHI value of the high-frequency words is very high, but they have no significant contribution to class distinctions. Therefore, the authors of [4, 5] presented two improved CHI formulas from different perspectives in order to make up for the lack of the original CHI method.

2.3 Synonym merging

With the development of text classification, some researchers start to propose a text classification system based on synonym merging to improve the accuracy of classification. The word similarity calculation methods based on "Tong YiCi Cilin" and "HowNet" were proposed in [17, 18] respectively. The work in [19] proposed a text feature selection method based on "TongYiCi Cilin" to reduce data's feature dimensions while ensuring data integrity and classification accuracy. A semantic kernel is used with SVM for text classification to improve the accuracy in [20, 21]. What is more, there are some other work in [22–24] that presents some excellent ideas, which is worth learning and reference when we are dealing with large-scale text classification. They can help us to speed up the calculation through big data technology.

The model proposed in this paper is a text classification model based on synonym merging, named SM-CHI. The difference with [19] is that we merge synonyms after feature selection based on CHI and we propose three improved weighting method for the merged feature words.

3 Text classification model based on semantic similarity

In this section, we mainly introduce the text classification model based on semantic similarity. Wherein, Section 3.1 describes the feature selection method based on χ^2 statistic. Section 3.2 introduces the method of synonym merging; Section 3.3 presents the traditional weight calculation method, TF-IDF.

3.1 Improved feature selection algorithm

In text classification, a feature word and its category tend to obey the CHI formula. Higher CHI value implies that a feature word has stronger ability to identify a category. The CHI value of word is calculated as follows [3]:

$$\chi^2(t,c) = \frac{N * (AD - BC)^2}{(A+C)(A+B)(B+D)(C+D)} \tag{1}$$

where N is the size of the training set; A is the number of documents that belong to class c and contain the word t; B is the number of documents that do not belong to class c but contain the word t; C is the number of documents that belong to the class c but do not contain the word t; and D is the number of documents that do not belong to class c and do not contain the word t.

Although the CHI formula has a relatively good performance in text classification, it also has some shortcomings [16]. First of all, high-frequency words that appear in all categories have higher CHI values, but they do not make much sense for class distinctions. Secondly, the CHI formula only considers the appearance of a word but not the frequency of the word in a document. Therefore, CHI formula also has "low frequency words flawed". For example, assuming word $t1$ appears in 99 documents, each appears 10 times; word $t2$ appears in 100 documents, each appears one time; obviously $t2$ has a higher CHI value, but in fact $t1$ is more representative for this category. There are many studies amended for its defects. The work in [4] proposed the multiplication by a log entry based on the original CHI to reduce the CHI value of high-frequency words. The formula is as follows:

$$chi\_imp\_1 = \log\left(\frac{N}{A+B}\right) * \chi^2(t,c) \tag{2}$$

where $A + B$ represents the number of documents that contain word t and N represents the total number of documents. In this case, the CHI value of the high-frequency words that appear in all categories are close to zero so that they would not be selected as a feature word.

In addition, the work in [5] has made a corresponding improvement to the word frequency, which is multiplied by term $\beta(t,c)$ on the basis of the original CHI formula. The calculation formula is as follows:

$$chi\_imp\_2 = \beta(t,c)\chi^2(t,c) \tag{3}$$

where $\beta(t,c)$ is calculated as follows:

$$\beta(t,c) = \frac{tf(t,c)}{\sum_{i=1}^{m} f(t,c_i)} \tag{4}$$

In the formula, m is the total number of categories and $tf(t,c)$ is the frequency of the word t in the category c.

3.2 Synonym merging algorithm based on "Tong YiCi Cilin"

Some of the feature words selected by CHI formula may be the same or have similar meaning. They have the same effect on class distinctions. If the synonym are merged, not only the classification accuracy will be improved, but also the dimension of the feature space can be reduced so that the efficiency of the algorithm can be improved. For example, "GanMao", "ZhaoLiang", "ShangFeng" are the synonym of "Cold" in Chinese. If the "Health Care" category articles contain these words respectively, then the feature words for the text classification contain too much redundant information. We use the method of synonym merging to deal with it.

In this paper, we use the "Tong YiCi Cilin" provided by Harbin Institute of Technology as the method of word similarity calculation [17]. Its structure has five layers. You can easily calculate the similarity between the two terms. The structure of "Tong YiCi Cilin" is shown in Fig. 2.

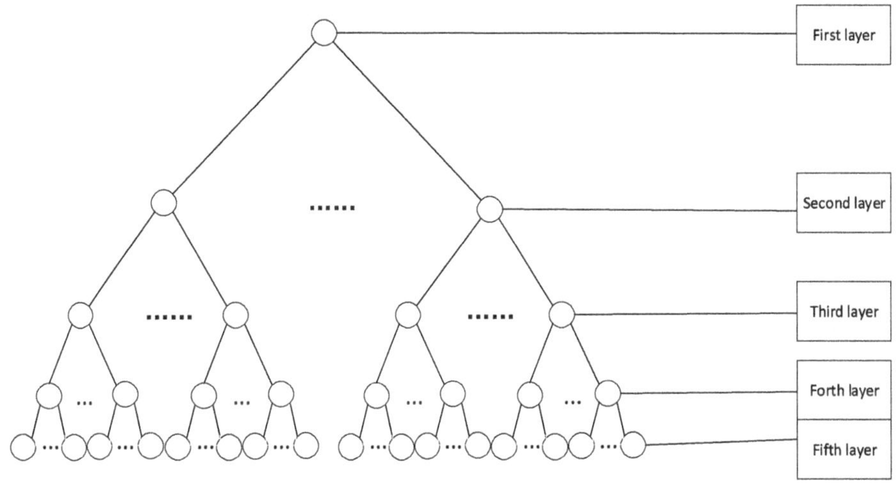

Fig. 2 Hierarchy structure of "Tong YiCi Cilin"

The concrete similarity calculation method is introduced in detail in [17]. When the similarity of two words is greater than threshold α, they will be regarded as a pair of synonym to merge. The optimal value of α will be discussed later in the experiment. In addition, all merged synonyms are stored in a list. Nested lists are used to store feature words so that all synonym information remains in the feature vector. To calculate the feature vector of each document, we propose three improved methods which will be discussed in Section 3.3.

3.3 Weight calculation method

Traditional TF-IDF weight calculation formula [1] is as follows:

$$weight_{t,d} = tf_{t,d} + idf_t \tag{5}$$

This formula represents the weight calculation method for word t in document d. Here, $tf_{t,d}$ denotes the frequency of occurrence of word t in d, and idf_t denotes the anti-document frequency of t, which is used to quantify the distribution of t in the training set. If n is used to denote the number of documents which contain t in the training set, the calculation formula of idf_t is as follows:

$$idf_t = \log\left(\frac{N}{n}\right) \tag{6}$$

As we mentioned above, the TF-IDF method is used to calculate the weight of each feature word in each text.

4 Algorithm description

4.1 Feature selection method based on the synonym merging

In this section, we introduce the feature selection algorithm (SM-CHI). This method firstly selects candidate feature words based on an improved CHI formula, and then merges synonyms to re-select those feature words that can represent the categories better and reduce dimension. The method is represented as the following formula:

$$SM - CHI = LF(t) * CHI(t) * SM(t) \tag{7}$$

where $LF(t)$ denotes whether the word t exists in the word bag or not, and is mainly decided according to the part of speech and stopping words. If the word t is a stopping word and in the part of speech that does not belong to verb, noun, and adjective, $LF(t) = 0$, otherwise $LF(t) = 1$. $CHI(t)$ represents the CHI value of the word t and is calculated by Eq. (2). $SM(t)$ indicates whether the word t contains synonym. If yes, it needs to merge all of its synonyms.

Firstly, all the texts in the training set are preprocessed, including Chinese word segmentation, part-of-speech tagging, and discarding stop words. The remaining words constitute the word bag of the training set. Secondly, we calculate the CHI value of each word. Choose the first 200 words from each category to form candidate sets of feature words. Note that the characteristic words selected for each category may be duplicated. The candidate set is stored using the HashSet (a data structure) and the de-emphasis is performed. After obtaining the candidate set, the similarity between each word is calculated according to "Tong YiCi Cilin" and threshold is set to α. The synonym merging is performed only when the word similarity is greater than α. We will experimentally determine the optimal value of hyper-parameter α. The pseudocode of SM-CHI is shown in Algorithm 1.

4.2 Improved method for calculating eigenvalue weight

In the scene of SM-CHI feature selection method presented in this paper, the traditional TF-IDF formula has some drawbacks. For the features after synonym merging,

Algorithm 1 SM-CHI feature selection algorithm

```
Input:a training set D
Output:a feature space F
(Initialization) A, B, TFIDF can all be a null dict
for each file in D:
    word_list=file.process()
    for word in word_list:
        A[file.class][word] += 1
        TFIDF[file.class][file.num][word] +=1
end for
Calculate B in the CHI formula from A
for cla in
    for word in cla:
        Calculate the CHI value according to formula 2
    Selects the first 200 words as the feature
end for
Combine features of the 9 categories to obtain word_features
for word1, word2 in word_features:
    sim=calcWordsSimilarity(word1, word2)
    if sim > α:
        Merge word1 and word2
end for
```

the original weight calculation formula will cause "unfairness". Because the merged feature words are stored in the nested list, so which word among them will be regarded as the feature is a question. For this problem, we present the following three solutions:

- Sum the weights of all items up in the feature list of each dimension as the weight of the list;
- Take the largest weight among the synonym as the weight value of the feature;
- Multiply the first item by 1.1 for times of the number of items in the feature list.

5 Experiments and results

In this section, three groups of experiments are designed to evaluate the performance of three CHI formulas, the optimal threshold α for synonym merging, and the performance of feature weight update method. We use the whole news data set from Sogou Lab [25] to test the accuracy of the experiment.

5.1 Performance evaluation and data set

The standard precision rate P, recall rate R, and $F1$ score are used to measure the classification performance. For the i_th category, the formula is as follows [26]:

$$P_i = \frac{TP}{TP+FN}, R_i = \frac{TP}{TP+FP}, F1_i = \frac{2*P_i*R_i}{P_i+R_i} \tag{8}$$

where TP is the number of documents correctly classified as class i, FP is the number of documents classified as class i but not actually i, and FN is the number of documents that is not classified as class i but is actually class i.

This article will use the whole network news data set provided by Sogou Lab to test our experiments. The corpus includes nine kinds of news types, such as Automobile, Finance, and IT. Each category contains thousands of documents. In this experiment, each category takes 400 documents, of which 280 are training set and 120 are test sets. Therefore, the training set contains 2520 documents, and the test set includes a total of 1080 documents.

The preprocessing module uses a third-party library for python, named jieba, to complete the work of word segmentation, part of speech tagging, and discarding of stop words. In addition, we use Naive Bayesian classifier provided in Python's NLTK library as the classifier.

5.2 Experiments and results

In this section, the following three groups of experiments are carried out to test the three innovation points of SM-CHI with control variable method . In Experiment I, we test the three feature selection algorithms without using synonym merges. In Experiment II, we use the first improved CHI method to select features and use grid search method to find the optimal threshold α. On the basis of Experiment I and II, we designed Experiment III to find the best weight update method.

5.2.1 *Experiment I*

In order to test the effect of three kinds of feature selection methods described in section 3.2, we conducted the following experiments. But we did not use synonym merging here. The results of experiment are shown in Fig. 3:

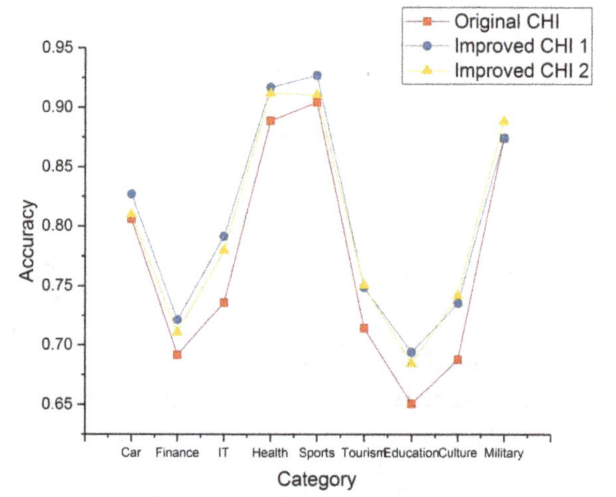

Fig. 3 Classification results comparing three kinds CHI formula

From the results, we can see that the two improved CHI formulas have a great effect on enhancing the value of F1 score of each category as compared to the original CHI formula, which means that the improved CHI formulas can select more representative words. They both make some improvement based on the original CHI. In addition, when the two improved CHI formulas are compared, the first improved method has a slight advantage, showing a better discrimination effect in the preceding categories. The result also shows that the log term successfully suppresses the CHI values of the high-frequency words appearing in all classes, which achieves relatively good results. Therefore, we will use the first improved CHI formula as our base feature selection method in the fellow experiment.

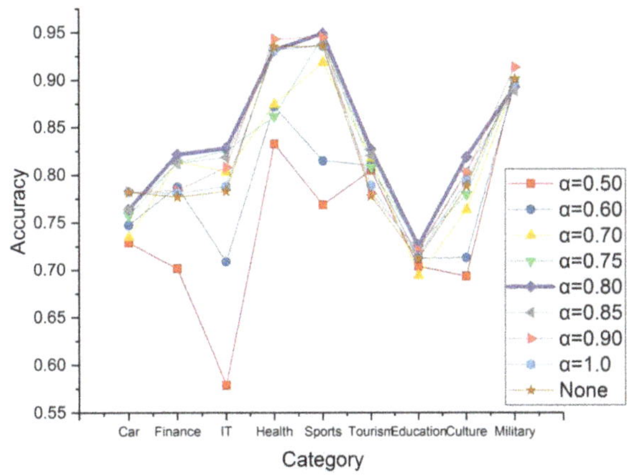

Fig. 4 F1 score of each category when α takes different values

5.2.2 *Experiment II*

In order to select the appropriate threshold α for synonymy merging, we designed the following experiment. The first improved CHI formula was used for feature selection, and the range of α is set to [0.5, 0.6, 0.7, 0.75, 0.8, 0.85, 0.9,1.0]. A total of nine experiments are conducted, including a comparative experiment that did not use synonym merging. The experimental results are shown in Table 1 and Fig. 4.

From the results in Figs. 4 and 5, we can draw the conclusion that the classification accuracy is the highest when $\alpha = 0.8$ and worst when $\alpha = 0.5$. The use of synonym merging improved classification accuracy by approximately 3 percentage points compared to use CHI only. By specific analysis of each category, we found that when we use synonym merging, only the first category has a low accuracy compared to no synonym merging. The reason is that the eigenvectors after synonym merging have reduced the text discrimination degree of the "car" category.

5.2.3 *Experiment III*

Based on the previous two experiments, we designed the following experiment to select the optimal weight update method introduced in Section 3.3. We designed three experiments, all with the first improved CHI formula and a threshold of 0.8. The experimental results are shown in Fig. 6:

According to Fig. 6, it can be seen that the classification accuracy of method 1 is the lowest. The reason is that this method adds the weights of all the synonymy words as the weight of the feature, but a word and its synonym words appear in more than one category. This simple superposition will make the feature differentiate the category worse. In contrast, methods 2 and 3 use the combined synonym as a one-dimensional feature and achieve better classification results and F1 scores. In contrast, method 2 is more effective which shows that the maximum value of the synonym is a better method because it can represent the maximum abilities of all synonyms to differentiate the text categories. By multiplying the power of 1.1 by the n, method 3 incorrectly increases the ability of the feature to distinguish text categories.

6 Conclusions

In this paper, we propose the SM-CHI feature selection method based on the common method used in Chinese text classification. This method mainly considers

Table 1 F1 score of each category when α takes different values

	Car	Finance	IT	Health	Sports	Tourism	Edu	Culture	Mil
$\alpha = 0.50$	0.729	0.701	0.579	0.833	0.769	0.805	0.704	0.693	0.900
$\alpha = 0.60$	0.747	0.787	0.709	0.872	0.815	0.810	0.712	0.713	0.895
$\alpha = 0.70$	0.734	0.814	0.803	0.874	0.919	0.817	0.694	0.764	0.896
$\alpha = 0.75$	0.757	0.813	0.825	0.861	0.946	0.808	0.726	0.779	0.900
$\alpha = 0.80$	0.763	0.821	0.828	0.931	0.949	0.827	0.726	0.819	0.903
$\alpha = 0.85$	0.763	0.812	0.818	0.931	0.937	0.821	0.721	0.804	0.889
$\alpha = 0.90$	0.782	0.783	0.808	0.943	0.945	0.781	0.720	0.802	0.913
$\alpha = 1.0$	0.782	0.781	0.788	0.932	0.936	0.789	0.714	0.794	0.897
None	0.782	0.776	0.783	0.935	0.936	0.778	0.711	0.788	0.901

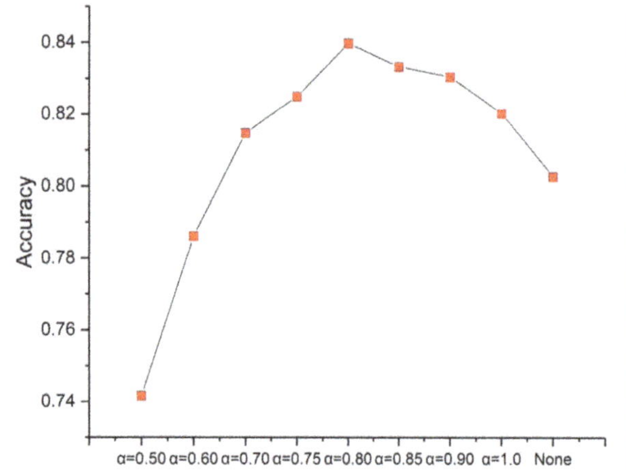

Fig. 5 Total accuracy when α takes different values

part of speech tagging, improved CHI formula and synonym merging. In addition, this paper proposes an update method for calculating the weight of feature words after synonym merging to obtain a more accurate vector representing of the text for classifier processing. The results of the experiment proves that the feature dimension can be reduced and the accuracy and effectiveness of text classification can be improved at the same time with this method.

In the future, we will focus on using synonym similarity calculation method based on "HowNet" instead of "Tong YiCi CiLin", because "HowNet" uses more than 1500 generics to build a unique knowledge description form and rich lexical semantic knowledge, so that we can more accurately calculate the similarity of words. What is more, SVM and neural networks have become the mainstream classifier in the text classification system because of its high accuracy, so we will use SVM or neural network instead of Naive Bayesian classification.

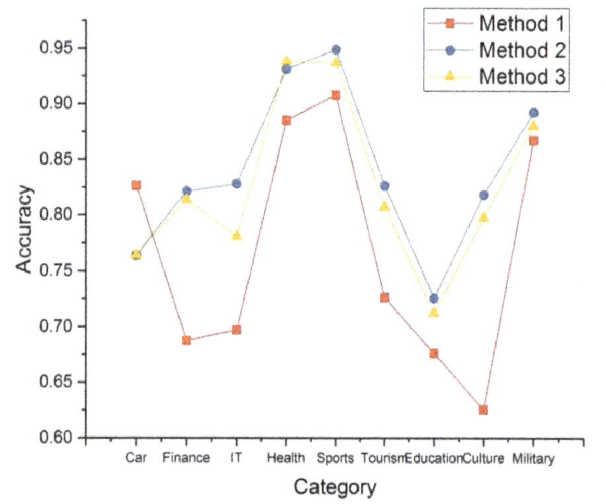

Fig. 6 Three kinds of special effects classification weight value updating method of comparison

Acknowledgements

This work is supported in part by the Shandong Provincial Natural Science Foundation, China (Grant No. ZR2014FQ018), in part by the BUPT-SICE Excellent Graduate Students Innovation Fund, in part by the National Natural Science Foundation of China (Grant No. 61471056), and in part by the China research project on key technology strategy of infrastructure security for information network development. The authors also gratefully acknowledge the helpful comments and suggestions of the reviewers, which have improved the presentation.

Authors' contributions

The idea was arisen from the discussions between HY and PZ. CL did the simulation and code implementation and wrote the Chinese version of the paper with the guide of PZ. HY wrote the Abstract and Conclusions. LW help us to translate the paper into English and made a lot of changes. All authors read and approved the final manuscript.

Competing interests

The authors declare that they have no competing interests.

Author details

[1]State Key Laboratory of Networking and Switching Technology, Beijing University of Posts and Telecommunications, Haidian District Xitucheng Road 10, Beijing 100876, People's Republic of China. [2]Advanced Innovation Center for Future Internet Technology, Beijing University of Technology, Chaoyang District Pingleyuan 100, Beijing 100124, People's Republic of China. [3]College of Computer & Communication Engineering, China University of Petroleum (East China), Changjiang West Road 66, Qingdao 266580, China.

References

1. wikipedia, tf-idf. https://en.wikipedia.org/wiki/Tf%E2%80%93idf. Accessed 01 Sept 2017
2. HT Ng, WB Goh, KL Low, in *International ACM SIGIR Conference on Research and Development in Information Retrieval*. Feature selection, perceptron learning, and a usability case study for text categorization (ACM, Philadelphia, 1997), pp. 67–73
3. Y Yang, JO Pedersen, in *Fourteenth International Conference on Machine Learning*. A comparative study on feature selection in text categorization (Morgan Kaufmann Publishers Inc., 1997), pp. 412–420
4. GC Feng, S Cai, in *Fourth International Conference on Computer, Mechatronics, Control and Electronic Engineering*. An Improved Feature Extraction Algorithm Based on CHI and MI (ICCMCEE, 2015)
5. Y Tang, T Xiao, in *International Conference on Computational Intelligence and Software Engineering*. An improved χ^2 (chi) statistics method for text feature selection (IEEE, 2009), pp. 1–4
6. DD Lewis, M Ringuette, in *Third Annual Symposium on Document Analysis & Information Retrieval*. A comparison of two learning algorithms for text categorization (ISRI, 1994), pp. 81–93
7. Y Yang, An evaluation of statistical approaches to text categorization. Inf. Retr. J. **1**(1), 69–90 (1999)
8. E Velasco, LC Thuler, CA Martins, LM Dias, VM Gonçalves, Automated learning of decision rules for text categorization. Acm Trans. Inf. Syst. **12**(3), 233–251 (1994)
9. T Joachims, *Text Categorization with Support Vector Machines: Learning with Many Relevant Features*. (Springer Berlin, Heidelberg, 1998)
10. J Bhimani, N Mi, M Leeser, Z Yang, in *IEEE International Conference on Cloud Computing*. Fim: Performance prediction model for parallel computation in iterative data processing applications (IEEE, 2017)
11. E Wiener, J Pedersen, AS Weigend, A neural network approach to topic spotting. Proc. Fourth Ann. Symp. Document Anal. Inf. Retr. (SDAIR). **92**(3), 482–487 (1995)
12. KW Church, P Hanks, Word association norms, mutual information, and lexicography. Comput. Linguist. **16**(1), 22–29 (1989)
13. JR Quinlan, Introduction of decision trees. Mach. Learn. **1**(1), 81-106

14. WJ Wilbur, K Sirotkin, The automatic identification of stop words. J. Inf. Sci. **18**(1), 45–55 (1992)
15. Y Yang, in *International ACM SIGIR Conference on Research and Development in Information Retrieval*. Noise reduction in a statistical approach to text categorization (ACM, 1995), pp. 256–263
16. T Dunning, Accurate methods for the statis15 of surprise and coincidence. Linguist. 74 Comput. Dirk Geeraerts Stefan Grondelaers. **19**(1), 61–74 (1993)
17. J Tian, W Zhao, Words similarity algorithm based on tongyici cilin in semantic web adaptive learning system. J. Jilin University. **28**(06), 602–608 (2010)
18. SJ Li, *Word Similarity Computing Based on How-net, The third Chinese mandarin semantics seminar*, (Taibei, 2002)
19. YH Zheng, DZ Zhang, A text feature selection method based on tongyici cilin. J. Xiamen University. **51**(2), 200–203 (2012)
20. B Altınel, MC Ganiz, B Diri, A corpus-based semantic kernel for text classification by using meaning values of terms. Eng. Appl. Artif. Intell. **43**(C), 54–66 (2015)
21. B Altınel, B Diri, MC Ganiz, A novel semantic smoothing kernel for text classification with class-based weighting. Knowl. Based Syst. **89**, 265–277 (2015)
22. J Wang, T Wang, Z Yang, Y Mao, N Mi, B Sheng, in *International Conference on Computing, NETWORKING and Communications*. Seina: A Stealthy and Effective Internal Sttack in Hadoop Systems (IEEE, 2017)
23. Z Yang, J Wang, D Evans, N Mi, in *International Workshop on Communication, Computing, and NETWORKING in Cyber Physical Systems*. Autoreplica: Automatic Data Replica Manager in Distributed Caching and Data Processing Systems (IEEE, 2016)
24. J Wang, T Wang, Z Yang, N Mi, B Sheng, in *IEEE International PERFORMANCE Computing and Communications Conference*. eSplash: Efficient Speculation in Large Scale Heterogeneous Computing Systems (IEEE, 2016)
25. sogou, Sogou data. http://www.sogou.com/labs/resource/ca.php. Accessed 01 Sept 2017
26. S Qin, J Song, P Zhang, Y Tan, in *International Conference on Fuzzy Systems and Knowledge Discovery*. Feature selection for text classification based on part of speech filter and synonym merge (IEEE, 2015), pp. 681–685

13

Heading estimation fusing inertial sensors and landmarks for indoor navigation using a smartphone in the pocket

Zhian Deng[1], Weijian Si[1], Zhiyu Qu[1], Xin Liu[2*] and Zhenyu Na[3*]

Abstract

Principal component analysis (PCA)-based approach for user heading estimation using a smartphone in the pocket suffers from an inaccurate estimation of device attitude, which plays a central role in both obtaining acceleration signals in the horizontal plane and the ultimate global walking direction extraction. To solve this problem, we propose a novel heading estimation approach based on two unscented Kalman filters (UKFs) fusing inertial sensors and landmarks. The first UKF is developed for the recalibration of device attitude estimation. We mathematically derive the measurement equation connecting observed user heading from landmarks with the quaternion vector representing device attitude. To decrease the nonlinearity of the measurement equation and make the filter more robust, we deploy the difference between user heading derived from the landmark and estimation result of PCA-based approach as the observation variable. The second UKF is developed for user heading estimation fusing estimation results of PCA-based approach and observed user headings from landmarks. Besides, we develop a robust landmark identification method by exploiting the acceleration and device pitch patterns, while noisy barometers are no longer required as previous methods. Experiments show that the proposed landmark-aided user heading estimation approach may improve accuracy performance significantly, which is very useful for continuous indoor navigation.

Keywords: Indoor navigation, User heading estimation, Unscented Kalman filter, Principal component analysis (PCA)

1 Introduction

Indoor positioning techniques have been paid increasing attentions from industry and academia due to the mass market for positioning applications. For outdoor environments, Global Navigation Satellite Systems (GNSSs) may provide reasonable accuracy performance. However, due to signal attenuations, they are always unavailable for indoor environments. Among various indoor positioning approaches [1, 2], pedestrian dead reckoning (PDR) using inertial sensor built-in smartphones is a promising solution, since it is self-contained and requires no extra infrastructures. There are two kinds of PDR, the strapdown approach [3] and the step-and-heading approach [4, 5]. The accumulated tracking errors of the strapdown approach may grow rapidly, since it involves a double integration of noisy acceleration signals. The strapdown approach is only feasible when continuous corrections are available, such as zero velocity updates for foot-mounted situations.

For unconstrained use of smartphones, such as a device put in the trouser pocket, it is more suitable to deploy the step-and-heading approach [6]. Step-and-heading approach infers the current pedestrian position sequentially by adding relative displacement to the position of previous step. The displacement is determined by estimated step length and user heading. User heading estimation is a central problem and the main error source of the step-and-heading approach. Moreover, user heading estimation may also be used in many other areas [7–9], such as human facing direction estimation in virtual reality and human computer interaction in smart environments. This paper focuses on the user heading estimation using a smartphone put in the pocket, which is one of the most popular device-carrying positions [10].

* Correspondence: liuxinstar1984@dlut.edu.cn; nazhenyu@dlmu.edu.cn
[2]School of Information and Communication Engineering, Dalian University of Technology, Dalian 116024, China
[3]School of Information Science and Technology, Dalian Maritime University, Dalian 116026, China
Full list of author information is available at the end of the article

Due to the changing device orientations caused by body locomotion, such as leg locomotion, it is inapplicable to compute user heading by the most commonly used device estimation approach [11], which adds heading offset to device heading. This is because the heading offset between device heading and user heading varies due to the changing device orientations and is difficult to be estimated. For a smartphone put in the pocket, uDirect approach [12] has been proposed by extracting walking direction in a specific region, where the walking direction dominates the acceleration vector. However, the specific region may be easily corrupted by body locomotion. In contrast, the PCA-based approaches [13] are more robust, since it exploits all samples in the walking step. Regardless of the changing device orientations, PCA may extract the walking direction along the maximum variations of the acceleration signals in the horizontal plane.

In order to obtain the horizontal accelerations more accurately, our previous work has proposed a PCA-based approach called RMPCA [14], combining rotation matrix (RM) with principal component analysis (PCA) for user heading estimation. Firstly, we continuously track the device attitude by developing extended Kalman filter (EKF) fusing inertial sensors. Then, we combine related rotation matrix to project the accelerations at local device coordinate system (DCS) into the global coordinate system (GCS). Finally, the global walking direction may be extracted by PCA over horizontal accelerations. Due to gyro and acceleration drifts, the accuracy performance of user heading estimation and positioning may degrade rapidly over a relatively short period.

Exploiting landmarks to aid pedestrian navigation is one of the most promising techniques to guarantee user heading estimation performance and limit accumulated tracking errors. Traditional landmark-based methods mainly rely on a pre-defined database and related infrastructures. For example, the densely deployed ultra-wide bandwidth (UWB) [15] and radio frequency identification (RFID) [16] anchors may provide distance information from the landmarks to the pedestrian, through time of arrival (ToA) and received signal strength (RSS) measurements, respectively. Wireless local area network (WLAN) or magnetic fingerprints [17, 18] can also be regarded as landmarks to aid pedestrian navigation. Traditional landmark-based methods may improve positioning accuracy significantly. However, these methods may increase the cost and disrupt self-containedness of the PDR system.

Recently, user motion states [19], including walking stairs and taking elevators and escalators, have been considered as indoor landmarks to aid indoor positioning. These landmarks [20] require neither extra infrastructures nor complex pre-defined database. Previous works [21, 22] have proposed these landmarks for both location estimation and direct user heading estimation recalibration. Significant user heading and positioning accuracy improvement has been reported. However, a re-estimation of the quaternion vectors describing device attitude is neglected. The accurate estimation of the quaternion vector is critical for ultimate user heading estimation, since it may directly affect the acceleration signal extraction in the horizontal plane and the ultimate global walking direction extraction by PCA. Besides, the previous device attitude estimation method fusing inertial sensors and magnetometers relies on EKF [21], which cannot adapt the nonlinearity of the measurement equation well.

In this paper, we propose a novel landmark-aided heading estimation approach based on two unscented Kalman filters (UKFs) and a recalibration of device attitude estimation. The main novelty is to fuse landmark information for device attitude recalibrations by constructing an explicit measurement equation in an UKF. The measurement equation relating landmarks with the quaternion vector describing device attitude is mathematically derived upon the principle of PCA-based approach. In order to reduce the nonlinearity of the measurement equation, we deploy the difference between user heading derived from the landmark and estimation result of RMPCA as the observation variable. For ultimate user heading estimation, we develop the second UKF fusing landmarks and estimation results of an improved RMPCA. The improved RMPCA may reduce the nonlinearity of the state equation of the second UKF, by extracting walking direction at a reference coordinate system. Besides, we develop a more robust user motion recognition method for landmark identification. Instead of requiring noisy barometers for vertical displacement detection as previous methods [19], we just deploy inertial sensors based on the acceleration and device pitch patterns.

Experiments demonstrate the accuracy performance improvement and reliability of the proposed landmark-aided heading estimation approach. In summary, our work makes the following contributions:

- We propose a novel heading estimation approach fusing inertial sensors and landmarks based on two developed UKFs and a recalibration of the device attitude estimation.
- We derive measurement equation of the first UKF mathematically upon the principle of the PCA-based approach. To decrease the nonlinearity of the measurement equation and make the filter more robust, the heading estimation difference is deployed as the observation variable.
- We develop the second UKF fusing landmarks and estimation results of an improved RMPCA, which

may reduce the nonlinearity of the state equation of the second UKF.
- We develop a robust landmark-identification method without requiring barometers, by only exploiting the inertial sensors and magnetometers.

The rest of the paper is organized as follows: Section 2 gives an overview of the proposed heading estimation approach. Section 3 presents the user motion recognition method for landmark identification. Section 4 describes the first UKF-based device attitude estimation module. Section 5 presents ultimate user heading estimation based on the second UKF. Section 6 provides experimental evaluations of the proposed approach. Finally, conclusions are presented in Section 7.

2 Overview of the proposed user heading estimation approach

Figure 1 overviews the proposed heading estimation approach fusing inertial sensors and landmarks using smartphones in the pocket. The proposed approach consists of three main modules: landmark identification, device attitude estimation, and user heading estimation.

The landmark identification module deploys a decision tree-based approach to recognize the motion states, including normal walking, standing, walking stairs, taking elevators, and taking escalators. Among these motion states, walking stairs, taking elevators, and taking escalators can be considered as landmarks to aid user heading estimation.

The device attitude estimation module deploys the first UKF to continuously estimate device attitude. The state model of the first UKF involves quaternion-based time evolution equation, while the measurement model involves measurement update from magnetic field values, accelerations under quasi-static situations, and identified landmarks. An explicit measurement equation relating the quaternion vector and user heading is mathematically derived upon the principle of PCA-based user heading estimation approach.

In order to describe the user heading estimation module, we define three coordinate systems, including global coordinate system (GCS), device coordinate system (DCS), and reference coordinate system (RCS). GCS consists of three axes X_G, Y_G, and Z_G, which point east, north, and the opposite direction of the gravity vector. We collect all raw inertial signals including acceleration and angular velocity samples at DCS. DCS consists of three axes X_D, Y_D, and Z_D. The two axes X_D and Y_D point rightward and forward, respectively, which are parallel with the phone screen. The axis Z_D is the cross

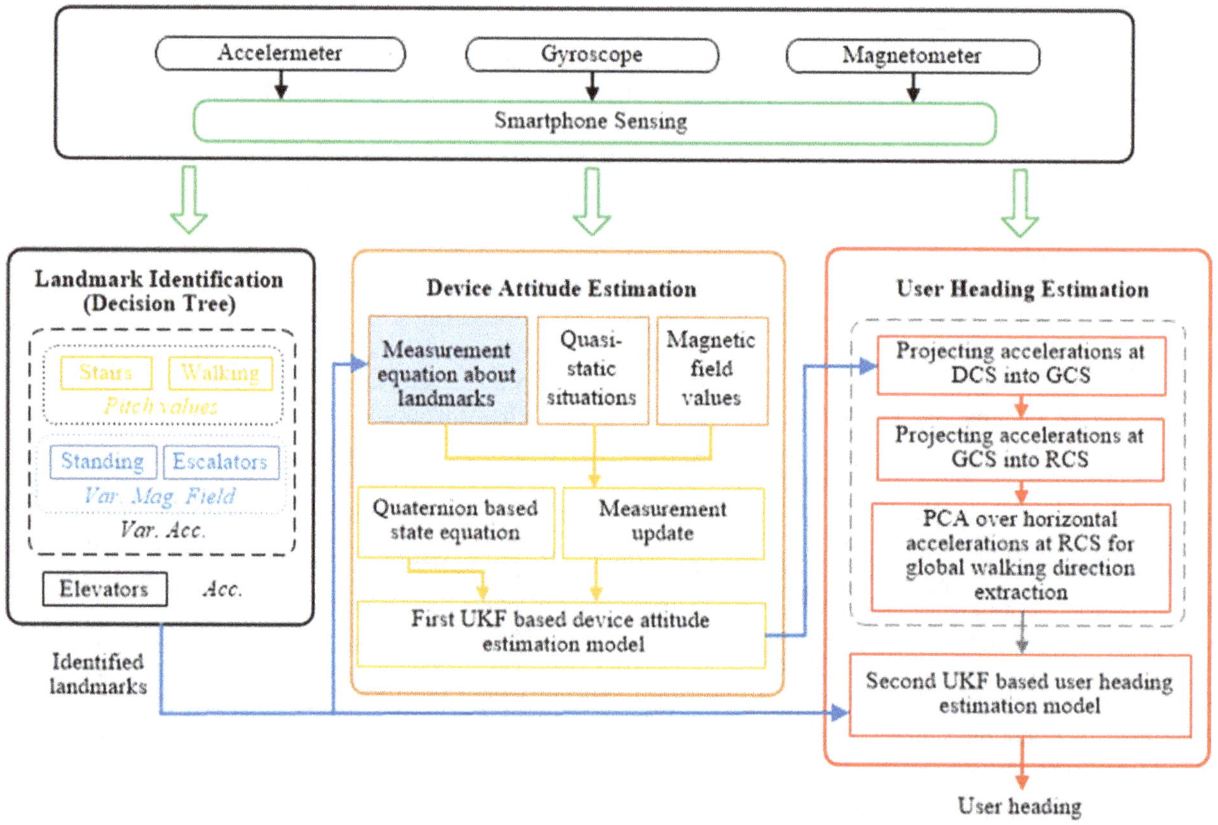

Fig. 1 Overview of the proposed user heading estimation approach fusing inertial sensors and landmarks

product of axes X_D and Y_D. To reduce nonlinearity of the measurement equation in the first UKF and state equation in the second UKF, we define RCS by rotating GCS UH_{RMPCA} radians around Z_G counterclockwise, which also includes three related axes X_R, Y_R, and Z_R. The angle UH_{RMPCA} is user heading initially estimated by RMPCA approach for each walking step. It should be noted that DCS and RCS may change with the body locomotion, while GCS is a fixed coordinate system.

User heading is defined as the angle that rotates from positive direction of the Y_G axis to the walking directions at GCS counterclockwise. The user heading estimation module deploys the second UKF to fuse identified landmarks with an improved RMPCA. For the improved RMPCA, rotation matrix obtained from the first UKF-based attitude estimation model is firstly used to project the accelerations at DCS into GCS. For each walking step, the accelerations at GCS are then projected into the related RCS. Finally, accelerations in the horizontal plane at RCS are obtained, and the global walking direction is extracted by PCA at RCS. The walking direction is extracted at RCS to reduce nonlinearity of the state equation.

3 Landmark identification module

The landmarks used in this paper include taking elevators, taking escalators, and walking upstairs/downstairs. When users enter or leave elevators, take escalators, and walk upstairs/downstairs, the user headings are always limited into a small region. Therefore, not only the landmarks can be used to recalibrate location estimation, but also can be used to recalibrate the heading estimation. We deploy a decision tree-based landmark identification method to detect and distinguish these landmarks from normal walking and standing motion states. The decision tree has three levels, as seen in Fig. 2.

In the first level, we firstly distinguish the elevator from the other motion states by exploiting the unique acceleration pattern of an elevator [23]. For the whole period of taking an elevator, the process includes standing still to wait for the elevator, entering the elevator, standing inside, and finally walking out of it. When standing inside the elevator for a short duration, a pair of positive/negative impulses of accelerations along the gravity direction occur, due to the related hyper-gravity/hypo-gravity effects. Between two impulses, there is a stationary duration, depending on the number of floors the elevator passes. In order to accurately capture the user heading information, when users enter or walk out of an elevator, we deploy the magnitude change of the magnetic field, since the total magnitude notably decreases or increases, respectively.

In the second level, we distinguish taking escalators/standing from walking stairs/walking by exploiting their acceleration variances. The acceleration variances of walking stairs/walking motion states are notably bigger than those of taking escalators/standing, since the former states involve the higher locomotion intensity. Furthermore, in the third level, we distinguish taking escalators from standing by exploiting the variances of the magnetic field values. The magnetic field values of taking escalators change rapidly due to the changed locations of moving escalators, while those of standing remain unchanged.

In the third level, to further distinguish between normal walking and walking upstairs/downstairs, we do not deploy barometers as previous methods [19, 24]. The barometers are only available in some relatively expensive smartphones, and the atmospheric pressure values measured by barometers may be influenced by many factors, such as the temperature, humidity, and opening/closing windows. Therefore, we exploit the pitch value pattern to detect walking upstairs/downstairs motion states. For the human leg, we define the leg pitch value as the angle leg rotates around the axis X_G, and the leg pitch value equals to zero when the leg is parallel to the gravity vector. The opening angle of the leg is defined as:

$$\Delta Pitch = Pitch_{max} - Pitch_{min} \quad (1)$$

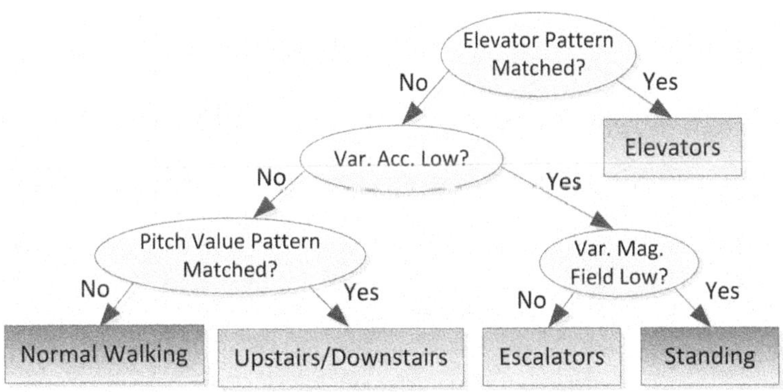

Fig. 2 Identification of landmarks using a decision tree

where $\text{Pitch}_{\max}$ and $\text{Pitch}_{\min}$ indicate the maximum and minimum value of leg pitch value. The leg pitch values show periodic change and thus make the maximum or the minimum leg pitch values occur only once per walking step. For each pedestrian, two observations may be seen during walking steps. Firstly, the opening angle of the leg for walking upstairs $\Delta\text{Pitch}^{\text{Upstairs}}$ is notably bigger than that of normal walking $\Delta\text{Pitch}^{\text{Normal}}$, while the opening angle of the leg for walking downstairs $\Delta\text{Pitch}^{\text{Downstairs}}$ is notably smaller than the opening angle $\Delta\text{Pitch}^{\text{Normal}}$, given as follows,

$$\Delta\text{Pitch}^{\text{Upstairs}} > \Delta\text{Pitch}^{\text{Normal}} > \Delta\text{Pitch}^{\text{Downstairs}} \quad (2)$$

Secondly, the maximum leg pitch value of walking upstairs $\text{Pitch}_{\max}^{\text{Upstairs}}$ is notably bigger than that of normal walking $\text{Pitch}_{\max}^{\text{Normal}}$, while the minimum leg pitch value of walking downstairs $\text{Pitch}_{\min}^{\text{Downstairs}}$ is notably bigger than that of normal walking $\text{Pitch}_{\min}^{\text{Normal}}$, given as follows,

$$\text{Pitch}_{\max}^{\text{Upstairs}} > \text{Pitch}_{\max}^{\text{Normal}}, \quad \text{Pitch}_{\min}^{\text{Downstairs}} > \text{Pitch}_{\min}^{\text{Normal}} \quad (3)$$

According to the observations, we may compute the average opening angle, the maximum and minimum leg pitch values for normal walking and walking upstairs/downstairs during offline phase. The related angle values can be seen as the pitch value pattern. During online phase, the pattern-matching process comparing the related angles as seen in (2) and (3) may be carried to distinguish between normal walking and walking upstairs/downstairs.

We collected 1000 test samples for each motion state, including taking elevators, taking escalators, walking stairs, normal walking, and standing. A total number of 5000 test samples were collected, 4000 samples collected in our office building, while the rest 1000 test samples of taking escalators were collected in a shopping mall. Each kind of samples are randomly divided into five parts, one part is used for testing and the rest four parts for training the related parameters of the decision tree. Table 1 shows the landmark identification results by a confusion matrix. The results show that almost all landmarks can be identified correctly, except for the negligible confusion between normal walking and walking stairs, due to some irregular leg locomotion. Furthermore, most of these wrong identified samples can be corrected by adjacent right identified samples. If the motion state of one walking step is different from that of the two adjacent steps, its motion state is assumed to be wrongly identified and considered as that of the two adjacent steps. Therefore, we assume that all the landmarks can be correctly identified and used in the proposed user heading estimation approach.

Table 1 Confusion matrix for landmark identification

Motion states	Stairs	Elevators	Escalators	Walking	Standing
Stairs	0.997	0	0	0.003	0
Elevators	0	1.0	0	0	0
Escalators	0	0	1.0	0	0
Walking	0.004	0	0	0.996	0
Standing	0	0	0	0	1.0

4 Device attitude module based on first UKF

We deploy quaternion vector as the state vector of UKF to describe the time evolution of device attitude. Firstly, we give the state and measurement models of UKF. Then, we derive the measurement equation relating quaternion vector with user heading mathematically upon the PCA-based approach. Finally, we describe unscented transformation and UKF equations.

4.1 Unscented Kalman filter design

Before designing UKF, we establish the relationship between device attitude and quaternion vector by deploying rotation matrix,

$$\mathbf{h}^{\text{DCS}}(t) = \left(\mathbf{R}_{\text{GCS}}^{\text{DCS}}(\mathbf{q}(t))\right)^T \mathbf{h}^{\text{GCS}}(t) \quad (4)$$

$$\mathbf{R}_{\text{GCS}}^{\text{DCS}}(\mathbf{q}) = \begin{bmatrix} q_0^2+q_1^2-q_2^2-q_3^2 & 2(q_1q_2-q_0q_3) & 2(q_1q_3+q_0q_2) \\ 2(q_1q_2+q_0q_3) & q_0^2-q_1^2+q_2^2-q_3^2 & 2(q_2q_3-q_0q_1) \\ 2(q_1q_3-q_0q_2) & 2(q_0q_1+q_2q_3) & q_0^2-q_1^2-q_2^2+q_3^2 \end{bmatrix} \quad (5)$$

where $\mathbf{R}_{\text{GCS}}^{\text{DCS}}(\mathbf{q}(t))$ is the rotation matrix from GCS to DCS at time t; $\mathbf{h}^{\text{GCS}}(t)$ and $\mathbf{h}^{\text{DCS}}(t)$ are the same 3×1 vectors with different representations at GCS and DCS, respectively; and $\mathbf{q} = [q_0 \ q_1 \ q_2 \ q_3]^T$ is the normalized quaternion vector with the scalar part q_0 and the vector part $\mathbf{e} = [q_1 \ q_2 \ q_3]^T$.

Based on the rigid body kinematic law [25], the state model of quaternion vector can be given as:

$$\mathbf{q}_{k+1} = F_k\mathbf{q}_k + \mathbf{w}_k^q \quad (6)$$

where state transition matrix $F_k = \exp(0.5^* T_s^* \Omega(\mathbf{w}_k))$,

$$\Omega(\mathbf{w}_k) = \begin{bmatrix} 0 & -w_k^x & -w_k^y & -w_k^z \\ w_k^x & 0 & w_k^z & -w_k^y \\ w_k^y & -w_k^z & 0 & w_k^x \\ w_k^z & w_k^y & -w_k^x & 0 \end{bmatrix} \quad (7)$$

where T_s is the system interval and $\mathbf{w}_k = \left[w_k^x \ w_k^y \ w_k^z\right]^T$ is the angular velocity vector measured at DCS at time instants kT_s. Process noise variable $\mathbf{w}_k^q$ and related covariance matrix W_k can be given as:

$$\mathbf{w}_k^q = \Xi_k \mathbf{w}_k^{\text{gyro}} = -\frac{T_s}{2}\begin{bmatrix} [\mathbf{e}_k \times] + q_0^k \mathbf{I} \\ -\mathbf{e}_k^T \end{bmatrix} \mathbf{w}_k^{\text{gyro}} \tag{8}$$

$$W_k = \mathbf{w}_k^q \left(\mathbf{w}_k^q\right)^T = \Xi_k W_k^{\text{gyro}} \Xi_k^T \tag{9}$$

where q_0^k is the scalar part of $\mathbf{q}_k$ and $\mathbf{e}_k = \begin{bmatrix} q_1^k & q_2^k & q_3^k \end{bmatrix}^T$ is the related vector part, $\mathbf{w}_k^{\text{gyro}}$ is the zero-mean white Gaussian noise of gyroscope outputs at time instants kT_s, $[\mathbf{e}_k \times]$ is a standard vector cross-product operator, W_k^{gyro} is the covariance matrix for gyroscope measurement noise with $W_k^{\text{gyro}} = \sigma_{\text{gyro}}^2 \mathbf{I}$, and $\mathbf{I}$ is an 3×3 identity matrix.

The measurement model of UKF is given as follows:

$$\mathbf{z}_{k+1} = \begin{bmatrix} \mathbf{a}_{k+1} \\ \mathbf{m}_{k+1} \\ \Delta \text{UH}_{k+1} \end{bmatrix} = \phi(\mathbf{q}_{k+1}) + \mathbf{v}_{k+1}$$

$$= \begin{bmatrix} \left(\mathbf{R}_{\text{GCS}}^{\text{DCS}}(\mathbf{q}_{k+1})\right)^T & \mathbf{0} & 0 \\ \mathbf{0} & \left(\mathbf{R}_{\text{GCS}}^{\text{DCS}}(\mathbf{q}_{k+1})\right)^T & 0 \\ \mathbf{0} & \mathbf{0} & 1 \end{bmatrix}$$

$$\begin{bmatrix} \mathbf{g} \\ \mathbf{h} \\ f(\mathbf{q}_{k+1}) \end{bmatrix} + \begin{bmatrix} \mathbf{v}_{k+1}^a \\ \mathbf{v}_{k+1}^m \\ \mathbf{v}_{k+1}^{\text{UH}} \end{bmatrix} \tag{10}$$

where $\mathbf{a}_{k+1}$ and $\mathbf{m}_{k+1}$ are the observed accelerations under quasi-static situations and magnetic field values at DCS, respectively; $\mathbf{g}$ and $\mathbf{h}$ are the local gravity vector and magnetic field values at GCS; ΔUH_{k+1} is the difference between user heading derived from a landmark and estimation result of RMPCA approach at GCS, $f(\mathbf{q}_{k+1})$ is the function relating user heading with the quaternion vector and accelerations upon PCA-based approach, as will be given in the next section; and $\mathbf{v}_{k+1}^a$, $\mathbf{v}_{k+1}^m$, and $\mathbf{v}_{k+1}^{UH}$ are the related zero mean white Gaussian measurement noise of the accelerometer, magnetometer, and landmark, respectively. The covariance of the measurement noise $\mathbf{R}_{k+1}$ can be given as follows:

$$\mathbf{R}_{k+1} = \begin{bmatrix} \mathbf{R}_{k+1}^a & 0 & 0 \\ 0 & \mathbf{R}_{k+1}^m & 0 \\ 0 & 0 & \mathbf{R}_{k+1}^{\text{UH}} \end{bmatrix} = \begin{bmatrix} {}^R\sigma_a^2 \mathbf{I}_3 & 0 & 0 \\ 0 & {}^R\sigma_m^2 \mathbf{I}_3 & 0 \\ 0 & 0 & {}^R\sigma_{UH}^2 \end{bmatrix} \tag{11}$$

where ${}^R\sigma\text{UH}^2$ is set during offline phase according to the style and related realistic environments of a landmark, as described in Section 3; and ${}^R\sigma a^2$ and ${}^R\sigma m^2$ are the parameters adaptively tuned according to the perturbation intensity of accelerations and magnetic field values:

$${}^R\sigma_a^2 = \begin{cases} \sigma_a^2, & \left| \|\mathbf{a}_{k+1}\|_2 - \|\mathbf{g}\|_2 \right| < \varepsilon_{a1} \cap \text{var}\left(\|\mathbf{a}_{k+1-Na/2}\|_2 : \|\mathbf{a}_{k+1+Na/2}\|_2\right) < \varepsilon_{a2} \\ \infty, & \text{otherwise} \end{cases} \tag{12}$$

$${}^R\sigma_m^2 = \begin{cases} \sigma_m^2, & \left| \|\mathbf{m}_{k+1}\|_2 - \|\mathbf{h}\|_2 \right| < \varepsilon_{m1} \cap \text{var}\left(\|\mathbf{m}_{k+1-Nm/2}\|_2 : \|\mathbf{m}_{k+1+Nm/2}\|_2\right) < \varepsilon_{m2} \\ \infty, & \text{otherwise} \end{cases} \tag{13}$$

where ε_{a1} represents allowed maximum difference between acceleration vector and the local one; ε_{m1} represents allowed maximum difference between magnetic field vector and the local one; ε_{a2} and ε_{m2} represent the related allowed maximum variances of signal samples, respectively; N_a and N_m represent the sizes of centered windows for acceleration and magnetic field samples, respectively; and $\text{var}(\cdot)$ is the function computing variance of samples in the centered window. Since the absolute static situations are always unavailable in realistic environments, we exploit the quasi-static situations, in which the magnitudes of the accelerations are similar to that of the static situations and the variances of the accelerations are assumed to be small.

4.2 Derivation of measurement equation relating quaternion vector with user heading

We derive the explicit measurement equation relating quaternion vector with user heading upon the principle of RMPCA approach. RMPCA firstly projects all acceleration samples within a walking step at DCS into the GCS and obtains the accelerations in the horizontal plane at GCS. In order to establish the relationship between the user heading and the quaternion vector at a specific time, we deploy a temporary rotation matrix between two DCSs to project all accelerations into the DCS at a specific time,

$$\mathbf{a}^{\text{DCS1}}(t2) = \left(\mathbf{R}_{\text{DCS2}}^{\text{DCS1}}\right)^T \mathbf{a}^{\text{DCS2}}(t2) \tag{14}$$

$$\mathbf{R}_{\text{DCS2}}^{\text{DCS1}} = \left(\mathbf{R}_{\text{GCS}}^{\text{DCS}}(\mathbf{q}(t2))\right)^T \left(\mathbf{R}_{\text{GCS}}^{\text{DCS}}(\mathbf{q}(t1))\right) \tag{15}$$

where DCS1 and DCS2 represent the DCS at time instant $t1$ and $t2$, $\mathbf{R}_{\text{DCS2}}^{\text{DCS1}}$ is the temporary rotation matrix from DCS2 to DCS1, $\mathbf{q}(t1)$ and $\mathbf{q}(t2)$ are the quaternion vectors at time instant $t1$ and $t2$, and $\mathbf{a}^{\text{DCS2}}(t2)$ and $\mathbf{a}^{\text{DCS1}}(t2)$ are the same accelerations measured at time instant $t2$ and represented at DCS1 and DCS2, respectively. The time interval between time instant $t1$ and $t2$ within a walking step is small enough, and thus, the accumulated error for computing related temporary rotation matrix $\mathbf{R}_{\text{DCS2}}^{\text{DCS1}}$ can be neglected. Derived from Eqs. (14) and (15), we project all accelerations within a walking step into the DCS at a specific time $t1$ within the walking step:

$$\mathbf{a}^{\mathrm{DCS1}}(j) = (\mathbf{R}(j))^T \mathbf{a}^{\mathrm{DCS}}(j), \quad j = 1, ..., N_{\mathrm{acc}}^{\mathrm{step}} \tag{16}$$

where $\mathbf{a}^{\mathrm{DCS1}}(j)$ and $\mathbf{a}^{\mathrm{DCS}}(j)$ are the representations of the j-th acceleration sample at DCS1 and its raw DCS, $\mathbf{R}(j)$ is the temporary rotation matrix from the DCS of the j-th acceleration sample to DCS1, and $N_{\mathrm{acc}}^{\mathrm{step}}$ is the number of acceleration samples within the walking step.

After projecting all accelerations within the walking step into a specific time $t1$, we may obtain the accelerations at GCS,

$$\mathbf{a}^{\mathrm{GCS}}(j) = \mathbf{R}_{\mathrm{GCS}}^{\mathrm{DCS}}(\mathbf{q}(t1))\mathbf{a}^{\mathrm{DCS1}}(j), \quad j = 1, ..., N_{\mathrm{acc}}^{\mathrm{step}} \tag{17}$$

where $\mathbf{a}^{\mathrm{GCS}}(j) = \begin{bmatrix} a_x^{\mathrm{GCS}}(j) & a_y^{\mathrm{GCS}}(j) & a_z^{\mathrm{GCS}}(j) \end{bmatrix}^T$ is the representation of the j-th acceleration sample at GCS, and $\mathbf{R}_{\mathrm{GCS}}^{\mathrm{DCS}}(\mathbf{q}(t1))$ is the temporary rotation matrix described by quaternion vector at time $t1$. Then, we can obtain the acceleration components in the horizontal plane as follows,

$$\begin{bmatrix} a_x^{\mathrm{GCS}}(j) \\ a_y^{\mathrm{GCS}}(j) \end{bmatrix} = Q(\mathbf{q}(t1))\mathbf{a}^{\mathrm{DCS1}}(j), \quad j = 1, ..., N_{\mathrm{acc}}^{\mathrm{step}} \tag{18}$$

$$Q(\mathbf{q}(t1)) = \begin{bmatrix} \hat{q}_0^2 + \hat{q}_1^2 - \hat{q}_2^2 - \hat{q}_3^2 & 2\left(\hat{q}_1\hat{q}_2 - \hat{q}_0\hat{q}_3\right) & 2\left(\hat{q}_1\hat{q}_3 + \hat{q}_0\hat{q}_2\right) \\ 2\left(\hat{q}_1\hat{q}_2 + \hat{q}_0\hat{q}_3\right) & \hat{q}_0^2 - \hat{q}_1^2 + \hat{q}_2^2 - \hat{q}_3^2 & 2\left(\hat{q}_2\hat{q}_3 - \hat{q}_0\hat{q}_1\right) \end{bmatrix} \tag{19}$$

where $\mathbf{q}(t1) = \begin{bmatrix} \hat{q}_0 & \hat{q}_1 & \hat{q}_2 & \hat{q}_3 \end{bmatrix}^T$ is the quaternion vector at time $t1$.

In order to make the nonlinear measurement equation converge fast, we take the difference between user heading derived from a landmark and estimation result of RMPCA approach as the observed variable,

$$\Delta \mathrm{UH} = \mathrm{UH}_{\mathrm{landmark}} \text{-} \mathrm{UH}_{\mathrm{RMPCA}} \tag{20}$$

where ΔUH is the observed difference value, $\mathrm{UH}_{\mathrm{landmark}}$ is the observed user heading derived from a landmark, and $\mathrm{UH}_{\mathrm{RMPCA}}$ is the estimation result of RMPCA approach for the current step. The acceleration components in the horizontal plane at RCS may be given as,

$$\begin{bmatrix} a_x^{\mathrm{RCS}}(j) \\ a_y^{\mathrm{RCS}}(j) \end{bmatrix} = C \begin{bmatrix} a_x^{\mathrm{GCS}}(j) \\ a_y^{\mathrm{GCS}}(j) \end{bmatrix}, \quad j = 1, ..., N_{\mathrm{acc}}^{\mathrm{step}} \tag{21}$$

$$C = \begin{bmatrix} \cos(\mathrm{UH}_{\mathrm{RMPCA}}) & \sin(\mathrm{UH}_{\mathrm{RMPCA}}) \\ -\sin(\mathrm{UH}_{\mathrm{RMPCA}}) & \cos(UH_{\mathrm{RMPCA}}) \end{bmatrix} \tag{22}$$

Then, derived from Eqs. (16) to (18), the acceleration components in the horizontal plane at RCS may be described by the raw measured accelerations represented at DCS1 directly,

$$\begin{bmatrix} a_x^{\mathrm{RCS}}(j) \\ a_y^{\mathrm{RCS}}(j) \end{bmatrix} = A\mathbf{a}^{\mathrm{DCS1}}(j), \quad j = 1, ..., N_{\mathrm{acc}}^{\mathrm{step}} \tag{23}$$

$$A = CQ(\mathbf{q}(t1)) \tag{24}$$

where $\mathbf{a}^{\mathrm{DCS1}}(j)$ is the j-th measured accelerations represented at DCS1.

According to the principle of PCA-based approach and pattern recognition [26, 27–29], the maximum energy of the accelerations in the horizontal plane may be obtained along the walking direction at RCS,

$$\Delta \mathrm{UH} = \max_{\Delta\theta}\left\{ \begin{bmatrix} -\sin\Delta\theta & \cos\Delta\theta \end{bmatrix} \sum_{j=1}^{N_{\mathrm{acc}}^{\mathrm{step}}} \begin{bmatrix} a_x^{\mathrm{RCS}}(j) \\ a_y^{\mathrm{RCS}}(j) \end{bmatrix} \begin{bmatrix} a_x^{\mathrm{RCS}}(j) \\ a_y^{\mathrm{RCS}}(j) \end{bmatrix}^T \begin{bmatrix} -\sin\Delta\theta & \cos\Delta\theta \end{bmatrix}^T \right\} \tag{25}$$

where $\Delta\theta$ is the angle variable that rotates from the axis Y_R to the walking direction counterclockwise. Substitute Eq. (23) into Eq. (25),

$$\Delta \mathrm{UH} = \max_{\Delta\theta}\left\{ \begin{bmatrix} -\sin\Delta\theta & \cos\Delta\theta \end{bmatrix} A \sum_{j=1}^{N_{\mathrm{acc}}^{\mathrm{step}}} \mathbf{a}^{\mathrm{DCS1}}(j)\mathbf{a}^{\mathrm{DCS1}}(j)^T A^T \begin{bmatrix} -\sin\Delta\theta & \cos\Delta\theta \end{bmatrix}^T \right\} \tag{26}$$

Define the following matrices described by quaternion vector at time $t1$,

$$\tilde{A}(\mathbf{q}(t1)) = \begin{bmatrix} \tilde{a}_{11} & \tilde{a}_{12} \\ \tilde{a}_{21} & \tilde{a}_{22} \end{bmatrix} = CQ(\mathbf{q}(t1)) \left[\sum_{j=1}^{N_{\mathrm{acc}}^{\mathrm{step}}} \mathbf{a}^{\mathrm{DCS1}}(j)\mathbf{a}^{\mathrm{DCS1}}(j)^T \right] Q(\mathbf{q}(t1))^T C^T \tag{27}$$

$$\tilde{f}(\mathbf{q}(t1), \Delta\theta) = \begin{bmatrix} -\sin\Delta\theta & \cos\Delta\theta \end{bmatrix} \tilde{A}(\mathbf{q}(t1)) \begin{bmatrix} -\sin\Delta\theta & \cos\Delta\theta \end{bmatrix}^T \tag{28}$$

Combine Eqs. (26), (27), and (28),

$$\frac{\partial\left(\tilde{f}(\mathbf{q}(t1), \Delta\theta)\right)}{\partial(\Delta\theta)} = 0 \tag{29}$$

$$\tilde{f}(\mathbf{q}(t1), \Delta\theta) = \begin{bmatrix} -\sin\Delta\theta & \cos\Delta\theta \end{bmatrix} \times \begin{bmatrix} \tilde{a}_{11} & \tilde{a}_{12} \\ \tilde{a}_{21} & \tilde{a}_{22} \end{bmatrix} \begin{bmatrix} -\sin\Delta\theta & \cos\Delta\theta \end{bmatrix}^T \tag{30}$$

We will get the following restriction equation about the observed variable ΔUH,

$$\tan(2 * \Delta \mathrm{UH}) = \frac{\tilde{a}_{12}}{\tilde{a}_{11} - \tilde{a}_{22}} \tag{31}$$

where $\tilde{a}_{11}$, $\tilde{a}_{12}$, and $\tilde{a}_{22}$ can be obtained from Eq. (27). Generally, we assume that the absolute difference between user heading derived from a landmark and from RMPCA approach is less than $\pi/4$. Therefore, we may obtain the measurement equation relating quaternion vector with observed variable as follows,

$$\Delta \mathrm{UH} = f(\mathbf{q}(t1)) = 0.5^{*} \arctan\left(\frac{\tilde{a}_{12}}{\tilde{a}_{11} - \tilde{a}_{22}}\right) \tag{32}$$

Usually, the absolute difference between user heading derived from a landmark and from RMPCA approach may be restricted into a small value, such as less than $\pi/8$. As a result, the arc tangent function in Eq. (25) may be approximated by a low-order polynomial function. If the absolute difference exceeds 45°, though the probability is rather low, we will not exploit the landmark to recalibrate the device attitude estimation module.

4.3 UKF equations for device attitude estimation

UKF [30, 31] is a Kalman filter based on unscented transformation (UT). UT provides an effective way to approximately calculate the change of the mean and covariance of a random variable when it undergoes a nonlinear transformation. As seen in Section 4.1, combining Eqs. (6) and (10), the state and measurement equations are given as follows:

$$\begin{cases} \mathbf{q}_{k+1} = F_k \mathbf{q}_k + \mathbf{w}_k^q \\ \mathbf{z}_{k+1} = \phi(\mathbf{q}_{k+1}) + \mathbf{v}_{k+1} \end{cases} \tag{33}$$

The state equation of the quaternion vector is linear, while the measurement equation is nonlinear. Therefore, we only need to deploy UT on the measurement equation.

Firstly, the same as that in Kalman filter, given the state estimation and its related covariance $\left(\widehat{\mathbf{q}}_k, , P_k\right)$, the state update equations may be given as follows:

$$\begin{cases} \widehat{\mathbf{q}}_{k+1}^{-} = F_k \widehat{\mathbf{q}} k \\ P_{k+1}^{-} = F_k P_k F_k^T + W_k \end{cases} \tag{34}$$

where $\widehat{\mathbf{q}}_{k+1}^{-}$ is the a priori state estimate and P_{k+1}^{-} is the related covariance matrix.

To calculate the statistic of the observed variable $\mathbf{z}_{k+1}$, UT designs a series of sigma points $\xi_{i,k+1}$ $(i = 0, 1, \cdots, 2L)$ with corresponding weights w_i using $\left(\widehat{\mathbf{q}}_{k+1}^{-}, P_{k+1}^{-}\right)$, according to the following:

$$\begin{cases} \xi_{0,,k+1} = \widehat{\mathbf{q}}_{k+1}^{-} \\ \xi_{i,,k+1} = \widehat{\mathbf{q}}_{k+1}^{-} + \left(\sqrt{(L+\lambda)P_{k+1}^{-}}\right)_i (i = 1, \cdots, L) \\ \xi_{i,,k+1} = \widehat{\mathbf{q}}_{k+1}^{-} - \left(\sqrt{(L+\lambda)P_{k+1}^{-}}\right)_i (i = L+1, \cdots, 2L) \end{cases} \tag{35}$$

$$\begin{cases} w_0^m = \dfrac{\lambda}{L+\lambda} \\ w_0^c = \dfrac{\lambda}{L+\lambda} + \left(1 - \alpha^2 + \beta\right) \\ w_i^m = w_i^c = \dfrac{\lambda}{2(L+\lambda)}, i = 1, \cdots, 2L \end{cases} \tag{36}$$

$$\lambda = \alpha^2 (L + \kappa) - L \tag{37}$$

where λ is a scaling factor, α is usually set to a small positive value (e.g., 1e-3) and L is set to 4 (the dimensionality of the state variable), κ is set to 0 and β is set to 2 for Gaussian distribution, and $\left(\sqrt{(L+\lambda)P_{k+1}^{-}}\right)_i$ is the i – th column of the square root matrix. w_i^m and w_i^c are the weight coefficients for calculating mean and covariance of $\mathbf{z}$.

After obtaining the sigma points, the measurement update equations are given as follows:

$$\begin{cases} \mathbf{z}_{i,k+1} = \phi(\xi_{i,k+1}) \\ \mathbf{z}_{k+1}^{-} = \sum_{i=0}^{2L} w_i^m \mathbf{z}_{i,k+1} \\ P_{\mathbf{zz},k+1} = \sum_{i=0}^{2L} w_i^c \left[\mathbf{z}_{i,k+1} - \mathbf{z}_{k+1}^{-}\right] \left[\mathbf{z}_{i,k+1} - \mathbf{z}_{k+1}^{-}\right]^T + \mathbf{R}_{k+1} \end{cases} \tag{38}$$

Then, the Kalman filter gain [32] may be obtained as follows:

$$\begin{cases} P_{\mathbf{qz},k+1} = \sum_{i=0}^{2L} w_i^c \left[\xi_{i,k+1} - \xi_{0,k+1}\right] \left[\mathbf{z}_{i,k+1} - \mathbf{z}_{k+1}^{-}\right]^T \\ K_{k+1} = P_{\mathbf{qz},k+1} \left(P_{\mathbf{zz},k+1}\right)^{-1} \end{cases} \tag{39}$$

The ultimate estimation of the state variable and its related covariance matrix are given as follows:

$$\begin{cases} \widehat{\mathbf{q}}_{k+1} = \widehat{\mathbf{q}}_{k+1}^{-} + K_{k+1}\left(\mathbf{z}_{k+1} - \mathbf{z}_{k+1}^{-}\right) \\ P_{k+1} = P_{k+1}^{-} - K_{k+1} P_{\mathbf{zz},k+1} K_{k+1}^T \end{cases} \tag{40}$$

5 User heading estimation module

This section describes the user heading estimation module based on the second UKF fusing heading estimation result of improved RMPCA and landmarks. The main difference between the proposed improved RMPCA and the original RMPCA is that attitude estimation is achieved by UKF rather than EKF. The state model of UKF is constructed upon the improved RMPCA. Firstly, as seen in Eq. (16), all acceleration samples measured at DCS within a walking step are projected into a specific time $t1$ during the same step. Then, deploying the quaternion vector at the specific time and related rotation matrix, the horizontal acceleration samples at GCS are obtained, as seen in Eq. (17). Similarly to Section 4.2, to reduce nonlinearity of the state equation, we also project accelerations at GCS into RCS and extract the walking direction at RCS. The state equation of user heading is given as follows,

$$\begin{cases} \mathrm{UH} = \mathrm{UH}_{\mathrm{ImRMPCA}} + f(\mathbf{q}(t1)) + w \\ f(\mathbf{q}(t1)) = 0.5^{*} \arctan\left(\frac{\tilde{a}_{12}}{\tilde{a}_{11} - \tilde{a}_{22}}\right) \end{cases} \tag{41}$$

where $\tilde{a}_{11}$, $\tilde{a}_{12}$, and $\tilde{a}_{22}$ can be seen as a function of the quaternion vector variable and obtained from Eq. (27), $\mathrm{UH}_{\mathrm{ImRMPCA}}$ is the user heading estimation result of improved RMPCA for current step, and w is the process noise variance of the state model.

If there is a landmark available, the user heading derived from the landmark can be used to recalibrate the ultimate user heading estimation. The measurement equation is given as follows,

$$\mathrm{UH}_{\mathrm{landmark}} = \mathrm{UH}_{\mathrm{ImRMPCA}} + \nu \tag{42}$$

where ν is the measurement noise of the user heading derived from the landmark. With the quaternion vector $\mathbf{q}(t1)$ and its covariance matrix calculated by attitude tracking model developed in Section 4, we can deploy UKF to calculate the ultimate user heading and its variance. If there is no landmark available, the ultimate user heading estimation can be calculated by unscented transformation on the state model, with the quaternion vector and its covariance matrix as input variable.

For different landmarks, the uncertainties of the derived user heading are different. User heading of taking an escalator or entering an elevator has a smaller variance than that of walking upstairs/downstairs. We set the measurement noise of each landmark at the target environment during the offline phase. Firstly, we define an interval covering all possible user headings of

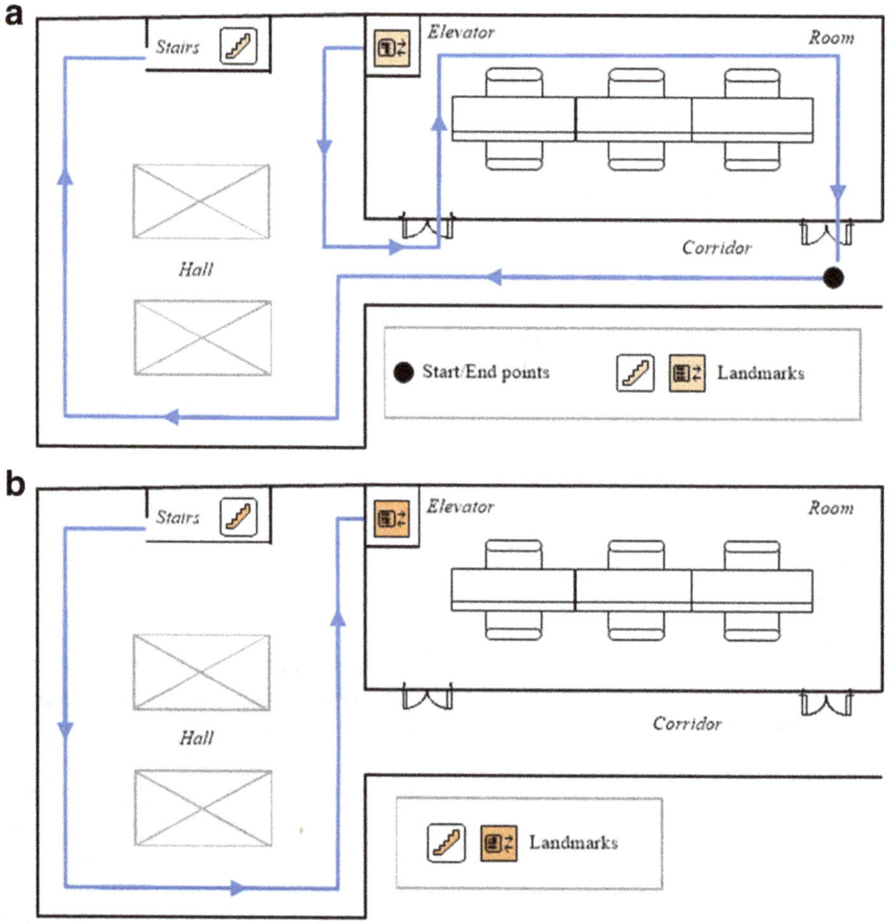

Fig. 3 Experimental environment and the true walking path including walking at the second floor, walking downstairs, walking at the first floor, taking the elevator, and walking at the second floor again. **a** Second floor. **b** First floor

a specific landmark. Secondly, for simplicity, the measurement noise is assumed to follow Gaussian distribution. We choose a three-sigma interval, where the probability of the user heading values lying in can reach up as high as 98.89%. Thirdly, the center value of the defined interval is set as the observed user heading of the landmark, and the standard deviation is set to one sigma. The parameter settings may vary with different types of the landmarks and realistic environments.

6 Evaluation

6.1 Experimental setup

Experiments were carried in a typical indoor building area covering two floors, including a hall, a corridor and a room for each floor, as seen in Fig. 3. The size of the experimental area for each floor is 18.5 m × 11.2 m. We deploy a Samsung Galaxy S4 smartphone as the device to collect gyroscope, accelerometer, and magnetometer data. The whole walking process for each participant includes starting on the walking path at the second floor along the corridor and the hall, walking downstairs, walking at the first floor along the hall, taking elevator, and walking at the second floor again along the hall and the room. The process was repeated by four individual participants with a total number of 100 times to test the user heading estimation performance of compared approaches. Each participant initially held the phone in hand, started the application, put it into the pocket and then started the walking. As in many other works [14, 33, 34], some parameters such as the initial walking direction are assumed to be known. To label the ground truth and compute the user heading estimation errors, we deployed a video to record the entire walking process of each participant.

6.2 Performance analysis

We compare the user heading estimation accuracy of various RMPCA-based approaches, including original RMPCA approach, landmark-aided RMPCA without device attitude recalibration, improved RMPCA using UKF, landmark-aided improved RMPCA without device attitude recalibration, and the proposed landmark-aided approach. As seen in Fig. 4, the proposed landmark-aided approach performs significantly better than the other compared approaches. Particularly, probability of absolute estimation error within 10° for proposed landmark-aided approach is 75.6%, while those of landmark-aided improved RMPCA without device attitude recalibration, improved RMPCA, landmark-aided RMPCA without device attitude recalibration, and RMPCA are 71.5, 69.1, 67.4, and 65.4%, respectively. Probability of absolute estimation error within 20° for proposed landmark-aided approach is 96.2%, while those of landmark-aided improved RMPCA without device attitude recalibration, improved RMPCA, landmark-aided RMPCA without device attitude recalibration, and RMPCA are 93.5, 91.4, 87.9, and 85.8%, respectively.

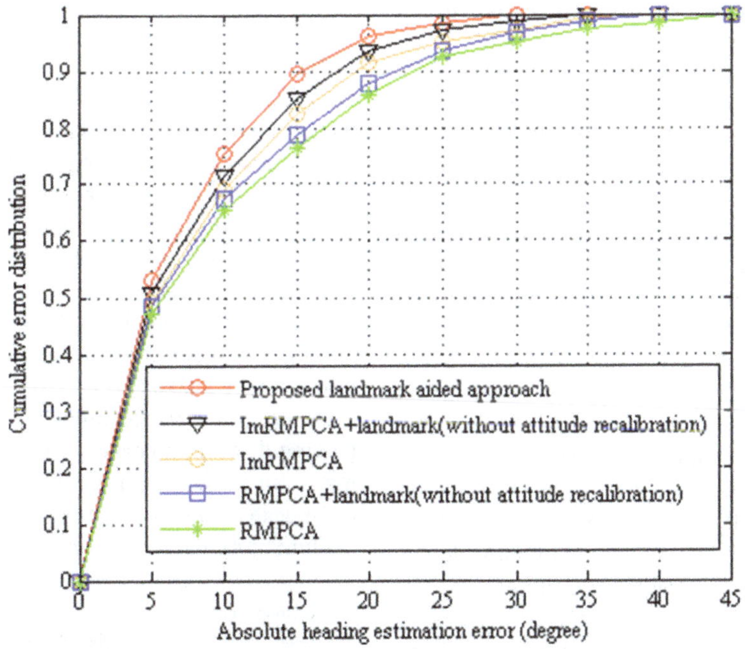

Fig. 4 Cumulative error distributions of absolute heading estimation error of compared user heading estimation approaches

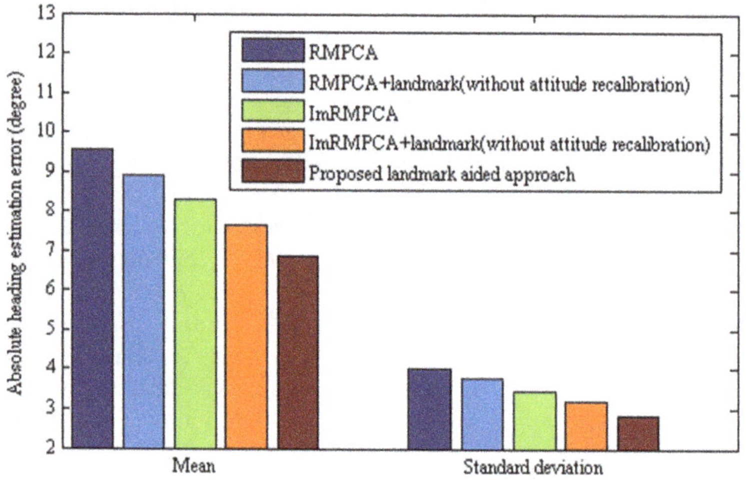

Fig. 5 Performance comparisons of mean and standard deviation of absolute heading estimation error

Figure 5 shows the performance comparisons of mean and standard deviation of absolute heading estimation error. The proposed landmark-aided approach obtains the smallest mean and standard deviation values of absolute heading estimation error. Particularly, mean absolute estimation error for proposed landmark-aided approach is 6.85°, while those of landmark-aided improved RMPCA without device attitude recalibration, improved RMPCA, landmark-aided RMPCA without device attitude recalibration, and RMPCA are 7.65°, 8.31°, 8.90°, and 9.56°, respectively. Standard deviation of absolute heading estimation error for proposed landmark-aided approach is 2.84°, while those of landmark-aided improved RMPCA without device attitude recalibration, improved RMPCA, landmark-aided RMPCA without device attitude recalibration, and RMPCA are 3.19°, 3.46°, 3.76°, and 4.03°, respectively. Compared with RMPCA approach, the proposed landmark-aided approach and previous landmark-aided RMPCA approach reduce the mean absolute heading estimation error by 28.3% (2.71°) and 6.9% (0.66°), respectively.

The proposed landmark-aided approach improves the heading estimation accuracy from two aspects. First, the improved RMPCA approach is developed for user heading estimation without landmarks. The improved RMPCA deploys UKF for quaternion-based device attitude estimation, which may better adapt the nonlinearity of the related measurement equation than previous RMPCA approach using EKF. This aspect can also be verified by the heading estimation result comparisons between the improved RMPCA and RMPCA approaches. Second, we not only deploy landmarks for direct user heading recalibrations, but also for device attitude estimation recalibrations, which may render more accurate extraction of the horizontal accelerations. This is also the reason why the proposed landmark-aided approach performs better than that of landmark-aided improved RMPCA without device attitude recalibration.

7 Conclusions

In this paper, we propose two UKF-based user heading estimation approach by fusing inertial sensors and landmarks. The second UKF fusing landmarks and estimation results of an improved RMPCA is developed for direct user heading estimation, while the first UKF is developed for device attitude estimation. The proposed approach not only exploits landmarks for direct user heading estimation recalibration, but also for device attitude recalibration, which is important for accurate walking direction extraction. Compared with previous RMPCA approach, instead of using EKF, the improved RMPCA using the first UKF may better adapt nonlinearity of measurement equation. Besides, we develop a robust user motion recognition method for landmark identification, without requiring noisy barometers. Experimental results show that the proposed landmark-aided approach may obtain significant user heading estimation accuracy improvement. Compared with previous landmark-aided RMPCA approach, the proposed approach may reduce the mean absolute heading estimation error by 23.0% (2.05°).

In our future works, more kinds of landmarks such as passing doors and corners will be used to aid user heading estimation and indoor navigation. Besides, more complicated and a large-scale indoor environment including more landmarks will be tested.

Acknowledgments

This research is supported by National Natural Science Foundation of China (Granted No. 61671168, 61301132, 61601221, and 61301131), and the Fundamental Research Funds for the Central Universities No. 3132017129.

Authors' contributions
ZD proposed the original idea and wrote this paper; WS and ZQ gave some valuable suggestions and improved the paper; ZN and XL supervised and revised the paper. All authors read and approved the final manuscript.

Competing interests
The authors declare that they have no competing interests.

Author details
[1]College of Information and Communication Engineering, Harbin Engineering University, Harbin 150001, China. [2]School of Information and Communication Engineering, Dalian University of Technology, Dalian 116024, China. [3]School of Information Science and Technology, Dalian Maritime University, Dalian 116026, China.

References

1. R. Harle, A survey of indoor inertial positioning systems for pedestrians. IEEE Commun. Surv. Tutorials. **15**(3), 1281–1293 (2013)
2. Y. Gu, A. Lo, I. Niemegeers, A survey of indoor positioning systems for wireless personal networks. IEEE Commun. Surv. Tutorials. **11**(1), 13–32 (2009)
3. D. H. Titterton, J. L. Weston, Strapdown inertial navigation technology. Institution of Electrical Engineers, 2004
4. C. Valérie, Christophe, magnetic, acceleration fields and gyroscope quaternion (MAGYQ)-based attitude estimation with smartphone sensors for indoor pedestrian navigation. Sensors (Switzerland) **14**, 22864–22890 (2014)
5. F. Zhao, X. Sun, H. Chen, R. Bie, Outage performance of relay-assisted primary and secondary transmissions in cognitive relay networks. EURASIP J. Wirel. Commun. Netw. **1**(60), 1–10 (2014)
6. L. Chen, E. Wu, M. Jin, G. Chen, Intelligent fusion of Wi-Fi and inertial sensor-based positioning systems for indoor pedestrian navigation. IEEE Sensors J. **14**(11), 4034–4042 (2014)
7. R. Atienza, A. Zelinsky, Active gaze tracking for human-robot interaction. IEEE Int. Conf. Multimodal Interfaces, ICMI 2002 (2002), pp. 261–266
8. F. Zhao, L. Wei, H. Chen, Optimal time allocation for wireless information and power transfer in wireless powered communication systems. IEEE Trans. Veh. Technol. **65**(3), 1830–1835 (2016)
9. M. Jia, L. Wang, Q. Guo, X. Gu, W. Xiang, A low complexity detection algorithm for fixed up-link SCMA system in mission critical scenario. IEEE Int Things J **1**(1), 99 (2017)
10. Z.A. Deng, G. Wang, D. Qin, Z. Na, Y. Cui, J. Chen, Continuous indoor positioning fusing WiFi, smartphone sensors and landmarks. Sensors (Switzerland) **16** (2016)
11. H. Lee, J. Lee, J. Cho, N. Chang, Estimation of heading angle difference between user and smartphone utilizing gravitational acceleration extraction. IEEE Sensors J. **16**(10), 3746–3755 (2016)
12. S.A. Hoseinitabatabaei, A. Gluhak, R. Tafazolli, W. Headley, Design, realization, and evaluation of uDirect-an approach for pervasive observation of user facing direction on mobile phones. IEEE Trans. Mob. Comput. **13**(9), 1981–1994 (2014)
13. K. Kunze, P. Lukowicz, K. Partridge, B. Begole, Which way am i facing: inferring horizontal device orientation from an accelerometer signal. Int. Symp. Wearable Comput. ISWC, 149–150 (2009)
14. Z.A. Deng, G. Wang, Y. Hu, D. Wu, Heading estimation for indoor pedestrian navigation using a Smartphone in the pocket. Sensors (Switzerland) **15**, 21518–21536 (2015)
15. F. Zampella, A.R. Jiménez, R.F. Seco, Light-matching: a new signal of opportunity for pedestrian indoor navigation. Int. Conf. Indoor Position. Indoor Navig. IPIN **2013** (2013)
16. A. R. J. Ruiz, F. S. Granja, J. C. P. Honorato, J. I. G. Rosas, Pedestrian indoor navigation by aiding a foot-mounted IMU with RFID signal strength measurements. Int. Conf. Indoor Position. Indoor Navig. IPIN 2010, 1–9 (2010)
17. S. Han, Z. Gong, W. Meng, C. Li, An indoor radio propagation model considering angles for WLAN infrastructures. Wirel. Commun. Mob. Comput. **15**(16), 2038–2048 (2015)
18. L. Zhang, S. Valaee, Y. Bin Xu, L. Ma, F. Vedadi, Graph-based semi-supervised learning for indoor localization using crowdsourced data. Appl. Sci. **7**(467), 1–24 (2017)
19. Z. Chen, H. Zou, H. Jiang, Q. Zhu, Y.C. Soh, L. Xie, Fusion of WiFi, smartphone sensors and landmarks using the Kalman filter for indoor localization. Sensors (Switzerland) **15**, 715–732 (2015)
20. H. Wang, A. Elgohary, and R. R. Choudhury, No need to war-drive: unsupervised indoor localization. Proc. 10th Int. Conf. Mob. Syst. Appl. Serv. (MobiSys '12), 197–210 (2012)
21. Y. Gu, Q. Song, Y. Li, M. Ma, Z. Zhou, An anchor-based pedestrian navigation approach using only inertial sensors. Sensors (Switzerland) **16**, 334–351 (2016)
22. F. Ichikawa, J. Chipchase, R. Grignani, Where's the phone? A study of mobile phone location in public spaces. International Conference on Mobile Technology, 1–8 (2009)
23. H. Abdelnasser, R. Mohamed, A. Elgohary, M.F. Alzantot, H. Wang, S. Sen, R.R. Choudhury, M. Youssef, SemanticSLAM: using environment landmarks for unsupervised indoor localization. IEEE Trans. Mob. Comput. **15**(7), 1770–1782 (2016)
24. K. Frank, E. Diaz, P. Robertson, F. Sanchez, Bayesian recognition of safety relevant motion activities with inertial sensors and barometer. Location and Navigation Symposium - PLANS 2014, 2014 IEEE/ION, pp. 174–184 (2014)
25. A.M. Sabatini, Quaternion-based extended Kalman filter for determining orientation by inertial and magnetic sensing. IEEE Trans. Biomed. Eng. **53**(7), 1346–1356 (2006)
26. F. Zhao, W. Wang, H. Chen, Q. Zhang, Interference alignment and game-theoretic power allocation in MIMO heterogeneous sensor networks communications. Signal Process. **126**, 173–179 (2016)
27. U. Steinhoff, B. Schiele, Dead reckoning from the pocket—an experimental study. Proceedings of 2010 IEEE International Conference on Pervasive Computing and Communications, 2010, pp. 162–170
28. F. Zhao, B. Li, H. Chen, X. Lv, Joint beamforming and power allocation for cognitive MIMO systems under imperfect CSI based on game theory. Wirel. Pers. Commun. **73**(3), 679–694 (2013)
29. M. Jia, X. Gu, Q. Guo, Broadband hybrid satellite-terrestrial communication systems based on cognitive radio toward 5G. IEEE Wirel. Commun. **23**(6), 96–106 (2013)
30. K. Xiong, H.Y. Zhang, C.W. Chan, Performance evaluation of UKF-based nonlinear filtering. Automatica **42**(2), 261–270 (2006)
31. H. Qasem, L. Reindl, Unscented and extended Kalman estimators for non linear indoor tracking using distance measurements. Positioning, Navigation and Communication, 2007. WPNC '07. 4th Workshop on Navigation and Communication, 177–181 (2007)
32. Y. Bar-Shalom, X. R. Li, T. B. T.-E. Kirubarajan, Estimation, tracking and navigation:theory, algorithms and software. John Wiley & Sons, 2002
33. F. Zhao, H. Nie, H. Chen, Group buying spectrum auction algorithm for fractional frequency reuses cognitive cellular systems. Ad Hoc Netw. **58**, 239–246 (2017)
34. M Jia, X Liu, X Gu, Joint cooperative spectrum sensing and channel selection optimization for satellite communication systems based on cognitive radio. Int. J. Satell. Commun. Netw. **23**(3), 139–150 (2015)

14

On the performance of the Code Division Duplex system using interference rejection codes

Yufang Yin[1], Gangjun Li[1], Li Li[2] and Hua Wei[2*]

Abstract

In this paper, we investigate the performance of the Code Division Duplex (CDD) system using the interference rejection codes, which exhibit zero correlation values on the relative delay-induced code offsets and can be applied to mitigate the mutual interference between the uplink and downlink. Specifically, we propose to employ loosely synchronous codes to effectively combat the interference through the interference-free window. Moreover, the simplified RAKE structure is proposed as the receiver of the CDD device, which can significantly reduce the complexity in the implementation. The simulation results demonstrate that our proposed method can achieve a near inference-free performance when the user load of the system is moderate.

1 Introduction

A duplex communication system [1] is a point-to-point system composed of two connected parties or devices that can communicate with one another in both directions, normally termed as uplink and downlink individually. To circumvent the mutual interference between the uplink and downlink, the Frequency Division Duplex (FDD) [1] and Time Division Duplex (TDD) [1] apparatuses are commonly deployed in communication systems. In the FDD system, the uplink and downlink signals are gapped in different frequency bands, while in the TDD system, the uplink and downlink transmissions are identified in different time slots. In this way, the uplink and downlink can be treated as being orthogonal in the FDD or TDD apparatus since they utilize different frequency or time resources.

Recently, No Division Duplex (NDD) [2, 3] techniques have captured growing interests for the 5G communication system, since they can significantly increase the capacity of the cellular system as well as simplify the infrastructure of the cellular network. The basic philosophy of the No Division Duplex [2, 3] is that a radio device can transmit and receive on the same frequency at the same time; more explicitly, it can operate in a full-duplex fashion. However, as the radio signals attenuate quickly in an exponential way over the distance, the signal from a local transmitting antenna is hundreds of thousands of times stronger than transmissions from other nodes. Therefore, the challenge in implementing a full-duplex system is to recover the desired signal from the excessively strong local interferers. Hence, a complicated interference canceller [3] must be deployed in the NDD device to combat the local interference, in both the analog RF component as well as the digital baseband component.

In Code Division Multiple Access (CDMA) systems, the spreading sequences characterize the properties of the associated Intersymbol Interference (ISI) as well as the Multiple Access Interference (MAI) [4]. Traditional spreading sequences, such as m-sequences [4], Gold codes [4], and Kasami codes [4], exhibit non-zero off-peak auto-correlations and cross-correlations, which result in a high MAI in the system. To circumvent this issue, the ideal solution is to design the optimal sequences whose auto- and cross-correlations are zero for the infinite code offsets, as shown in Fig. 1a. However, this kind of perfect sequence does not exist in theory. Therefore, considerable research has been invested on the interference rejection code which can exhibit zero correlation values on the relative delay-induced code offsets. The area which has the value of zero correlation is the so-called Zero Correlation Zone (ZCZ) or Interference-Free Window (IFW) of the spreading code [5], as shown in Fig. 1d. The auto- and

*Correspondence: weihua@cuit.edu.cn
[2]Chengdu University of Information and Technology, Chengdu, China
Full list of author information is available at the end of the article

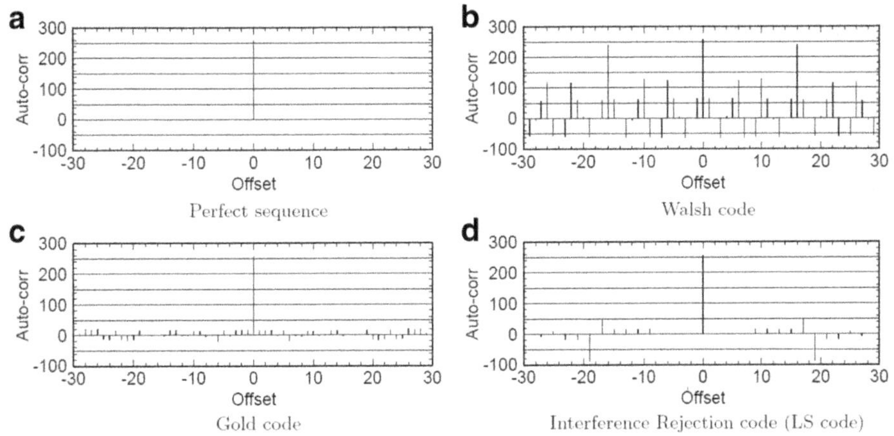

Fig. 1 Cross-correlations of four different codes. **a** Perfect sequence. **b** Walsh code. **c** Gold code. **d** Interference rejection code (LS code)

cross-correlation properties of four spreading codes are exhibited in Fig. 1. In this figure, we can observe that the correlation value of the interference rejection code is zero in IFW. Loosely Synchronous (LS) code [6, 7] is one of these specific interference rejection codes which exhibit an IFW, where the off-peak values of aperiodic auto- and cross-correlations are zero, resulting in zero ISI and zero MAI in the IFW. Given these conditions, a major benefit of the interference rejection codes is that they are capable of combating the mutual interference between the uplink and downlink without a complicated interference canceller. Hanzo, Wei, and Ni [8–10] investigated the performance of the CDMA system in the context of the deployment the LS codes. However, in their work, the power control [11] scheme is employed to avoid the near-far effect, and only the cellar communication scenarios are evaluated in previous work. In [9, 12], the interference rejection codes are deployed in the multi-carrier CDMA system and the performance is analyzed. The authors in [13] investigated the performance of the CDMA system in conjunction with the multiple antenna technique and the interference rejection code. Various chip waveforms [14] are investigated for the LS codes, and the network capacity in the cellular scenario is analyzed in [15].

In this paper, to circumvent the implementation of complicated interference canceller, Code Division Duplex (CDD) is proposed for a synchronous communication system. As shown in Fig. 2, the basic concept of the CDD is that the downlink and uplink signals are distinguished by their unique spread sequence C_i, i.e., the CDD radio can transmit and receive simultaneously at the same frequency band with different spreading signatures. The *state of the art* CDD device can operate in the same fashion as the NDD device. However, the advantage of the CDD receiver is to employ a simple matched filter or RAKE receiver rather than a complicated interference canceller in the NDD device owing to the cross-correlation properties of the interference rejection code, which is beneficial in the engineering implementation and power saving.

In this paper, we will investigate the performance of CDD communication systems in the context of two practical scenarios. The first scenario is a cellular-like communication system, where a base station node operates as the centric node and all the other CDD nodes must communicate with the center node. In this specific scenario, the center node has a much stronger transmission power than other nodes. The second scenario is an ad hoc communication system where each CDD device transmits on the same power in the network and can communicate with each other. In this contribution, we will investigate the Bit Error Rate (BER) performance and capacity of CDD systems in both scenarios when communicating over a Nakagami-m channel.

This paper is organized as follows. In Section 2, we will introduce the specific interference rejection code, namely LS code, while in Section 3, we will describe the model

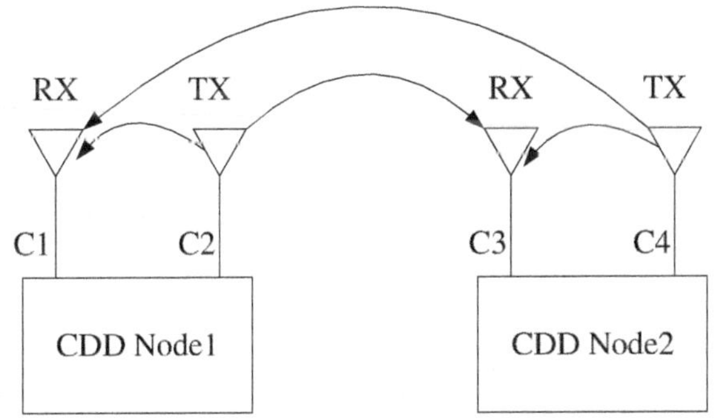

Fig. 2 Code Division Duplex system

of the CDD system and fading channel. In Section 4, we will characterize the BER performance of the CDD communication, and in Section 5, we will discuss our findings. Finally, in Section 6, we will offer our conclusion.

2 Interference rejection codes

There exists a specific family of spreading codes, which exhibit an IFW; more specifically, Loosely Synchronized (LS) codes [6] exploit the properties of the so-called orthogonal complementary sets [6, 7]. Since the detailed construction method of binary LS codes was described in [6, 9, 10], here, we only focus our attention on the properties and the applications of the LS codes [16, 17].

Generally, an LS code can be denoted as LS(N, P, W_0), which denotes the family of LS codes generated by applying a $(P \times P)$-dimensional Walsh-Hadamard (WH) matrix to an orthogonal complementary code set of length N and inserting W_0 number of zeros between two orthogonal complementary sets [6, 9, 10]. The constructed LS(N, P, W_0) codes can be constructed for any arbitrary length of the codes by selecting the proper parameter N, and then, the generated LS codes can exhibit an IFW of interference rejection code as shown in Fig. 1; more explicitly, the spreading signature function $c_i(t), c_j(t)$ of the constructed LS codes satisfy:

$$\int_{-\infty}^{-\infty} c_i(t)c_j(t-\tau)dt = 0, \quad \tau < W_I, \quad i \neq j \tag{1}$$

where $c_i(t)$ and $c_j(t)$ are the signature functions in the generated LS(N, P, W_0) code set and W_I is the width of the IFW. More explicitly, the constructed LS(N, P, W_0) codes are still orthogonal to each other even in the scenarios of the multipath environment provided that the delay spread of the multipath is less than the width of IFW. Hence, the LS code with inherent IFW can effectively suppress the Multipath Interference (MPI) and multiuser access interference (MAI), which also significantly simplify the design of receiver as well as reduce the complexity of the receiver in CDMA system.

For example, the LS(N, P, W_0) codes can be generated based on the complementary pair of [6, 16], and the total number of available codes in the family of LS(N, P, W_0) is given by NP. The total number of NP codes can be classified by N sets of P number of LS codes through their Walsh-Hadamard Matrix, and each set has P number of LS codes. The LS codes in the same set exhibit an IFW length of $[-W_I, +W_I]$, where we have $W_I = \min\{W_0, N-1\}$. The aperiodic auto-correlation and cross-correlation function $\rho_{kk}(\tau)$, $\rho_{jk}(\tau)$ of the codes belonging to the same set will be zero, provided that we have $\tau \leq W_I T_c$. Furthermore, the LS codes belonging to the N different sets are still orthogonal to each other at zero offset, namely in a perfectly synchronous environment. However, the LS codes belonging to the N different sets will lose their orthogonality, when they have a non-zero code offset.

To combat the interference incurred by its own transmission, the spreading codes in the CDD device have the following assignment policy: the spreading codes of the uplink and downlink are wisely chosen to belong to different sets of Walsh-Hadamard. The benefit of this assignment is that the spreading codes between the uplink and downlink have the maximum width of the IFW, which can guarantee the orthogonality between the uplink and downlink in one CDD node so that their mutual interference can be negligible.

3 System model

3.1 System model

Assume that the system supports K synchronous users and each user is assigned two unique spreading signature waveforms for its uplink and downlink, respectively. The spreading signature waveform can be noted as $\mathbf{c}_k(t) = \sum_{i=0}^{G-1} c_{ki}\psi_{T_c}(t - iT_c)$, where G is the spreading gain and $\psi_{T_c}(t)$ is the rectangular chip waveform, which is defined over the interval $[0, T_c)$. Consequently, when the K users' signals are transmitted over the frequency-selective fading channel, the complex low-pass equivalent signal received at a given RX antenna can be expressed as:

$$R(t) = \sum_{k=1}^{K}\sum_{l=0}^{L_p-1} \sqrt{2P_k}\mathbf{c}_k(t - lT_c - \tau_k)b_k(t - lT_c - \tau_k) \times h_{kl}\exp(j\theta_{kl}) + N(t), \tag{2}$$

where $N(t)$ is the complex-valued low-pass-equivalent AWGN (additive white Gaussian noise) having a double-sided spectral density of N_0 and τ_k is the propagation delay between the transmitter and receiver for user k. In the same CDD device, τ_k is assumed to be 0, while in different CDD devices, τ_k is determined by the propagation distance: without loss of generality, $\tau_k = \frac{d_k}{c}$, where c is the speed of light. L_p is the total number of resolvable paths. P_k is the received power, which is determined by the large-scale fading of the wireless channel.

3.2 Wireless large-scale fading model

The complexity of wireless signal propagation makes it difficult to obtain a single model that characterizes path loss accurately across a range of different environments. In the textbook [1], various propagation models of the wireless channel are proposed and investigated in wireless communications, such as the well-known free-space channel model, Okumura Model [1], Hata Model [18], and COST207 [1, 18] which are traditionally used in system analysis. However, for the convenience of analysis of various system performance, we simplify the channel model as well as capture the essence of signal propagation without resorting to complicated path loss models, which is sufficiently accurate to approximate the real channel model. In this contribution, the following simplified models [1] for path loss as a function of distance are commonly used for system design:

$$P_r = P_t G_a \left[\frac{d}{d_0}\right]^{-\gamma}, \quad (3)$$

where d_0 is the wavelength of the radio signal, d is the distance between the transmitter and receiver, and G_a is the constant which is related to the characteristic of the antenna, such as the height and direction. Parameter γ is the exponential attenuation factor which is related to the practical environment. Generally, the attenuation factor γ is about 2 to 4 according to the free-space, indoor, urban, or hilly environments.

3.3 Channel model

The DS-CDMA signal experiences independent frequency-selective Nakagami-m fading. The complex low-pass equivalent representation of the Channel Impulse Response (CIR) encountered by the kth user is given by [19]:

$$h_k(t) = \sum_{l=0}^{L_p-1} h_{kl}\delta(t - lT_c)\exp\left(j\theta_{kl}\right), \quad (4)$$

where h_{kl} represents the Nakagami-distributed fading envelope, lT_c is the relative delay of the lth path of user k with respect to the main path, while L_p is the total number of resolvable multipath components. Furthermore, θ_{kl} is the uniformly distributed phase-shift of the lth multipath component of the channel and $\delta(t)$ is the Kronecker delta function. More explicitly, the L_P multipath attenuations $\{h_{kl}\}$ are independent Nakagami-distributed random variables with a Probability Density Function (PDF) of [20–22]:

$$\begin{aligned} p(h_{kl}) &= M(h_{kl}, m_{kl}, \Omega_{kl}), \\ M(R, m, \Omega) &= \frac{2m^m R^{2m-1}}{\Gamma(m)\Omega^m} e^{(-m/\Omega)R^2}, \end{aligned} \quad (5)$$

where $\Gamma(\cdot)$ is the gamma function [19] and m_{kl} is the Nakagami-m fading parameter, which characterizes the severity of the fading for the lth resolvable path of user k [23] . Specifically, $m_{kl} = 1$ represents Rayleigh fading, $m_{kl} \to \infty$ corresponds to the conventional Gaussian scenario, and $m_{kl} = 1/2$ describes the so-called one-sided Gaussian fading, i.e., the worst-case fading condition. The Rician and log-normal distributions can also be closely approximated by the Nakagami distribution in conjunction with values of $m_{kl} > 1$. The parameter Ω_{kl} in Eq. 5 is the second moment of h_{kl}, i.e., we have $\Omega_{kl} = E[(h_{kl})^2]$. We assume a negative exponentially decaying Multipath Intensity Profile (MIP) given by $\Omega_{kl} = \Omega_{k0}e^{-\eta l}$, $\eta \geq 0, l = 0, \ldots, L_p - 1$, where Ω_{k0} is the average signal strength corresponding to the first resolvable path and η is the rate of average power decay.

4 BER performance analysis

As shown in Fig. 3, let the first user be the user of interest and consider a receiver using de-spreading as well as multipath diversity combining. The conventional matched filter-based RAKE receiver using maximum ration combining (MRC) can be invoked for detection, where we assume that the RAKE receiver combines a total of L_r number of diversity paths, which may be more or possibly less than the actual number of resolvable components at the current chip rate. It is advocated that the system has achieved perfect time synchronization and perfect estimates of the channel tap weights. Then, after appropriately delaying the outputs of the individual matched filter, in order to coherently combine the L_r number of path signals with the aid of the RAKE receiver, the output Z_{kl} of the RAKE receiver's lth finger sampled at $t = T + lT_c + \tau_k$ can be expressed as:

$$Z_{kl} = D_{kl} + I_{kl}, \quad (6)$$

where D_{kl} represents the desired direct component, which can be expressed as:

$$D_{kl} = \sqrt{2P}T_s b_k[0]\, h_{kl}^2, \quad (7)$$

where T_S is the symbol duration.

The MRC's decision variable Z_k, which is given by the sum of all the RAKE fingers' outputs, can be expressed as:

$$Z_k = \sum_{l=0}^{L_r-1} Z_{kl}. \quad (8)$$

The RAKE fingers' output signal Z_{kl} is a Gaussian distributed random variable with a mean of D_{kl}.

The term I_{kl} in Eq. 6 represents the total interference incurred by the Multipath Interference (MPI) as well as the Multiple Access Interference (MAI), which can be expressed as:

$$I_{kl} = I_{kl}[S] + I_{kl}[M] + N_k, \quad (9)$$

where $I_{kl}[S]$ represents the multipath interference imposed by the user of interest, which can be expressed as:

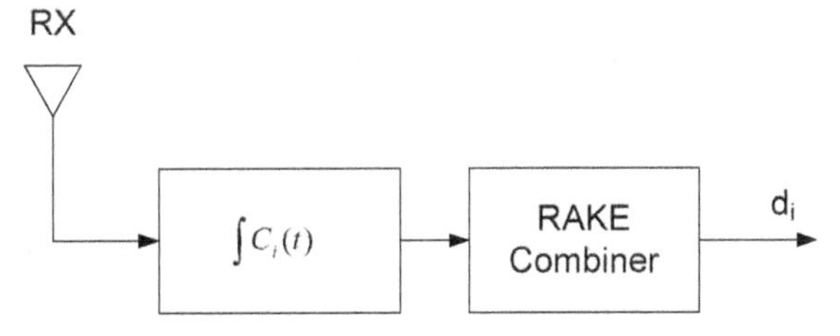

Fig. 3 CDD receiver structure

$$I_{kl}[S] = \sqrt{2P}T_s h_{kl} \sum_{\substack{l_p=0 \\ l_p \neq l}}^{L_p-1} \frac{h_{kl_p} \cos\theta_{kl_p}}{T_s}$$

$$\times \int_0^{T_s} b_k\left[t-(l_p-l)\,T_c\right] \cdot c\left[t-(l_p-l)\,T_c\right]c[t]dt. \quad (10)$$

Furthermore, $I_{kl}[M]$ represents the multiuser interference inflicted by the $K-1$ number of interfering signals, which can be expressed as:

$$I_{kl}[M] = \sqrt{2P}T_s h_{kl} \sum_{\substack{k'=1 \\ k' \neq k}}^{K} \sum_{l_p=0}^{L_p-1} \frac{h_{k'l_p} \cos\theta_{k'l_p}}{T_s}$$

$$\cdot \int_0^{T_s} b_{k'}[t-(l_p-l)T_c-(\tau_{k'}-\tau_k)] \times$$

$$c_{k'}[t-(l_p-l)T_c-(\tau_{k'}-\tau_k)]\, c[t]\, dt. \quad (11)$$

In Eqs. 10 and 11, the $\cos(\cdot)$ terms are contributed by the phase differences between the incoming carrier and the locally generated carrier used in the demodulation. Finally, the noise term in Eq. 9 can be expressed as:

$$N_{kl} = h_{kl} \int_0^{T_s} n(t)c[t]\cos(2\pi f_c t + \theta_{kl})dt, \quad (12)$$

which is a Gaussian random variable with a zero mean and a variance of $N_0 T_s h_{kl}^2$, where $\{h_{kl}\}$ represents the path attenuations.

Furthermore, $I_{kl}[S]$ represents the MPI imposed by the user of interest. Without loss of generality, we can assume that the width of the IFW is longer than the channel spread time $L_p T_c$, and in this case, the MPI becomes zero since the auto-correlation of the spreading code is zero.

Hence, the performance of the CDD system is mainly determined by the MAI term $I_{kl}[M]$, which is affected by channel delay profile, distance amongst the nodes, and the number of users supported in a cell.

Again, the uplink TX and downlink RX have different spreading sequences for the kth user in one CDD device, and recall that the system can assign different sets of the spreading codes for the uplink and downlink; more specifically, the spreading codes of the uplink and downlink have the maximum width of IFW in the generated LS codes. And the distance between the TX antenna and RX antenna is very close, i.e., $\tau_k = 0$. Hence, after the matched filter and RAKE receiver, the interference introduced by its own transmission can be ignored due to the inherent orthogonality of the spreading codes. Thus, the MAI are mainly introduced by other CDD device nodes.

Since the power control is difficult to be implemented in the CDD system, the near-far effect must be taken into account for the BER performance analysis. Assume that the power of the interested signals is P_k while the power of the interference is P_j, and P_k and P_j can be obtained from Eq. 3.

As shown in Fig. 4, if one path of the channel impulse response of the interference user is located within the IFW, this specific path will not introduce any interference to the interested signals. More explicitly, only these paths which are located outside the interference-free window will introduce the MAI to the desired signal. Without loss of generality, we can assume that the multipath intensity profile of all users obeys the same distribution, i.e., $\Omega_{k0} = \Omega_0, k = 1, 2, ...K$, and $E_b = PT_s$ is the energy per bit. Thus, the variance of the lth RAKE finger's output samples Z_{kl} for a given set of channel amplitudes $\{h_{kl}\}$ may be approximated as [23, 24]:

$$\sigma_{kl}^2 = 2PT_s^2 \left[\frac{\chi_{kl}}{3G} + \left(\frac{2\Omega_0 E_b}{N_0}\right)^{-1}\right] \cdot \Omega_0 h_{kl}^2, \quad (13)$$

$$\chi_{kl} = \sum_{j=1}^{K} \sum_{l=0}^{L_p-1} \frac{P_j}{P_k} e^{-\eta l} u\left(W_I - |\tau_j - \tau_k| - l\right) \quad (14)$$

where W_I is the width of the IFW; τ_j and τ_k are the propagation delay of the interference user j and the interested user k, respectively; L_p is the total number of resolvable paths; and $u(x)$ is the step function, i.e., $u(x) = 1, x \geq 0$, otherwise $u(x) = 0, x < 0$. According to the large-scale fading model, P_j and P_k are the strength of the received signals, which can be derived in Eq. 3 as $P_j = P_{tj} G_a \left[\frac{d_j}{d_0}\right]^{-\gamma}, P_k = P_{tk} G_a \left[\frac{d_k}{d_0}\right]^{-\gamma}$. Moreover, the propagation delay obeys $\tau_j = d_j/c, \tau_k = d_k/c$.

Hence, the MRC's output sample Z_k can be approximated by a Gaussian random variable with a mean of

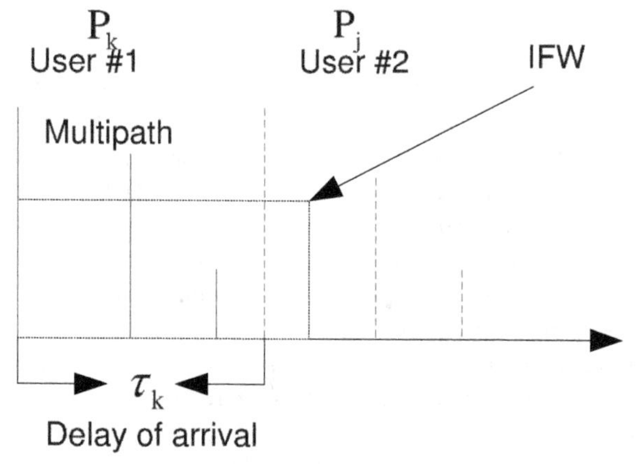

Fig. 4 MAI effect with IFW

$\mathrm{E}[Z_k] = \sum_{l=0}^{L_r-1} D_{kl}$ and a variance of $\mathrm{Var}[Z_k] = \sum_{l=0}^{L_r-1} \sigma_{kl}^2$ [21, 24], where we have:

$$\mathrm{E}[Z_k] = \sum_{l=0}^{L_r-1} \sqrt{2PT_s} b_k[0]\, h_{kl}^2, \qquad (15)$$

$$\mathrm{Var}[Z_k] = \sum_{l=0}^{L_r-1} \sigma_{kl}^2. \qquad (16)$$

Hence, the BER using BPSK modulation conditioned on a set of fading attenuations $\{h_{kl}, l = 0, 1, \ldots, L_r - 1\}$ can be expressed as [1, 18]:

$$P_b(\gamma) = Q\left(\sqrt{\frac{(\mathrm{E}[Z_k])^2}{\mathrm{Var}[Z_k]}}\right) = Q\left(\sqrt{\sum_{l=0}^{L_r-1} 2\gamma_l}\right), \qquad (17)$$

where $Q(x)$ represents the Gaussian Q-function which can also be simplified as [1, 23]:

$$Q(x) = \frac{1}{\pi}\int_0^{\pi/2} \exp\left(-\frac{x^2}{2\sin^2\theta}\right) d\theta, x \geq 0. \qquad (18)$$

Furthermore, $2\gamma_l$ in Eq. 17 represents the output Signal to Interference plus Noise Ratio (SINR) at the lth finger of the RAKE receiver.

Let us now substitute Eqs. 13, 15, and 16 into Eq. 17, and the expressions under the square-root functions must be equal, which allows us to express γ_l as follows:

$$\gamma_l = \left[\frac{\chi_{kl}}{3G} + \left(\frac{\Omega_0 E_b}{N_0}\right)^{-1}\right]^{-1} \cdot \frac{h_{kl}^2}{\Omega_0}. \qquad (19)$$

For the convenience of the BER derivation, we define a term γ_c as:

$$\gamma_c = \left[\frac{\chi_{kl}}{3G} + \left(\frac{\Omega_0 E_b}{N_0}\right)^{-1}\right]^{-1}. \qquad (20)$$

With the aid of Eq. 20, we have $\gamma_l = \gamma_c \cdot \frac{(h_l)^2}{\Omega_0}$ and h_l obeys the Nakagami-m distribution characterized by Eq. 5, and the PDF of γ_l can be formulated as:

$$p_{\gamma_l}(\gamma_l) = \left(\frac{m_l}{\overline{\gamma}_l}\right)^{m_l} \frac{\gamma^{m_l-1}}{\Gamma(m_l)} \exp\left(-\frac{m_l\gamma_l}{\overline{\gamma}_l}\right), \quad \gamma_l \geq 0, \qquad (21)$$

where $\overline{\gamma}_l = \gamma_c e^{-\eta l}$ for $l = 0, 1, \ldots, L_r - 1$. The average BER, $P_b(E)$, can be obtained by the weighted averaging of the conditional BER expression of Eqs. 17 and 21 over the joint PDF of the instantaneous SNR values.

5 Simulation performance of CDD

As we mentioned before, two scenarios are investigated in this paper. The first scenario is a cellular system which has a center node, and all other nodes besides the center node only communicate with the center one. The second scenario is a relay and ad hoc communication environment, where any two of the nodes can communicate with each other. For the CDD performance evaluation, we assume that the system has a chip rate of 1.2288 M chips which is the same as the CDMA2000 and EVDO systems [25]. The bandwidth of our system is 1.25 MHz. The carrier frequency of the system is $f_c = 2$ GHz, and the large-scale fading factor $\gamma = 2.5$. Moreover, the radius of a cell is configured to 2000 m. All the CDD nodes are operated in synchronous transmission mode, i.e., all nodes transmit simultaneously. We also assume that the maximum delay spread of the channel is $\tau = 3\ \mu s$ [11]. Thus, the number of resolvable paths is considered as $L_p = \lfloor \frac{\tau}{T_c} \rfloor + 1 = 4$, where $\tau = 3\ \mu s$, and the channel's delay spread profile is negatively exponentially distributed with the factor η in the range of $[0.3, 3]\ \mu s$. The RAKE receiver combines $L_r = 3$ number of fingers for BER performance evaluation, and the channel's exponential decay factor is $\eta = 0.2$.

In our simulation, the LS(4,32,4) codes are deployed as the spreading codes. More explicitly, the total length of the LS(4,32,4) code is $L = NP + 2W_0 = 136$ chips, and its spreading factor may be calculated by simply noting that each bit to be transmitted is spread by one of the constituent LS(4,32,4) codes. Although the length of this LS code is $L = NP + 2W_0 = 136$, the effective spreading gain of the LS code is identical to $G_{\text{eff}} = 128$ since the inserting zeros must not be taken into account. The LS(4,32,4) codes are constituted by a total of 128 spreading codes. The width of the IFW W_I and the number of supported user K satisfies that:

$$\begin{aligned} W_I &= 4, K \leq 16; \\ W_I &= 2, 16 < K \leq 32; \\ W_I &= 1, 32 < K \leq 64; \\ W_I &= 0, 64 < K \leq 128. \end{aligned}$$

First, we investigated the BER performance of CDD when there existed a strong interference CDD node. In this scenario, a cell has three CDD nodes, the source node, the destination node, and the interference node. In our simulation, the interference node has a transmission power 20 dB higher than the source node. It is obvious that the distance between the interference node and destination node dramatically affects the performance of the RAKE receiver since the strength of interference is inversely proportional to the distance. Figure 5 exhibits the BER performance of the destination node in conjunction with the distance between the interference node and destination node. More explicitly, we can observe that when the width of IFW W_I is below 2 and the distance of the interference node is less than 300 m, the effect of the interference node cannot be completely rejected by the interference spreading code so that the MAI leak inevitably in the RAKE receiver due to the insufficient width of the IFW. However, when the W_I increases to 3

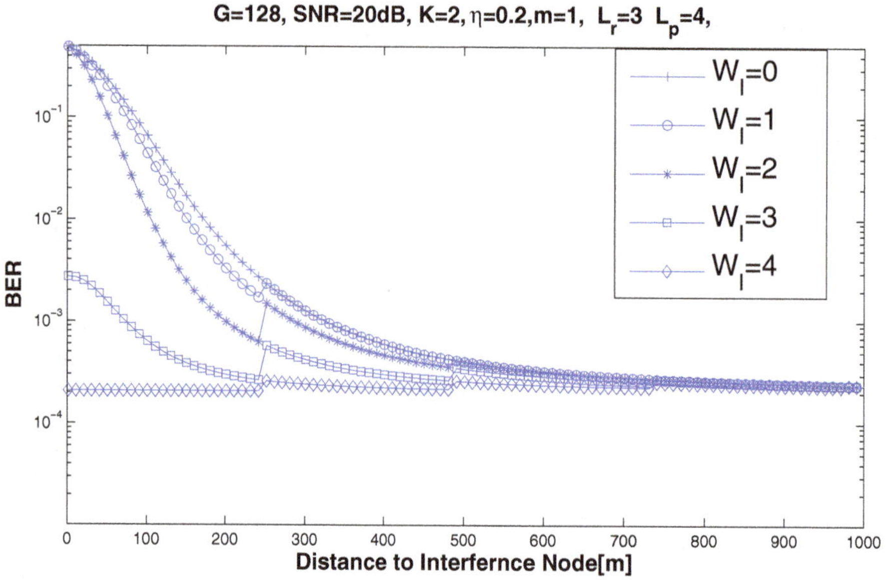

Fig. 5 BER vs distance of interference node in the cellular CDD communication system

or 4, the interference is suppressed significantly and the effect of the interference node is negligible.

The performance of CDD in two common scenarios is shown in Fig. 6. In the cellular scenario, a center node or Base Station (BS) exists in the network, which has a high transmission power than other nodes. In our simulation, we assume that the base station node has a 20-dB higher transmission power. The second is an ad hoc scenario where each node has the same transmission power. Figure 6 plots the BER performance of the CDD communication system in the context of a various number of users K. In Fig. 6, we can observe that the performance in the ad hoc scenario is superior than that in the cellular scenario. This phenomenon is reasonable since the cellular environment exists as a strong base station which also acts as a strong interference node to other destination nodes and then degrades the BER performance in the cellular condition.

Figures 7 and 8 exhibit the performance of the CDD system in conjunction with various supported users K. From Fig. 7, we can observe that the near interference-free performance can be achieved when the number of users K is less than 16 since the spreading codes have the maximum width of the IFW in this condition. As

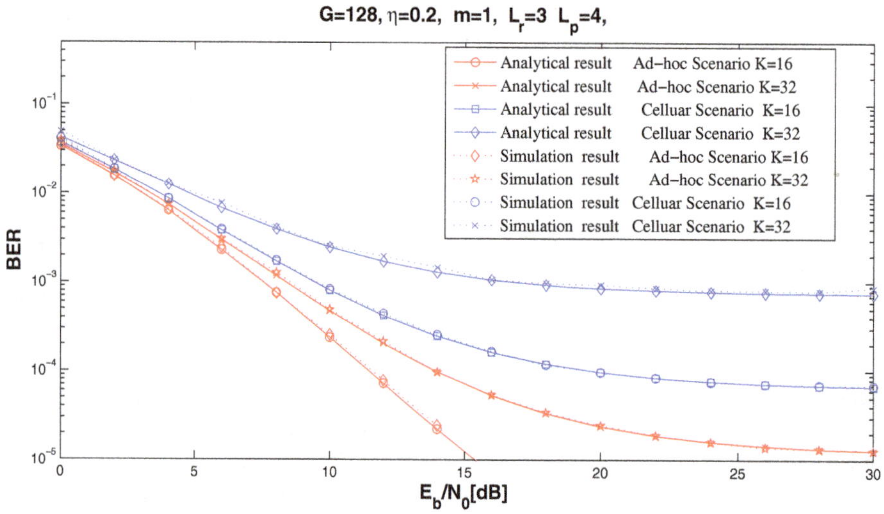

Fig. 6 BER vs E_b/N_0 performance in the context of the ad hoc and cellular scenarios

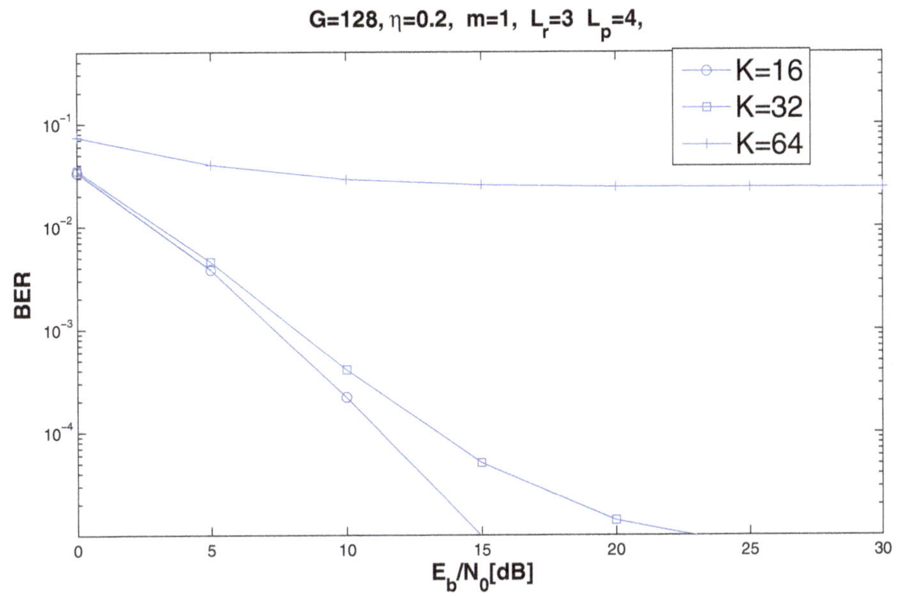

Fig. 7 BER performance of the CDD communication system with various number of users *K*

the number of users is increased to 32, the performance of the CDD system is degraded moderately, since the width of the IFW is still maintained in $W_I \leq 2$. However, when the number of users is increased to more than 64, the performance is degraded significantly since the width of IFW is decreased to $0 \sim 1$, which reduces the capability of combating the MAI. Figure 8 portrays the BER performance of the CDD in conjunction with various users *K*. Two different SNR parameters are configured in our simulation, i.e., SNR = 10 dB and SNR = 20 dB, respectively. From Fig. 8, we can conclude that the CDD system exhibits a near interference-free performance in the case $K \leq 32$. In Figs. 7 and 8, it is advocated that the CDD system should be deployed in a relatively-low

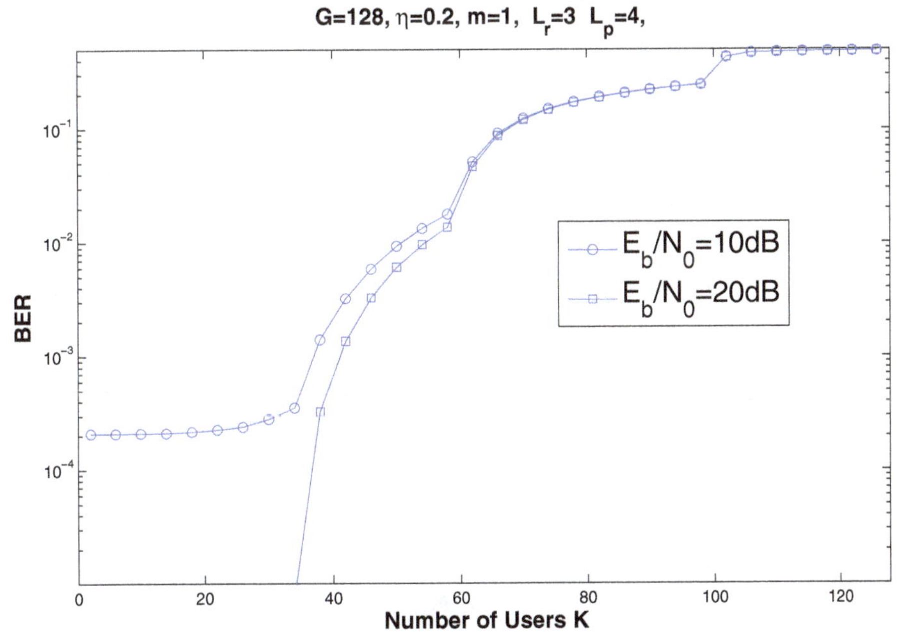

Fig. 8 BER performance vs. the number of supported users *K* in a cell

load of users. For a spreading gain of 128 systems, the CDD system can support $K = 40$ user communications simultaneously without the mutual interference; this load is promising since the linear Minimum Mean Square Estimator (MMSE) multiuser detector with a much higher complexity can only support to $K = 40$ to achieve an ideal performance. Similarly, the performance of the linear MMSE multiuser detector [26, 27], the Parallel Interference Canceller (PIC) [26, 27], and Serial Interference Canceller (SIC) [26, 27] deteriorates significantly for the high load scenario in a CDMA system; furthermore, the traditional MMSE detector, PIC, and SIC suffer the serious near-far effect and degrade the performance in our CDD scenarios. The ideal Maximum Likelihood detector [26, 27] can solve this issue; however, the complexity of the ML detector is excessively high so that it is impossible to implement in practical communication systems.

Finally, Fig. 9 plots the BER performance of different channel fading models. The Nakagami-m channel is a generic model for the wireless channel, where the choice of the parameter m determines the fading type of the channel model. In Fig. 9, three different fading models are simulated and analyzed in our investigation. More explicitly, the Rayleigh fading [1, 18], Rician fading [1, 18], and Gaussian Fading [1, 18] are evaluated in the simulation. In the practical applications in a cellular communication, the Rayleigh fading and Rician fading model are commonly adopted for the performance evaluation. Hence, in this detection, we evaluate the CDD performance in conjunction with the real scenario of the cellar communications. The simulation results in Fig. 9 conform to the mathematic analysis, and we can conclude that the performance of Rayleigh fading is the worst and the performance of Gaussian fading is the best amongst these three fading models.

6 Conclusions

In this paper, we investigated the BER performance of the CDD communication system, which can exhibit the same operation fashion as the NDD device. Moreover, the advantage of the CDD is that the receiver can be simplified as the matched filter or RAKE receiver by carefully selecting our proposed interference rejection codes—LS codes. Hence, our proposed method can significantly reduce the complexity and difficulty in the implementation of the NDD fashion. Our simulation results show that the CDD system can achieve a near interference-free performance in a low-user load scenario due to the limited number of available LS codes having a certain IFW width; more explicitly, our simulation and analysis show that the CDD system can achieve an ideal performance when the number of user $K \leq 32$, and in this condition, a simple matched filter or RAKE receiver is sufficient to combat all the interference. For future research, the method of the smart assignment of the spreading codes will be investigated to reduce interference between neighboring nodes and increase the capacity and spectrum efficiency of the CDD network.

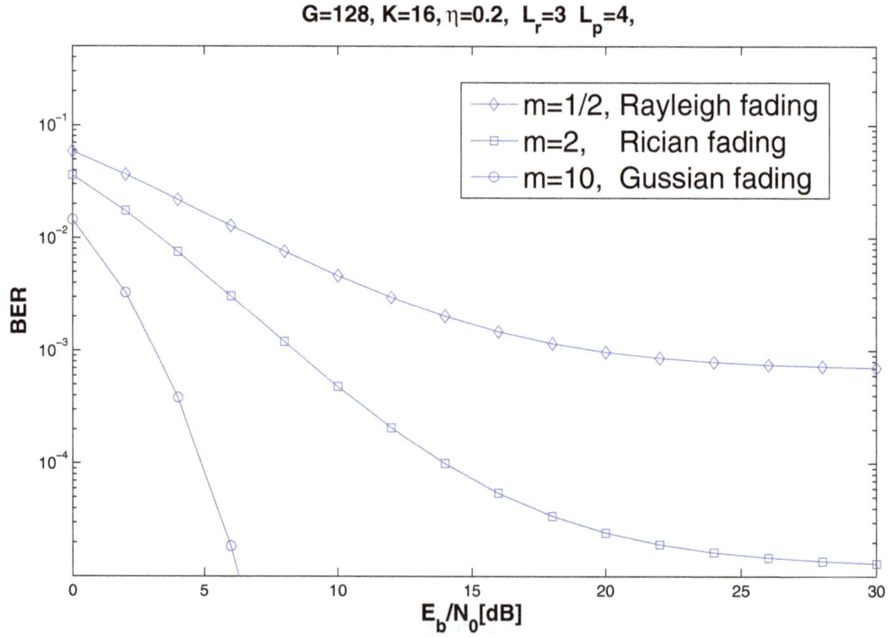

Fig. 9 BER performance of the CDD communication system in adjunction with different fading channel models

Acknowledgements
This work is supported by the Scientific Research Foundation of Chengdu Technological University (grant no. 2017RC004) and the Scientific Research Foundation of CUIT (KYTZ201501, KYTZ201502); the Sichuan Provincial Department of Science and Technology in granted program nos. 2015GZ0340, 2015GZ0286, 2015GZ0290, and 2015RZ0060; and the National Defense Innovation Technology Program 173-162-11-ZT-003-012-02. Finally, great thanks to the anonymous reviewers for the their valuable suggestions to improve the quality of this paper.

Funding
The work is sponsored by Chengdu Technological University (grant no. 2017RC004); Chengdu University of Information and Technology (KYTZ201501, KYTZ201502); Sichuan Provincial Department of Science and Technology in granted program nos. 2015GZ0340, 2015GZ0286, 2015GZ0290, and 2015RZ0060; and the National Defense Innovation Technology Program 173-162-11-ZT-003-012-02.

Authors' contributions
YY and HW contributed to the conception and design of the study. YY, LL, GL, and HW contributed to the acquisition of data and simulation. YY, GL, LL, and HW contributed to the analysis and interpretation of the simulation data. All authors read and approved the final manuscript.

Authors' information
Dr. Yufang Yin obtained her Ph.D. degree in 2007 from the EE department, the University of Texas at San Antonio. From January 2009 to June. 2012, she worked as the specialist and research scientist in Nokia Networks. From July 2012 to October 2014, she worked as the staff engineer in Spread Spectrum Communication. Since November 2016, she joined Chengdu Technological University. Her research area includes statistical signal processing and applied machine learning.
Gangjun Li received the B.S. degree in Mechanical Engineering from the University of Electronic Science and Technology of China in 1987 and the M.S. degree in Mechanics from the Harbin Institute of Technology in 1990. He received the Ph.D. degree in Robotics Engineering from the Southwest Jiaotong University in 2002. Since 2010, he is a professor of Chengdu Technological University (CDTU). His research interests are in areas of robotics, simulation, modeling of complex systems, and nonlinear control.
Dr. Li (Alex) Li obtained his Ph.D. degree in electronics engineering from the University of York in 2014, and now he is working in Chengdu University of Information Technology, China. His research interests include fields of wireless communication, MIMO-OFDM, adaptive filtering, and ASIC design.
Hua Wei obtained his Ph.D. degree from the ECS department, University of Southampton in September 2005. From January 2006 to September 2014, he worked as the principle engineer in Spread spectrum Communication for wireless modem design. Since September 2014, he joined in Chengdu University of information Technology. His research area includes adaptive equalization, signal processing, and VLSI design for wireless baseband processor.

Competing interests
The authors declare that they have no competing interests.

Author details
[1]Chengdu Technological University, Chengdu, China. [2]Chengdu University of Information and Technology, Chengdu, China.

References
1. A GoldSmith, *Wireless communications*. (Cambridge University Press, 2005)
2. T Riihonen, S Werner, R Wichman, Mitigation of loopback self-interference in full-duplex MIMO relays. IEEE Trans. Sign. Process. **59**(12), 5983–5993 (2011)
3. M Duarte, C Dick, A Sabharwal, Experiment-driven characterization of full-duplex wireless systems. IEEE Trans. Wirel. Commun. **1**(12), 4296–4307 (2012)
4. L Hanzo, LL Yang, EL Kuan, K Yen, *Single- and multi-carrier DS-CDMA*. (John Wiley and IEEE Press, 2003), p. 1060
5. P Fan, L Hao, Generalized orthogonal sequences and their applications in synchronous CDMA systems. IEICE Trans. Fundam. **E83-A**, 2054–2069 (2000)
6. S Stańczak, H Boche, M Haardt, in *GLOBECOM '01*. Are LAS-codes a miracle? vol. 1 (IEEE Global Telecommunications Conference, San Antonio, 2001), pp. 589–593
7. CC Tseng, CL Liu, Complementary sets of sequences. IEEE Trans. Inf. Theory. **18**, 644–652 (1972)
8. H Wei, L Hanzo, On the uplink performance of asynchronous LAS-CDMA. IEEE Trans. Wirel. Commun. **5**, 1187–1196 (2006)
9. H Wei, L-L Yang, L Hanzo, Interference-free broadband single- and multi-carrier DS-CDMA. IEEE Commun. Mag. **43**(2), 68–73 (2005)
10. S Ni, H Wei, JS Blogh, L Hanzo, Network performance of asynchronous UTRA-like FDD/CDMA systems using loosely synchronised spreading codes. IEEE Veh. Technol. Conf. **2**, 1359–1363 (2003)
11. WCY Lee, *Mobile communications engineering*, 2nd ed. (McGraw-Hill, New York, 2008)
12. L-L Yang, W Hua, L Hanzo, Multiuser detection assisted time- and frequency-domain spread multicarrier code-division multiple-access. IEEE Trans. Veh. Technol. **55**(1), 397–404 (2006)
13. H Wei, L-L Yang, L Hanzo, Downlink space–time spreading using interference rejection codes. IEEE Trans. Veh. Technol. **55**(6), 1838–1847 (2006)
14. H Wei, L Hanzo, in *2005 6th IEE International Conference on 3G and Beyond*. LAS-CDMA Using Various Time Domain Chip-Waveforms (IET Conference Publications, 2005)
15. X Liu, H Wei, L Hanzo, Analytical bit error rate performance of DS-CDMA ad hoc networks using large area synchronous spreading sequences. IET Commun. **1**(4), 760–764 (2007)
16. RL Frank, Polyphase complementary codes. IEEE Trans. Inf. Theory. **26**, 641–647 (1980)
17. R Sivaswamy, Multiphase complementary codes. IEEE Trans. Inf. Theory. **24**, 546–552 (1978)
18. D Tse, P Viswanath, *Fundamentals of wireless communication*. (Cambridge University Press, 2005)
19. JG Proakis, *Digital communications*, 3rd edn. (Mc-Graw Hill International Editions, 1995)
20. N Nakagami, in *Statistical methods in radio wave propagation*, ed. by WG Hoffman. The m-distribution, a general formula for intensity distribution of rapid fading (Pergamon, Oxford, 1960)
21. T Eng, LB Milstein, Coherent DS-CDMA performance in Nakagami multipath fading. IEEE Trans. Commun. **43**, 1134–1143 (1995)
22. V Aalo, O Ugweje, R Sudhakar, Performance analysis of a DS/CDMA system with noncoherent M-ary orthogonal modulation in Nakagami fading. IEEE Trans. Veh. Technol. **47**, 20–29 (1998)
23. M-S Alouini, AJ Goldsmith, A unified approach for calculating error rates of linearly modulated signals over generalized fading channels. IEEE Trans. Commun. **47**, 1324–1334 (1999)
24. MK Simon, M-S Alouini, A unified approach to the probability of error for noncoherent and differentially coherent modulation over generalized fading channels. IEEE Trans. Commun. **46**, 1625–1638 (1998)
25. 3GPP2, Overviews of the CDMA2000 standards. Website: www.3gpp2.org
26. X Wang, VH Poor, *Wireless communication systems: advanced techniques for signal reception*. (Prentice Hall, 2011)
27. HV Poor, L Tong, *Signal processing for wireless communication system*. (Springer, 2012)

15

Joint ABS and user grouping allocation for HetNet with picocell deployment in downlink

Wei-Chen Pao[1], Jhih-Wei Lin[2], Yung-Fang Chen[2*] and Chin-Liang Wang[3]

Abstract

In order to resolve the co-channel inter-cell interference problem in heterogeneous networks (HetNet), the feature of almost blank subframes (ABS) in the time domain of the enhanced inter-cell interference coordination (eICIC) is utilized. In this paper, an ABS configuration design is developed on downlink in HetNet and the associated resource allocation problem for maximizing the system performance with fairness among user equipments (UEs) is considered. Compared to conventional problems, the resource assignment problems include the configuration of ABS pattern and the resource allocation for macro UEs and pico UEs, which aims to maximize the downlink throughput and balance the traffic offloading in intra-frequency HetNet deployments. First, this paper introduces an ABS pattern design by using the channel condition, which is developed in terms of the time domain resource. Subframes are categorized as protected or normal subframes for reducing interference impact to pico UEs. Based on the configuration of the ABS pattern, we develop a grouping strategy to determine which pico UEs use either protected or normal subframes. Besides, the assignment of resource blocks with respect to the resource in the frequency domain is developed along with the fairness among UEs. The proposed joint allocation scheme takes the system throughput and the fairness into account, and has better performance than the existing schemes. Simulation results also reveal that the performance of the proposed joint allocation scheme approximates the optimal solution with the full search scheme.

Keywords: Inter-cell interference coordination, Almost blank subframe, Heterogeneous network, Proportional fairness, Resource allocation

1 Introduction

Orthogonal frequency division multiple access (OFDMA) system has been chosen for the next-generation broadband wireless system standards [1] in order to satisfy the growing demands on the high data traffic in the mobile communication systems. The Third Generation Partnership Project (3GPP) Long Term Evolution (LTE) has been regarded as a promising mobile technology with increased system sum rates [2]. The LTE-Advanced (LTE-A) is an evolution of LTE, which achieves International Mobile Telecommunications (IMT)-Advanced requirements [3–5]. With more data traffic demand in the future, enhancing the system spectral efficiency by the deployment of traditional macro eNodeBs (eNBs) has high cost. Therefore, heterogeneous networks (HetNet) have been widely discussed in 3GPP LTE-A standards [6, 7]. HetNet includes high-power macro eNBs and low-power nodes, such as femto eNBs, pico eNBs, and relays [8–10].

In order to offload user equipments (UEs) from a macro eNB to a pico eNB more efficiently, cell range expansion (CRE) has been introduced in 3GPP LTE-A [11]. In the CRE method, a bias value is introduced and added to reference signal received power (RSRP) of pico eNBs. A CRE region is introduced where the system pretends that UEs have better signal quality from the pico eNB. Therefore, the system may tend to offload UEs from the macro eNB to the pico eNB. More UEs may connect to pico eNBs for load balancing. The control of the bias and the related offloading is discussed in [12, 13]. However, the UE which is handed over from the macro eNB to the pico eNB with the CRE technique may suffer severe interference from the macro eNB since the received signal from

* Correspondence: yfchen@ce.ncu.edu.tw
[2]Department of Communication Engineering, National Central University, Taoyuan, Taiwan, Republic of China
Full list of author information is available at the end of the article

pico eNB is still weak. A major problem in HetNet [14] is the cross-tier inter-cell interference (ICI) because the low-power nodes such as pico eNBs share the same frequency band with the macro eNBs. In order to improve the performance and reduce the cross-tier interference, a major feature of enhanced inter-cell interference coordination (eICIC) [10, 15–18] is to coordinate inter-cell interference in time domain by implementing almost blank subframe (ABS). The method of time-domain multiplexing (TDM) using ABS [19, 20] is introduced to avoid heavy ICI on both data and control channels of the downlinks. When the ABS scheme is employed, subframes will be further configured as either normal subframes or protected subframes. For normal subframes, UEs served by macro eNBs (macro UEs) and served by pico eNBs (pico UEs) are all allowed to use these subframes. For protected subframes, only pico UEs are allowed to use those subframes. The designs of the CRE region and the ABS pattern can be developed to have gain and benefit for the whole system, such as traffic offloading and throughput.

Regarding the above discussion, we focus on the challenges for maximizing the system performance with fairness among UEs in HetNet which include the configuration design of the ABS pattern and the resource allocation for macro UEs and pico UEs in terms of the time domain resource, i.e., subframe configuration including protected subframe and normal subframe where the resource is located in the time domain, and the frequency domain resource, i.e., subcarrier allocation to multiple UEs where this resource is located in the frequency domain. Due to the design of the CRE region and ABS to protect Pico UEs, the resource allocation problem becomes more complicated. The ABS pattern needs to be configured and coordinated among eNBs for the purpose of the maximization of system capacity or throughput. Also, subcarriers should be properly assigned to macro UEs and Pico UEs. The associated configuration, i.e., a normal or a protected subframe of a particular subcarrier will determine the amount of the suffered interference. We also consider fairness among Pico UEs. Consequently, the proposed resource allocation scheme comprises the ABS configuration, the Pico UE grouping, and the subcarrier allocation.

First, this paper focuses on the design for the configuration of an ABS pattern. The problems of cell selection combined with ABS density are investigated in [21–23]. The channel state is usually assumed to be time invariant in the period of an ABS pattern [24, 25]. Thus, most of papers only discuss the ABS density without determining which subframes should be configured as protected subframes. The system performance, e.g., throughput, is affected by the ABS configuration since each subcarrier or each UE experiences different channel condition. In the view of the time domain, the ABS pattern design strategy is developed. We utilize the channel condition to design an evaluation function as an indicator which aims to maximize the sum rate of the system. The indicator can efficiently determine which subframes should be configured as protected subframes. Different from the previous work [22], the ABS pattern is dynamically adjusted, instead of fixed.

Second, this paper develops a pico UE grouping strategy based on an optimization technique to determine which pico UEs use protected subframes or normal subframes. Various schemes of pico UE grouping [24–27] have been investigated in the HetNet with CRE. The simplest way is that all pico UEs are assigned to use protected subframes [26], which means pico eNBs only work in protected subframes. This scheme may cause the reduced performance in pico eNBs due to the limited resources. In [27], some pico UEs which are handovered from a macro eNB to a pico eNB with the CRE technique are assigned to the protected UE set and others are assigned to the normal UE set. In [25], "brute force search" by employing integer programming is used to get the optimal solution for the problem of the time-domain resource partitioning for enhanced inter-cell interference coordination. However, this scheme has a high computational complexity. In order to reduce the computational complexity, one scheme [24] using Nash bargaining solution (NBS) is proposed to reduce the complexity, but the NBS scheme only finds a sub-optimal solution in some sense. Therefore, it raises our motivation to develop a pico UE grouping scheme with a low computational complexity while approximating the optimal solution. We also develop a strategy for the pico UE grouping to determine which pico UEs use either the normal subframes or the protected subframes. Deviated from the tradition UE grouping methods [28–30], Pico UEs will be redistributed by using a fast adjustment in a group basis and a refinement mechanism on a per-UE basis. Accompanied by the re-distribution, the radio resource blocks in the frequency domain are allocated jointly per subframe.

Finally, in this paper, we propose joint allocation scheme to maximize the system performance while considering fairness among UEs by appending some processing procedures. A dynamic ABS pattern design is introduced. Based on the ABS pattern, UE grouping strategies including the fast adjustment and the refinement mechanism are introduced. Meanwhile, the radio resource allocation is executed as well. The proposed scheme outperforms the existing schemes [24, 27] and approaches the full search scheme [25] while the computational complexity is greatly reduced. The proposed algorithm is different from other works, such as (1) time-domain resource portioning [23, 24, 27] without user grouping or subcarrier allocation, (2) fair scheduling [31, 32] without ABS configuration and user grouping, or (3) user selection and resource allocation algorithm [33] without ABS configuration.

Our contributions and new ideas include (a) a joint allocation scheme is first developed in the time domain and the frequency domain, including the consideration of the proportional fairness among UEs; (b) a new evaluation function based on an optimization technique is proposed to determine the configuration of subframes, i.e., normal and protected subframes. A low complexity associated strategy is thus proposed; (c) another new evaluation function is derived for UE grouping, i.e., in either the normal or the protected UE set; (d) the proposed scheme outperforms the existing schemes [24, 27], and approaches the optimal solution with the full search scheme [25].

2 System model and problem formulation

We consider a downlink transmission scenario of the two-tier cellular network in an OFDMA system, which consists of one macro eNB and one pico eNB. The HetNet deployment including a pico eNB within the coverage of a macro eNB is depicted in Fig. 1. In the cell range expansion (CRE) [11] technique, a bias value is introduced and added to reference signal received power (RSRP) of pico eNBs. The numbers of UEs served by the macro eNB and the pico eNB are denoted by $K^{'}$ and $K^{''}$, respectively. The corresponding UE sets are $\Omega^{(\text{Macro})}$ and $\Omega^{(\text{Pico})}$. All K UEs ($K = K^{'} + K^{''}$) are uniformly distributed within the coverage of the macro eNB. Each UE is only allowed to connect a single eNB. The serving eNB for the kth UE is determined by comparing the values of RSRP [29, 30]. The kth UE chooses the pico eNB with a bias if the following holds true:

$$\text{RSRP}_k^{(\text{Pico})} + \text{bias} \geq \text{RSRP}_k^{(\text{Macro})} \tag{1}$$

where $\text{RSRP} = p_r \cdot |G|^2$. G means the channel gain and p_r means the reference signal power per resource element [29]. bias (in dB) is chosen to be a positive value toward pico eNBs. Otherwise, the kth UE is served by the macro eNB.

The offloaded UEs in the CRE region may suffer from ICI of macro eNBs. In normal subframes, macro UEs and pico UEs are all allowed to use those subframes. In this paper, pico UEs will be further classified into two sets, a normal UE set $\Omega^{(\text{normal})}$ and a protected UE set $\Omega^{(\text{protected})}$. UEs in the normal UE set use the normal subframes, and UEs in the protected UE set use the protected subframes. We also assume that the same ABS patterns are configured for all eNBs under consideration, which is the synchronous configuration of ABS pattern [34] as shown in Fig. 1. An ABS pattern means the pattern of the protected subframes in a frame. The channel state information is known to eNBs and the information of ABS configuration exchanges through X2 interface between all eNBs in LTE [19].

We assume that the system has N resource blocks (RBs) and each RB includes Q subcarriers. For pico UEs, the signal to interference plus noise ratio (SINR) for the

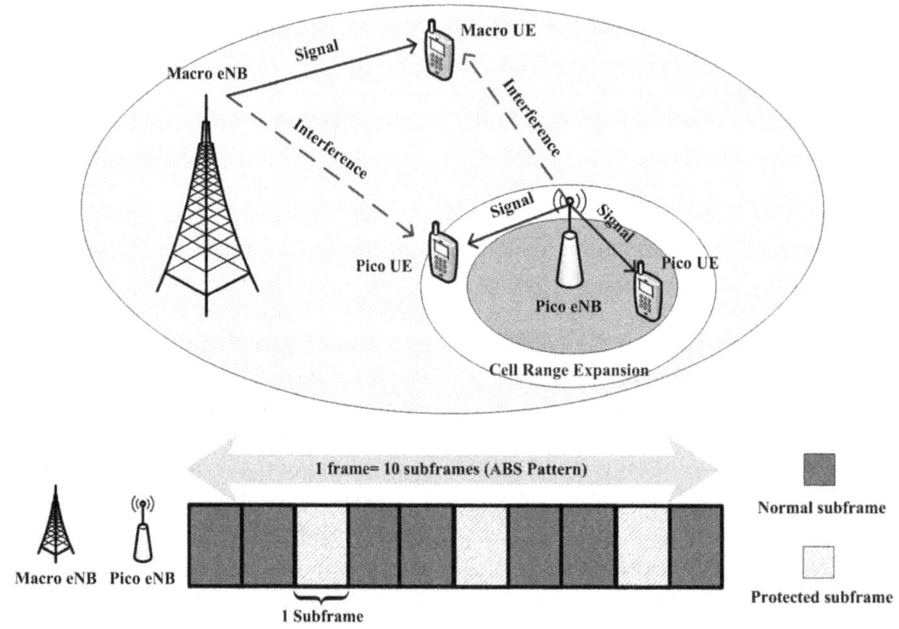

Fig. 1 Heterogeneous network with inter-cell interference and cell range expansion

k''th pico UE on the qth subcarrier of the nth RB at the ith subframe can be expressed as:

$$\mathrm{SINR}_{k'',n,q}^{(\mathrm{Pico})}[i] = \frac{\left|G_{k'',n,q}^{(\mathrm{Pico})}[i]\right|^2 p_{k'',n,q}^{(\mathrm{Pico})}[i]}{N_0\Delta f + \left|G_{k'',n,q}^{(\mathrm{Macro})}[i]\right|^2 \varepsilon[i] p_{k',n,q}^{(\mathrm{Macro})}[i]} \quad (2)$$

where i is the subframe index; $G_{k'',n,q}^{(\mathrm{Pico})}[i]$denotes the channel gain between the pico eNB and the k''th pico UE on the qth subcarrier of the nth RB; $p_{k'',n,q}^{(\mathrm{Pico})}[i]$ is the amount of power for the k''th pico UE on the qth subcarrier of the nth RB; N_0 is the AWGN noise power spectral density; $\Delta f$is the subcarrier spacing; $G_{k'',n,q}^{(\mathrm{Macro})}[i]$ denotes the channel gain between the macro eNB and the k''th pico UE on the qth subcarrier of the nth RB; $p_{k',n,q}^{(\mathrm{Macro})}[i]$ is the amount of power for the k'th macro UE on the qth subcarrier of the nth RB; $\varepsilon[i]$denotes the binary indicator; $\varepsilon[i] = 1$represents that the ith subframe is the normal subframe; $\varepsilon[i] = 0$represents that the ith subframe is the protected subframe.

In the frequency domain, the basic resource allocation unit for scheduling is the RB unit. In the time domain, resource allocation scheduling is performed based on the subframe unit. Therefore, the data rate for the k''th pico UE on the nth RB at the ith subframe can be expressed as:

$$r_{k'',n}^{(\mathrm{Pico})}[i] = \sum_{q=1}^{Q} \Delta f \log_2\left(1 + SINR_{k'',n,q}^{(\mathrm{Pico})}[i]\right) = \begin{cases} r_{k''}^{(\mathrm{normal})}[i], k'' \in \Omega^{(\mathrm{normal})} \\ r_{k''}^{(\mathrm{protected})}[i]\ k'' \in \Omega^{(\mathrm{protected})} \end{cases} \quad (3)$$

where $r_{k'',n}^{(\mathrm{normal})}[i]$and $r_{k'',n}^{(\mathrm{protected})}[i]$ are the data rate for the k''th pico UE on the nth RB in the normal UE set and the protected UE set, respectively.

The sum of the data rate for the k''th pico UE at the ith subframe can be expressed as:

$$\begin{aligned} R_{k''}^{(\mathrm{Pico})}[i] &= \sum_{n=1}^{N} \rho_{k'',n}^{(\mathrm{Pico})}[i] r_{k'',n}^{(\mathrm{Pico})}[i] \\ &= \sum_{n=1}^{N} \rho_{k'',n}^{(\mathrm{Pico})}[i] \sum_{q=1}^{Q} \Delta f \log_2\left(1 + SINR_{k'',n,q}^{(\mathrm{Pico})}[i]\right) \\ &= \begin{cases} R_{k''}^{(\mathrm{normal})}[i], k'' \in \Omega^{(\mathrm{normal})} \\ R_{k''}^{(\mathrm{protected})}[i]\ k'' \in \Omega^{(\mathrm{protected})} \end{cases} \end{aligned} \quad (4)$$

where $\rho_{k'',n}^{(\mathrm{Pico})}[i]$ denotes the binary indicator. $\rho_{k'',n}^{(\mathrm{Pico})}[i] = 1$ represents that the nth RB is assigned to the k''th pico UE. $R_{k''}^{(\mathrm{normal})}[i]$ and $R_{k''}^{(\mathrm{protected})}[i]$ are the sum of data rate for the k''th pico UE in the normal UE set and the protected UE set, respectively. Note that, if the ith subframe is normal, $R_{k''}^{(\mathrm{normal})}[i]$ is evaluated and $R_{k''}^{(\mathrm{protected})}[i]$ equals 0. Similarly, for macro UEs, the SINR for the k'th macro UE on the qth subcarrier of the nth RB at the ith subframe can be expressed as:

$$\mathrm{SINR}_{k',n,q}^{(\mathrm{Macro})}[i] = \frac{\left|G_{k',n,q}^{(\mathrm{Macro})}[i]\right|^2 \varepsilon[i] p_{k',n,q}^{(\mathrm{Macro})}[i]}{N_0\Delta f + \left|G_{k',n,q}^{(\mathrm{Pico})}[i]\right|^2 p_{k'',n,q}^{(\mathrm{Pico})}[i]} \quad (5)$$

The data rate for the k'th macro UE on the nth RB at the ith subframe can be expressed as:

$$r_{k',n}^{(\mathrm{Macro})}[i] = \sum_{q=1}^{Q} \Delta f \log_2\left(1 + SINR_{k',n,q}^{(\mathrm{Macro})}[i]\right). \quad (6)$$

The sum of the data rate for the k'th macro UE at the ith subframe can be expressed as:

$$\begin{aligned} R_{k'}^{(\mathrm{Macro})}[i] &= \sum_{n=1}^{N} \rho_{k',n}^{(\mathrm{Macro})}[i] r_{k',n}^{(\mathrm{Macro})}[i] \\ &= \sum_{n=1}^{N} \rho_{k',n}^{(\mathrm{Macro})}[i] \sum_{q=1}^{Q} \Delta f \log_2\left(1 + SINR_{k',n,q}^{(\mathrm{Macro})}[i]\right) \end{aligned} \quad (7)$$

where $\rho_{k',n}^{(\mathrm{Macro})}[i]$ denotes the binary indicator. $\rho_{k',n}^{(\mathrm{Macro})}[i] = 1$ represents that the nth RB is assigned to the k'th macro UE. Therefore, the sum rate of the system is known as $\sum_{i=1}^{I}\left\{\sum_{k'=1}^{K'} R_{k'}^{(\mathrm{Macro})}[i] + \sum_{k''=1}^{K''} R_{k''}^{(\mathrm{Pico})}[i]\right\}$.

Our goal is to maximize the system performance subject to the transmit power constraints while considering the fairness among all UEs [31, 32]. The resource allocation problem for the time domain eICIC is formulated as:

$$\begin{aligned} &\max_{\varepsilon[i],\varepsilon^*[i],C_{k''},C^*_{k''}} \left\{\sum_{i=1}^{I}\left(\sum_{k'=1}^{K'} \frac{\varepsilon[i] R_{k'}^{(\mathrm{Macro})}[i]}{\overline{R}_{k'}^{(\mathrm{Macro})}[i]} + \sum_{k''=1}^{K''} \frac{\varepsilon[i] C_{k''} R_{k''}^{(\mathrm{normal})}[i]}{\overline{R}_{k''}^{(\mathrm{Pico})}[i]} + \sum_{k''=1}^{K''} \frac{\varepsilon^*[i] C^*_{k''} R_{k''}^{(\mathrm{protected})}[i]}{\overline{R}_{k''}^{(\mathrm{Pico})}[i]}\right)\right\} \\ &= \left\{\sum_{i=1}^{I}\left(\sum_{k'=1}^{K'} \frac{\varepsilon[i]}{\overline{R}_{k'}^{(\mathrm{Macro})}[i]} \sum_{n=1}^{N} \rho_{k',n}^{(\mathrm{Macro})}[i] r_{k',n}^{(\mathrm{Macro})}[i] + \sum_{k''=1}^{K''} \frac{\varepsilon[i] C_{k''}}{\overline{R}_{k''}^{(\mathrm{Pico})}[i]} \sum_{n=1}^{N} \rho_{k'',n}^{(\mathrm{Pico})}[i] r_{k'',n}^{(\mathrm{normal})}[i] + \sum_{k''=1}^{K''} \frac{\varepsilon^*[i] C^*_{k''}}{\overline{R}_{k''}^{(\mathrm{Pico})}[i]} \sum_{n=1}^{N} \rho_{k'',n}^{(\mathrm{Pico})}[i] r_{k'',n}^{(\mathrm{protected})}[i]\right)\right\} \end{aligned} \quad (8)$$

subject to

$$\sum_{k' \in \Omega^{(\mathrm{Macro})}} \sum_{n=1}^{N} \sum_{q=1}^{Q} p_{k',n,q}^{(\mathrm{Macro})}[i] \leq P^{(\mathrm{Macro})}[i] \tag{9}$$

$$\sum_{k'' \in \Omega^{(\mathrm{Pico})}} \sum_{n=1}^{N} \sum_{q=1}^{Q} p_{k'',n,q}^{(\mathrm{pico})}[i] \leq P^{(\mathrm{Pico})}[i] \tag{10}$$

$$\sum_{k'=1}^{K'} \rho_{k',n}^{(\mathrm{Macro})}[i] = 1, \quad \text{for } n = 1, \ldots, N \tag{11}$$

$$\sum_{k''=1}^{K''} \rho_{k'',n}^{(\mathrm{Pico})}[i] = 1, \quad \text{for } n = 1, \ldots, N \tag{12}$$

$$\varepsilon[i] + \varepsilon^{*}[i] = 1 \tag{13}$$

$$C_{k''} + C_{k''}^{*} = 1 \tag{14}$$

where the first term of Eq. (8) is the feasible rate relative to its current average data rate of the macro UEs coupled with the average data rate $\overline{R}_{k'}^{(\mathrm{Macro})}[i]$ of the k'th macro UE; the second term of Eq. (8) is the feasible rate relative to its current average data rate of the pico UEs which are in the normal UE set coupled with the average data rate $\overline{R}_{k''}^{(\mathrm{Pico})}[i]$ of the k''th pico UE; the third term of Eq. (8) is the feasible rate relative to its current average data rate of the pico UEs which are in the protected UE set coupled with the average data rate $\overline{R}_{k''}^{(\mathrm{Pico})}[i]$ of the k''th pico UE; the average data rate for macro UEs and pico UEs is defined as:

$$\overline{R}^{k'(\mathrm{Macro})}[i+1] = \left(1 - \frac{1}{JI}\right) \overline{R}_{k'}^{(\mathrm{Macro})}[i] + \frac{1}{JI} R_{k'}^{(\mathrm{Macro})}[i] \tag{15}$$

and

$$\overline{R}_{k''}^{(\mathrm{Pico})}[i+1] = \left(1 - \frac{1}{JI}\right) \overline{R}_{k''}^{(\mathrm{Pico})}[i] + \frac{1}{JI} R_{k''}^{(\mathrm{Pico})}[i]. \tag{16}$$

$1/\overline{R}_{k'}^{(\mathrm{Macro})}[i]$ and $1/\overline{R}_{k''}^{(\mathrm{Pico})}[i]$ in Eq. (8) are the proportional fairness (PF) factors which are introduced to strike a balance between the system performance and the fairness among UEs [32]; i is the subframe index; I is the number of subframes in one frame; J is the window size which is the number of the frames in a PF period; $\varepsilon[i]$ and $\varepsilon^{*}[i]$ are the binary indicators; $\varepsilon[i] = 1$ ($\varepsilon^{*}[i] = 0$) represents that the ith subframe belongs to the normal subframe; $\varepsilon^{*}[i] = 1$ ($\varepsilon[i] = 0$) represents that the ith subframe belongs to the protected subframe; $C_{k''}$ and $C_{k''}^{*}$ denote the indicators for the UE k'' whether it is assigned to the normal UE set or the protected UE set. $C_{k''} = 1$ ($C_{k''}^{*} = 0$) represents that the k''th pico UE is assigned to the normal UE set; $C_{k''}^{*} = 1$ ($C_{k''} = 0$) represents that the k''th pico UE is assigned to the protected UE set; Eq. (9) represents that the sum of the allocated power $p_{k',n,q^{(\mathrm{Macro})}}[i]$ is less than the total power of the macro eNB $P^{(\mathrm{Macro})}[i]$; Eq. (10) represents the transmit power constraint of the pico eNB $P^{(\mathrm{Pico})}[i]$; Eq. (11) denotes that each RB in the macro eNB is only allocated to one macro UE; each RB in one eNB is not allowed to be shared among UEs served in the eNB; Eq. (12) denotes that each RB in the pico eNB is only allocated to one pico UE; Eq. (13) means that each subframe is only classified as either the normal subframe or the protected subframe; the sum of $\varepsilon[i]$ and $\varepsilon^{*}[i]$ for a particular subframe i is equal to 1; Eq. (14) implies that one pico UE can only be in the normal UE set or the protected UE set.

This resource allocation considered in this paper contains two main challenges: (a) *configuration of the ABS pattern,* i.e. determines the values of $\varepsilon[i]$ and $\varepsilon^{*}[i]$; and (b) *design of pico UE grouping,* i.e., determines the values of $C_{k''}$ and $C_{k''}^{*}$. So far, there is no research for considering these two problems jointly. We first consider the design of the ABS pattern, including how many protected subframes per ABS pattern should be configured in the system and which subframes should be configured as protected subframes. In this paper, we will present a joint solution for this resource allocation problem to maximize the sum rate of the system while maintaining the fairness among UEs. The proposed scheme can achieve balance between the system performance and the computational complexity, further comprising (a) *an ABS pattern design in Section 3;* (b) *a Pico UE grouping design* in Section 4; and (c) *a joint allocation scheme* in Section 5. The ABS pattern determination is executed at the macro eNB while the UE grouping and the joint allocation is executed at the pico eNB. The backhaul signaling among eNBs is required for the communication of ABS patterns. The following is a summary of symbols used in the paper along with their explanations.

$G^{(Pico)}_{k'',n,q}$	the channel gain between the pico eNB and the k''th pico UE on the qth subcarrier of the nth RB
$p^{(Pico)}_{k'',n,q}$	the amount of power for the k''th pico UE on the qth subcarrier of the nth RB
$G^{(Macro)}_{k'',n,q}$	the channel gain between the macro eNB and the k''th pico UE on the qth subcarrier of the nth RB
$p^{(Macro)}_{k',n,q}$	the amount of power for the k'th macro UE on the qth subcarrier of the nth RB
$\varepsilon[i]$	the binary indicator. Subframe configuration
$r^{(normal)}_{k'',n}$	the data rate for thek''th pico UE on the nth RB in the normal UE set
$r^{(protected)}_{k'',n}$	the data rate for thek''th pico UE on the nth RB in the protected UE set
$\rho^{(Pico)}_{k'',n}$	the binary indicator. RB assignment
$\Omega^{(normal)}$	the normal UE set
$\Omega^{(protected)}$	the protected UE set
$R^{(normal)}_{k''}$	the sum of data rate for thek''th pico UE in the normal UE set
$R^{(protected)}_{k''}$	the sum of data rate for thek''th pico UE in the protected UE set
$\overline{R}$	the average data rate
I	the number of subframes in one frame
J	the windows size which is the number of the frames in a PF period
$C_{k''}$	the indicators for the UE k'' assigned to the normal UE set
$C^*_{k''}$	the indicators for the UE k'' assigned to the protected UE set
$F^{(Macro)}[i]$	denotes that the sum rate of all macro UEs in the ith subframe
$F^{(Pico)}_{(protected)}[i]$	denotes that the sum rate of pico UEs in the protected UE set in the ith subframe
$F^{(Pico)}_{(normal)}[i]$	denotes that the sum rate of pico UEs in the normal UE set in the ith subframe
$\alpha,\beta,\mu,\psi_{k''},\varphi_{k''},\zeta_{k''},\varsigma_{k''},\lambda_{k''}$	non-negative Lagrangian multipliers
N	the number of resource block in a system
Q	the number of subcarriers per resource block

3 Proposed ABS pattern design

As mentioned in the introduction section, two important issues need to be considered for the design: (a) the number of the protected subframes per ABS pattern should be determined in the system and (b) which subframes should be categorized as protected subframes. The time-invariant channel during a subframe is assumed because of the slow time-varying channel [24]. Therefore, we would design the ABS pattern by using subframe as the processing unit. We assume that all RBs are available and shared among UEs. Initially, without considering the UE grouping and resource allocation in this design phase, the channel information of all subcarriers in each subframe toward two eNBs in an ABS period is used. Our goal is to compare the performance index of each subframe which would be determined as normal or protected. Three variables are defined as:

$$F^{(Macro)}[i] = \sum_{k'=1}^{K'}\sum_{n=1}^{N}\sum_{q=1}^{Q}\Delta f\log_2\left(1+SINR^{(Macro)}_{k',n,q}[i]\right); \tag{17}$$

$$F^{(Pico)}_{(normal)}[i] = \sum_{k''=1}^{K''}\sum_{n=1}^{N}\sum_{q=1}^{Q}\Delta f\log_2\left(1+SINR^{(normal)}_{k'',n,q}[i]\right); \tag{18}$$

$$F^{(Pico)}_{(protected)}[i] = \sum_{k''=1}^{K''}\sum_{n=1}^{N}\sum_{q=1}^{Q}\Delta f\log_2\left(1+SINR^{(protected)}_{k'',n,q}[i]\right); \tag{19}$$

where $F^{(Macro)}[i]$denotes that the sum rate of all macro UEs in the ith subframe; $F^{(Pico)}_{(normal)}[i]$ denotes that the sum rate of pico UEs in the normal UE set in the ith subframe; $F^{(Pico)}_{(protected)}[i]$ denotes that the sum rate of pico UEs in the protected UE set in the ith subframe. Therefore, the aggregate sum rate can be rewritten as:

$$\max_{\varepsilon[i],\varepsilon^*[i]} \sum_{i=1}^{I}\left(\varepsilon[i]F^{(Macro)}[i] + \varepsilon[i]F^{(Pico)}_{(normal)}[i] + \varepsilon^*[i]F^{(Pico)}_{(protected)}[i]\right) \tag{20}$$

where the objective function (20) is subject to Eqs. (9–10, 13); the maximization is achieved by considering the values of $\varepsilon[i]$and $\varepsilon^*[i]$ according to the channels of each subframe. By using an optimization technique, we obtain the Lagrangian function with the relaxation [35] as:

$$\begin{aligned} L[i] = &-\sum_{i=1}^{I}\varepsilon[i]F^{(Macro)}[i] - \varepsilon[i]F^{(Pico)}_{(normal)}[i] - \varepsilon^*[i]F^{(Pico)}_{(protected)}[i] \\ &+\sum_{i=1}^{I}\alpha[i]\left(\sum_{k'\in K'}\sum_{n=1}^{N}\sum_{q=1}^{Q}p_{k',n,q}[i] - P^{(Macro)}[i]\right) \\ &+\sum_{i=1}^{I}\beta[i]\left(\sum_{k''\in K''}\sum_{n=1}^{N}\sum_{q=1}^{Q}p_{k'',n,q,l}[i] - P^{(Pico)}[i]\right) \\ &+\sum_{i=1}^{I}\mu[i]\left(\varepsilon[i] + \varepsilon^*[i] - 1\right) \end{aligned} \tag{21}$$

where $\alpha[i]$, $\beta[i]$, and $\mu[i]$ are non-negative Lagrangian multipliers. $\varepsilon[i]$ is initially relaxed to be assumed as a real-valued number, and it will be used to determine its binary value. After differentiating $L[i]$ with respect to $\varepsilon[i]$and $\varepsilon^*[i]$, respectively, we have

$$\partial L[i]/\partial \varepsilon[i] = F^{(\text{Macro})}[i] + F^{(\text{Pico})}_{(\text{normal})}[i] - \mu[i] = 0 \tag{22}$$

and

$$\partial L[i]/\partial \varepsilon^*[i] = F^{(\text{Pico})}_{(\text{protected})}[i] - \mu[i] = 0. \tag{23}$$

Therefore, we can determine whether the ith subframe should be classified as the normal subframe or the protected subframe by considering the difference between Eq. (22) and Eq. (23).

$$\Delta F[i] = F^{(\text{Macro})}[i] + F^{(\text{Pico})}_{(\text{normal})}[i] - F^{(\text{Pico})}_{(\text{protected})}[i]. \tag{24}$$

Based on the derived result, $\Delta F[i]$ implies that the sum rate difference of the ith subframe configured as the normal subframe or the protected subframe. If the value of $\Delta F[i]$ is greater than zero, the ith subframe is configured as the normal subframe because it may achieve higher sum rate. Otherwise, the ith subframe is categorized as the protected subframe. After the development of the configuration of an ABS pattern, the UE grouping design will be shown in the next section.

4 Proposed UE grouping strategy

The resource allocation problem in the view of frequency domain is considered, especially for pico UEs. An UE grouping strategy is developed based on an optimization technique to determine which pico UEs use either the normal subframes or the protected subframes. Besides, the issue of proportional fairness among UEs is also studied. In this section, by using the ABS pattern designed in the previous section, we develop a strategy to determine the pico UE grouping while considering the fairness among UEs. The pico UE grouping problem for the maximization of the sum rate of pico eNBs can be formulated by using the second and the third term of Eq. (8) as:

$$U^{(\text{Pico})} = \max_{C_{k''}, C^*_{k''}} \sum_{i=1}^{I} \sum_{k''=1}^{K''} \left\{ \frac{C_{k''}}{\overline{R}^{(\text{Pico})}_{k''}[i]} \sum_{n=1}^{N} \rho^{(\text{Pico})}_{k'',n}[i] r^{(\text{normal})}_{k'',n}[i] + \frac{C^*_{k''}}{\overline{R}^{(\text{Pico})}_{k''}[i]} \sum_{n=1}^{N} \rho^{(\text{Pico})}_{k'',n}[i] r^{(\text{protected})}_{k'',n}[i] \right\} \tag{25}$$

where the function (25) is subject to Eqs. (9–12, 14); when $C_{k''} = 1$ represents that UE k'' is assigned to the normal UE set; when $C^*_{k''} = 1$ represents that UE k'' is assigned to the protected UE set on the pico eNB. However, this optimization problem is NP-hard and has a high computational complexity [24]. In this paper, we would analyze the pico UE grouping problem by using an optimization technique. The optimization problem (25) is transformed into the dual domain by forming its Lagrangian dual with the relaxation [35]. The Lagrangian function is shown as:

$$\begin{aligned} L'[i] = &-\sum_{i=1}^{I} \sum_{k''=1}^{K''} \frac{1}{\overline{R}^{(\text{Pico})}_{k''}[i]} \left(C_{k''} R^{(\text{normal})}_{k''}[i] + C^*_{k''} R^{(\text{protected})}_{k''}[i] \right) \\ &+ \sum_{k''=1}^{K''} \psi_{k''} \left(\sum_{k'=1}^{K''} \rho^{(\text{Macro})}_{k',n}[i] - 1 \right) + \sum_{k''=1}^{K''} \phi_{k''} \left(\sum_{k''=1}^{K''} \rho^{(\text{Pico})}_{k'',n}[i] - 1 \right) \\ &+ \sum_{k''=1}^{K''} \zeta'_{k''} \left(\sum_{k'' \in K''} \sum_{n=1}^{N} \sum_{q=1}^{Q} p^{(\text{Macro})}_{k',n,q}[i] - P^{(\text{Macro})}[i] \right) \\ &+ \sum_{k''=1}^{K''} \zeta_{k''} \left(\sum_{k'' \in K''} \sum_{n=1}^{N} \sum_{q=1}^{Q} p^{(\text{Pico})}_{k'',n,q}[i] - P^{(\text{Pico})}[i] \right) \\ &+ \sum_{k''=1}^{K''} \lambda_{k''} \left(C_{k''} + C^*_{k''} - 1 \right) \end{aligned} \tag{26}$$

where $\psi_{k''}$, $\phi_{k''}$, $\zeta_{k''}$, $\zeta'_{k''}$, and $\lambda_{k''}$ are non-negative Lagrangian multipliers for the constraints (9–12, 14). The optimal solution Eq. (25) is achieved when the value of Eq. (26) is maximized. Therefore, the expression of the parameter $C_{k''}$ and $C^*_{k''}$ should be derived, which implies one pico UE is assigned to one of the UE sets.

First, by setting the partial derivative of Eq. (26) with respect to $C_{k''}$ to zero, we have

$$\frac{\partial L'[i]}{\partial C_{k''}} = 0 \Rightarrow \sum_{i=1}^{I} \frac{1}{\overline{R}^{(\text{Pico})}_{k''}[i]} \sum_{n=1}^{N} \rho_{k'',n,l}[i] r^{(\text{normal})}_{k'',n,l}[i] - \lambda_{k''} = 0. \tag{27}$$

Then, according to Eq. (8), we can get

$$\sum_{n=1}^{N} \rho_{k'',n}[i] r^{(\text{normal})}_{k'',n}[i] = R^{(\text{normal})}_{k''}[i] / C_{k''}. \tag{28}$$

By Eqs. (27) and (28), we can get

$$C_{k''} = \sum_{i=1}^{I} R^{(\text{normal})}_{k''}[i] / \left(\lambda_{k''} \overline{R}^{(\text{Pico})}_{k''}[i] \right). \tag{29}$$

Similarly, by setting the partial derivative of Eq. (26) with respect to $C^*_{k''}$ to zero, we can get

$$C^*_{k''} = \sum_{i=1}^{I} R^{(\text{protected})}_{k''}[i] / \left(\lambda_{k''} \overline{R}^{(\text{Pico})}_{k''}[i] \right). \tag{30}$$

By taking $C_{k''}$ (29) and $C^*_{k''}$ (30) into Eq. (14), we obtain

$$\sum_{i=1}^{I}\left(R_{k''}^{(\text{normal})}[i]+R_{k''}^{(\text{protected})}[i]\right)/\left(\lambda_{k''}\overline{R}_{k''}^{(\text{Pico})}[i]\right)=1$$
$$\Rightarrow\lambda_{k''}=\sum_{i=1}^{I}\left(R_{k''}^{(\text{normal})}[i]+R_{k''}^{(\text{protected})}[i]\right)/\overline{R}_{k''}^{(\text{Pico})}[i]. \tag{31}$$

By taking the expression of $\lambda_{k''}$ (31) into Eq. (29) and Eq. (30), respectively, we obtain

$$C_{k''}=\sum_{i=1}^{I}\frac{R_{k''}^{(\text{normal})}[i]}{\overline{R}_{k''}^{(\text{Pico})}[i]}\cdot\sum_{i=1}^{I}\frac{\overline{R}_{k''}^{(\text{Pico})}[i]}{R_{k''}^{(\text{normal})}[i]+R_{k''}^{(\text{protected})}[i]} \tag{32}$$

and

$$C_{k''}^{*}=\sum_{i=1}^{I}\frac{R_{k''}^{(\text{protected})}[i]}{\overline{R}_{k''}^{(\text{Pico})}[i]}\cdot\sum_{i=1}^{I}\frac{\overline{R}_{k''}^{(\text{Pico})}[i]}{R_{k''}^{(\text{normal})}[i]+R_{k''}^{(\text{protected})}[i]}. \tag{33}$$

Therefore, we can determine UE k'' which is assigned to the normal UE set or the protected UE set by comparing the difference of $C_{k''}$ (32) and $C_{k''}^{*}$ (33) which are related to the data rate. The difference of $C_{k''}$ (32) and $C_{k''}^{*}$ (33) is defined as:

$$\Delta C_{k''}=C_{k''}-C_{k''}^{*}. \tag{34}$$

$\Delta C_{k''}$ is the indicator difference of the normal UE set and the protected UE set when UE k'' is assigned to them. If the value of $\Delta C_{k''}$ is more than zero, UE k'' should be assigned to the normal UE set; otherwise, UE k'' should be assigned to the protected UE set.

5 Proposed joint allocation scheme

In this section, the joint ABS configuration and UE grouping scheme is described to resolve the resource allocation problem while achieving high system performance and fairness among all UEs. The proposed scheme by utilizing the designed functions includes the designs of the ABS pattern, the pico UE grouping, and the RB allocation. The procedures of the proposed scheme are as follows:

The pico UE index is $k''\in\{1,\ldots,K''\}$.*i*and*j* are the subframe index and frame index, respectively. (i,j)denotes the *i*th subframe of the *j*th frame. $\overline{R}_{k'}^{(\text{Macro})}[i]$ and $\overline{R}_{k''}^{(\text{Pico})}[i]$ are the average date rate of macro UE and pico UE, respectively. One starts from $j=1$, $j\in\{1,\ldots,J\}$, J is the window size which is the number of the frames in a PF period.

Step 1: ABS Configuration. Start from$i=1$,$i\in\{1,\ldots,I\}$, $I$is the number of subframes in one frame. Each subframe is determined as the normal subframe or the protected subframe in one frame. The designed function $\Delta F[i]$ (24) determines which subframes should be configured as protected subframes in one frame. In order to find a better ABS pattern, we may sort the values of the designed function (24) in a descending order. And then, the number of protected subframes is limited to an upper bound of ABS density [22]. Therefore, the ABS pattern for a frame would be configured.

Step 2: Pico UE Grouping. After the ABS pattern is configured for a frame, we will assign pico UEs to the normal UE set or the protected UE set and allocate the RBs for each subframe. The initial UE set is determined by using Eq. (1). Pico UEs which are handed over from macro eNB to pico eNB because of bias values are assigned to the protected UE set $\Omega^{(\text{protected})}$, i.e., $C_{k''}^{*}=1$. The others are assigned to the normal UE set $\Omega^{(\text{normal})}$, i.e., $C_{k''}=1$.

Step 3: RB Allocation, for $n=1,\ldots,N$. After the protected UE set $\Omega^{(\text{protected})}$and the normal UE set $\Omega^{(\text{normal})}$are determined, the RB allocation of each subframe is performed. For the normal subframes, only pico UEs which are assigned to the normal UE set can use the RBs; similarly, the RBs in the protected subframes can only be used by UEs in the protected UE set. We assume that power is equally distributed to each RB in the view of base station. In the initial stage, equal power is assumed for each subcarrier of each UE. A simple strategy is to assign RBs to UEs with a higher value of data rate. Since it is the first run of the solution, a better solution is achieved after iterations by the proposed scheme.

For the macro UEs, the *n*th RB of the *i*th subframe is allocated to the k'th macro UE with the maximum value of the function:

$$\hat{k}'=\arg\max_{k'\in\Omega^{(\text{Macro})}}\left\{r_{k',n}^{(\text{Macro})}[i]/\overline{R}_{k'}^{(\text{Macro})}[i]\right\}. \tag{35}$$

The *n*th RB is assigned to the k'th macro UE denoted as $\rho_{k',n}^{(\text{Macro})}[i]=1$. For the pico UEs in the normal UE set, the *n*th RB of the *i*th subframe which belongs to the normal subframe is allocated to the k''th pico normal UE with the maximum value of the function:

$$\hat{k}''=\arg\max_{k''\in\Omega^{(\text{normal})}}\left\{r_{k'',n}^{(\text{normal})}[i]/\overline{R}_{k''}^{(\text{Pico})}[i]\right\}. \tag{36}$$

Similarly, for the pico UEs in the protected UE set, the *n*th RB of the *i*th subframe is allocated to the pico UE with the maximum value of the function:

$$\hat{k}'' = \underset{k'' \in \Omega^{(\text{protected})}}{\arg \max} \left\{ r_{k'',n}^{(\text{protected})}[i] / \overline{R}_{k''}^{(\text{Pico})}[i] \right\}. \quad (37)$$

The nth RB is determined as $\rho_{k'',n}^{(\text{Pico})}[i] = 1$ for the k''th pico UE. Repeat *Step 3* until all the RBs, for n = 1, ..., N, are allocated in the ith subframe; then, repeat the same process until $i = I$. The initial sum rate denoted as $U_{(\text{initial})}^{(\text{Pico})}$ is calculated according to Eq. (25).

Step 4: Fast Adjustment. We would re-distribute pico UEs into the two UE sets by using $C_{k''}$ (32) and $C_{k''}^{*}$ (33) in a group basis. If the k''th UE is assigned to the normal UE set, we can get the $C_{k''}$ according to Eq. (32); then, we temporarily move the k''th UE from the normal UE set to the protected UE set; and repeat *Step 3* to allocate the RBs for each subframe. We can get the $C_{k''}^{*}$ according to Eq. (33). Similarly, if the k''th UE is assigned to the protected UE set, the same approach is used to get $C_{k''}$ and $C_{k''}^{*}$. After that, $\Delta C_{k''}$ (34) is obtained for each pico UE. The value of $\Delta C_{k''}$ would be used to determine if the k''th UE is re-assigned to the normal UE set or the protected UE set. If $\Delta C_{k''} \geq 0$, the pico UE k'' is assigned to the normal UE set; otherwise, the pico UE k'' is assigned to the protected UE set.

$$\begin{cases} C_{k''} = 1 \text{ and } k'' \in \Omega^{(\text{normal})}, & \text{if } \Delta C_{k''} \geq 0 \\ C_{k''}^{*} = 1 \text{ and } k'' \in \Omega^{(\text{protected})}, & \text{if } \Delta C_{k''} < 0 \end{cases} \quad (38)$$

After the pico UE re-assignment, the updated UE sets, i.e., $\Omega^{(\text{normal})}$ and $\Omega^{(\text{protected})}$, are obtained. The sum rate $U^{(\text{Pico})}$ according to Eq. (25) is calculated correspondingly. If the value of $U^{(\text{Pico})}$ is larger than that of $U_{(\text{initial})}^{(\text{Pico})}$, the normal UE set $\Omega^{(\text{normal})}$ and the protected UE set $\Omega^{(\text{protected})}$ are updated. Then, update $U_{(\text{initial})}^{(\text{Pico})}$ as $U^{(\text{Pico})}$ and repeat *Steps 3–4* until the value of $U^{(\text{Pico})}$ is no more increased.

Step 5: Refinement Mechanism. Based on the result of the fast adjustment, the refinement mechanism is performed to exchange or move pico UEs between two UE sets on a per-UE basis. The main concept of the refinement mechanism is to re-assign only one UE to the normal UE set $\Omega^{(\text{normal})}$ or the protected UE set $\Omega^{(\text{protected})}$ in one iteration. In the exchanging operation, originally two pico UEs, e.g., $x$and$y$, are assigned to$\Omega^{(\text{normal})}$and$\Omega^{(\text{protected})}$, i.e., $x \in \Omega^{(\text{normal})}$ and$y \in \Omega^{(\text{protected})}$. After the exchanging operation, two pico UEs are exchanged between two pico UE sets, i.e., $y \in \Omega^{(\text{normal})}$ and $x \in \Omega^{(\text{protected})}$. In this fashion, two pico UEs are exchanged in each time and the number of pico UEs in each set is not changed. The sum rate (25) of all combinations is calculated correspondingly. The sum rate increment can be defined as:

$$\Delta U_{(\text{exchange})}^{(\text{Pico})} = \left\{ U_{s'}^{(\text{Pico})} - U_{(\text{initial})}^{(\text{Pico})} \right\}_{s' \in S'} \quad (39)$$

where S' denotes all combination cases in the exchanging operation; s' is one of combination cases; $U_{s'}^{(\text{Pico})}$ is the sum rate of the s'th combination case. In the moving operation, one pico UE is moved from $\Omega^{(\text{normal})}$ to $\Omega^{(\text{protected})}$ or from $\Omega^{(\text{protected})}$ to$\Omega^{(\text{normal})}$. After moving, the corresponding sum rate (25) is calculated. The sum rate increment can be defined as:

$$\Delta U_{(\text{move})}^{(\text{Pico})} = \left\{ U_{s''}^{(\text{Pico})} - U_{(\text{initial})}^{(\text{Pico})} \right\}_{s'' \in S''} \quad (40)$$

where S'' denotes all combination cases in the moving operation; s'' is one of combination cases; $U_{s''}^{(\text{Pico})}$ is the sum rate of the s''th combination case. Among all combinations $S' \cup S''$, one combination with the maximum sum rate increment would be selected to perform the corresponding operation.

$$s^{*} = \arg \max_{s^{*} \in S' \cup S''} \left\{ \Delta U_{(\text{exchange})}^{(\text{Pico})} \right\} \cup \left\{ \Delta U_{(\text{move})}^{(\text{Pico})} \right\}. \quad (41)$$

Then, update the sets of $\Omega^{(\text{normal})}$ and $\Omega^{(\text{protected})}$ and the initial value $U_{(\text{initial})}^{(\text{Pico})}$. Repeat *Step 5* until no sum rate increment can be achieved.

Step 6: Update the average data rates$\overline{R}_{k'}^{(\text{Macro})}[i]$and $\overline{R}_{k''}^{(\text{Pico})}[i]$ for the current frame. The proposed joint scheme is ready to be performed for the next frame$j = j + 1$. Return to Step 1 until the window size of frames $j = J$ is met.

In summary, Fig. 2a is a flow chart of the proposed joint allocation scheme. The proposed joint allocation scheme is executed on a per frame duration basis. First, ABS configuration for a frame determines each subframe as the normal subframe or the protected subframe at the macro eNB. The periodicity is a frame basis. The current channel state information may not be instantly estimated, so that we can use the channel information in the previous frame to design the ABS pattern because of the slow time-varying channel application. After that, the information of ABS configuration exchanges through the X2 interface between the macro eNB and the pico eNB. The X2 interface would be ideal backhaul as the latency may be less than 2.5 μs [36]. Therefore, the predicted sum rate can be calculated in the proposed ABS pattern design. The backhaul signaling may comprise the channel information of Pico UEs and the ABS configuration. The proposed Pico UE

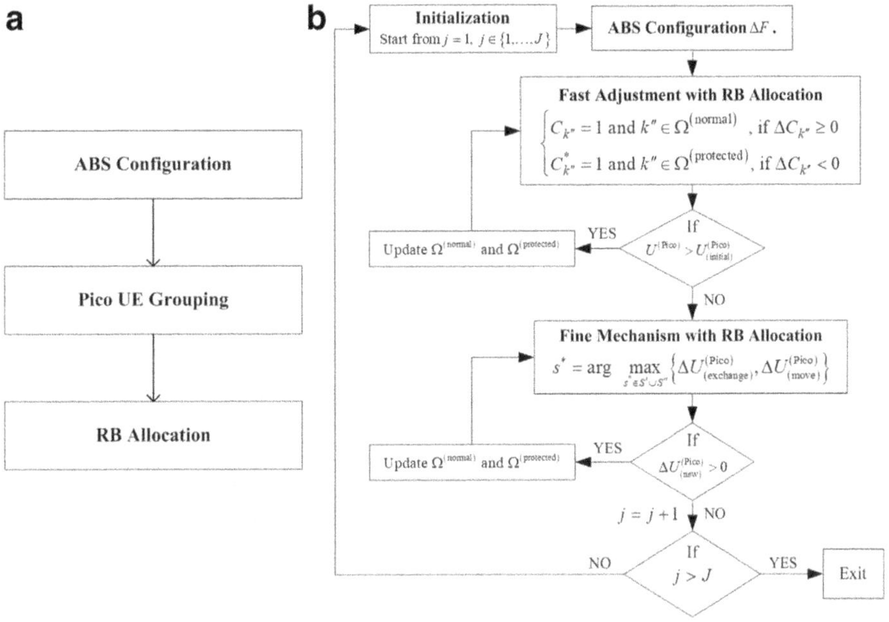

Fig. 2 a Flow chart of the proposed joint allocation scheme. **b** Procedures of the proposed joint allocation scheme

grouping strategy is performed locally for each subframe. i.e., pico UEs are assigned into two sets, i.e., the normal UE set or the protected UE set. In the step of Pico UE grouping, fast adjustment introduced in Section 4 and refinement mechanism in Section 5 are utilized to separate pico UEs into two groups for achieving better performance. The corresponding RB allocation in Section 5 is performed in the macro eNB and the pico eNB respectively. The proposed joint allocation scheme in Section 5 combines the above operations to maximize the sum rate of the system while maintaining the fairness among UEs.

We incorporate the computation and evaluation of related equations into Fig. 2a to complete the allocation procedures. Figure 2b illustrates the procedures of the proposed joint allocation scheme across the frequency domain and the time domain.

6 Computational complexity analysis

In this section, we focus on the computational complexity analysis of the RB allocation and UE grouping strategies in one frame in terms of Big-Oh notation. All compared schemes are based on the same system model, so the computational complexity of calculating the data rate associated with an SINR is the same, which is represented as $O(Q)$ by referring to Eq. (3) and Eq. (6). The complexities of the processing steps in the algorithm are the focus, in terms of calculating the SINR and the data rate.

Step 1 of the proposed joint allocation scheme is the ABS configuration for one frame. Regarding the proposed ABS pattern design, the calculation of the sum rate for each subframe in one frame is needed. Therefore, the complexity of the proposed ABS pattern design is $O(IK^{''}NQ)$. Step 2 uses Eq. (1) to determine the initial UE set. In that, $K^{''}$ UEs are considered which requires $O(K^{''})$. RB allocation is operated in *Step 3*. For the macro eNB, N RBs are considered for $K^{'}$ UEs per subframe. One frame contains I subframes. The computational complexity of the RB allocation needs $O(INK^{'}Q)$. For the pico eNB, we assume that $K^{''}$ UEs are equally distributed in each set. Each set needs the complexity of $O(INK^{''}Q/2)$ for the RB allocation. Therefore, $O(INK^{''}Q)$ is required for RA allocation of the pico UEs. *Step 4* focuses on the moving operation of the pico UEs. $K^{''}$ pico UEs are evaluated along with the RB allocation for one iteration. The complexity is calculated as $O((INK^{''}Q)K^{''}T_1)$. T_1 denotes the number of iterations. The refinement mechanism in *Step 5* comprises the exchanging operation and the moving operation. We also assume that the average number of pico UEs in each set is approximately $K^{''}/2$. For the exchanging operation, there are $(K^{''}/2)\cdot(K^{''}/2) \approx (K^{''})^2$ possible combinations per iteration. For the moving operation, the number of combinations is the number of total pico UEs $K^{''}$. Therefore, the complexity of the refinement mechanism along with the RB allocation is $O(INK^{''}Q((K^{''})^2 + K^{''})T_2) \approx O(IN(K^{''})^3QT_2)$. T_2 is the number of iterations. In summary, the complexity of the joint allocation scheme for one frame is $O(K^{''}INQ + K^{''} + INQ(K^{'} + K^{''}) + INQ(K^{''})^2T_1 + INQ(K^{''})^3T_2) \approx O(INQK^{'} + INQ(K^{''})^2T_1 + INQ(K^{''})^3T_2)$.

For the compared schemes, the computational complexity of the fixed ABS pattern [22] is $O(1)$ where it fixes the ABS density and determines an ABS pattern at random. Regarding pico UE grouping strategies, the full search scheme [25] is to search all UE grouping combinations of pico UEs in each ABS pattern. The complexity of all UE grouping combinations is $O\left(\sum_{j=1}^{K''} C_j^{K''}\right) \approx O\left(2^{K''}\right)$. So, the total computational complexity along with RB allocation is $O\left(INQK' + INQK'' \sum_{j=1}^{K''} C_j^{K''}\right) \approx O\left(INQK' + INQK'' 2^{K''}\right)$, including K' macro UEs and K'' pico UEs. $O(INQK')$ is the complexity of RA allocation for macro UEs. In [27], pico UE grouping is determined by CRE bias, which requires $O(K'')$. Therefore, the computational complexity along with RB allocation is $O(K'' + INQK' + INQK'')$. In [24], an indicator which involves the data rate is used to determine that each UE may be assigned to either the normal UE set or the protected UE set. The computational complexity of pico UE grouping is $O(K'' T_3)$. T_3 denotes the number of iterations. So, the total computational complexity along with RB allocation is $O(INQK' + INQK'' K'' T_3) = O(INQK' + INQ(K'')^2 T_3)$.

7 Simulation results

The simulation results will demonstrate the performance of the proposed scheme compared to those of the existing schemes [22, 24, 25, 27]. The network topology consists of one macro eNB and one pico eNB based on the 3GPP case 1 [37]. The radius of the macro eNB is 289 m, and the pico eNB is randomly distributed with a minimum distance of 75 m to the macro eNB. The frequency selective wireless channel model [38] is employed. We adopt Jake's model to generate the Rayleigh fading channel. The mobile speed is 4 km/h. The standard deviation of shadowing is 8 dB. $N_0$is −174 dBm/Hz. The channel power of the received signal for each UE is varied because of the various path losses at the different locations. The macro path loss model [39] is 128.1 + 37.6*$log(d1)$ in decibels. $d1$ is in kilometers. The pico path loss model [39] is 38 + 30*$log(d2)$ in decibels. $d2$ is in meters. The carrier frequency is 2 GHz. UEs are uniformly distributed within the coverage of the macro cell with the numbers from 12 to 36. Transmit power is 46 dBm for the macro eNB and 30 dBm for the pico eNB. Eight-decibel bias is considered for cell range expansion. The downlink FDD system is used for simulation with a bandwidth of 10 MHz comprising 50 RBs. Each RB has 12 subcarriers. Subcarrier spacing $\Delta f$is 15 kHz. The ABS pattern period is set to be 10 ms, i.e., 10-subframe duration. All results are the average values from 600 frames.

The first simulation is revealed to demonstrate the outperformance of the proposed ABS pattern design strategy. In order to have fair comparison, the fixed ABS approach is cooperated with a proper setting along with the proposed user grouping strategy and the proposed RB allocation scheme. The comparison baseline is the fixed ABS pattern method [22] with 50% ABS muting ratio, i.e., five ABS subframes per frame. The fixed ABS pattern [22] is also cooperated with the proposed user grouping strategy and the proposed RB allocation scheme. The result of Fig. 3 is conducted in the scenario with 24 UEs, which shows the results of the sum rates in the system versus different numbers of RBs. Figure 4 is conducted in the scenario with 50 RBs, which shows the results of the sum rates in the system versus different numbers of UEs. The simulation results indicate that the proposed ABS pattern design scheme outperforms the fixed ABS pattern method [22]. The enhancement is increased with the increment of the number of RBs because of more flexible resource scheduling. With user diversity, the performance is slightly improved with the number increment of UEs.

Based on the proposed ABS pattern design, the following simulation focuses on pico UE grouping strategies, including the full search scheme [25], the proposed joint allocation scheme, and the existing schemes [24, 27]. In Figs. 5 and 6, simulations are conducted in the scenario with 24 UEs at the bias value of 8 dB. In Fig. 5, it shows the results of the objective function (8) versus the number of RBs. It illustrates that the values of the objective function (8) are in the order of [Full search [23] > Proposed Joint Allocation Scheme > Ref. [24] > Ref. [27]] from large to small. The proposed joint allocation scheme approximates the full search scheme [25] while the computational complexity

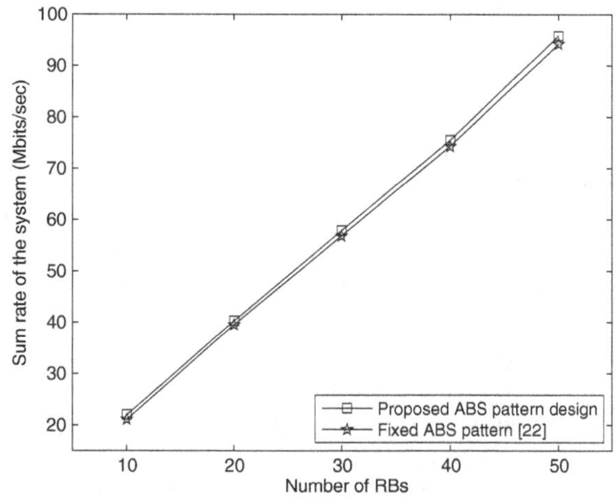

Fig. 3 Comparison of the proposed ABS pattern design with the fixed ABS pattern. The sum rate of the system versus the number of RBs with the number of UEs $K = 24$ and the bias value = 8 dB

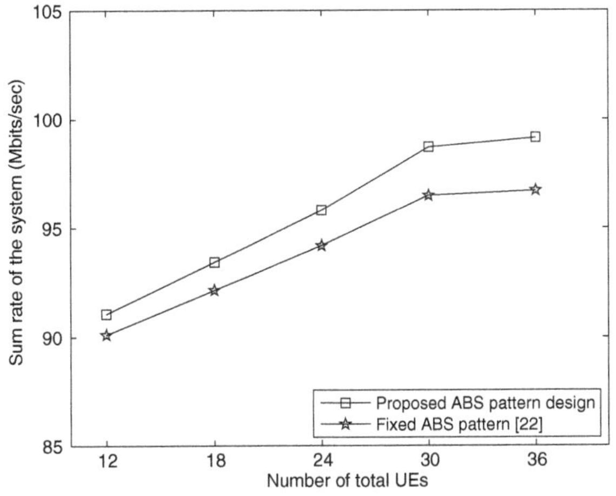

Fig. 4 Comparison of the proposed ABS pattern design with the fixed ABS pattern. The sum rate of the system versus the number of UEs with the number of RBs $N = 50$ and the bias value = 8 dB

is greatly reduced. The simulation results also indicate that the proposed joint allocation scheme outperforms the existing schemes [24, 27]. Figure 6 shows the performance comparison in terms of the sum rate of the system. The simulation results indicate that the proposed joint allocation scheme still outperforms the existing schemes [24, 27]. The performance of the proposed joint allocation scheme is much closer to that of the full search scheme [25].

In Figs. 7 and 8, the simulations are in the scenario with 50 RBs by varying the number of total UEs. With the same performance trend, the proposed joint allocation scheme outperforms the existing schemes [24, 27] in terms of the objective function and the sum rate. The proposed scheme dynamically determines the pico UE

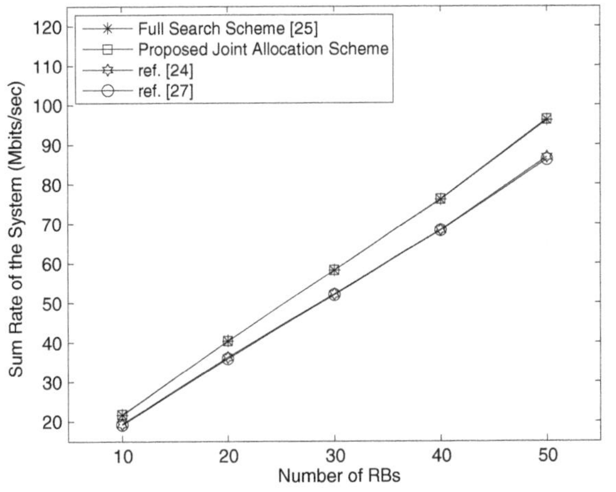

Fig. 6 Comparison of the sum rate of the system, in respect to Fig. 5

grouping according to the functions (Eqs. 25 and 34), which would indicate the related rate performance of pico UEs. Thus, the proposed joint allocation scheme has better performance than the existing schemes [24, 27]. Compared to the full search scheme [25], the performance loss of the proposed joint allocation scheme is very tiny. Due to the multi-user diversity, the sum rate of the system is increased as the number of UEs increases. However, under the proportional fairness scheduler, the increase of the sum rate would become slow if the numbers of UEs are greater.

Figure 9 shows the fairness index (42). Jain's fairness index (FI) is a real number in the interval [33] denoted as

$$\mathrm{FI} = \left(\sum_{k''=1}^{K''} R_{k''}^{(\mathrm{Pico})}\right)^2 \Big/ \left(K'' \sum_{k''=1}^{K''} \left(R_{k''}^{(\mathrm{Pico})}\right)^2\right). \qquad (42)$$

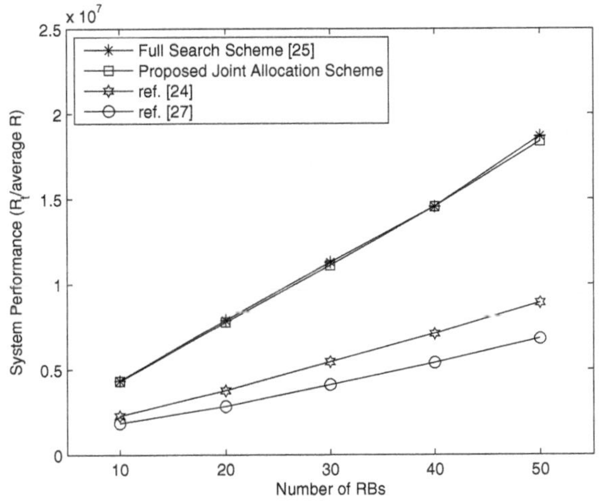

Fig. 5 Comparison of the proposed joint allocation scheme with the existing schemes. The value of the objective function versus the number of RBs with the number of UEs $K = 24$ and the bias value = 8 dB

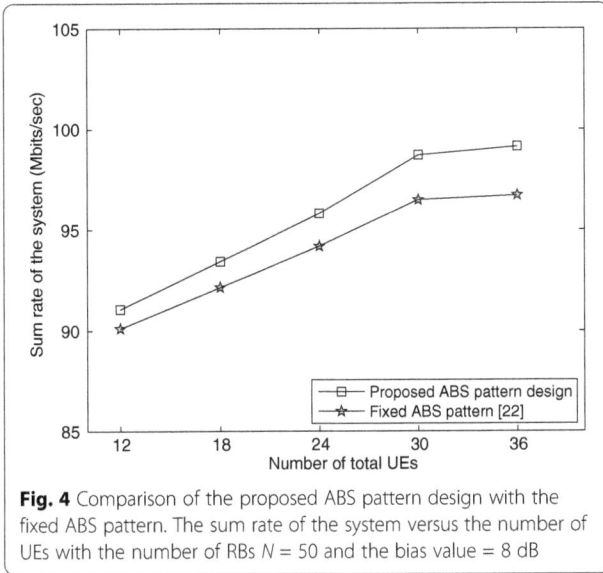

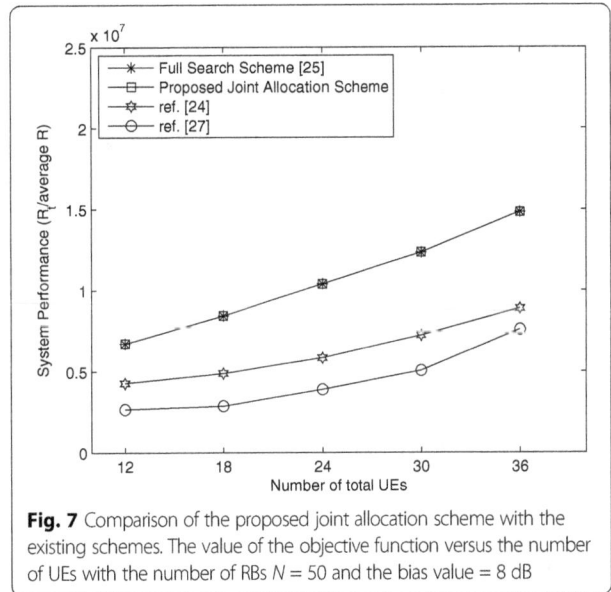

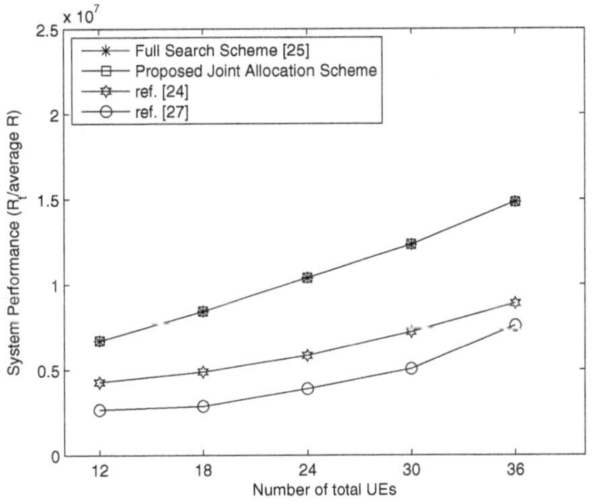

Fig. 7 Comparison of the proposed joint allocation scheme with the existing schemes. The value of the objective function versus the number of UEs with the number of RBs $N = 50$ and the bias value = 8 dB

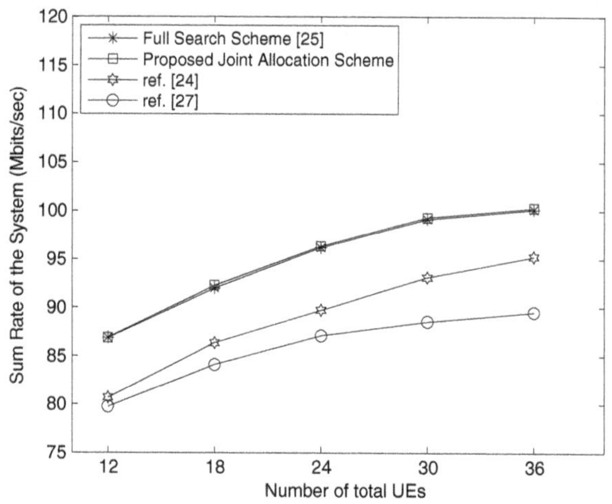

Fig. 8 Comparison of the sum rate of the system, in respect to Fig. 7

when FI = 1, it means that all UEs have the same performance with the highest fairness. When there are more UEs in the system, the fairness index would be descent. By considering the proportional fairness, the fairness index of the proposed joint allocation scheme would be almost equal to that of the full search scheme [25]. Note that the scheme [24] is designed to achieve the fairness of all UEs. Therefore, the value of the objective function would be better than that of the scheme [27] as shown in Figs. 5 and 7. Although slightly higher fairness is revealed as shown in Fig. 9, the sum rate is much worse than the proposed joint allocation scheme as shown in Figs. 6 and 8. From the results in Fig. 9, the difference between the scheme [24] and the proposed joint allocation scheme is negligible. Regarding the numbers of iterations required for the proposed joint allocation scheme and the existing scheme [24], almost the same numbers of iterations are revealed in our simulation. The average number of iterations is less than 6.

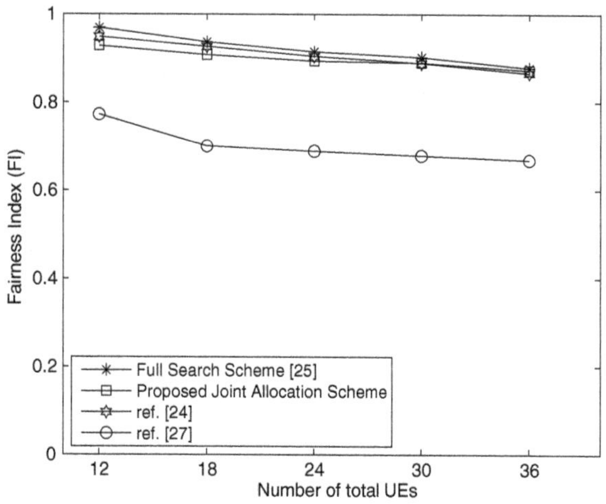

Fig. 9 Comparison of the fairness index, in respect to Fig. 7

8 Conclusions

A strategy with the help of the derived evaluation function to determine the ABS pattern is designed in this paper. With a proper design of ABS pattern, the performance would be further improved in term of sum rates. The novel UE grouping strategies are presented to improve the performance of the system without a prohibitive computational complexity. The proposed joint allocation scheme obtains the balance in the computational complexity and the performance, and also considers the fairness among UEs. The proposed joint allocation scheme outperforms the existing schemes about 10% gains in term of sum rate, and approaches 99.5% of the full search scheme while having a much lower complexity.

Acknowledgements

There is no other person who contributed toward the article who does not meet the criteria for authorship.

Funding

This work was supported by the Information and Communications Research Laboratories, Industrial Technology Research Institute (ITRI), Hsinchu, Taiwan, under Grant 2017–41–5G-0301.

Authors' contributions

W-CP is responsible for the development of the most parts of the algorithms and conducting simulations. J-WLin is responsible for the development of some parts of the algorithms and conducting simulations. Y-FC is responsible for the development of the algorithms, verification of the derivations and simulation results, writing the paper for the whole idea, and supervising the process of the research. C-LW is responsible for providing the whole concept of the algorithms to develop and supervising the process of the research. All authors read and approved the final manuscript.

Competing interests

The authors declare that they have no competing interests.

Author details

[1]Industrial Technology Research Institute, Zhudong, Taiwan, Republic of China. [2]Department of Communication Engineering, National Central University, Taoyuan, Taiwan, Republic of China. [3]Department of Electrical Engineering and Institute of Communications Engineering, National Tsing Hua University, Hsinchu, Taiwan, Republic of China.

References

1. E. Dahlman, S. Parkvall, J. Skold, *4G LTE/LTE-Advanced for Mobile Broadband* (Elsevier Ltd., UK, 2011)
2. *3GPP, Technical Specification Group Radio Access Network; Physical layer aspects for evolved Universal Terrestrial Radio Access (UTRA)*, TR 25.814, 2006
3. 3GPP, *Mobile Broadband Innovation Path to 4G: Release 9, Release 10 and Beyong: HSPA+, LTE/SAE and LTE-Advance*, 2010
4. 3GPP, *Requirements for further advancements for Evolved Universal Terrestrial Radio Access (E-UTRA) (LTE-Advanced)*, TR 36.913, 2011

5. 3GPP, *Further advancements for E-UTRA physical layer aspects*, TR 36.814, 2010
6. A. Khandekar, N. Bhushan, J. Tingfang, V. Vanghi, in *Proc. European Wireless Conf.* LTE-Advanced: heterogeneous networks (2011), pp. 978–982
7. A. Prasad, O. Tirkkonen, P. Lunden, O.N.C. Yilmaz, L. Dalsgaard, C. Wijting, Energy-efficient inter-frequency small cell discovery techniques for LTE-advanced heterogeneous network deployments. IEEE Commun. Mag. **51**(5), 72–81 (2013)
8. V. Chandrasekhar, J.G. Andrews, Femtocell networks: a survey. IEEE Commun. Mag. **46**(9), 59–67 (2008)
9. A. BouSaleh, S. Redana, B. Raaf, J. Hämäläinen, in *Proc. IEEE Vehicular Technology Conf.-Fall.* Comparison of relay and pico eNB deployments in LTE-advanced (2009), pp. 1–5
10. 3GPP, *Evolved Universal Terrestrial Radio Access (E-UTRA); Mobility enhancements in heterogeneous networks (Release 11)*, TR 36.839, 2012
11. R1-100701, in *3GPP TSG RAN WG1 Meeting#59*. Importance of Serving Cell Selection in Heterogeneous Networks (2010)
12. Y. Song, P.Y. Kong, Y. Han, Minimizing Energy Consumption through Traffic Offloading in a HetNet with 2-class Traffic. IEEE Commun. Lett. **19**(8), 1394–1397 (2015)
13. P.Y. Kong and G. K. Karagiannidis, "Backhaul-Aware Joint Traffic Offloading and Time Fraction Allocation for 5G HetNets", *IEEE Transactions on Vehicular Technology, DOI:* https://doi.org/10.1109/TVT.2016.2517671, 2016
14. A. Damnjanovic et al., A survey on 3GPP heterogeneous networks. IEEE Wirel. Commun. **18**(3), 10–21 (2011)
15. D. Luo, B. Li, D. Yang, in *Proc. IEEE Vehicular Technology Conf.-Fall.* Performance evaluation with range expansion for heterogeneous networks (2011), pp. 1–5
16. R1-112543, "Scenarios for eICIC evaluations," *3GPP TSG-RAN WG1 Meeting#66*, 2011
17. R1-106143, "Details of eICIC in macro-pico case," *3GPP TSG-RAN WG #63*, 2010
18. R1-110175, "Remaining issues of Rel-10 eICIC," *3GPP TSG-RAN WG1#63bis*, 2011
19. D. Lopez-Perez, I. Guvenc, G.D.L. Roche, M. Kountouris, T.Q.S. Quek, J. Zhang, Enhanced intercell interference coordination challenges in heterogeneous networks. IEEE Wirel. Commun. **18**(3), 22–30 (2011)
20. 3GPP, *Evolved Universal Terrestrial Radio Access (E-UTRA) and Evolved Universal Terrestrial Radio Access Network (E-UTRAN); Overall description; Stage 2 (Release 13)*, TS 36.300, 2015
21. J. Oh, Y. Han, in *Proc. IEEE Int. Symposium on Personal Indoor and Mobile Radio Commun.* Cell selection for range expansion with almost blank subframe in heterogeneous networks (2012), pp. 653–657
22. Y. Wang, K.I. Pedersen, in *Proc. IEEE Vehicular Technology Conf.-Spring.* Performance analysis of enhanced Inter-cell Interference Coordination in LTE-Advanced heterogeneous networks (2012), pp. 1–5
23. S. Deb, P. Monogioudis, J. Miernik, J.P. Seymour, Algorithms for enhanced Inter-cell Interference Coordination (eICIC) in LTE HetNets. IEEE/ACM Trans. Networking **99**, 1 (2013)
24. L. Jiang, M. Lei, in *Proc. IEEE Int. Symposium on Personal, Indoor and Mobile Radio Commun.* Resource allocation for eICIC scheme in heterogeneous networks (2012), pp. 448–453
25. J. Pang, J. Wang, D. Wang, G. Shen, Q. Jiang, J. Liu, in *Proc. IEEE Wireless Commun. and Networking Conf.* Optimized time-domain resource partitioning for enhanced inter-cell interference coordination in heterogeneous networks (2012), pp. 1613–1617
26. R1-100142, *"System performance of heterogeneous networks with range expansion," 3GPP TSG-RAN WG1 Meeting#59bis*, 2010
27. R1-112411, *"Scenarios for further enhanced non ca-based icic for lte," 3GPP TSG RAN WG1 Meeting#66*, 2011
28. A. Weber, O. Stanze, in *Proc. IEEE Int. Conf. Commun.* Scheduling strategies for HetNets using eICIC (2012), pp. 6787–6791
29. 3GPP, *Evolved Universal Terrestrial Radio Access (E-UTRA); Physical channels and modulation,* TR 36.211, 2011
30. 3GPP, *Evolved Universal Terrestrial Radio Access (E-UTRA); Physical layer – Measurements,* TR 36.214, 2010
31. H. Seo, B.G. Lee, in *Proc. IEEE Global Telecommun. Conf.* A proportional-fair power allocation scheme for fair and efficient multiuser OFDM systems, vol 6 (2004), pp. 3737–3741
32. H. Kim, Y. Han, A proportional fair scheduling for multicarrier transmission systems. IEEE Commun. Lett. **9**(3), 210–212 (2005)
33. V.D. Papoutsis, I.G. Fraimis, S.A. Kotsopoulos, User selection and resource allocation algorithm with fairness in MISO-OFDMA. IEEE Commun. Lett. **14**(5), 411–413 (2010)
34. R1-105406, in *3GPP TSG RAN WG1 meeting#62bis*. Support of time domain icic in rel-10 (2010)
35. S. Boyd, L. Vandenberghe, *Convex Optimization* (Cambridge University Press, 2004)
36. 3GPP, *Technical Specification Group Radio Access Network; Scenarios and requirements for small cell enhancements for E-UTRA and E-UTRAN (Release 12)*, TR 36.932, 2013
37. 3GPP, *Evolved Universal Terrestrial Radio Access (E-UTRA); Further advancements for E-UTRA physical layer aspects (Release 9)*, TS 36.814, 2010
38. L. Dong, G. Xu, H. Ling, in *Proc. IEEE Global Telecommun. Conf.* Prediction of fast fading mobile radio channels in wideband communication systems, vol 6 (2001), pp. 3287–3291
39. 3GPP, *LTE; Evolved Universal Terrestrial Radio Access (E-UTRA); Radio Frequency (RF) requirements for LTE Pico Node B*, TR 36.931, 2011

16

Reconfiguration time and complexity minimized trust-based clustering scheme for MANETs

Sunho Seo[1], Jin-Won Kim[1,2], Jae-Dong Kim[1,2] and Jong-Moon Chung[1*]

Abstract

A trust management mechanism for mobile ad hoc networks (MANETs) is proposed to cope with security issues that MANETs face due to time constraints as well as resource constraints in bandwidth, computational power, battery life, and unique wireless characteristics. The trust-based reputation scheme *GlobalTrust* is a reliable trust management mechanism. In this paper, a clustering algorithm is applied to the *GlobalTrust* scheme (named Cluster-based GlobalTrust (*CGTrust*)) to find the optimal group size to minimize the configuration time, which consists of trust information computational time and complexity, while having to satisfy the trust reliability requirements. The optimal number of clusters is derived from the minimizing point of the computation complexity function. Simulation results show that the computational time and complexity of *CGTrust* are controllable and can be used effectively in time critical network operations that require trust analysis.

Keywords: MANET, Trust, Cluster, GlobalTrust, CGTrust

1 Introduction

Mobile ad hoc networks (MANETs) consist of distributed wireless mobile nodes. In MANETs, it is critical to make a decision of assessing the trustworthiness of participating nodes accurately by a trust authority. Trust evaluations are made on node reputation, which is the perception of a node evaluated by other nodes. A node's reputation is evaluated based on the collection of trust evaluations made on that node by other nodes [1, 2]. Cho et al. [1] analyzed the concepts and properties of trust. In addition, they surveyed MANET trust management schemes for secure routing, authentication, and intrusion detection, etc. Aberer and Despotovic [3] proposed a trust-based reputation management scheme. The authors of [4, 5] proposed distributed methods for reputation management. However, results in [3–5] are vulnerable to collusive attacks. Reputation management schemes in [6–9] subjectively evaluate reputation using direct observation that disturbs the global view. Quorum-based reputation management schemes are proposed based on *k-out-of-n* threshold signatures [10, 11]. The *k*-means-based reputation scheme is presented in [12, 13], which does not consider conflicting recommendation attacks. Chen et al. [2] proposed *GlobalTrust* for accurate decision-making of a trusted authority (TA) under various attacks, which shows an outstanding performance in trust on MANETs. The TA needs to collect trust-related data and compute trust values and set up optimal routing paths, which is a very computationally burdening task. Thus, a common issue would be the time constraints and required computing resources of the TA and MANET nodes. The computational time delay and complexity of *GlobalTrust* significantly increases as the number of nodes in the MANET increases. In addition, in a MANET, communicating nodes are mobile. Therefore, route reconfiguration may need to be conducted frequently. As a result, reconfiguration time minimization would be necessary. In this paper, a Cluster-based GlobalTrust (*CGTrust*) scheme is proposed with the objective to minimize the computational complexity and reconfiguration time delay experienced by a TA while supporting the required trust reliability requirements. *CGTrust* consists of the following three unique features.

*Correspondence: jmc@yonsei.ac.kr
[1] School of Electrical and Electronic Engineering, Yonsei University, Seoul, Republic of Korea
Full list of author information is available at the end of the article

- *CGTrust* evaluates trust at cluster heads (CHs) and at the TA in contrast to *GlobalTrust* that focuses on the TA. This approach has the benefit of requiring less time and uses less computing resources at the TA in the network.
- *CGTrust* provides a computational complexity minimized mechanism to form MANET clustering, where the complexity considers both intra and inter cluster computations. This algorithm helps to drastically reduce the complexity associated with the trust evaluation and computation.
- *CGTrust* provides a mechanism to evaluate the trust of both non-CH nodes and CHs to prevent false trustworthiness decisions of non-CH nodes by a non-trustworthy CH and minimize the setup and reconfiguration time.

2 CGTrust

The proposed scheme is for MANETs that use the *GlobalTrust* trust-based reputation scheme [2]. It is assumed that MANETs consist of multiple nodes communicating via multiple hops. The objective of *CGTrust* is to provide a trust-based clustering algorithm that derives the optimal number of clusters k (i.e., $k_{optimal}$) to minimized network setup and reconfiguration time delay based on the constraints of the required trust level that needs to be satisfied. *CGTrust* forms clusters by using a trust-based reputation algorithm, where the TA makes decisions based on trust values. The TA is a system that can determine the trust level of all nodes in the network and manage the security and performance of the network [2]. Therefore, it is assumed that the TA does not perform the role of a node in the network, rather the TA exists independently and manages the network because the TA requires a large amount of resources to perform the trustworthiness evaluation functions. The TA collects evidence periodically to assess the trust of participating nodes through CHs and makes a trustworthy decision. In the initial setup stage, if no CHs have been preselected, then the TA will select CHs and provide initial cluster information (where this initial formation may not be an optimal setting). If The number of nodes changes, the TA recomputes the optimal number of clusters, and then reassigns CHs among the nodes with the highest reputation values. Figure 1 is a diagram

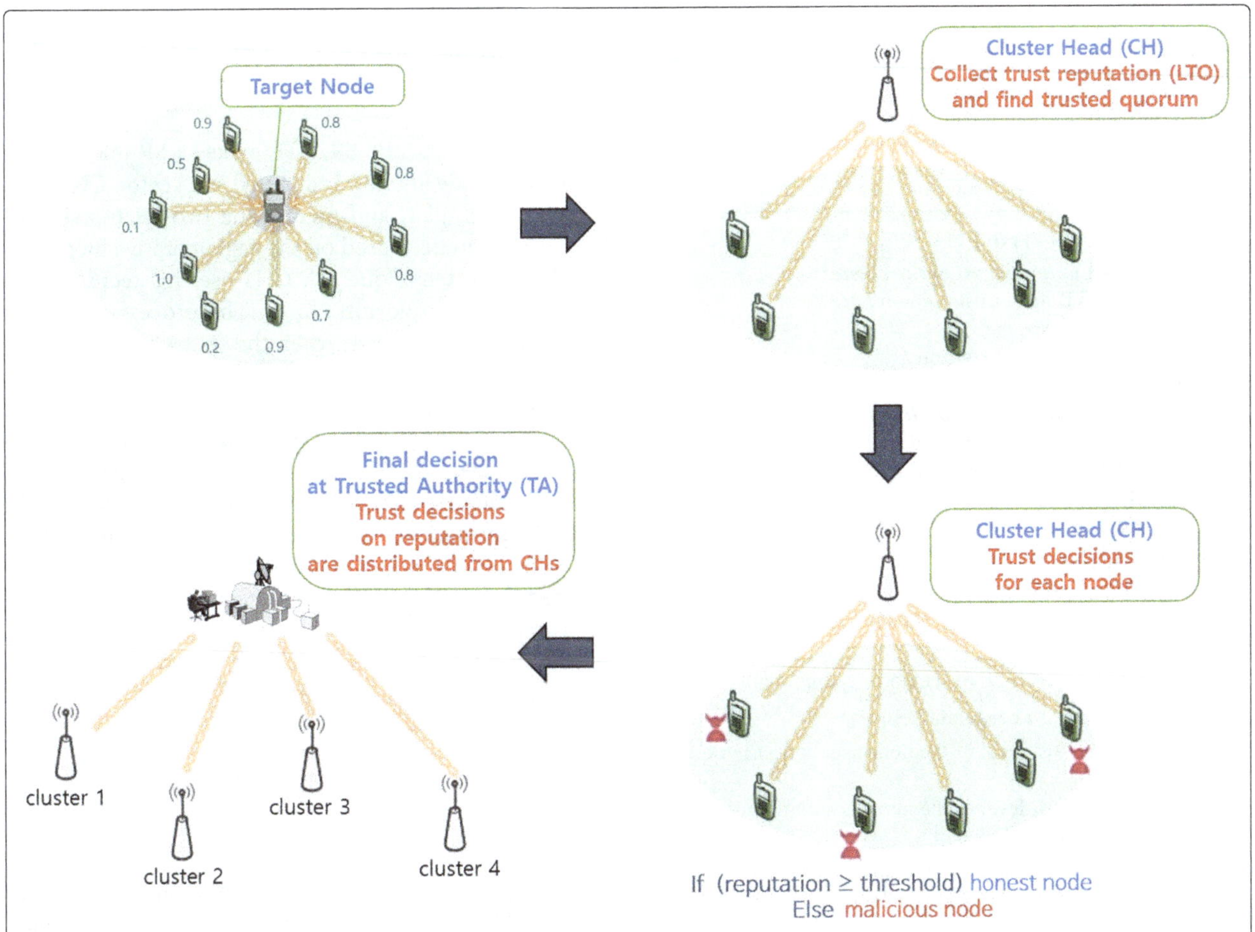

Fig. 1 *CGTrust* network operations

that illustrates the operations of *CGTrust*. The *CGTrust* process is described in the following steps (Fig. 2).

Step 1: Non-CH nodes that belong to the *i*th cluster transmit local trust opinions (LTOs) of their neighbor nodes to the local CH. $\mathrm{LTO}_{w,u}$ is node *w*'s trust opinion towards node *u* based on direct observations. $\mathrm{LTO}_{w,u}$ is calculated as below

$$\mathrm{LTO}_{w,u} = \frac{p_{w,u}}{p_{w,u} + n_{w,u}} \tag{1}$$

where $p_{w,u}$ and $n_{w,u}$ are respectively the total number of positive events and the total number of negative events. If there are no events between node *w* and node *u*, the LTO is null. The sum of positive and negative events $(p_{w,u} + n_{w,u})$ is the sum of the number of events sent by node *w* to node *u*. Positive events represented by $p_{w,u}$ are events that node *w* is determined that a packet was transmitted well to node *u*. Negative events represented by $n_{w,u}$ are events that node *w* is determined that a packet was not delivered to node *u* for some reasons (noise, wrong decision, intentional packet drop, etc.). The transmission can be determined by an acknowledgement packet, etc. As can be seen from Eq. (1), the higher the positive events, the closer to 1 the LTO and the closer to 0 the negative events become. The higher the LTO, the trust opinion of node *u* approaches the reputation of an honest node.

An honest node correctly sends a packet to a predefined route when it receives a packet from another node, except for errors that occur during the transmission process. In addition, when the LTO is reported to the TA or CH, the reporting of the LTO is performed without distortion.

Algorithm: CGTrust

At the CH – Compute trust of the cluster

1: **COMPUTE** local trust opinion *LTO*
2: **COMPUTE** *Sim* and *SR*
3: **FIND** CH's trusted quorum *D*
4: **COMPUTE** *BR*, *CR* and *GR*

At the TA – Compute trust of all nodes

5: **IF** cluster is using encrypted packet mode (β)
  COLLECT encrypted *LTOs* from non-CH nodes
6: **ELSE**
  COLLECT *GR* from CH
7: **COMPUTE** *Sim* and *SR*
8: **FIND** TA's trusted quorum *D*
9: **COMPUTE** *BR*, *CR*, and *GR*
10: **IF** *GR* is unknown
  SET the trust level of the node as unknown
11: **ELSEIF** *GR* ≥ decision factor θ
  SET the trust level of the node as a honest node
12: **ELSE**
  SET the trust level of the node as a malicious node
14: **IF** *N* has changed
  COMPUTE $k_{optimal}$ using (15)
  Assign nodes to the $k_{optimal}$ clusters and select CHs

Fig. 2 *CGTrust* algorithm

A malicious node may drop a packet when it is received from another node, or intentionally send a packet to a different route. In addition, information may be distorted during the transmission and reporting process of the LTO to the TA or CH.

The *i*th CH (CH_i) aggregates trust-related evidences and computes the subjective reputation (SR) of node *u* as it is assessed by node *w* ($\mathrm{SR}_{w,u}$) using

$$\mathrm{SR}_{w,u} = \sum_{j \in S_u} \mathrm{LTO}_{j,u} \frac{\mathrm{HR}_j \mathrm{Sim}(w,j)}{\sum_{j \in S_u} \mathrm{HR}_j \mathrm{Sim}(w,j)} \tag{2}$$

where $\mathrm{Sim}(w,j)$ is the *cosine* similarity of the LTOs between node *w* and node *j*, S_u is the set of nodes that have non-null LTOs over node *u* (including *w* if *w* has one), and HR_j is the hierarchical rank of node *j*. HR_j can specify that administrators have different values, or they can all have the same value, depending on how trustworthy the node's opinion is [2]. The TA should have a hierarchy rank higher than that of any CH or node, because the TA's opinion is more strongly reflected than the CHs or nodes as it is where the reputation judgment is made. In this paper, the hierarchy rank is designated as TA = 3, CH = 2, and node = 1.

After calculating the SR, CH_i makes a SR matrix. The SR tuple in node *w*'s view is denoted as a vector. CH_i compares all SR tuples, and merges the two SR tuples with the least difference based on the agglomerative hierarchical clustering technique [2]. CHs use this technique to form its trusted quorum (D_{CH}). The process of finding the trusted quorum ends when the set having the largest number of nodes has more than half of the total number of network nodes. When the total number of nodes is *N*, the number of nodes belonging to the trusted quorum may have a value between $[N/2, N]$, depending on how it is calculated. Since all SR tuples are compared and a trusted quorum is found, the complexity does not change depending on the size of the trusted quorum.

Based on the trusted quorum, CH_i obtains the behavioral reputation (BR) of node *u* ($\mathrm{BR}_{\mathrm{CH}_{i,u}}$). $\mathrm{BR}_{\mathrm{CH}_{i,u}}$, reflecting how CH_i views node *u*'s network behavior, is computed by averaging the SR tuples using (3).

$$\mathrm{BR}_{\mathrm{CH}_{i,u}} = \frac{\sum_{w \in D} \mathrm{SR}_{w,u}}{|D_{CH}|} \tag{3}$$

Then CH_i computes the credibility reputation (CR) of node *u* ($\mathrm{CR}_{\mathrm{CH}_{i,u}}$) using (4). $\mathrm{CR}_{\mathrm{CH}_{i,u}}$ indicates how trustworthy *u*'s reported LTOs are. This is computed based on the difference between u's reported LTOs and BRs of the nodes that node u has reported LTOs over.

$$CR_{CH_{i,u}} = 1 - \sqrt{\frac{\sum_{j\in\{LTO_{u,j}\neq null\}} \left(LTO_{u,j} - BR_{CH_{i,j}}\right)^2}{|j \in \{LTO_{u,j} \neq null\}|}} \quad (4)$$

Next CH_i computes the global reputation (GR) of node u ($GR_{CH_{i,u}}$) using the normalized factor γ selected from the range in [0,1].

$$GR_{CH_{i,u}} = \gamma BR_{CH_{i,u}} + (1-\gamma)CR_{CH_{i,u}} \quad (5)$$

Step 2: Once a CH computes all GRs of its cluster's nodes, it sends the result information to the TA. The TA computes the SR from GR based on the resultant data as in (6). The cosine similarity ($Sim(CH_i, CH_k)$) in Eq. (6) is calculated on the basis of the LTO between the CH communication. This allows the TA to determine if a CH is infected and can make a final decision. Step 2 is performed when the number of clusters is two or more. If the network is not divided into clusters, the TA directly makes all trust decisions of all nodes in the network.

$$SR_{CH_{i,u}} = \sum_{CH_k\in S'_u} GR_{CH_k,u} \frac{HR_{CH_k} Sim(CH_i, CH_k)}{\sum_{CH_k\in S'_u} HR_{CH_k} Sim(CH_i, CH_k)} \quad (6)$$

After (6), the TA calculates the BR of node u in (7), and the CR of node u (i.e., $CR_{TA,u}$) in (8).

$$BR_{TA,u} = \frac{\sum_{CH_i\in D} SR_{CH_{i,u}}}{|D_{TA}|} \quad (7)$$

$$CR_{TA,u} = 1 - \sqrt{\frac{\sum_{u\in\{GR_{CH_{i,u}}\neq null\}} \left(GR_{CH_{i,u}} - BR_{TA,u}\right)^2}{|u \in \{GR_{CH_{i,u}} \neq null\}|}} \quad (8)$$

Then the TA computes the GR of node u (i.e., $GR_{TA,u}$) using

$$GR_{TA,u} = \gamma' BR_{TA,u} + (1-\gamma')CR_{TA,u} \quad (9)$$

Step 3: After the TA computes the GR, the TA decides the trustworthiness of each node using

$$Decision(u) = \begin{cases} \text{unknown} & \text{if } GR_{TA,u} = \text{unknown} \\ \text{honest} & \text{if } GR_{TA,u} \geq \theta \\ \text{malicious} & \text{if } GR_{TA,u} < \theta \end{cases} \quad (10)$$

where θ is a decision factor selected from the range in [0, 1]. The detection errors can be reduced by selecting the most appropriate θ value.

The TA evaluates each CH's trustworthiness using direct trust computation, which can be conducted using the encrypted packet mode, which encrypts packets exchanged between non-CH nodes and the TA. In this mode, non-CH nodes send (encrypted) information packets to their CH and the CH forwards these packets to the TA without trust computation. Using the encrypted packet mode, the TA computes the CHs' trustworthy level periodically considering β, which is the ratio of nodes that use encrypted packet mode in the cluster. In this computation process, β is a variable that represents the possibility that a CH is a malicious node. As the value of β increases, the number of nodes that the TA needs to directly compute increases, making it difficult to reduce the computational complexity. Considering the computational complexity of trust computation and the worst case where the majority of CHs are infected, the suitable value of β is [0, 0.5].

3 Cluster-based network analysis

3.1 Computational complexity analysis

The proposed scheme computes the GR of each node based on *GlobalTrust* that uses a trusted quorum *D*. *GlobalTrust* uses the *agglomerative hierarchical clustering technique* to find a minimum dominating cluster [2].

For an accurate complexity analysis, the method of [14] is applied to *CGTrust*, where the complexity of the pseudocode steps is computed. Each cluster has *N/k* nodes. Every node will collect the SRs of all other nodes in its cluster, which are ((*N/k*)-1) SRs. Two nodes in a cluster will pair up and compare their collected SRs, but will exclude the SR of the paired node in this comparison process. Therefore, ((*N/k*)-2) SRs will be compared by the node pair. In addition, since there are $\binom{N/k}{2}$ combinations of possible node pairs in each cluster, the computational complexity of one cluster is O[((*N/k*)- 2)(*N/k*)((*N/k*)-1)/2] = O((*N/k*)^3) (step 3).

In case of inter-clusters, the TA uses $((1-\beta)k+\beta N-1)$ SRs for each node since the TA computes the GR directly at the rate of β. The minimum pair complexity becomes $O(((1-\beta)k+\beta N-1)^2)$. The computational complexity for all steps is $O(((1-\beta)k+\beta N-1)((1-\beta)k+\beta N-2)/2))$ and based on this the computational complexity of each node is $O(((1-\beta)k+\beta N-1)^2) + O(((1-\beta)k+\beta N-1)((1-\beta)k+\beta N-2)/2))$. Therefore, to compute the TA's trusted quorum the required complexity is $O[((1-\beta)k+\beta N-1)^2 + ((1-\beta)k+\beta N-1)((1-\beta)k+\beta N-2)/2)(1-\beta)k+\beta N]$ (step 9). After organizing the terms, the computational complexity can be expressed as $O((1-\beta)^3k^3 + (\beta N)^3)$. In *CGTrust*, the complexity of both intra-cluster and inter-clusters are considered. Therefore, the total computational complexity (C_{total}) becomes

$$C_{total} = O\left((N/k)^3 + (1-\beta)^3k^3 + (\beta N)^3\right) \quad (11)$$

3.2 Trust information computation time

In order to minimize the time required to evaluate the trust profile of a large MANET, the optimal cluster size (k_{optimal}) that minimizes the computational complexity used in evaluating the trust profile of all nodes in the network is derived. As shown in Fig. 3, the computational complexity is directly proportional to the computational time required in evaluating the trust profile of all nodes of the MANET.

3.3 Minimization of computational complexity

The main objective of the *CGTrust* scheme is to minimize the computational complexity that results in a minimized network reconfiguration time delay. The optimization statement and constraints are established as below.

$$\begin{aligned} \underset{k}{\text{minimize}} \quad & C_{\text{total}}(k) \\ \text{subject to} \quad & 0 \leq \beta \leq 0.5, \\ & N, k \in \mathbb{Z}_{>0}. \end{aligned}$$

The total computational complexity in the TA is a function of k in the form of C_{total}, where the computational complexity minimizing optimal number of clusters can be obtained from

$$\begin{aligned} k' &= \arg\left[\frac{\partial C_{\text{total}}(k)}{\partial k} = 0\right] \\ &= \arg\left[-\frac{3\left(N^3 + (\beta-1)^3 k^6\right)}{k^4} = 0\right] \end{aligned} \tag{12}$$

which results in the following candidate solutions.

$$k' = \left\{ \begin{array}{lll} \frac{\sqrt{N(1-\beta)}}{1-\beta}, & \frac{-\sqrt{N(1-\beta)}}{1-\beta}, & -\sqrt{\frac{N(i\sqrt{3}+1)}{2(1-\beta)}}, \\ \sqrt{\frac{N(i\sqrt{3}+1)}{2(1-\beta)}}, & -\sqrt{\frac{N(i\sqrt{3}-1)}{2(1-\beta)}}, & \sqrt{\frac{N(i\sqrt{3}-1)}{2(1-\beta)}} \end{array} \right\} \tag{13}$$

There are 6 solutions of (13), where N is a positive number and the range of β is [0, 0.5]. Considering the fact that k needs to be a positive integer, the only feasible solution is $\frac{\sqrt{N(1-\beta)}}{1-\beta}$ which is a positive real number. In addition, the second derivative of the objective function is

$$\frac{\partial^2}{\partial k^2} C_{\text{total}}(k) = \frac{12N^3}{k^5} + 6(1-\beta)^3 k \tag{14}$$

in which $k' = \frac{\sqrt{N(1-\beta)}}{1-\beta}$ results in a positive value of the convex objective function $C_{\text{total}}(k)$. The optimal number of clusters (k_{optimal}) has to be a positive integer, and therefore, the nearest integer function (i.e., $Nint(\cdot)$) is applied to result in (15).

$$k_{\text{optimal}} = Nint\left(\frac{\sqrt{N(1-\beta)}}{1-\beta}\right) \tag{15}$$

Figure 4 presents the C_{total} profile and the k_{optimal} values for the range of interest based on N and k.

4 Performance evaluation

Malicious attack patterns investigated for the FN and FP performance evaluation include the following five attack patterns [2].

- Naive malicious attack (NMA): a malicious node provide improper services with probability α. However, it reports its LTOs honestly.
- Collusive rumor attack (CRA): In addition to providing improper services with probability α, malicious nodes collude to report false LTOs. Malicious nodes report LTOs of 1 to malicious node and LTOs of 0 to honest nodes.
- Non-collusive rumor attack (NRA): a malicious node can report a false LTO that is opposite to the observed evidence. For example, if an LTO is evaluated as p, the malicious node may report $(1-p)$ as the LTO.
- Malicious spy attack (MSA): some malicious nodes misbehave with probability α. Other malicious nodes behave honestly. These malicious nodes may collude

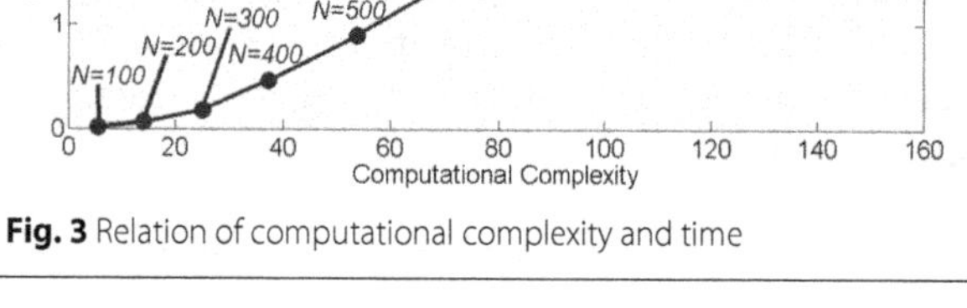

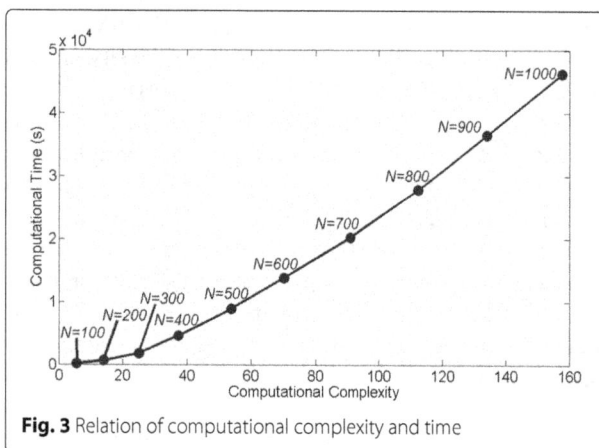

Fig. 3 Relation of computational complexity and time

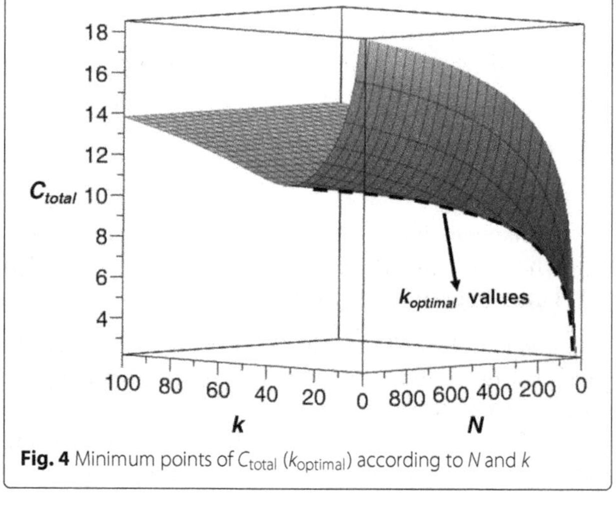

Fig. 4 Minimum points of C_{total} (k_{optimal}) according to N and k

and report LTOs of 1 to malicious node and LTOs of 0 to honest nodes to confuse the trust and reputation system.

- Conflicting behavior attack (CBA): malicious nodes can collude to confuse the trust and reputation system such as CRA and MSA. However, they misbehave only to some of honest nodes, and report LTOs of 1 to malicious node and LTOs of 0 to honest nodes to confuse the trust and reputation system. This attack causes LTO disagreement among honest nodes, which makes it difficult to find malicious nodes.

CBA is considered the most demanding type of attack because it makes it difficult to distinguish malicious nodes by confusing LTO information of honest nodes with respect to other nodes. For the above reasons, CBA was selected and evaluated.

The simulation based performance analysis of *CGTrust* and *GlobalTrust* was conducted using Matlab with N nodes randomly distributed with a uniform density in a 2×2 km^2 square area. Simulation parameters were set same to the experiments in [2], where the ratio of malicious nodes was set to 0.3 and every node randomly requests of its neighbor nodes to send a packet (which is multihop relayed) 100 times per minute, and β is in the range in [0, 0.5]. Honest nodes were made to drop packets based on a 0.05 packet error rate (PER) and the detection error probability of the monitoring system was set to 0.05. Each node was made to transmit trust data packets every 30 s and the TA computes the GRs based on the accumulated data of the past 30 min.

In addition, the probability that a malicious node drops a packet was set to 0.5, $\gamma = 0.7$, $\theta = 0.7$, and the upper bound probability of FN and FP were set to 0.1 as used in [2].

In the simulation, the TA is not a target of a malicious node. If the TA is infected, trust decisions on network nodes will not be correct. It is assumed that CH and other nodes can be malicious nodes, based on the restriction that the malicious ratio is not more than 0.5. If the malicious node ratio is greater than 0.5, the malicious nodes can take control of all the opinions in the network and the trust decision cannot be determined correctly. Although it is assumed that the overall ratio of malicious nodes is less than 0.5, the proportion of malicious nodes in a cluster is not limited. Therefore, in some clusters, malicious nodes may not properly report to the TA because they have taken control of the cluster.

Figure 5 compares C_{total} of *GlobalTrust* and *CGTrust* based on β, where it can be observed that C_{total} of *CGTrust* decreases to less than 1/1000 compared to *GlobalTrust*. The results show that *CGTrust* can significantly reduce the required computations and thereby reduce the network's trust evaluation time. Figure 6 shows the detection error probability of FN and FP for *CGTrust*, where for the number of nodes of interest (i.e., $100 \leq N \leq 1000$), the probability of FN and FP are always lower than the upper bound 0.1. More significantly, the FN probability is always below 0.00012. In addition to being much faster than *GlobalTrust*, having a very low FN probability is a very advantageous feature of *CGTrust* because (compared to FP) FN is a significantly more critical security problem due to the fact that FN represents the probability that a malicious node has been not detected and still remains operational in the network. On the other hand, an erroneous FP decision on a node can be easily corrected by additional checking of the node. Figure 7 investigates influence of changes in malicious node ratios, where the results show that the FN probability of *CGTrust* is an approximate 0.1 times lower compared to *GlobalTrust*, while the FP probability of the two are similar. Abrupt changes in the performance can be observed in Figs. 4 and 5, which are due to the positive integer rounding effect applied to k_{optimal}.

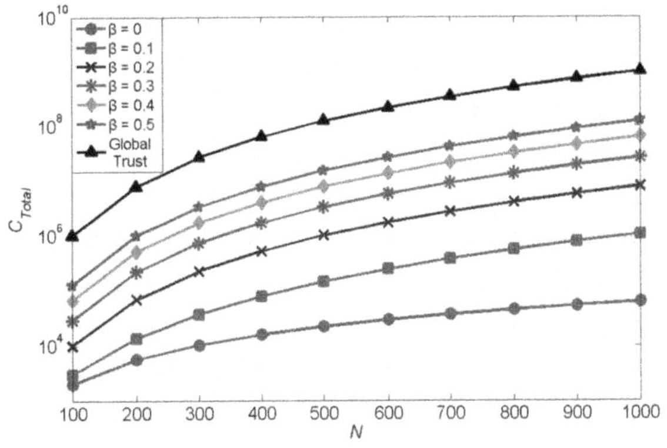

Fig. 5 C_{total} comparison of *GlobalTrust* and *CGTrust* based on N and β

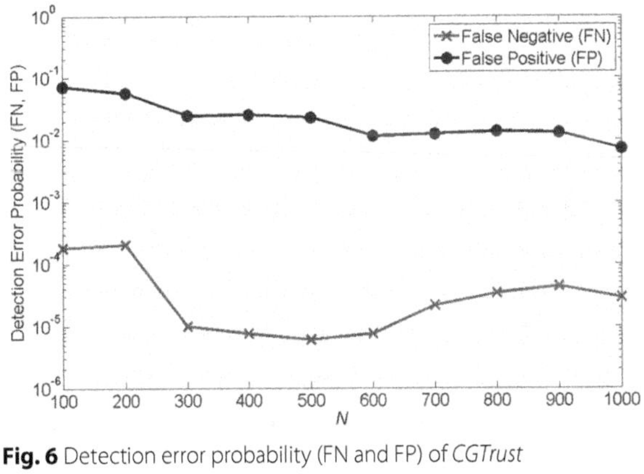

Fig. 6 Detection error probability (FN and FP) of *CGTrust*

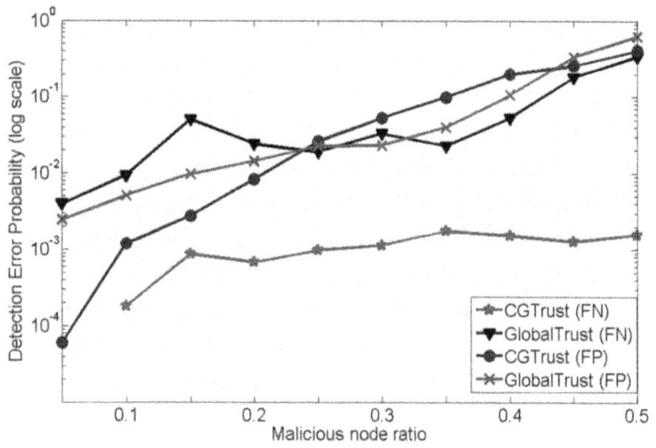

Fig. 7 Detection error probability (FN and FP) comparison of *GlobalTrust* and *CGTrust* based on various malicious node ratios

5 Conclusion

Mission supportive MANETs require fast updates on node conditions in order to properly support command and control operations. To support this objective, *CGTrust* was designed to minimize the time required to evaluate the trust profile of a MANET through optimal cluster size control applied to *GlobalTrust*. The simulation results show that for the number of nodes and malicious node ratios of practical interest (based on $\beta = 0.1$ and 0.5), *CGTrust* can be approximately 1000 and 10 times faster compared to *GlobalTrust*, respectively. In addition, the results also show that the FN probability is approximately 0.1 times lower when *CGTrust* is used instead of *GlobalTrust* for the malicious node ratio range of 0.05 to 0.5.

Acknowledgements
This work was supported by the ICT R&D program of MSIT/IITP, Republic of Korea (B0101-17-1276, Access Network Control Techniques for Various IoT Services).

Competing interests
The authors declare that they have no competing interests.

Author details
[1]School of Electrical and Electronic Engineering, Yonsei University, Seoul, Republic of Korea. [2]Republic of Korea Air Force, Gyeryong, Republic of Korea.

References

1. J-H Cho, A Swami, I-R Chen, A Survey on Trust Management for Mobile Ad Hoc Networks. IEEE Commun. Surv. Tutor. **13**(4), 562–583 (2011)
2. X Chen, J Cho, S Zhu, in *Proceedings of the IEEE International Conference on Sensing, Communication and Networking (SECON) 2014*. GlobalTrust: An Attack-Resilient Reputation System for Tactical Networks, (Singapore, 2014), pp. 275–283
3. K Aberer, Z Despotovic, in *Proceedings of the 2001 ACM International Conference on Information and Knowledge Management (CIKM)*. Managing trust in a peer-2-peer information system, (Atlanta, GA, USA, 2001), pp. 310–317
4. SD Kamvar, MT Schlosser, H Garcia-Molina, in *Proceedings of the 2003 ACM International Conference on World Wide Web (WWW)*. The eigentrust algorithm for reputation management in p2p networks, (Budapest, Hungary, 2003), pp. 640–651
5. L Xiong, L Liu, Peertrust: Supporting reputation-based trust for peer-to-peer electronic communities. IEEE Trans. Knowl. Data Eng. **16**(7), 843–857 (2004)
6. S Buchegger, JY Le Boudec, *A Robust Reputation System for Mobile ad hoc Networks*. Technical Report IC/2003/50, EPFL-DI-ICA, (Lausanne, 2003)
7. Q He, D Wu, P Khosla, in *Proceedings of the IEEE Wireless Communications and Networking Conference (WCNC) 2004*. Sori: a secure and objective reputation based incentive scheme for ad-hoc networks, (Atlanta, GA, USA, 2004), pp. 825–830
8. WL Teacy, J Patel, NR Jennings, M Luck, Travos: Trust and reputation in the context of inaccurate information sources. Auton. Agent Multi Agent Syst. **12**(2), 183–198 (2006)
9. A Jsang, R Ismail, in *Proceedings of the Electronic Commerce Conference 2002*. The beta reputation system, (Bled, Slovenia, 2002), pp. 2502–2511
10. H Chan, VD Gligor, A Perrig, G Muralidharan, On the distribution and revocation of cryptographic keys in sensor networks. IEEE Trans. Dependable Secure Comput. **2**(3), 233–247 (2005)
11. M Raya, MH Manshaei, M Felegyhazi, J-P Hubaux, in *Proceedings of ACM Conference on Computer and Communications Security (CCS) 2008*. Revocation games in ephemeral networks, (Alexandria, VA, USA, 2008), pp. 199–210
12. S Reidt, M Srivatsa, S Balfe, in *Proceedings of the ACM Conference on Computer and Communications Security (CCS) 2009*. The fable of the bees: incentivizing robust revocation decision making in ad hoc networks, (Chicago, IL, USA, 2009), pp. 291–302
13. X Chen, H Patankar, S Zhu, M Srivatsa, J Opper, in *Proceedings of the IEEE International Conference on Sensing, Communication and Networking (SECON) 2013*. Zigzag: Partial mutual revocation based trust management in tactical ad hoc networks, (New Orleans, LA, USA, 2013), pp. 131–139
14. S Kim, J-M Chung, Message Complexity Analysis of Mobile Ad Hoc Network Address Autoconfiguration Protocols. IEEE Trans. Mobile Comput. **7**(3), 358–371 (2008)

Permissions

All chapters in this book were first published in EURASIP JWCN, by Springer International Publishing AG.; hereby published with permission under the Creative Commons Attribution License or equivalent. Every chapter published in this book has been scrutinized by our experts. Their significance has been extensively debated. The topics covered herein carry significant findings which will fuel the growth of the discipline. They may even be implemented as practical applications or may be referred to as a beginning point for another development.

The contributors of this book come from diverse backgrounds, making this book a truly international effort. This book will bring forth new frontiers with its revolutionizing research information and detailed analysis of the nascent developments around the world.

We would like to thank all the contributing authors for lending their expertise to make the book truly unique. They have played a crucial role in the development of this book. Without their invaluable contributions this book wouldn't have been possible. They have made vital efforts to compile up to date information on the varied aspects of this subject to make this book a valuable addition to the collection of many professionals and students.

This book was conceptualized with the vision of imparting up-to-date information and advanced data in this field. To ensure the same, a matchless editorial board was set up. Every individual on the board went through rigorous rounds of assessment to prove their worth. After which they invested a large part of their time researching and compiling the most relevant data for our readers.

The editorial board has been involved in producing this book since its inception. They have spent rigorous hours researching and exploring the diverse topics which have resulted in the successful publishing of this book. They have passed on their knowledge of decades through this book. To expedite this challenging task, the publisher supported the team at every step. A small team of assistant editors was also appointed to further simplify the editing procedure and attain best results for the readers.

Apart from the editorial board, the designing team has also invested a significant amount of their time in understanding the subject and creating the most relevant covers. They scrutinized every image to scout for the most suitable representation of the subject and create an appropriate cover for the book.

The publishing team has been an ardent support to the editorial, designing and production team. Their endless efforts to recruit the best for this project, has resulted in the accomplishment of this book. They are a veteran in the field of academics and their pool of knowledge is as vast as their experience in printing. Their expertise and guidance has proved useful at every step. Their uncompromising quality standards have made this book an exceptional effort. Their encouragement from time to time has been an inspiration for everyone.

The publisher and the editorial board hope that this book will prove to be a valuable piece of knowledge for researchers, students, practitioners and scholars across the globe.

List of Contributors

Qichao Song and Baisen Lui
College of Electrical and Information Engineering, Heilongjiang Institute of Technology, Harbin 150050, China

Xiaoming Zhu
College of Electrical and Information Engineering, Heilongjiang Institute of Technology, Harbin 150050, China
College of Information and Communication Engineering, Harbin Engineering University, Harbin 150001, China

Xiaodong Yang
College of Information and Communication Engineering, Harbin Engineering University, Harbin 150001, China
Collaborative Research Center, Meisei University, Tokyo 1918506, Japan

Ioannis-Prodromos Belikaidis, Stavroula Vassaki, Andreas Georgakopoulos, Aristotelis Margaris and Kostas Tsagkaris
WINGS ICT Solutions, Athens, Greece

Federico Miatton
Sistelbanda, Valencia, Spain

Uwe Herzog
EURESCOM, Heidelberg, Germany

Panagiotis Demestichas
University of Piraeus, Piraeus, Greece

Jesper H. Sørensen and Elisabeth de Carvalho
Department of Electronic Systems, Aalborg University, Fredrik Bajers Vej 7, 9220 Aalborg, Denmark

Manaf Zghaibeh and Najam Ul Hassan
Department of Electrical and Computer Engineering, Dhofar University, P. O. Box 2509, 211, Salalah, Oman

Tshiamo Sigwele and Yim F. Hu
Faculty of Engineering and Informatics, University of Bradford, BD7 1DP Bradford, UK

Prashant Pillai
Faculty of Technology, Design and Environment, Oxford Brookes University, Oxford, UK

Atm S. Alam
Institute of Communication Systems, 5G Innovation Centre University of Surrey, Düsternbrooker Weg 20, GU2 7XH Guildford, UK

Xuan-Xinh Nguyen and Dinh-Thuan Do
Wireless Communications Research Group, Faculty of Electrical and Electronics Engineering, Ton Duc Thang University, 19 Nguyen Huu Tho Street, Ho Chi Minh City, Vietnam

Tao Jing, Zhen Li, Yan Huo and Fan Zhang
School of Electronics and Information Engineering, Beijing Jiaotong University, 100044 Beijing, China

Kaiwei Jiang
School of Electronics and Information Engineering, Beijing Jiaotong University, 100044 Beijing, China
School of Electric and Information Engineering, Taizhou Vocational and Technical College, 318000 Taizhou, China

Liang Li
Department of Electrical Engineering and Computer Science, the University of Tennessee, Knoxville, Knoxville, USA

Ju Bin Song
Sequans Communications, 1732, Deogyoung Road, 446701 Giheung, Yongin, South Korea

Husheng Li
Department of Electrical Engineering and Computer Science, the University of Tennessee, Knoxville, Knoxville, USA
Sequans Communications, 1732, Deogyoung Road, 446701 Giheung, Yongin, South Korea

Shuping Gong
Department of Electronic Engineering, Kyung Hee University, 90 Washington Valley Road, 07921 Bedminster, NJ, USA

Yuki Kasai, Masahiro Sasabe and Shoji Kasahara
Graduate School of Information Science, Nara Institute of Science and Technology, 8916-5 Takayama-cho, Ikoma, 630-0192 Nara, Japan

Anum L. Enlil Corral-Ruiz and Felipe A. Cruz-Perez
Electrical Engineering Department, CINVESTAV-IPN, Mexico City, Mexico

S. Lirio Castellanos-Lopez and Genaro Hernandez-Valdez
Electronics Department, UAM, Mexico City, Mexico

Mario E. Rivero-Angeles
Communication Networks Laboratory, CIC-Instituto Politécnico Nacional, Mexico City, Mexico

Tanee Demeechai and Pratana Kukieattikool
National Electronics and Computer Technology Center, Phahonyothin Road, 12120 Pathum Thani, Thailandw

Haipeng Yao and Chong Liu
State Key Laboratory of Networking and Switching Technology, Beijing University of Posts and Telecommunications, Haidian District Xitucheng Road 10, Beijing 100876, People's Republic of China

Peiying Zhang
State Key Laboratory of Networking and Switching Technology, Beijing University of Posts and Telecommunications, Haidian District Xitucheng Road 10, Beijing 100876, People's Republic of China
College of Computer & Communication Engineering, China University of Petroleum (East China), Changjiang West Road 66, Qingdao 266580, China

Luyao Wang
Advanced Innovation Center for Future Internet Technology, Beijing University of Technology, Chaoyang District Pingleyuan 100, Beijing 100124, People's Republic of China

Zhian Deng, Weijian Si and Zhiyu Qu
College of Information and Communication Engineering, Harbin Engineering University, Harbin 150001, China

Xin Liu
School of Information and Communication Engineering, Dalian University of Technology, Dalian 116024, China

Zhenyu Na
School of Information Science and Technology, Dalian Maritime University, Dalian 116026, China

Yufang Yin and Gangjun Li
Chengdu Technological University, Chengdu, China

Li Li and Hua Wei
Chengdu University of Information and Technology, Chengdu, China

Wei-Chen Pao
Industrial Technology Research Institute, Zhudong, Taiwan, Republic of China

Jhih-Wei Lin and Yung-Fang Chen
Department of Communication Engineering, National Central University, Taoyuan, Taiwan, Republic of China

Chin-Liang Wang
Department of Electrical Engineering and Institute of Communications Engineering, National Tsing Hua University, Hsinchu, Taiwan, Republic of China

Sunho Seo and Jong-Moon Chung
School of Electrical and Electronic Engineering, Yonsei University, Seoul, Republic of Korea

Jin-Won Kim and Jae-Dong Kim
School of Electrical and Electronic Engineering, Yonsei University, Seoul, Republic of Korea
Republic of Korea Air Force, Gyeryong, Republic of Korea

Index

www.ingramcontent.com/pod-product-compliance
Lightning Source LLC
Chambersburg PA
CBHW082033190326
41458CB00010B/3350
* 9 7 8 1 6 8 2 8 5 5 9 5 9 *